American River College Library
4700 College Oak Drive
Sacramento, CA 95841

Nutritional Ergogenic Aids

Edited by
IRA WOLINSKY
JUDY A. DRISKELL

CRC PRESS

Boca Raton London New York Washington, D.C.

Library of Congress Cataloging-in-Publication Data

Nutritional ergogenic aids / edited by Ira Wolinsky, Judy A. Driskell.
p. cm.
Includes bibliographical references and index.
ISBN 0-8493-1626-X
1. Athletes—Nutrition. 2. Dietary supplements. 3. Sports—Physiological aspects. I. Wolinsky, Ira II. Driskell, Judy A. (Judy Anne)
RC1235.N88 2004
613.2'024'139–dc22   2004041371

This book contains information obtained from authentic and highly regarded sources. Reprinted material is quoted with permission, and sources are indicated. A wide variety of references are listed. Reasonable efforts have been made to publish reliable data and information, but the author and the publisher cannot assume responsibility for the validity of all materials or for the consequences of their use.

Neither this book nor any part may be reproduced or transmitted in any form or by any means, electronic or mechanical, including photocopying, microfilming, and recording, or by any information storage or retrieval system, without prior permission in writing from the publisher.

All rights reserved. Authorization to photocopy items for internal or personal use, or the personal or internal use of specific clients, may be granted by CRC Press LLC, provided that $1.50 per page photocopied is paid directly to Copyright Clearance Center, 222 Rosewood Drive, Danvers, MA 01923 USA. The fee code for users of the Transactional Reporting Service is ISBN 0-8493-1626-X/04/ $0.00+$1.50. The fee is subject to change without notice. For organizations that have been granted a photocopy license by the CCC, a separate system of payment has been arranged.

The consent of CRC Press LLC does not extend to copying for general distribution, for promotion, for creating new works, or for resale. Specific permission must be obtained in writing from CRC Press LLC for such copying.

Direct all inquiries to CRC Press LLC, 2000 N.W. Corporate Blvd., Boca Raton, Florida 33431.

Trademark Notice: Product or corporate names may be trademarks or registered trademarks, and are used only for identification and explanation, without intent to infringe.

Visit the CRC Press Web site at www.crcpress.com

© 2004 by CRC Press LLC

No claim to original U.S. Government works
International Standard Book Number 0-8493-1626-X
Library of Congress Card Number 2004041371
Printed in the United States of America 2 3 4 5 6 7 8 9 0
Printed on acid-free paper

Preface

Never was there a subject in the field of sports nutrition that has commanded more attention, interest and speculation than the topic of nutritional ergogenic aids. Often, pursuit of the intricacies of this topic has generated more heat than light. There is a vast number of ergogenic aids to consider; however, we have focused on the more common ones and assembled a roster of the best authors to discuss individual nutritional ergogenic aids. The first chapter in the book deals expertly with definitions and regulatory issues. The nutritional ergogenic aids are then divided into three broad categories to facilitate logical coverage: amino acid derivatives, lipid derivatives and other substances in foods not classified as essential. A chapter on the systematic and critical evaluation of benefits and possible risks of nutritional ergogenic aids follows. Hopefully, this volume will provide the reader and researcher with a reference book on the subject, a summary of the latest results and a possible basis for consensus. Sports nutritionists, sports medicine and fitness professionals, researchers, students, health practitioners and the informed layman will find this book timely and useful.

This book is part of a grouping of books on nutrition in exercise and sport that we have edited. Other related titles are: *Sports Nutrition: Vitamins and Trace Elements; Macroelements, Water and Electrolytes in Sports Nutrition; Energy-Yielding Macronutrients; Energy Metabolism in Sports Nutrition; Nutritional Applications in Exercise and Sport;* and *Nutritional Assessment of Athletes.* Also related are *Nutrition in Exercise and Sport*, Third Edition, edited by Ira Wolinsky, and *Sports Nutrition*, authored by Judy Driskell.

Ira Wolinsky, Ph.D.
University of Houston

Judy A. Driskell, Ph.D., R.D.
University of Nebraska

Dedication

We appreciate the opportunity to have worked with the chapter authors, experts all, on this book and on our other books in the area of sports nutrition. We learned from them and dedicate this book to them.

Dedication

The Editors

Ira Wolinsky, Ph.D., is Professor of Nutrition at the University of Houston. He received his B.S. degree in Chemistry from the City College of New York and his M.S. and Ph.D. degrees in Biochemistry from the University of Kansas. He has served in research and teaching positions at the Hebrew University, the University of Missouri and The Pennsylvania State University, as well as conducting basic research in NASA life sciences facilities and abroad.

Dr. Wolinsky is a member of the American Society of Nutritional Sciences, among other honorary and scientific organizations. He has contributed numerous nutrition research papers in the open literature. His major research interests relate to the nutrition of bone and calcium and trace elements, and to sports nutrition. He has been the recipient of research grants from both public and private sources. He has been the recipient of several international research fellowships and consultantships to the former Soviet Union, Bulgaria, Hungary and India. He merited a Fulbright Senior Scholar Research Fellowship to Greece in 1999.

Dr. Wolinsky has co-authored a book on the history of the science of nutrition, *Nutrition and Nutritional Diseases*. He co-edited *Sports Nutrition: Vitamins and Trace Elements; Macroelements, Water, and Electrolytes in Sports Nutrition; Energy-Yielding Macronutrients and Energy Metabolism in Sports Nutrition; Nutritional Applications in Exercise and Sport; Nutritional Assessment of Athletes* and the current book *Nutritional Ergogenic Aids*, all with Judy Driskell. Additionally, he co-edited *Nutritional Concerns of Women* with Dorothy Klimis-Zacas, *The Mediterranean Diet: Constituents and Health Promotion* with his Greek colleagues, and *Nutrition in Pharmacy Practice* with Louis Williams. He edited three editions of *Nutrition in Exercise and Sport*. He served also as the editor, or co-editor, for the CRC Series on *Nutrition in Exercise and Sport*, the CRC Series on *Modern Nutrition*, the CRC Series on *Methods in Nutrition in Nutrition Research* and the CRC Series on *Exercise Physiology*.

Judy Anne Driskell, Ph.D., R.D. is Professor of Nutritional and Health Sciences at the University of Nebraska. She received her B.S. degree in Biology from the University of Southern Mississippi in Hattiesburg. Her M.S. and Ph.D. degrees were obtained from Purdue University. She has served in research and teaching positions at Auburn University, Florida State University, Virginia Polytechnic Institute and State University and the University of Nebraska. She has also served as the Nutrition Scientist for the U.S. Department of Agriculture/Cooperative State Research Service and as a Professor of Nutrition and Food Science at Gadjah Mada and Bogor Universities in Indonesia.

Dr. Driskell is a member of numerous professional organizations including the American Society of Nutritional Sciences, the American College of Sports Medicine, the Institute of Food Technologists and the American Dietetic Association. In 1993 she received the Professional Scientist Award of the Food Science and Human Nutrition Section of the Southern Association of Agricultural Scientists. In addition, she was the 1987 recipient of the Borden Award for Research in Applied Fundamental Knowledge of Human Nutrition. She is listed as an expert in B-Complex Vitamins by the Vitamin Nutrition Information Service.

Dr. Driskell co-edited the CRC books *Sports Nutrition: Minerals and Electrolytes* with Constance V. Kies. In addition, she authored the textbook *Sports Nutrition* and co-authored the advanced nutrition book *Nutrition: Chemistry and Biology*, both published by CRC. She co-edited *Sports Nutrition: Vitamins and Trace Elements*; *Macroelements, Water, and Electrolytes in Sports Nutrition*; *Energy-Yielding Macronutrients and Energy Metabolism in Sports Nutrition*; *Nutritional Applications in Exercise and Sport*; *Nutritional Assessment of Athletes* and the current book *Nutritional Ergogenic* Aids, all with Ira Wolinsky. She has published more than 100 refereed research articles and 10 book chapters as well as several publications intended for lay audiences, and has given numerous professional and lay presentations. Her current research interests center around vitamin metabolism and requirements, including the interrelationships between exercise and water-soluble vitamin requirements.

Contributors

Martha A. Belury, Ph.D., R.D. Department of Human Nutrition, Ohio State University, Columbus, Ohio

Christina Beer, Ph.D. Weider Nutrition International, Salt Lake City, Utah

Luke R. Bucci, Ph.D., C.C.N., A.S.C.P., C.N.S. Weider Nutrition International, Salt Lake City, Utah

Timothy P. Carr, Ph.D. Department of Nutrition and Health Sciences, University of Nebraska, Lincoln, Nebraska

Sara A. Chelland, M.S. Department of Nutrition, Food and Exercise Science, Florida State University, Tallahassee, Florida

Veda Diwadkar-Navsariwala, Ph.D. Department of Human Nutrition, University of Illinois at Chicago, Chicago, Illinois

Hans C. Dreyer, P.T., M.S. Department of Biokinesiology and Physical Therapy, University of Southern California, Los Angeles, California

Judy A. Driskell, Ph.D., R.D. Department of Nutrition and Health Sciences, University of Nebraska, Lincoln, Nebraska

David J. Dyck, Ph.D. Department of Human Biology and Nutritional Science, University of Guelph, Guelph, Ontario, Canada

Jeff Feliciano, B.S. Weider Nutrition International, Salt Lake City, Utah

Mike Greenwood, Ph.D., C.S.C.S., F.A.C.S.M. Department of Health, Physical Education and Recreation, Baylor University, Waco, Texas

James R. Guest, M.D., F.A.A.P. University Health Center, University of Nebraska, Lincoln, Nebraska

Jean E. Guest, M.S., R.D., L.M.N.T. Department of Nutrition and Health Sciences, University of Nebraska, Lincoln, Nebraska

Mark D. Haub, Ph.D. Department of Human Nutrition, Kansas State University, Manhattan, Kansas

Catherine G.R. Jackson, Ph.D., F.A.C.S.M.   Department of Kinesiology, California State University at Fresno, Fresno, California

Satya Jonnalagadda, Ph.D.   Department of Nutrition, Georgia State University, Atlanta, Georgia

Douglas S. Kalman, M.S., R.D., F.A.C.N.   Miami Research Associates, Miami, Florida

Jay Kandiah, Ph.D., R.D., C.D.   Department of Family and Consumer Sciences, Ball State University, Muncie, Indiana

Stavros A. Kavouras, Ph.D.   Department of Nutrition and Dietetics, Harokopio University, Athens, Greece

Celeste G. Koster, M.A. Ed.   Department of Human Nutrition, Ohio State University, Columbus, Ohio

Richard B. Kreider, Ph.D., F.A.C.S.M.   Department of Health, Physical Education and Recreation, Baylor University, Waco, Texas

Brian C. Leutholtz, Ph.D., F.A.C.S.M.   Department of Health, Human Performance and Recreation, Baylor University, Waco, Texas

Nancy M. Lewis, Ph.D., R.D., F.A.D.A., L.M.N.T   Department of Nutrition and Health Sciences, University of Nebraska, Lincoln, Nebraska

Henry C. Lukaski, Ph.D.   U.S. Department of Agriculture, Grand Forks Human Nutrition Research Center, Grand Forks, North Dakota

Faidon Magkos, M.S.   Department of Nutrition and Dietetics, Harokopio University, Athens, Greece

Susan H. Mitmesser, Ph.D.   MDS Pharma Services, Lincoln, Nebraska

Robert J. Moffatt, Ph.D   Department of Nutrition and Exercise, Florida State University, Tallahassee, Florida

Steven L. Nissen, D.V.M., Ph.D.   Department of Animal Science, Iowa State University, Ames, Iowa

Tausha D. Robertson, M.S.   Student Health Services, University of North Carolina at Chapel Hill, Chapel Hill, North Carolina

Rob Skinner, M.S., R.D., L.D., C.S.C.S.   Georgia Tech Athletic Association, Georgia Institute of Technology, Atlanta, Georgia

Neal Spruce Apex Fitness Group, Carmarillo, California

Maria Stacewicz-Sapuntzakis, Ph.D. Department of Human Nutrition, University of Illinois at Chicago, Chicago, Illinois

Shawn Talbott, Ph.D. Pharmanex LLC, Provo, Utah

C. Alan Titchenal, Ph.D., C.N.S. Department of Human Nutrition, Food and Animal Sciences, University of Hawaii at Manoa, Honolulu, Hawaii

Amy A. Turpin, B.S. Weider Nutrition International, Salt Lake City, Utah

Robert E.C. Wildman, Ph.D., R.D. Bally Total Fitness Corporation, Chicago, Illinois

Robert Wiswell, Ph.D. Department of Biokinesiology and Physical Therapy, University of Southern California, Los Angeles, California

Ira Wolinsky, Ph.D. Department of Health and Human Performance, 110 Cameron Building, University of Houston, Houston, Texas

Shi Zhou, Ph.D. School of Exercise Science and Sport Management, Southern Cross University at Lismore, Lismore, New South Wales, Australia

Contents

Part I Introduction

1   Nutritional Ergogenic Aids: Introduction, Definitions
    and Regulatory Issues 3
    Amy A. Turpin, Jeff Feliciano and Luke R. Bucci

Part II Amino Acid Derivatives

2   Arginine 21
    Jean E. Guest, Nancy M. Lewis and James R. Guest

3   Aspartate 37
    Douglas S. Kalman

4   Branched-Chain Amino Acids 47
    Robert E.C. Wildman

5   Carnitine 61
    Robert J. Moffatt and Sara A. Chelland

6   Creatine 81
    Richard B. Kreider, Brian C. Leutholtz and Mike Greenwood

7   Gelatin 105
    Douglas S. Kalman

8   Glucosamine and Chondroitin Sulfate 115
    Catherine G.R. Jackson

9   Glutamine 129
    Satya S. Jonnalagadda and Rob Skinner

10 β-Hydroxy-β-Methylbutyrate 147
    Steven L. Nissen

11 Lysine 171
   Neal Spruce and C. Alan Titchenal

12 Ornithine, Ornithine Alpha-Ketoglutarate and Taurine 197
   Tausha D. Robertson

Part III Lipid Derivatives

13 Conjugated Linoleic Acid 209
   Celeste G. Koster and Martha A. Belury

14 Medium-Chain Triglycerides and Glycerol 221
   Timothy P. Carr

15 Wheat Germ Oil and Octacosanol 247
   Susan H. Mitmesser

Part IV Other Substances in Foods Not Classified As Essential

16 Buffers: Bicarbonate, Citrate and Phosphate 257
   Mark D. Haub

17 Caffeine 275
   Faidon Magkos and Stavros A. Kavouras

18 Carotenoids 325
   Maria Stacewicz-Sapuntzakis and Veda Diwadkar-Navsariwala

19 Coenzyme Q_{10} 355
   Shi Zhou

20 Ginseng 379
   Luke R. Bucci, Amy A. Turpin, Christina Beer
   and Jeff Feliciano

21 Lipoic Acid 411
   Henry C. Lukaski

22 Myo-Inositol and Pangamic Acid 431
    Robert A. Wiswell and Hans C. Dreyer

23 Pyruvate and Dihydroxyacetone 445
    David J. Dyck

24 Tannins 455
    Jay Kandiah

Part V Evaluation of Effectiveness

25 **Systematic and Critical Evaluation of Benefits and
    Possible Risks of Nutritional Ergogenic Aids** 469
    *Amy A. Turpin, Shawn M. Talbott, Jeff Feliciano
    and Luke R. Bucci*

Part VI Summary

26 **Summary and Implications — Nutritional Ergogenic
    Aids**... 507
    Ira Wolinsky and Judy A. Driskell

Index 519

Part I
Introduction

1
Nutritional Ergogenic Aids: Introduction, Definitions and Regulatory Issues

Amy A. Turpin, Jeff Feliciano and Luke R. Bucci

CONTENTS

I. Introduction 3
II. Working Definitions 4
III. Sports Nutrition Marketplace Update 5
IV. Regulatory Aspects 6
   A. Laws Regulating Foods and Dietary Supplements 7
      1. NLEA: Pertinent Issues with Ergogenic Aids 7
      2. DSHEA: Pertinent Issues with Ergogenic Aids 8
      3. Controlled Substances Act 11
      4. Fair Trade Practices: FTC Guidelines for Advertising 12
   B. Dietary Supplement Industry Self-Regulation 12
      1. Industry Trade Organizations 12
   C. Sports Governing Agencies 13
      1. Doping Issues 13
   D. Certification Programs 14
V. Summary and Conclusions 14
References 16

I. Introduction

Nutritional ergogenic aids have enjoyed recent attention and notoriety in the past decade more so than ever before. New regulations, more scientific study, more usage, more controversy, more media focus and more public scrutiny have enormously increased the awareness of nutritional ergogenic aids, but not necessarily an understanding. As with any emerging topic, there are many hidden agendas, misperceptions, dogmas and beliefs surrounding ergogenic aids. This chapter will clarify definition of dietary supplements, with an update on the United States marketplace for these products and the framework of laws within which ergogenic aids are regulated, and will address several recent controversies.

This chapter is written by supplement industry insiders with academic science roots, both in teaching and research. Unlike most reviewers of sports nutrition, we live daily with fiduciary and regulatory obligations different from those that academicians, clinicians and government employees must observe. The perspective of the dietary supplement industry has largely been ignored or disregarded, partly due to economic competition with much larger groups and partly due to a tradition of being labeled as alternative. However, the supplement industry is a significant, viable and growing segment of the U.S. economy that affects the lives of more than 59% of citizens daily. What may appear to many to be a sensational industry is actually quite steadfast and committed to promoting health. Thus, this chapter will present controversial issues from a rational and objective viewpoint that is seldom heard, but with a healthy respect for all opinions.

II. Working Definitions

Nutritional ergogenic aids are perceived as many things. Synonyms include "sports nutrition," "sports supplements," "sports drinks," "performance enhancers," "anabolics," "weight loss aids," "hydration drinks," and many other monikers. In actuality, a comprehensive definition for nutritional ergogenic aids would be: dietary manipulations to improve physical and sport performance. Dietary manipulations encompass three major efforts:

1. Alteration of food choices
2. Addition of macronutrients for specific uses in sports and exercise
3. Addition of micronutrients for specific uses in sports and exercise

Abundant scientific information exists for dietary and macronutrient manipulations, but only for some micronutrients.

Alteration of food choices concerns dietary habits and choices — i.e., what a person eats. This topic pertains to overall caloric intake, ratio of protein to carbohydrates to fat and how eating affects overall health as well as physical performance. Since everyone eats, this category is the largest type of nutritional ergogenic aid, but also the most unrecognized. Most persons who routinely exercise adjust their caloric demands via simple food choices. In the context of sports-nutrition products, entire categories of meal replacement, protein sources, refined carbohydrate sources, energy bars, snacks, low-fat foods and low-carbohydrate foods have merged with traditional food distribution. Foods and many sports-nutrition products are widely available at grocery stores and other common retail outlets. Many products originally designed for use by exercising individuals (i.e., sports energy bars) have crossed over into mainstream use as food choices for sedentary individuals or nonexercise uses.

Addition of macronutrients includes drinking water or water/electrolyte/carbohydrate drinks to extend or maintain performance, carbohydrate loading, increased protein intake, fat loading, hyperhydration and other manipulations of intake from macronutrients. Many investigators have reviewed the primary benefits of macronutrient manipulations,[1-10] and some sports governing agencies have issued guidelines for use, especially for hydration and electrolyte replacement.[11-15] Usually, macronutrient manipulations are accomplished by specific sports nutrition food products (i.e., sports drinks, energy bars, protein powders). Considerable overlap between alteration of food choices and macronutrient manipulation exists as more products replace foods.

The most controversial category is the micronutrients.[2-5, 16-20] This category contains what most people commonly think of when nutritional ergogenic aids or sports performance enhancers are discussed. Usually, products in this category are dietary supplements. Usually, but not exclusively, dietary supplements are pills (tablets, capsules and softgels). Dietary supplements in nonpill forms overlap with the other categories and some products contain macronutrients and micronutrients simultaneously. An example would be a protein powder with added vitamins, minerals, herbs and metabolites (such as creatine).

This chapter will focus primarily on marketing and regulatory issues surrounding nutritional ergogenic aids, so that other chapters in this volume might be seen in the context of current realities.

III. Sports Nutrition Marketplace Update

After passage of the Dietary Supplement Health Education Act (DSHEA) in 1994, the dietary supplement industry enjoyed yearly sales growth of 15–20% until 2000, when sales flattened. In 1999, according to a report presented to the FDA by the Research Triangle Institute Center for Economics Research, the sports nutrition category was estimated at $927 million dollars annually and 9% of dietary supplement sales.[21] However, of the $10.4 billion in dietary supplement sales, some portion of other categories (vitamins, herbals and botanicals, meal supplements, minerals and specialty supplements) were undoubtedly used for physical performance. To get an idea of the size of the industry, in 1997 only 16 supplement manufacturers had sales over $100 million annually and only eight raw vitamin material suppliers had sales over $50 million annually, while only eight herbal raw material suppliers had sales of over $20 million. This indicates that the dietary supplement industry was not monopolized and had many smaller players. Interestingly, the largest companies from each category were mostly pharmaceutical company subsidiaries.

In 2001, the U.S. nutrition industry was valued at $53 billion, of which $18.5 billion was classified as functional foods and $17.7 billion was classified as supplements.[22] Sports nutrition sales were estimated to be $1.7 billion and meal supplements were estimated at $2.3 billion. Growth rates for these categories were the highest across all supplement categories at around 10% annually. Again, an unknown percentage of supplements from other categories were undoubtedly used to assist performance.

The *Nutrition Business Journal* has classified the diet, weight loss, energy and sports supplements under the sports nutrition heading, as it is almost impossible to ascertain which products are used for each classification and many people combine diet and exercise to lose weight. Their estimates for 2001 indicated that the broad category of sports nutrition plus weight loss is in the neighborhood of $9.94 billion annually.[23] Sports and energy drinks (Gatorade®, AllSport®, Red Bull®) were estimated to bring in $2.9 billion annually. Weight loss meal supplements (e.g., SlimFast®) had $2.0 billion in sales, while weight loss pills (e.g., Metabolife®) had $1.9 billion in sales. Nutrition bars (e.g., PowerBar®, Clif Bar®) had $1.4 billion in sales. Sports supplements had $1.7 billion in sales, mostly from powders ($1.5 billion). Sports-specific drinks ($120 million) and pills ($100 million) were a much smaller portion of the sports nutrition supplement category. The large majority (78%) of sports nutrition and weight loss products (bars, meals, drinks and powders) were labeled and sold as foods, not dietary supplements. The segment that generates the most controversy (pills) brought in $2.0 billion in sales, mostly for weight loss. Pills for performance had 1% of sports nutrition product sales.

Dietary supplements in general ($17.6 billion annual U.S. sales in 2001) were sold about equally from mass merchandise retailers (club, drug, grocery, mass merchandise) (34%), natural health food outlets (33%) and other avenues (33%) in 2000.[23,24] Multilevel companies led other categories with 19% of total supplement sales, while Internet sales were only 1.5% of total supplement sales. Thus, it appears that most supplements are bought in familiar retail outlets. Data on where sports nutrition supplements are sold are less clear, but it follows that most are also purchased in familiar retail outlets.

IV. Regulatory Aspects

Few industries have generated as much controversy and misinterpretation over regulation as the dietary supplement industry and the broader sports nutrition segment that includes weight loss, in particular. In recent years, senior FDA officials have publicly stated that the industry is "unregulated," when it is their duty to enforce the existent regulations. The media has also perpetrated the incorrect notion that the dietary supplement industry is unregulated. Any allusion to the dietary supplement industry's being unregulated is not true.

A. Laws Regulating Foods and Dietary Supplements

Sports-nutrition products are subject to regulation as foods under either one of two amendments to the Food, Drug and Cosmetic Act, depending upon classification as a food or dietary supplement. Food products (those not labeled as dietary supplements) are subject to the Nutrition Labeling Education Act of 1990 (NLEA),[25] and dietary supplements are subject to the DSHEA of 1994.[26] By these legislative acts, the Food and Drug Administration of the United States Human Health Services (FDA) is responsible for enforcement of regulations for sports-nutrition products. The Center for Food Safety and Nutrition (CFSAN) is the department of the FDA that focuses on these categories. In addition, the National Institutes of Health (NIH) has two departments related to sports nutrition issues: Office of Dietary Supplements (ODS — mandated by DSHEA) and Office of Alternative Medicine (OAM). Thus, the U.S. government has considerable resources devoted to the science and enforcement of regulations covering sports-nutrition products.

1. NLEA: Pertinent Issues with Ergogenic Aids

Sports-nutrition products that are not labeled as dietary supplements are considered foods, the same as an apple or a hamburger. As such, food-like sports-nutrition products are regulated by the NLEA.[25] Pertinent issues with the NLEA are a prohibition for making disease treatment or prevention claims, except for specific and allowed Health Claims approved by the FDA. A nutrition facts panel describing content of calories, macronutrient amounts and essential vitamin and mineral amounts are required, in addition to an ingredients listing. Statements of facts are permitted. Sports performance is not considered a disease state or abnormal condition contributing to a disease state (such as inflammation, which has been so ruled) and thus, claims about sports-nutrition products are allowable as long as they do not aver or imply to treat, cure, prevent or mitigate a disease condition.

a. Health Claims

To account for a scientific consensus concerning the effects of foods or food components on disease states, eight health claims were permitted for foods and prevention of disease. Standards for what constitutes a scientific consensus were set very narrowly until recently, when court cases determined that expert scientific opinions on the weight of evidence favoring a claim are sufficient for FDA review of health claims. Nevertheless, health claims have been limited by court cases to permit only disease prevention or risk reduction claims, and claims about symptom reduction or disease treatment, regardless of veracity, are prohibited. At first, health claims did not apply to dietary supplements, but recently, health claims for dietary supplements have been allowed.

2. DSHEA: Pertinent Issues with Ergogenic Aids

In 1994, a grass roots campaign culminated in passage of the DSHEA, which defined dietary supplements as a separate subcategory of foods.[26] The DSHEA states "Except for purposes of section 201(g), a dietary supplement shall be deemed to be a food within the meaning of this Act."[26] Thus, supplements are legally considered foods and are not drugs. Therefore, any representations of dietary supplements as "drugs" are false. Both proponents and detractors of dietary supplements have used drug and supplement appellation interchangeably. In fact, peer-reviewed medical and scientific publications have allowed food and supplement components to be called drugs numerous times since 1995, in direct contradiction to the law. This common practice shows a lack of knowledge, misunderstanding or intentional disregard of current regulations by academicians and health care professionals and thus, the media and public.

a. Definition of Dietary Supplements

DSHEA was passed to assist the health status of the U.S. population and to take advantage of increasing documentation on the preventive and health-promoting effects of certain nutrients.[26] To integrate dietary supplements into improving health in the U.S., Congress deemed it necessary to give dietary supplements a separate definition and set of regulations. To be a dietary supplement, a product must be labeled as "dietary supplement" and be intended for ingestion.

As defined by DSHEA, a dietary supplement is a product other than tobacco intended to supplement the diet containing one or more of the following: (1) vitamins; (2) minerals; (3) herb or other botanical; (4) amino acid; (5) dietary substance; and (6) a concentrate, metabolite, constituent, extract or combination of any of (1) through (5).[26] DSHEA grandfathered all ingredients in commercial use as of October 14, 1994 as dietary supplements.

b. FDA Authority

i. Safety — The FDA has the burden of proof in U.S. courts to show a dietary supplement "… presents a significant or unreasonable risk of illness or injury …" or poses "… an imminent hazard to public health or safety …"[26] Safety risks are reported to U.S. attorneys and an individual is held responsible. Such person has 10 days to respond to proceedings initiated by the FDA. The FDA also has authority to remove products that are adulterated, contain a new dietary ingredient with insufficient safety data or are mislabeled. This chain of events to remove dietary supplement products represents a considerable departure from FDA practices prior to DSHEA and was specifically included to prevent unnecessary persecution of products knowing that, in general, dietary supplements had a low level of risk for harm.

Since DSHEA, the FDA has continued to remove individual products with success, in spite of a general perception that FDA enforcement powers have been limited by DSHEA. Another way the FDA can effectively limit product usage is to issue letters for consumers. The FDA has issued warning letters against using ephedrine-containing products, as well as other ingredients shown to have known risk factors (chaparral or gamma hydroxybutyrate, for instance). Thus, consumer safety has demonstrably been upheld under current regulations.

One obvious breach of DSHEA regulations easily found among dietary supplements is products labeled as sublingual or intended for buccal or nasal mucosal absorption. These products are clearly not intended to deliver their contents by ingestion, automatically making them unapproved drugs and adulterated dietary supplements. In spite of successful FDA actions in removing such products, many still appear.

Another condition of mislabeling is calling a dietary supplement "pharmaceutical grade" or "prescription strength." Comparison of dietary supplements to drugs is prohibited, since comparison implies they are drugs. Nevertheless, this practice still persists among a few dietary supplement manufacturers. Combinations of nutrients and drugs in the same package or pill are prohibited and FDA has sent letters asking manufacturers not to market such combinations until their legal status is clarified. Companies that proceeded to market such products have had actions taken against them.

ii. Premarket Notification — New ingredients for dietary supplements must have safety information submitted to the FDA before they are able to be marketed, or else the supplement is considered adulterated.[26] New dietary ingredient (NDI) submissions have been numerous and some have not been approved due to a lack of sufficient information to verify safety. Thus, since passage of DSHEA, the FDA has been able to approve new ingredients or remove those that did not follow the correct process of premarket notification. In other words, new ingredients cannot simply be put into the marketplace without FDA oversight. Other restrictions were that if a new substance had investigational-new-drug or new-drug status before being sold as a dietary supplement, it was ineligible as a dietary supplement.

iii. Good Manufacturing Practices — One provision of DSHEA was for the FDA to set forth good manufacturing processes (GMPs) specifically for dietary supplements at a later date.[26] The FDA was accepting public responses to its proposed GMPs in early 2003, with finalization and enactment to follow. At present, proposed dietary supplement GMPs resemble nonprescription pharmaceutical regulations instead of food or previous supplement practices. Regardless of outcome, dietary supplement GMPs will be a welcome change for consumers and larger supplement manufacturers, and a drastic change for some supplement manufacturers. Quality control and manufacturing procedures will be standardized throughout the industry, more evenly ensuring that label contents are met and no adulteration has occurred. It is expected that GMPs will provide a more level playing field and weed out less scrupulous manufacturers.

c. *Structure/Function Claims and Health Claims*

Another major intent of DSHEA was to allow legitimate claims about health for dietary supplements, as long as disease treatment claims were not made.[26] Statements of nutritional support for dietary supplements are allowed for:

- Benefits related to classical nutrient deficiency diseases (such as scurvy)
- Descriptions of the role of a nutrient "... intended to affect the structure or function in humans, ..."
- Descriptions of the mechanisms to maintain structures or functions
- Provision of well-being.[26]

Structure/function claims must have "... substantiation that such statement is truthful and not misleading ..." and an obvious disclaimer on the package.[26] The disclaimer is "This statement has not been evaluated by the Food and Drug Administration. This product is not intended to diagnose, treat, cure or prevent any disease."[26] Thus, accurate statements about how a nutrient works are allowed, but effects on diseases or abnormal health conditions leading to disease symptoms are prohibited, even if true. Manufacturers are required to file claims with the FDA and keep substantiation on file for inspection by the FDA. The FDA does not approve claims, but a new industry has sprung up to advise supplement companies on such matters. The period from 1994 to 2000 saw a gradual understanding of the limits of structure/function claims and the FDA has issued numerous courtesy and warning letters to dietary supplement manufacturers.

i. FDA Final Ruling on Structure/Function Claims — On January 6, 2000, the FDA issued a final ruling on structure/function claims that more clearly demarcated disease claims.[27] Pregnancy, premenstrual syndrome, menopause, aging and some gastrointestinal symptoms were considered normal nondisease conditions for which claims could be made, as long as substantiation existed. Other claims about reducing symptoms of disease, or conditions associated with or leading to disease, were considered the domain of drugs. For example, anti-inflammatory effects are prohibited because inflammation is considered an abnormal condition by the FDA, because pain is a symptom of inflammation that is also a symptom of disease or need for medical attention. Pain reduction or relief, except for noninjury exercise-induced muscle and joint soreness, is prohibited. For the purposes of sports nutrition, physical performance and its sequelae are not considered abnormal or disease conditions. Thus, structure/function claims for dietary supplement ergogenic aids have not attracted much scrutiny from the FDA.

To bridge the gap between the weight of scientific evidence and inability to improve health of Americans by being unable to convey disease prevention or treatment claims to consumers, court cases have allowed health claims for food and dietary supplements to be based on a preponderance of evidence rather than the more lengthy scientific consensus. Health claim submissions can now be sent to the FDA, which has to review the evidence in a relatively short time frame and approve or disapprove the claims. Nevertheless, in spite of favorable evidence, court rulings have not allowed claims for reduction of disease or disease symptoms. Health claims can only describe disease prevention or risk reduction, continuing the practice as for foods (low fat diets, calcium, fiber, soy protein). For example, proposed Health claims for saw palmetto stating reduction of benign prostatic hypertrophy symptoms were not allowed, since they were disease-treatment claims. In the time since court cases have redefined how health claims are submitted, health claims for prevention of some cancers by selenium, for cardiovascular disease risk reduction by sterols and reduction of risk for dementia from phosphatidylserine have been approved. Health claims for sports nutrition are not likely since such claims are already allowed as statements of fact or structure/function claims.

3. Controlled Substances Act

Another controversial issue in sports nutrition is use of prohormones as dietary supplements. This category is not included in this volume, but their mention is important because these products are available and have raised doping and drug testing issues for athletes that have received international media attention and public awareness. DSHEA allowed some prohormones to be marketed as dietary supplements because of grandfathering (DHEA, pregnenolone) and documented presence in the diet (androstenedione). Other prohormones (norandrostenediols, norandrostenediones and others) have relied on the definition of dietary supplements and presence in the diet in order to be introduced into the marketplace. The Drug Enforcement Agency (DEA) has jurisdiction over regulation of anabolic agents, including steroid hormones. The DEA is investigating whether prohormones can be classified as anabolic agents. If classified as anabolic agents, those prohormones would automatically become Schedule III compounds, in the same category as testosterone and anabolic steroids. This classification would immediately convert these compounds into controlled drug substances, thus prohibiting them for sale as dietary supplements, under DEA oversight.

The issue of whether (or which) prohormone products can be continued for sale as dietary supplements is a dynamic and complicated issue, with an uncertain outcome that hinges on interpretation of the current regulations. These issues will continue to evolve and their outcome will have an effect on how DSHEA is interpreted.

4. Fair Trade Practices: FTC Guidelines for Advertising

The Federal Trade Commission (FTC) regulates advertising claims. As with any industry, advertising claims should be truthful, nonmisleading and substantiated. The FTC has jurisdiction to enforce these guidelines. In 1998, the FTC produced a booklet with guidelines for advertising dietary supplements.[28] Guideline booklets are readily available online and from the FTC. Guidelines are similar to the long-standing Enforcement Policy Statement on Food Advertising, Food Policy Statement (information available online from www.ftc.gov). The FTC stressed that implication and "net impression" are used to assess truthfulness and accuracy of claims.[28] So far, the only guidelines for what constitutes adequate substantiation is "competent and reliable" evidence. Claims supported by scientific research studies are stronger if the actual product was tested and if study design is acknowledged by scientists to be well controlled. Research evidence can be borrowed if identity and dosage are equivalent for intended use. In other words, scientific expert opinions determine the veracity and applicability of substantiation for advertising claims for sports-nutrition products.

The FTC has given priority to products with weight loss claims, which are now classified in the realm of sports nutrition. The FTC has enjoined many companies from making weight loss claims, including some highly visible cases. In addition, the FTC has issued consent decrees for misrepresentation of prohormone supplements. Thus, the FTC is active in enforcement of claims for sports-nutrition products.

B. Dietary Supplement Industry Self-Regulation

As is true for any successful industry that reaches mass market acceptance and is faced with consumer health issues, industry self-policing becomes more necessary. The dietary supplement industry has reached the point where self-regulation is possible. Although no group claims total responsibility for self-regulation, several industry trade organizations are taking such steps.

1. Industry Trade Organizations

Several industry trade organizations are attempting to set guidelines for conduct for dietary supplement manufacturers. Membership is voluntary and entails considerable effort and expense. Primary industry organizations are:

- Council for Responsible Nutrition (CRN)
- National Nutritional Foods Association (NNFA)

- American Herbal Products Association (AHPA)
- Dietary Supplement Education Alliance (DSEA)

Industry trade organizations set standards of conduct, interact with media and regulatory agencies, lobby legislators and are intimately involved in current issues with dietary supplements. All have websites that describe their roles and interests. For example, CRN has a Sports Nutrition section and includes a scientific symposium on sports nutrition at their annual meeting. CRN has also issued guidelines on supplement use for teens and adolescents, using a green-, yellow- and red-light designation for supplement ingredients.[29] Thus, industry trade organizations have been taking active roles in sports-nutrition products.

C. Sports Governing Agencies

Numerous sports governing agencies also affect sports-nutrition products. Such agencies include the International Olympic Committee (IOC), United States Olympic Committee (USOC), professional sports associations, collegiate sports associations, coach and trainers professional organizations and medical associations (American College of Sports Medicine, in particular). These organizations set policies concerning use of diet, dietary supplements and drugs. They also enforce doping policies, testing athletes for use of substances banned from use, dependent on each governing agency. Most agencies have websites that outline their roles and functions.

1. Doping Issues

Dietary supplements have come under considerable attention for containing banned substances (for a current list, consult the United States Anti-Doping Agency website).[30] Of particular interest are the prohormones, all of which are banned substances for most sports competitions.[30] While dietary supplements with such prohormones as known ingredients pose an obvious hazard for athletes undergoing testing, athletes have accused dietary supplements in general of containing prohormones as undeclared trace contaminants. However, identifying banned substances in supplements not labeled to contain such has been extremely rare, in spite of intense efforts, although 7 out of 75 products purchased over the Internet contained unlabeled prohormones or stimulants.[31]

The issue of contamination of dietary supplements with banned substances was made possible by general ignorance of usual manufacturing practices. As a response, until implementation of dietary supplement GMPs, independent testing of products for banned substances has been sought and advocated. As outlined in the next section of this chapter, independent certification programs are not evolved sufficiently to provide comprehensive product testing. The World Anti-Doping Agency and United States Anti-

Doping Agency have no such product-testing programs, nor the desire to initiate such programs. Neither does any sports governing agency conduct such testing, although products from various companies have been screened for IOC banned substances by ConsumerLabs.com's Athletic Banned Substances Screening Program. Other efforts to independently test products for banned substances are in progress.

At present, if a dietary supplement manufacturer wishes to show its products contain no banned substances, or that taking its products will not cause positive doping tests, another option is to collaborate with academic institutions to perform such testing with IOC-sanctioned testing labs. One such example has been conducted and presented for a single herbal product.[32] Thus, options, albeit unwieldy and costly for manufacturers with a line of products, exist for independent verification of doping testing after human use of sports-nutrition products.

D. Certification Programs

A perceived need for greater public acceptance of product quality has led to several groups' offering independent testing programs of dietary supplement contents for label claims and lack of contaminants (heavy metals and pathogenic microbes). Major retailers have also expressed interest in product certification, and some actually require testing of products by independent laboratories before acceptance for sale. United States Pharmacopeia (USP) and the National Sanitation Foundation (NSF) have product certification programs under way. NNFA certifies manufacturing sites for GMPs that have turned out to be similar to proposed FDA GMPs for dietary supplements. Product content certification programs mean that validated assays for active ingredients must be developed and agreed upon. Such assays have been daunting from a technical standpoint and slow efforts to certify many products. Nevertheless, as independent certification programs become more extensive, consumers and retailers will have increased assurances that products are manufactured properly with high quality.

V. Summary and Conclusions

Nutritional ergogenic aids include dietary manipulations, macronutrient manipulations by functional foods or dietary supplements and micronutrient supplementation by dietary supplements. Quantitatively, more nutritional ergogenic aids are used for endurance exercise and are composed mostly of diet and macronutrient manipulations, as reflected by the sales for sports drinks spread among a few brands. However, an almost equally large number of sales are spread among many products and outlets for building muscle and losing body fat. In fact, the distinction between weight-loss and sports-nutrition products has become so blurred that industry followers consider them as one category. Indeed, many individuals exercise for their health or weight loss instead of practicing specific sports, and almost all desire leanness. In addition, there remains a segment of sports-nutrition products that is targeted toward anabolism, and these seem to generate the most controversy among sports supplements.

Nutritional ergogenic aids have an advantage over other food or dietary supplement categories in being able to make factual claims without running afoul of making disease or drug claims. This is because physical performance is not regarded as a disease or abnormal condition and thus, nutritional ergogenic aid product claims are mostly concerned with being accurate and nonmisleading. For example, it is not allowable under DSHEA to explain how a glucosamine dietary supplement has efficacy for reducing pain and improving mobility in mild-to-moderate osteoarthritis. However, products intended for weight loss purposes are vigorously scrutinized for claims and substantiation. As scientific research continues to examine the efficacy of ergogenic aids, ingredients fall into tiers of substantiation. For example, the sheer number of human-outcome studies on creatine have generated a large amount of substantiation for specific claims concerning short-term, repetitive, anaerobic exercise performance and increased body weight.

Another trend in dietary supplements as ergogenic aids is rapid turnover of ingredients and sophistication of proposed mechanisms of action for products targeted for anabolism. Even a brief perusal of current products in "muscle magazines" and Internet sites finds many ingredients intended for other uses being applied to anabolism. For example, many companies are utilizing substances that affect glucose and insulin metabolism in an effort to feed muscles. These product ingredient trends indicate a shift away from obvious mimics of anabolic steroids and their accompanying controversies to a more acceptable mechanism of action.

A seemingly eternal issue with ergogenic aids is the extent and quality of research into their efficacy and safety. There will never be enough timely research to test most substances or actual products. Thus, the ability to craft claims from available substantiation remains the major means of communicating potential product benefits.

Issues of active concern include acceptance of product quality assurance, doping, regulatory status of prohormones, safety of stimulants, certification of products and industry credibility. These issues are all being actively addressed by industry, regulatory agencies and other groups. In spite of controversies, nutritional ergogenic aids are a growing market. Segments of sports drinks, energy bars and meal replacements/snacks for control of caloric intake are popular and supported by large amounts of substantiation. Thus, nutritional ergogenic aids are a somewhat schizophrenic category of well-entrenched and accepted, but mostly invisible, products with a tiny subset of controversial products. Continued research and addressing of controversial issues with the goal of consumer acceptance is shaping nutritional ergogenic aids toward greater utility and acceptance.

References

1. Anonymous, Joint position statement: Nutrition and athletic performance. American College of Sports Medicine, American Dietetic Association and Dietitians of Canada, *Med. Sci. Sports Exerc.*, 32, 2130, 2000.
2. Antonio, J. and Stout, J.R., *Sports Supplements*, Lippincott Williams & Wilkins, Philadelphia, PA, 2001.
3. Applegate, E., Effective nutritional ergogenic aids, *Int. J. Sport Nutr.*, 9, 229, 1999.
4. Brouns, F., Nieuwenhoven, M., Jeukendrup, A. and Marken Lichtenbelt, W., Functional foods and food supplements for athletes: From myths to benefit claims substantiation through the study of selected biomarkers, *Br. J. Nutr.*, 88 Suppl. 2, S177, 2002.
5. Bucci, L., *Nutrients as Ergogenic Aids for Sports and Exercise*, CRC Press, Boca Raton, FL, 1993.
6. Driskell, J.A. and Wolinsky, I., *Energy-Yielding Macronutrients and Energy Metabolism in Sports Nutrition*, CRC Press, Boca Raton, FL, 2000.
7. Hargreaves, M., Pre-exercise nutritional strategies: Effects on metabolism and performance, *Can. J. Appl. Physiol.*, 26 Suppl, S64, 2001.
8. Ivy, J.L. Dietary strategies to promote glycogen synthesis after exercise, *Can. J. Appl. Physiol.*, 26 Suppl, S236, 2001.
9. Jacobs, K.A. and Sherman, W.M., The efficacy of carbohydrate supplementation and chronic high-carbohydrate diets for improving endurance performance, *Int. J. Sport Nutr.*, 9, 92, 1999.
10. Maughan, R., Sports nutrition: An overview, *Hosp. Med.*, 63, 136, 2002.
11. American College of Sports Medicine, Position stand on the prevention of thermal injuries during distance running, *Med. Sci. Sports Exer.*, 17, ix, 1985.
12. Sawka, M.N. and Coyle, E.F., Influence of body water and blood volume on thermoregulation and exercise performance in the heat, *Exerc. Sport Sci. Rev.*, 27, 167, 1999.
13. Maughan, R.J., Food and fluid intake during exercise, *Can. J. Appl. Physiol.*, 26 Suppl., S71, 2001.
14. Kay, D. and Marino, F.E., Fluid ingestion and exercise hyperthermia: Implications for performance, thermoregulation, metabolism and the development of fatigue, *J. Sports Sci.*, 18, 71, 2000.
15. Buskirk, E.R. and Puhl, S.M., *Body Fluid Balance: Exercise and Sport*, CRC Press, Boca Raton, FL, 1996.
16. Kreider, R.B., Dietary supplements and the promotion of muscle growth with resistance exercise, *Sports Med.*, 27, 97, 1999.
17. Lawrence, M.E. and Kirby, D.F., Nutrition and sports supplements: Fact or fiction, *J. Clin. Gastroenterol.*, 35, 299, 2002.
18. Nissen, S.L. and Sharp, R.L., Effect of dietary supplements on lean mass and strength gains with resistance exercise: A meta-analysis, *J. Appl. Physiol.*, 94, 651, 2003.
19. Schwenk, T.L. and Costley, C.D., When food becomes a drug: Nonanabolic nutritional supplement use in athletes, *Am. J. Sports Med.*, 30, 907, 2002.
20. Talbott, S.M., Dietary supplements in sport — a practical approach, *Curr. Topics Nutraceutical Res.*, in press, 2004.

21. Muth, M.K., Domanico, J.L.K. anderson, D.W., Siegel, P.H. and Bloch, L.J., Dietary supplement sales information. Final report, Contract No. 223-96-2290: Task Order 4, Research Triangle Institute, Research Triangle Park, NC, 1999.
22. National Nutritional Foods Association, Facts and stats, www.nnfa.org/facts/ Newport Beach, CA, May 2003.
23. Rea, P., Overview of the sports nutrition market, presented at Sports Nutrition 2002. Strategies for Profiling the New Sports Nutrition Consumer and Introducing New Product and Ingredient Concepts, *Nutrition Business Journal*, San Diego, CA, 2002. (contact Nutrition Business Journal at newhope.com)
24. www.Supplementinfo.org/industry/marketplace.htm, May 2003.
25. Nutrition Labeling and Education Act of 1990 (NLEA), Public Law 101-535, *Federal Register*, 55, 101, 1990.
26. Dietary Supplement Health and Education Act of 1994 (DSHEA), Public Law 103-417, *Federal Register*, 59, 4325, 1994.
27. Regulations on Statements Made for Dietary Supplements Concerning the Effect of the Product on the Structure and Function of the Body, *Federal Register*, 65, 1000, January 6, 2000.
28. Federal Trade Commission, Dietary supplements: An advertising guide for industry, Bureau of Consumer Protection, Washington, DC, 1998.
29. Council for Responsible Nutrition, Guidelines for young athletes: Responsible use of sports nutrition supplements, Council for Responsible Nutrition, Washington, DC, 2002.
30. United States Anti-Doping Agency, Prohibited classes of substances and prohibited methods, www.usantidoping.org/prohibited_sub/, May 2003.
31. Kamber, M., Baume, N., Saugy, M. and Rivier, L., Nutritional supplements as a source for positive doping cases?, *Int. J. Sport Nutr. Exerc. Metab.*, 11, 258, 2001.
32. Fomous, C.M., Costello, R.B. and Coates, P.M., Symposium: Conference on the science and policy of performance-enhancing products, *Med. Sci. Sports Exerc.*, 34, 1685, 2002.

Part II
Amino Acid Derivatives

2

Arginine

Jean E. Guest, Nancy M. Lewis and James R. Guest

CONTENTS

I. Introduction 21
II. Chemical Structure and Classification 22
   A. Structure 22
   B. Classification 22
III. Metabolic Functions 22
   A. Arginine 23
   B. Nitric Oxide 23
IV. Nutrient Assessment 24
   A. Body Reserves 24
   B. Requirements 25
   C. Dietary Supplementation 25
   B. Toxicity 27
V. Clinical Applications 27
   A. Hormone Secretion 27
   B. Immune Functions 28
   C. Physical Activity 29
VI. Summary 31
References 31

I. Introduction

Athletes of all performance levels strive to maximize physical abilities. Historically, ancient Greek Olympians ate mushrooms to enhance physical performance. In modern times, athletes have focused on dietary supplements as ergogenic aids to increase work output. Consumption of dietary supplements in the form of amino acids, individually or in combination, has become commonplace among athletes. In particular, athletes desiring androgenic effects have targeted the amino acid arginine as an ergogenic aid. This chapter discusses arginine's potential as an ergogenic aid.

II. Chemical Structure and Classification

A. Structure

Arginine is a basic, positively charged amino acid with neutral polarity containing two amino groups (NH_2), and one carboxyl group (COOH) (Figure 2.1). Like most of the 20 or 21 amino acids found in animal proteins, with the exception of glycine, arginine is of the L-optical configuration.

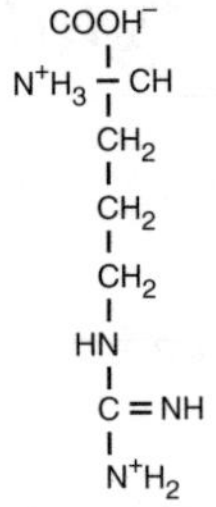

FIGURE 2.1
Structure of arginine.

B. Classification

In humans, amino acids have traditionally been classified as essential or nonessential, as described by Rose in 1957.[1] The essentiality classification of amino acids was based on the ability of the human body to synthesize proteins in sufficient quantities to meet metabolic requirements. Laidlaw and Kopple[2] in 1987 expanded the classification of amino acids by coining the terms " truly dispensable" (nonessential) and "conditionally indispensable" (essential).

Truly dispensable amino acids can be adequately synthesized in the human body under normal physiologic conditions. Conditionally indispensable amino acids may not be able to meet metabolic needs during specific physiological conditions.[3-6] Arginine is a truly dispensable amino acid in healthy children[7] and adults.[1] However, in premature infants[8] and individuals experiencing catabolic stress[3] arginine becomes a conditionally indispensable amino acid.

III. Metabolic Functions

Arginine is a complex amino acid essential in numerous physiologic and metabolic functions. Most human tissues utilize arginine for cytoplasmic and nuclear protein biosynthesis. In addition to protein synthesis, arginine has been identified as an essential nutrient in renal, cardiovascular, endocrine and immune functions.

A. Arginine

The primary site of arginine synthesis is the liver by way of the urea cycle. Ammonia produced from metabolism combines with carbon dioxide to form carbamoyl phosphatate. Carbamyl phosphate synthetase I is required for

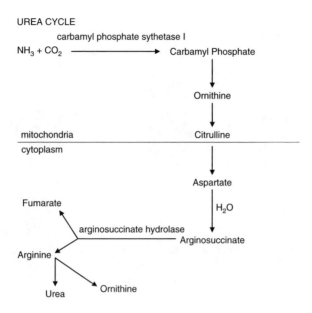

FIGURE 2.2
Synthesis of arginine via hepatocytes.

this reaction in the mitochondria of hepatocytes. A reaction between carbamyl phosphate and ornithine then produces citrulline. After citrulline is transported into the cytoplasm it reacts with aspartate to become argininosuccinate. Argininosuccinate via arginase then forms arginine and fumarate. Ornithine returns to hepatocyte mitochondria to continue the cycle of arginine synthesis.

Arginase activity is quite high in the liver to facilitate detoxification functions (Figure 2.2). Although arginase activity occurs in other tissues and organs, it is low by comparison with hepatic levels. Despite lower arginase activity, arginine is synthesized in the kidney and other tissues (i.e., endothelium, skeletal, muscle and brain).

B. Nitric Oxide

In 1980, Furchgott and Zawadzki[9] identified endothelium-derived relaxing factor (EDRF) as a humoral agent mediating the action of some vasodilators. These researchers demonstrated that endothelium was required for the vasodilator action of acetylcholine (ACh) in isolated animal vascular strips and arterial rings. In fact, the relaxant effect of ACh without endothelium resulted in contraction. Subsequently, Furchgott suggested that EDRF was actually nitric oxide (N0). In 1987, Palmer and co-workers[10] confirmed this hypothesis by measuring biological activity of EDRF and NO from endothelial cells in culture.

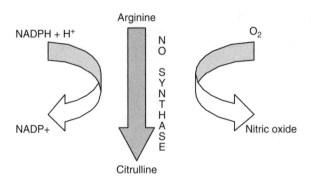

FIGURE 2.3
Nitric oxide synthesis.

In humans, NO is a naturally occurring gas produced by vascular endothelial cells.[10] NO is synthesized from nitric oxide synthase (NOS) interacting with the amino acid arginine through what is termed the arginine-nitric oxide pathway[11] (Figure 2.3). NOS has at least two families in this enzyme system. One, which is calcium- and calmodulin-dependent, is present in neurons, endothelial cells, platelets, mast cells, mesangial cells, endocardial and myocardial cells.[12–14] Another NOS family is not calcium- or calmodulin-dependent. This family is present in vascular smooth muscle, tumor cells, hepatocytes, macrophages, Kupffer cells, neutraphils, myocardial cells and fibroblasts.[14–16]

NO is important in numerous physiological functions in the human body. It increases cyclic guanosine monophosphate by activating guanlyate cyclase in vascular smooth muscle leading to smooth muscle relaxation. Relaxation of smooth muscle facilitates maintenance of regional blood flow and arterial blood pressure. NO is also important in neurotransmitter, gastrointestinal, respiratory, genitourinary and host defense functions.

IV. Nutrient Assessment

A. Body Reserves

Proteins are complex organic nitrogen-containing compounds. Most proteins consist of 20–22 α-amino acids in peptide linkages. Each protein has a unique shape and function determined by genetically embedded amino acid sequences. Proteins are important in cellular structure, enzymes, transport and storage, antibodies, blood clotting factors, visual pigments, hormones and some neurotransmitters. Protein stores can be utilized as an energy source if necessary.

The primary function of protein in the human body is to build and maintain body tissues. This occurs by a process known as protein turnover in which protein is continuously synthesized and degraded. Protein turnover in adults is approximately 250 g/d[17] based on a daily protein intake of 70–100 g. Rate of protein turnover varies at different stages of life. Whole body protein synthesis is highest in premature infants (17.4 g/kg/d) and lowest in the elderly (1.9 g/kg/d).[18]

Body protein status is determined by nitrogen balance involving dietary protein intake and protein losses. Body protein provides protein stores available to meet demands during times of increased protein need or stress. The majority of protein in the adult human body is present as skeletal muscle (about 43%).[19] The remaining protein is contained in other structural tissue (15%), metabolically active tissue (10%) and organs such as brain, lung, heart and bone (34%).

Protein turnover is a dynamic process involving the continuous exchange of body protein and free amino acids as well as *de novo* synthesis in cells. Free amino acid pools are found in plasma, tissue and body fluids. The exchange between body protein and free amino acid pools results in what is often referred to as "labile" protein stores.

Arginine is incorporated in tissue proteins. During times of homeostasis, arginine can be synthesized in adequate amounts from ornithine. It is only during times of physiological stress or in children with inborn errors of urea cycle metabolism that arginine may become an indispensable amino acid. Short-term body protein loss from dietary changes occurs mainly from the viscera,[20] while loss associated with chronic physiological stress comes from skeletal muscle.[21] Arginine body reserves have not been determined, and may actually vary by individual metabolic demands at any point in time.

B. Requirements

According to the FAO/WHO/UNU[22] technical report, adults require 117/mg/kg/d of arginine. Normal adult dietary intake of arginine has been reported to be 5.4 g/100g mixed dietary proteins.[23] A tolerable upper intake limit has not been determined for arginine due to a lack of sufficient data.

C. Dietary Supplementation

Arginine doses utilized in clinical research have ranged from 500 mg/d to 30 g/d over a period of 3 to 80 days. Reported dose-related effects of orally administered arginine encompassed the full spectrum from no observed effects to potentially adverse effects. No changes in liver function, BUN, creatinine or blood glucose were reported by Barbul and co-workers[24] in a group of healthy volunteers receiving 30 g/d of arginine hydrochloride over 7 to 21 days. Hurson et al.[25] reported no effects observed in 30 elderly adults supplemented with 30 g/d of arginine aspartate (17 g free arginine) over 14 days.

In another study, Siani et al.[6] studied the effect of a single blind, controlled, cross-over dietary intervention with two different arginine-containing diets and a control diet. In this study, six healthy volunteers aged 39 ± 4 years with a body mass index of 26 ± 1 kg/m² randomly received each of three diets for 1 week. Diet 1 was the control diet with a mean of 4.08 g/d of arginine. Diet 2 was identical to Diet 1 with the addition of arginine-containing foods to provide a mean arginine intake of 9.65 g/d. Diet 3 was identical to Diet 1 with the addition of 30 g/d of an oral arginine supplement. Dietary intake was controlled to provide a steady intake of calories and sodium (180mmol/d).

A significant decrease in systolic and diastolic blood pressure was seen in both Diet 2 ($p < 0.03$ and $p < 0.002$, respectively) and Diet 3 ($p < 0.001$ and $p < 0.006$, respectively) in this study. Other significant results demonstrated in this study include increased creatinine clearance ($P = 0.07$) and a fall in fasting blood glucose ($P = 0.06$) during Diet 3 when compared with Diet 2. However, Diet 2 provided a decrease in total serum cholesterol ($P = 0.06$) and triglyceride ($P = 0.009$) levels, while high-density lipoprotein cholesterol levels increased ($P = 0.04$) compared with Diet 3.

The results of this study are interesting in that they suggest numerous benefits to the health status of healthy adults resulting from increased arginine intake. The beneficial effects were different depending on arginine dose and source. Further studies in healthy humans are needed to realize whether the benefits observed in increased arginine intakes are reproducible and sustainable.

On the other hand, side effects from arginine supplementation have been reported. Barbul and colleagues[26] studied 36 healthy volunteers evenly divided into two treatment groups and a control group. Over 14 days, the treatment groups received 30 g/d of arginine as either hydrochloride (24.8 g free arginine) or as aspartate (17 g free arginine). Results of this study indicated that the arginine hydrochloride group experienced mild hyperchloremic acidosis. Participants in both treatment groups and the control group reported experiencing mild anorexia and gastrointestinal upset.

More serious effects of oral arginine supplementation have been reported by Park et al.[27] in breast cancer patients. These researchers administered 30 g/d of arginine to 10 breast cancer patients over 3 days just prior to surgery. A control group of 10 breast cancer patients received no arginine supplementation prior to surgery. Comparison of the two groups indicated that the arginine-supplemented patients had a median tumor protein synthesis rate that was double that of the control group. Additionally, a marked stimulation in the expression of the activation antigen Ki67 was measured in the arginine-supplemented group when compared with the control group. The results from this study suggest that large doses of oral arginine may stimulate tumor growth in certain cancers in humans.

Paradoxically, in breast cancer, arginine has been demonstrated to both promote tumor growth and stimulate host defenses. Brittenden et al.[28] demonstrated in breast cancer patients that a supplement of 30 g/d of arginine for 3 days prior to chemotherapy resulted in mitigation of chemotherapy-induced immune suppression and actually stimulated host defenses. This paradoxical nature of arginine provides some insight into the complexity of this amino acid.

B. Toxicity

Arginine is generally considered safe and nontoxic in humans. Large doses of arginine (0.5 g/kg to 30.0 g total) infused intravenously over a 20–30-minute period to assess pituitary function resulted in no reported side effects.[29,30] Despite arginine's being considered safe in humans, Tiwary[31] and co-workers in 1973 reported two cases of anaphylactic reactions to intravenous arginine infusion. However, purity of the amino acid solutions utilized in these anaphylactic reactions has been questioned.

V. Clinical Applications

A. Hormone Secretion

Arginine has been demonstrated to be a potent stimulator of endocrine hormones. These include human growth hormone,[32] insulin[32] and prolactin.[53] Human growth hormone has been associated with body mass accretion. In particular, athletes desiring androgenic effects have become interested in the ability of arginine to enhance human-growth-hormone[29] release, implying an ergogenic effect. In 1981, Isidori and co-workers[32] studied the effect of orally administered amino acids on the secretion of growth factors in 15 healthy male volunteers aged 15 to 20 years. Their results demonstrated that, in combination, lysine and arginine provoked a release of human growth hormone and insulin. The authors stated that this phenomenon was reproducible and did not reduce with time.

The suggestion that arginine has ergogenic properties was bolstered by two studies reported in the late 1980s by Elam et al.[34,35] Arginine and ornithine (1 g/d each) supplements in combination with a weight-training program resulted in reduced body fat, increased lean body mass and increased strength over a 5-week period. It should be noted that these studies have been criticized for being poorly controlled, using inappropriate statistical analyses and perhaps drawing conclusions not fully supported by data.[36]

In contrast, Fogelholm and co-workers[37] reported no significant effect on serum growth-hormone levels following ingestion of oral arginine, ornithine and lysine (2 g/d each) in competitive male weightlifters after 4 days. Similarly, Lambert et al.[38] reported that acute ingestion of either 2.4 g arginine-lysine or 1.85 g ornithine in male bodybuilders had no significant effect on serum growth hormone levels.

Another study casting doubt on the effect of arginine as an ergogenic aid was reported in 1997 by Suminski and colleagues.[39] These researchers conducted a factorial design study on the effects of arginine-lysine (1.5 g each) during resistance exercise. Sixteen young males were randomly assigned to one of four study periods consisting of exercise and amino acids, exercise and placebo, amino acids alone and placebo alone. All participants completed each study period.

Results of this study indicated that acute secretion of human growth hormone increased after amino acid administration. Plasma concentrations of human growth hormone were elevated at 30, 60 and 90 minutes during the exercise period compared with the no-exercise period ($p < 0.05$). Plasma growth-hormone concentrations were not significantly different among exercise periods. However, plasma growth-hormone concentrations were significantly elevated at 60 minutes after amino acid ingestion in the amino-acid-only period compared with the placebo-only period. This study infers that, although arginine-lysine supplementation elevates growth hormone secretion and circulating plasma levels it does not necessarily result in increased performance from muscle mass and strength.

Most of the research reported in the area of arginine and hormone stimulation has been in athletes involved with resistance-type exercise. Mixed reports regarding the effectiveness of arginine supplementation in combination with other amino acids in stimulating human growth hormone may reflect the low doses of arginine provided. On the other hand, exercise itself can stimulate human growth hormone, resulting in muscle mass accretion. The use of arginine as an anabolic ergogenic aid has not been supported by these studies.

B. Immune Functions

Arginine has been described as an immunonutrient for its ability to up-regulate immune function. Supplemental arginine has been demonstrated to have positive effects on immune responses in patients experiencing burns,[40] HIV/AIDS,[41] certain cancers,[28,40] surgery/trauma,[4] renal[42] and heart failure.[43]

Throughout the 1990s, studies in surgery/trauma patients demonstrated the beneficial effects of arginine supplementation on immune function. Reduction in length of hospital stay and infection rate,[21] fewer multiple-organ failures and intra-abdominal abscesses[44] and increased lymphocyte, CD3 and CD4 T-helper cells[45] were some of the significant outcomes attributed to supplemental arginine.

However, not all results of supplemental arginine studies have reported positive outcomes. In 1997, Saffle et al.[40] found no difference in incidence of wound infections in arginine-supplemented burn patients. These findings were in contrast to an earlier study by Gottschlich and co-workers,[3] who not only reported reduced wound infections but a shorter length of hospital stay in arginine-supplemented burn patients. In addition to burn patients, reports of no differences in immune function with arginine supplementation have been reported in HIV/AIDS,[5] and certain cancer[20,46,47] patients.

There is little information regarding the benefit of arginine supplementation and immune function in healthy athletes. Dietary arginine has been demonstrated to enhance T-cell-mediated immune function.[45] T-lymphocytes are important in cell-mediated immunity. The perception that T-lymphocytes are reduced in both moderate- and high-intensity exercise has been supported by some authors.[48-50] Despite reports that high-intensity exercise reduced B-lymphocytes,[51] natural killer cells[52] and neutrophils,[53] there is little information that these reductions are of clinical importance. Likewise, information supporting immune-function enhancement from arginine supplementation in athletes is lacking.

C. Physical Activity

The effect of arginine on physical activity in humans is not entirely understood. Beaumier et al.[54] investigated the effects of various levels of arginine intake in five healthy college-aged males participating in regular physical activity. A normal arginine diet (56.1mg/kg/d) or an arginine-supplemented diet (561 mg/kg/d) over a 6-day period was ingested by study participants. Intake of total energy, nitrogen and nutritionally indispensable amino acids were maintained at the same level during both diet periods. Tracer arginine, ornithine, leucine and urea were infused via nasogastric tube on day 7 of each study period.

Study results indicated that plasma arginine and ornithine fluxes increased significantly ($p < 0.05$) with arginine supplementation, as did the rate of plasma-labeled arginine to ornithine ($p < 0.05$) and rate of ornithine oxidation ($p < 0.001$). Urea production and excretion were significantly ($p < 0.05$) reduced, while plasma insulin levels increased significantly ($p < 0.05$) by approximately 25% during the arginine-supplemented diet period when compared with the normal diet period. In addition to the metabolic changes observed, participants lost a mean of 2.06 ± 1.08 kg ($p < 0.02$) during the arginine-supplemented diet period. Most weight loss occurred during days 2 and 3. Weight decrease was associated with increased daily urine output and total urinary sodium loss. The authors concluded that the kinetics observed in this study indicate a highly compartmentalized structure in whole body arginine and ornithine metabolism and their interrelationships.

The importance of arginine in physical activity and exercise may be as the precursor of NO. The physiologic response to exercise by NO has been attributed to endothelium-dependent vascular reactivity. NO inhibits platelet aggregation as well as platelet[55] and leukocyte[56] adhesion to the endothelium. The effect of NO on cardiovascular function has been suggested to play a role in increased endurance during exercise, and faster recovery after intense training.

Rodriguez-Plaza et al.[57] evaluated the relationship between physical activity and NO by measuring urinary excretion of NO metabolites (nitrates + nitrites) in 50 males. Participants were divided into four groups based on activity level. Group 1 consisted of 14 highly trained marathon runners (90 km/week), Group 2 had 11 male runners (64 km/week), Group 3 was composed of 12 sedentary males and Group 4 had 13 males with coronary artery disease who were participating in a 12-week cardiac rehabilitation program. Urinary NO metabolites were collected at baseline and after exercise in Groups 1 (a 42.2-km marathon), 2 (a 15-km race) and 4 (a 6-km walk), while only baseline urinary NO metabolites were collected in Group 3.

The results in this study showed baseline levels of urinary NO metabolites were highest in Group 1 (10.10 mmol/g creatinine), with Group 4 having the lowest (0.35 mmol/g creatinine). Only Group 1 experienced a significant (P = 0.0001) reduction (80%) in urinary NO metabolite excretion after exercise. However, Group 4 had been prospectively evaluated before and after their 12-week cardiac rehabilitation program. Urinary excretion of NO metabolites (mmol/g creatinine) were 157% higher when compared with baseline (P = 0.034) with a positive, significant correlation (P = 0.006) between the increases in exercise and urinary NO metabolite excretion induced by the 12-week program. According to these authors, the findings suggest that increased NO production may be a major adaptive mechanism by which chronic aerobic exercise training benefits the cardiovascular system. The marked increase in NO production induced by long-term high levels of aerobic exercise may be protective in athletes undertaking strenuous levels of exercise.

Jungersten et al.[58] theorized that individuals who exercise regularly have a higher resting level of NO formation. To test their hypothesis, 51 healthy nonsmoking individuals were examined to determine whether resting plasma nitrate (an end product of NO metabolism) levels were higher in athletes than nonathletes. Additionally, plasma nitrate levels were collected after exercise on a bicycle ergometer. Results of this study indicated that plasma nitrate was significantly ($p < 0.01$) higher in athletic participants than in nonathletic participants, and resting plasma nitrate level ($p < 0.01$) and urinary excretion of nitrate at rest ($p < 0.01$) correlated with participants' peak work rate. The authors of this study concluded that physical fitness and formation of NO at rest are positively linked, and that this relationship may help explain the beneficial effects of physical exercise on cardiovascular health.

Conventional wisdom accepts cardiovascular benefits of exercise as routine. The role of NO in endothelium-dependent vasodilation as a potential adaptive mechanism in chronic exercise was explored by Kingwell et al.[59] Endurance-trained male athletes were age matched with sedentary controls to assess vascular reactivity to infused Ach. Forearm vascular resistance to ACh was significantly greater in athletes (P = 0.03) when compared with sedentary controls. Enhanced vasodilation response to ACh was associated with lower total cholesterol (P = 0.03) in endurance-trained athletes. These authors concluded that enhanced endothelium-dependent dilator reserve was due to altered lipoprotein levels in athletes. This result implies that an adaptive mechanism from enhanced vascular reactivity might increase functional capacity in endurance-trained athletes. Although further studies are needed to confirm this finding, potential applications could be important to the general population as well as other athletes.

VI. Summary

Arginine supplementation has been demonstrated to be valuable in specific clinical applications, especially those involving physiological stress. These beneficial functions (hormone stimulation, immune regulation and endothelium reactivity) provoked interest in arginine as a potential ergogenic aid. Indeed, studies in healthy human beings and athletes demonstrated that arginine supplementation evoked metabolic and physiologic responses with few side effects. However, there is little evidence from the limited studies in the literature that these responses have true clinical significance. Therefore, the use of arginine as an ergogenic aid cannot be recommended at this time.

Future research needs to include dose-response effects of arginine supplementation using adequate duration of study length as well as large numbers of participants.

References

1. Rose, W.C., The amino acid requirements of adult man, *Nutr. Abstr. Rev.*, 27, 631, 1957.
2. Laidlaw, S.A. and Kopple, J.D., Newer concepts of the indispensable amino acids, *Am. J. Clin. Nutr.*, 46(4), 593, 1987.
3. Gottschlich, M.M., Jenkins, M., Warden, G., Boumer, T., Havens, P., Snook, J. and Alexander, J.W., Differential effects of three enteral dietary regimens on selected outcome variables in burn patients, *J. Parenter. Enteral. Nutr.* 14, 225, 1990.
4. Daly, J.M., Reynolds, J., Thom, A., Kinsley, L, Dietrick-Gallagher, M., Shon, J. and Ruttieri, B., Immune and metabolic effects of arginine in the surgical patient, *Ann. Surg.*, 208, 512, 1988.
5. Pichard, C., Sudre, P., Karsegard, V., Yerly, S., Slosman, D.O., Delley, V., Perrin, L. and Hirschel, B., A randomized double-blind controlled study of 6 months of oral nutrition, the Swiss HIV cohort study, supplementation with arginine and omega-3 fatty acids in HIV-infected patients, *AIDS*, 12, 53, 1998.
6. Siani, A., Pagano, E., Iacone, R., Iacoviello, L., Scopacasa, F. and Strazzullo, P., Blood pressure and metabolic changes during dietary L-arginine supplementation in humans, *Am. J. Hypertens.*, 13, 547, 2000.
7. Nakagawa, J., Takahashi, T., Suzuki, T. and Kobayashi, K., Amino acid requirements of children: minimal needs of tryptophan, arginine and histidine based on nitrogen balance method. *J. Nutr.*, 80, 305, 1963.
8. Heird, W.C., Dell, R.B., Driscoll, J.M., Grebin, B. and Winter, R.W., Metabolic acidosis resulting from intravenous alimentation mixtures containing synthetic amino acids, *New Eng. J. Med.*, 287, 943, 1972.

9. Furchgott, R.F. and Zawadzki, J.V., The obligatory role of endothelial cells in the relaxation of arterial smooth muscle by acetylcholine, *Nature*, 288, 373, 1980.
10. Palmer, R.M.J., Ferrige, A.G. and Moncada, S., Nitric oxide release accounts for the biological activity of endothelium-derived relaxing factor, *Nature*, 327, 524, 1987.
11. Moncada, S., Palmer, R.M.J. and Higgs, E.A., Biosynthesis of nitric oxide from L-arginine: A pathway for the regulation of cell function and communication, *Biochem. Pharmacol.*, 38, 1709, 1989.
12. Garthwaite, J., Garthwaite, G. and Palmer, R.M.J., NMDA receptor activation induces nitric oxide synthesis from Arginine in rat brain slices, *Eur. J. Pharmacol.*, 172, 413, 1989.
13. Moncada, S., Radomski, M.W. and Palmer, R.M.J., Endothelium-derived relaxing factor: Identification as nitric oxide and role in the control of vascular tone and platelet functions, *Biochem. Pharmacol.*, 37, 2495, 1988.
14. Palmer, R.M.J., The discovery of nitric oxide in the vessel wall. A unifying concept in the pathogenesis of sepsis. *Arch. Surg.*, 128, 396, 1993.
15. Kilbourn, R.G. and Belloni, P., Endothelial cell production of nitrogen oxides in response to interferon-Δ in combination with tumor necrosis factor, interleukin-1 or endotoxin. *J. Natl. Cancer Inst.*, 82, 772, 1990.
16. Nussler, A.K., Di Silvio, M., Billiar, T.R., Hoffman, R.A., Gellar, D.A., Selby, R., Madariaga, J. and Simmons, R.L, Stimulation of nitric oxide synthesis in human hepatocytes by cytokines and endotoxins, *J. Exp. Med.*, 169, 1467, 1989.
17. Waterlow, J.C., Protein turnover with special reference to man, *Quart. J. Exp. Physiol.*, 69, 409, 1984.
18. Hansen, R.D., Raja, C. and Allen, B.J., Total body protein in chronic disease and in aging, *Ann. N. Y. Acad. Sci.*, 904, 345, 2000.
19. Young, V.R., Steffee, W.P., Pencharz, P.B., Winterer, J.C. and Scrimshaw, N.S., Total human body protein synthesis in relation to protein requirements at various ages, *Nature*, 253, 192,1975.
20. McCarter, M.D., Genilini, O.D., Gomez, M.E. and Daly, J.M., Preoperative oral supplement with immunonutrients in cancer patients, *J. Parenter. Enteral Nutr.*, 22, 206, 1998.
21. Bower, R.H., Cerra, F.B., Bershadsky, B.,.Licari, J.J., Hoyt, D.B., Jensen, G.L., Van Buren, C.T., Rothkopf, M.M., Daly, J. M. and Adelsberg, B.R., Early enteral administration of a formula (Impact) supplemented with arginine, nucleotides and fish oil in intensive care patients: Results of a multicenter, prospective, randomized, clinical trial, *Crit. Care Med.*, 23, 436, 1995.
22. World Health Organization, Food and Agriculture Organization, United Nations University, Energy and protein requirements, *WHO Tech. Rep. Ser.*, 724, 1985.
23. National Research Council, National Academy of Sciences, *Dietary Reference Intakes for Energy, Carbohydrates, Fiber, Fat, Protein and Amino Acids*, Washington, D.C.: National Academy Press, 2002.
24. Barbul, A., Sisto, D.A., Wasserkrug, H.L. and Effron, G., Arginine stimulates lymphocytic immune response in healthy human beings, *Surgery*, 90, 244, 1981.
25. Hurson, M., Regan, M.C., Kirk, S.J., Wasserkrug, H.L. and Barbul, A., Metabolic effects of arginine in a healthy elderly population, *J. Parenter. Enteral Nutr.*, 19(3), 329, 1995.

26. Barbul, A., Lazarou, S.A., Efron, D.T., Wasserkrug, H.L. and Efron, G., Arginine enhances wound healing and lymphocyte immune response in humans, *Surgery*, 108, 331, 1990.
27. Park, K.G. M., Hays, S.D., Blessing, K., Kelley, P., NcNurlan, M.A., Eremin, O. and Garlick, P.J., Stimulation of human breast cancers by dietary L-arginine, *Clin. Sci.*, 82, 413, 1992.
28. Brittenden, J., Park, K.G. M., Heys, S.D., Ross, C., Ashby, J. and Eremin, O., L-Arginine stimulates host defenses in patients with breast cancer, *Surgery*, 115, 205, 1994.
29. Merimee, T.J., Lillicrap, D.A. and Rabinowitz, D., Effect of arginine on serum-levels of human growth-hormone, *Lancet*, 2, 668, 1965.
30. Merimee, T.J., Rabinowitz, D., Riggs, L., Burgess, J.A., Rimoin, D.L. and McKusick, V.A., Plasma growth hormone after arginine infusion: Clinical experiences. *New Engl. J. Med.*, 276, 434, 1967.
31. Tiwary, C.D., Rosenbloom, A.L and Julius, R.L., Anaphylactic reaction to arginine infusion, *New Engl. J. Med.*, 288, 218, 1973.
32. Isidori, A., Monaco, A.L. and Cappa, M., A study of growth hormone release in man after oral administration of amino acids, *Curr. Med. Res. Opin.*, 7, 475, 1981.
33. Rakoff, J.S., Siler, T.M., Sinha, Y.N. and Yen, S.S.C., Prolactin and growth hormone in response to sequential stimulation by arginine and synthetic TRF, *J. Clin. Endocrinol. Metab.*, 137, 641, 1973.
34. Elam, R.P., Hardin, D.H., Sutton, R.A. and Hagen, L., Effects of arginine and ornithine on strength, lean body mass and urinary hydroxyproline in adult males, *J. Sports Med. Phys. Fitness*, 29, 52, 1989.
35. Elam, R.P., Morphological changes in adult males from resistance exercise and amino acid supplementation, *J. Sports Med. Phys. Fitness*, 28, 35, 1988.
36. Williams, B.D., Chinkes, D.L. and Wolfe, R.R., Alanine and glutamine kinetics at rest and during exercise in humans, *Med. Sci. Sports Exerc.*, 30, 1053, 1998.
37. Fogelholm, G.M., Naveri, H.K., Kiilavuori, K.T. and Horman, M.H., Low-dose amino acid supplementation: No effects on serum human growth hormone and insulin in male weightlifters, *Int. J. Sports Nutr.*, 3, 290, 1993.
38. Lambert, M.I., Hefer, J.A., Millar, R.P. and Macfarlane, P.W., Failure of commercial oral amino acid supplements to increase serum growth hormone concentrations in male body-builders, *Int. J. Sports Nutr.*, 3, 298, 1993.
39. Suminski, R.R., Robertson, R.J., Goss, F.L., Arslanian, S.,Kang, J., DaSilva, S., Utter, A.C. and Metz, K.F., Acute effect of amino acid ingestion and resistance exercise on plasma growth hormone concentration in young men, *Int. J. Sports Nut.*, 7, 48, 1997.
40. Saffle, J.R., Wiebke, G., Jennings, K., Morris, S.E. and Barton, R.G., Randomized trial of immune-enhancing enteral nutrition in burn patients, *J. Trauma*, 42, 793, 1997.
41. Swanson, B., Keithley, J.K., Zeller, J.M. and Sha, B.E., A pilot study of the safety and efficacy of supplemental arginine to enhance immune function in persons with HIV/AIDS, *Nutrition*, 18, 688, 2002.
42. Annuk, M, Fellstrom, B. and Lind, L., Cyclooxygenase inhibition improves endothelium-dependent vasodilation in patients with chronic renal failure, *Nephrol. Dial. Transplant*, 17m 2159, 2002.

43. Watanabe, G., Tomiyama, H. and Doba, N., Effects of oral administration of L-arginine on renal function in patients with heart failure, *J. Hypertens.*, 18, 229, 2000.
44. Moore, F.A., Moore, E.E., Kudsk, K.A., Brown, R.O., Bower, R.H., Koruda, M.J., Baker, C.C. and Barbul, A., Clinical benefits of an immune-enhancing diet for early postinjury enteral feeding, *J. Trauma*, 37, 607, 1994.
45. Kirk, S.J., Hurson, M., Regan, M.C., Holdt, D.R., Wasserkrug, H.L. and Barbul, A., Arginine stimulates wound healing and immune function in elderly human beings, *Surgery*, 114, 155, 1993.
46. Park, K.G.M., Hays, S.D., Blessing, K., Kelley, P., NcNurlan, M.A., Eremin, O. and Garlick, P.J., Stimulation of human breast cancers by dietary L-arginine, *Clin. Sci.*, 82, 413, 1992.
47. Heys, S.D., Park, K.G.M. and McNurlan, M.A., Stimulation of protein synthesis in human tumors by parenteral nutrition: Evidence for modulation of tumor growth, *Br. J. Surg.*, 78, 483, 1991.
48. Frisina, J., Gaudieri, S., Cable, T., Keast, D. and Palmer, T., Effect of acute exercise on lymphocyte subsets and metabolic activity, *Int. J. Sports Med.*, 15, 36, 1994.
49. Mitchell, J., Pizza, F., Paquet, A., Davis, B., Forrest, M. and Braun, W., Influence of carbohydrate status on immune responses before and after endurance exercise, *J. Appl. Physiol.*, 84, 1917, 1998.
50. Nieman, D.C., Henson, D.A., Johnson, A., Lebeck L., Davis, J.M. and Nehlsen-Cannarella, S.L., Effects of brief, heavy exertion on circulating lymphocyte subpopulations and proliferative response, *Med. Sci. Sports Exerc.*, 24, 1339, 1992.
51. Gabriel, H., Schwarz, L., Born, P. and Kindermann, W., Differential mobilization of leukocyte and lymphocyte subpopulations into the circulation during endurance exercise, *Eur. J. Appl. Physiol. Occup. Physiol.*, 65, 529, 1992.
52. Nieman, D.C., Miller, A.R., Henson, D.A., Warren, B.J., Gusewitch, G., Johnson, R.L., Davis, J.M., Butterworth, D.E., Herring, J.L. and Nehlsen-Cannarella, S.L., Effects of high vs. moderate-intensity exercise on natural killer cell activity, *Med. Sci. Sports Exerc.*, 25,1126, 1993.
53. Hack, V., Strobel, G., Rau, J. and Weicker, H., The effect of maximal exercise on the activity of neutrophil granulocytes in highly trained athletes in a moderate training model, *Eur. J. Appl. Physiol. Occup. Physiol.*, 65, 520,1992.
54. Beaumier, L., Castillo, L., Ajami, A.M. and Young, V.R., Urea cycle intermediate kinetics and nitrate excretion at normal and "therapeutic" intakes of arginine in humans, *Am. J. Physiol.*, 269, E884, 1995.
55. Radomski, M.W., Palmer, R.M.J. and Moncada, S., Comparative pharmacology of endothelium-derived relaxing factor, nitric oxide and prostacyclin in platelets, *Br. J. Pharmacol.*, 92, 181, 1987.
56. Kubes, P., Suzuki, M. and Granger, D.N., Nitric oxide: an endogenous modulator of leukocyte adhesion, *Proc. Natl. Acad. Sci.*, 86, 4651, 1991.
57. Rodriguez-Plaza, L. G., Alfieri, A.B. and Cubeddu, L.X., Urinary excretion of nitric oxide metabolites in runners, sedentary individuals and patients with coronary artery disease: Effects of 42 km marathon, 15 km race and a cardiac rehabilitation program, *J. Cardiovasc. Risk*, 4(5-6), 367, 1997.

58. Jungersten, L., Ambring, A., Wall, B. and Wennmalm, A., Both physical fitness and acute exercise regulate nitric oxide formation in healthy humans, *J. Appl. Physiol.*, 82(3), 760, 1997.
59. Kingwell, B.A., Tran, B., Cameron, J.D., Jennings, G.L. and Dart, A.M., Enhanced vasodilation to acetylcholine in athletes is associated with lower plasma cholesterol, *Am. J. Physiol.*, 270, H2008, 1996.

3
Aspartate

Douglas S. Kalman

CONTENTS
I. Introduction 37
II. Aspartate's Role as a Nonessential Amino Acid 38
III. Aspartate as an Ergogen in Animals 38
IV. Aspartate as an Ergogen in Humans 41
V. The Malate-Aspartate Shuttle 42
VI. The Link to Liver Function 43
VII. Aspartate as a Delivery Agent 43
VIII. An Overview of Athletic Uses (Real and Potential) 43
IX. Summary 44
References 44

I. Introduction

L-aspartate is an amino acid naturally found in all living organisms including mammals and plants. L-aspartate is a dicarboxylic amino acid. Although most L-aspartate is in proteins, small amounts of free L-aspartate are found in body fluids and in plants. The normal diet contains about 2 grams of L-aspartate daily. It is also in the alternative dipeptide sweetener aspartame; the amount of L-aspartate from sweetener is a small fraction of the total L-aspartate consumed.

L-aspartate is also known as L-amino succinate. The physical form of commercially sold aspartate is a solid, with an acid form also available that is slightly soluble in water, and with additional salt forms that are more water soluble. Available salts include magnesium, calcium, potassium, zinc and combinations thereof. L-aspartate is used interchangeably with the term aspartic acid. The biological form of this substance, however, is the anion of aspartic acid, L-aspartate.

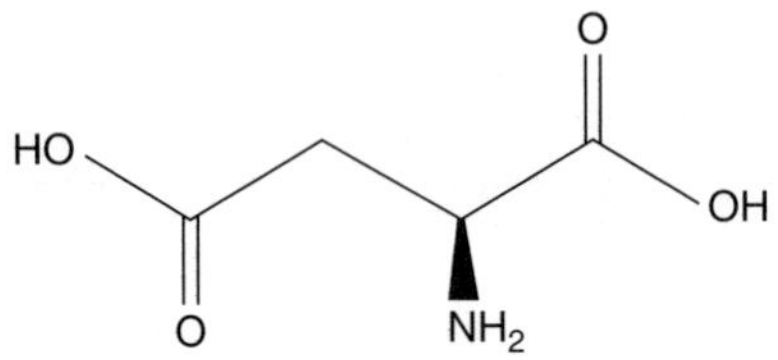

FIGURE 3.1
Structure of aspartic acid

Following ingestion, L-aspartate is absorbed from the small intestine by an active transport process. Following absorption, L-aspartate enters the portal circulation and from there is transported to the liver, where much of it is metabolized to protein, purines, pyrimidines and L-arginine, and is catabolized as well. L-aspartate is not metabolized in the liver; it enters the systemic circulation, which distributes it to various tissues of the body. The cations associated with L-aspartate independently interact with various substances in the body and participate in various physiological processes.[1]

The structure of aspartic acid is given in Figure 3.1.

II. Aspartate's Role as a Nonessential Amino Acid

L-aspartate is considered a nonessential amino acid, meaning that, under normal physiological conditions, sufficient amounts of the amino acid are synthesized in the body to meet the body's requirements. L-aspartate is formed by the transamination of the Krebs cycle intermediate oxaloacetate. The amino acid serves as a precursor for synthesis of proteins, oligopeptides, purines, pyrimidines, nucleic acids and L-arginine. L-aspartate is a glycogenic amino acid that can also promote energy production via its metabolism in the Krebs cycle. From aspartate's metabolic roles, nutritional theorists claim that supplemental aspartate has an antifatigue effect on skeletal muscle. This claim has been the source of many clinical investigations and may underpin the claims of aspartate as an ergogenic aid.[2]

III. Aspartate as an Ergogen in Animals

It is hypothesized that exogenous aspartate could spare muscle glycogen and/or enhance the resynthesis of ATP during exercise. To test this hypothesis, six groups of rats were studied.[3] The groups either received intraperitoneal (IP) injections of aspartate, saline or nothing. Some of the groups performed swimming exercise, while others did nothing. All animals were sacrificed after a predetermined length of time. The groups of rats that performed the swimming exercise swam until exhaustion (voluntary). There were no significant differences in exercise performance or duration when comparing the aspartate-supplemented group to the control (saline). However, the rats that received the aspartate had a significantly lower amount of plasma free fatty acids (FFA) after the exhaustive swims as compared with the saline-treated rats. Postmortem glycogen analysis did not reveal any differences between the groups. The IP dose of aspartate was 1 gm/kg body weight.

More recently, the potential for aspartate combined with asparagines to stave off exercise-induced fatigue was analyzed utilizing a multiple-day dosing period.[4] The aim of the study was to determine whether aspartate and asparagines could affect hydrogen ion accumulation and lactate concentration in rats swimming above the lactate threshold. Prior to the 7-day supplementation period, all the test rats performed 20-minute daily acclimation swimming exercise. The swimming was weighted, meaning that the rats performed progressive overload training utilizing weights for intervals of 3-minute swims and 1-minute rest. Lactate threshold was established. One group of rats received the aspartate/asparagines combination, while the other group received distilled water. After the acclimation period, the rats were further divided into two groups that would perform the exercise-to-exhaustion swimming and two groups that would be immediately sacrificed. In both sets, there was one supplement group and one control group. In the animals that were sacrificed after 5 days of supplementation, there were no changes in any physiological markers (muscle and liver glycogen, blood glucose, lactate, alanine and glutamine concentrations, as compared with the control group. However, in the rats that swam to exhaustion at a level above their lactate threshold, the supplement group was able to exercise significantly longer (by approximately 27 minutes) and had lower lactate levels (by about 2.5 mmol/L). Upon being sacrificed and examined, it was also noted that the supplement group had less glycogen used (better glycogen preservation) in various muscles of the leg (gastronemius and extensor digitorius). The results strongly suggest that the combination of aspartate and asparagines may positively influence the contribution of oxidative metabolism and delay fatigue during exercise performed above the lactate threshold. Further, the results suggest that alterations in the malate-aspartate shuttle were responsible for the reduced lactate formation from pyruvate (via reduced cytosolic hydrogen ion formation; see Figure 3.2). More research is needed to see if the same effect can be duplicated by a separate laboratory and in humans.

In high doses, aspartate can be toxic, as animal studies have demonstrated that it can cause brain lesions. This same effect has not been observed in humans.[2]

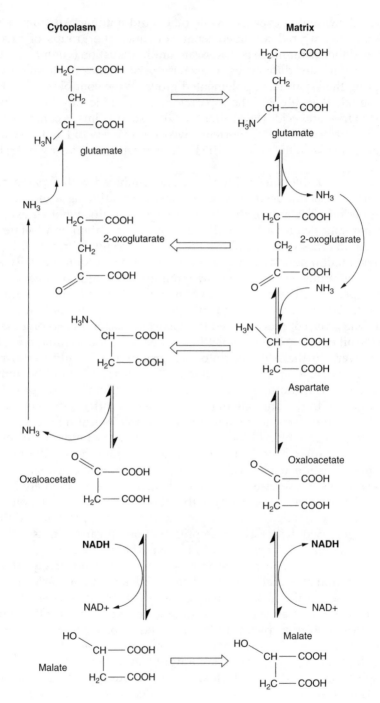

FIGURE 3.2
Malate-aspartate shuttle.

IV. Aspartate as an Ergogen in Humans

Aspartate is often delivered as a salt (magnesium, potassium, calcium and zinc), though it can be orally taken in its base form (L-aspartic acid). The most common dietary supplement salt form is a combination of potassium and magnesium. To test whether the popular combination of potassium/magnesium aspartate (K+/Mg+) could have any effect on ventilation, oxygen uptake, carbon dioxide production, respiratory quotient, heart rate or blood pressure, 7.2 grams of the salt was administered to seven healthy and fit adults. Study participants underwent control testing, placebo treatment and the K+/Mg+ aspartate treatment.[5] All subjects ingested their respective products or placebo 1 day prior to doing a sub-maximal treadmill test (90 minutes of walking at 62% VO_2 maximum). The study found that loading with the K+/Mg+ aspartate salt for 24 hours pre-exercise at a dose of 7.2 grams had no effect on any physiological parameter when compared with placebo or control within the confines tested. The possible reason for lack of any physiologic effect of supplementation may be due to the low level of exercise (lack of a stressful test) or the supplement's not being ingested the day of exercise.

Aspartate and asparagine are precursors to the Krebs cycle intermediate, oxaloacetate. The amino acid carnitine is the product of lysine and methionine (di-peptide) as well as being supplied by the diet. Carnitine is used to transport free fatty acids into the mitochondria for beta-oxidation. In a study where the researchers used a combination of aspartate, asparagines and carnitine and compared it with a control group, those receiving the supplement were able to exercise 40% longer.[6] In addition, serum analysis of the blood revealed greater glycogen preservation and free fatty acid utilization by those receiving the dietary supplement. The study has many weaknesses and was not well described in the literature; however, it is feasible that "priming" oxaloacetate production may have an effect on exercise endurance. This study needs modern replication where the combination of amino acids is compared with placebo rather than control. In addition, the replication study should employ a common lactate threshold/time-to-exhaustion trial format.

The effects of aspartate supplementation on ammonia production were examined during and after intense resistance-training workouts. Twelve weightlifters were randomly assigned to either receive the aspartate (150 mg/kg) or vitamin C (placebo) on the days of exercise testing (7). Subjects received the supplement or placebo over a 2-hour period prior to the exercise. Exercise consisted of bench, incline, shoulder and triceps presses along with bicep curls. Subjects lifted at 70% of their one-repetition maximum (RM). After the workout sessions, subjects performed a bench press test (repetitions to failure at 65% 1-RM). Blood was withdrawn pre-exercise, 20 minutes and 40 minutes into the workout, at the conclusion of training and 15 minutes post-exercise.

There were no significant differences between the placebo and aspartate group for ammonia production pre or post exercise. Thus, acute aspartate loading has no apparent buffering effect during or post exercise. The study did not examine whether supplementation had any effect on workload; while it may not have had one on ammonia production, the overall exercise performance may have been impacted. A repeat study is warranted to examine whether aspartate has any effect on weightlifting performance.

Arginine is an amino acid that is thought to be somatotrophic in physiologic concentrations. In can be combined with aspartate for better absorption (arginine aspartate). To test if arginine aspartate had any general metabolic effect in runners, a chronic dosing study was undertaken. Fourteen marathon runners participated in a double-blind crossover clinical trial.[8] Study participants were supplemented with either 15 grams of the arginine aspartate or a carbohydrate solution for 14 days prior to running a marathon. Blood samples were collected before the run, after 31 kilometers (~19 miles), at the end of the run and again 2 hours after finishing the marathon. The blood samples were analyzed for amino acids; hormones and the respiratory ratio were calculated during the run. Those receiving the carbohydrate supplement had no significant changes in hormones, creatine kinase or the respiratory exchange ratio. However, those who received the arginine aspartate supplementation had significant changes (increases) in somatotrophic hormone (growth hormone), glucagons, urea and arginine, while most of the other amino acids were reduced in the plasma. There was no effect on RQ from supplementation. The potential benefit or metabolic consequence of alteration of growth hormone in runners is unknown. Thus, the applicability of this study for translation into potential for ergogenic enhancement is unknown. It should be noted that aspartate was the delivery agent for arginine and most likely had no ergogenic effect by itself.

V. The Malate-Aspartate Shuttle

Most of the energy derived from the oxidation of glucose is not extracted directly as adenosine tri-phosphate (ATP), but as reduced NADH that transfers high-energy electrons to the electron transport chain in the inner mitochondrial membrane. The malate-aspartate shuttle (Figure 3.2) occurs in specific tissues. First, oxaloacetate (OAA) on the cytoplasmic side is reduced by NADH, creating malate and NAD+. The malate then crosses the matrix membrane via a dicarboxylic acid carrier. It is then converted back to OAA in the matrix, converting matrix NAD+ to NADH. Once inside, the energy in malate is extracted again by reducing NAD+ to make NADH, regenerating oxaloacetate. This NADH is then free to transfer its high-energy electrons to the electron transport chain. The oxaloacetate is transaminated with glutamate to make aspartate and alpha-ketoglutarate. Aspartate is returned to the cytosol by the aspartate-glutamate transporter, which moves glutamate into the mitochondria as it transports aspartate out. This aspartate then crosses the matrix membrane via an amino acid carrier, and is duly converted by the same transamination reaction in reverse back to OAA. The amino group from the aspartate is transferred to 2 oxoglutarate, giving glutamate, which crosses back to the matrix on the same carrier as the aspartate. When glutamate is deaminated in the matrix it becomes 2 oxoglutarate, which is then transferred to the matrix to accept the amino group off aspartate. This process generates three ATP per every molecule of NADH (glycolysis). The malate-aspartate shuttle is active within heart, kidney and liver cells (9).

VI. The Link to Liver Function

Aspartate undergoes metabolism in the liver. Aspartate aminotransferase (AST) is a liver enzyme. It is similar to alanine aminotransferase (ALT) but less specific for liver disease as it is also produced in muscle and can be elevated in other conditions (for example, early in the course of a heart attack). AST is produced by hepatocytes and myocytes. Dietary or supplemental aspartate has no effect on serum AST readings. The only commonality is aspartate itself.[10] AST catalyzes the aspartate to OAA reaction by transferring an amino acid to an alpha-keto acid (i.e., alpha ketogluturate).

VII. Aspartate as a Delivery Agent

Aspartate is an amino acid. It is readily available in the body as aspartic acid. Dietary supplement forms use L-aspartate. It is commercially available in the magnesium, potassium, calcium and zinc salts. Thus, aspartate is used in the creation of salt forms of many minerals and can be chelated with other amino acids (i.e., arginine). It is typically supplied in 600 mg doses.[1]

VIII. An Overview of Athletic Uses (Real and Potential)

The studies for performance benefit have been inconclusive. Many utilizing the time-to-exhaustion method show no benefit, while others demonstrate a physiological effect of aspartate supplementation. The potential exists that the right study question has not been asked in order to elucidate a definitive ergogenic property of aspartate.

Currently, it appears that there may be some benefits for high-intensity runners and people involved in high-intensity repetitive sports. More research is needed.

IX. Summary

The evidence for aspartate to be an ergogenic acid is equivocal. Some animal studies demonstrate benefit while others do not. The same pattern is true in human studies. Aspartate may have potential as an ergogenic aid under the right conditions (i.e., anaerobic metabolism testing exercise endurance), but more evidence is needed. It is apparent that aspartate offers no benefit to the athlete performing sub-maximal exercise, nor does it exert any body-composition-altering properties. Aspartate is considered an excitatory amino acid, as AST activity is common in neural tissue.[2] However, no one has examined whether aspartate supplementation (acute or chronic) has any effect on central nervous system activity.

There is currently not enough evidence to suggest that aspartate has ergogenic effects; more research is needed under the conditions mentioned. The most common dose of aspartate used (magnesium/potassium moiety) is 10 grams over a 24-hour period.[11]

References

1. Hendler, S.S. *The PDR® for Nutritional Supplements*. March 2001. Thomson Healthcare. Montvale, NJ.
2. Groff, J.L., Gropper, S.S. and Hunt, S.M. *Advanced Nutrition and Human Metabolism*. 2nd ed. 1995. PP 172, 181, 499.West Publishing Company, St. Paul, MN.
3. Trudeau, F. and Murphy, R. Effects of potassium-aspartate salt administration on glycogen use in rat during swimming stress. *Physiol. Behav.* 1993; 54(1):7–12.
4. Marquezi, M.L., Roshel, H.A., Costa, A., Sawada, L.A. and Lancha, A.H. Jr. Effect of aspartate and asparagines supplementation on fatigue determinants in intense exercise. *Int. J. Sport Nutr. Exerc. Metab.* 2003; 13:65–75.
5. Hagan, R.D., Upton, S.J., Duncan, J.J., Cummings, J.M. and Gettman, L.R. Absence of effect of potassium-magnesium aspartate on physiologic responses to prolonged work in aerobically trained men. *Int. J. Sports Med.* 1982; 3(3):177–181.
6. Lancha, A.H., Recco, M.B., Abdalla, D.S. and Curi, R. Effect of aspartate, asparagines and carnitine supplementation in the diet on metabolism of skeletal muscle during a moderate exercise. *Physiol. Behav.* 1995; 57(2):367–371.
7. Tuttle, J.L., Potteiger, J.A., Evans, B.W. and Ozmun, J.C. Effect of acute potassium-magnesium aspartate supplementation on ammonia concentrations during and after resistance training. *Int. J. Sport Nutr.* 1995; 5(2):102–109.
8. Colombani, P.C., Bitzi, R., Frey-Rindova, P., Frey, W., Arnold, M., Langhans, W. and Wenk, C. Chronic arginine aspartate supplementation in runners reduced total plasma amino acid level at rest and during marathon run. *Eur. J. Nutr.* 1999; 38(6):263–270.
9. McArdle, W.D., Katch, F.I. and Katch, V.L. *Exercise Physiology: Energy, Nutrition and Human Performance*. 5th ed. 2001, Page 143. Lippincott, Williams & Wilkens. Baltimore, MD.

10. Worman, H.J. Common Laboratory Tests in Liver Disease. Available on: http://cpmcnet.columbia.edu/dept/gi/labtests.html. Accessed June 20, 2003.
11. Williams MH. *The Ergogenics Edge: Pushing the Limits of Sports Performance.* pp 135. 1998 Human Kinetics, Champaign, IL.

4
Branched-Chain Amino Acids

Robert E.C. Wildman

CONTENTS

I. Introduction 47
II. Branched-Chain Amino Acid Supplementation 49
III. Food and Supplementation of Branched-Chain Amino Acids 49
IV. Absorption of Branched-Chain Amino Acids 50
V. Tissue Distribution and Metabolism of Branched-Chain
   Amino Acids 50
VI. Branched-Chain Amino Acids and Exercise-Induced Fatigue 52
VII. Branched-Chain Amino Acid Supplementation and
   Exercise-Induced Changes in Immune Status 55
VIII. Branched-Chain Amino Acid Supplementation and Muscle
   Fiber Hypertrophy and Pathology in Response to Exercise 56
IX. Conclusions 57
Acknowledgments 57
References 57

I. Introduction

For centuries, man has searched for means to improve physical performance. The list of substances tried is long and the purported benefits varied. The branched-chain amino acids (BCAA) emerged as a potential ergogenic aid a couple of decades ago and have remained popular since. Figure 4.1 presents an overview of the major purported ergogenic benefits of BCAA supplementation. These amino acids, namely leucine, isoleucine and valine (Figure 4.2), have been touted to improve physical performance by reducing peripherial and/or central fatigue associated with sustained moderately high-intensity activity.[1] Supplementation of BCAA has also been speculated to have a beneficial influence on immune status following strenuous exercise and to

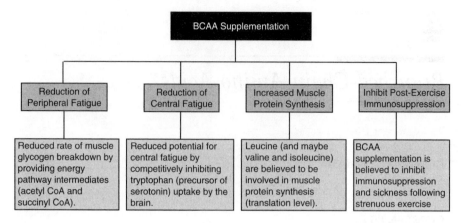

FIGURE 4.1
Major purported ergogenic benefits of BCAA supplementation.

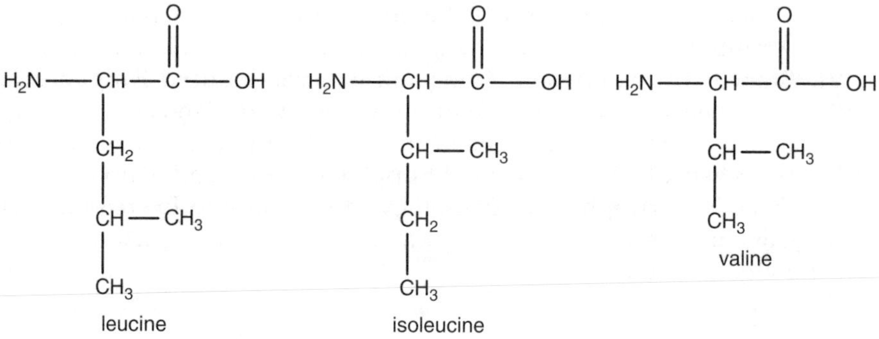

FIGURE 4.2
Branched-chain amino acids.

possibly reduce the incidence of sickness.[1] In addition, BCAA, and more specially leucine, have been identified as important factors involved in regulating the production of key muscle proteins that might promote positive protein balance and muscle hypertrophy. Hydroxymethyl butyric acid (HMB), a derivative of leucine, is also a popular supplement that is touted to increase strength and lean body mass.[1]

This chapter will review the basic structure, metabolism and properties of BCAA in relation to exercise and sport performance. In addition, a brief overview of the current industry trends related to BCAA, alone and as part of the amino acid preparations and intact protein, will be provided. HMB, mentioned briefly in this chapter, will be covered in more detail in another chapter in this volume.

II. Branched-Chain Amino Acid Supplementation

The nutrition supplement industry continues to grow in the United States and other countries. For instance, the sport supplement market in the United States grew by 8.7% between 2000 and 2001 as sales increased from $1.6 billion to $1.74 billion, with 32.4% of those sales in protein and muscle growth supplements.[1] Meanwhile, the sport drink industry raked in $2.4 billion during 2001.[1] BCAA (and HMB) can be found in classic supplemental form (pills, capsules, etc.) as well as formulated into sport drinks, bars and shakes (powder and ready-to-drink (RTD)). Supplementation recommendations and practices vary and can include five grams of leucine, four grams of valine and two grams of isoleucine. It is sometimes recommneded that the BCAA be taken together and in a balance of 2 to 2 to 1 (leucine:valine:isoleucine) although supportive information is scarce. Recommended levels of consumption seem to depend on training intensity, and manufacturers generally advise that individuals take from 1 gram/day on rest days and easy days to about 3 grams on heavy training days.[2]

III. Food and Supplementation of Branched-Chain Amino Acids

Amino acid supplements are often viewed by the public as safe and natural alternatives for those individuals wishing to avoid many of the hormone-based supplements on the market today. Like many nutrition supplements marketed for ergogenic purposes (such as creatine, carnitine and glycerol), BCAA are found naturally in nutritional significant levels in the diet. Therefore, both food and supplementation should be considered together when designing research studies and developing recommendations for supplementation. BCAA are found in all protein-containing foods, but egg and poultry, meat and milk (particularly whey) have notable BCAA content (BCAA to total protein ratio). β-lactoglobulin, the most abundant whey protein, is leucine rich. Whey protein has become one of the most popular types of protein used in powders and high-protein sport bars and shakes. Among plant food sources, oats, soy, lentils and wheat germ have respectable leucine levels. Isoleucine is found in especially high amounts in meats, fish, cheese, most seeds and nuts, eggs, chicken and lentils. Important sources of valine include soy flour, cottage cheese, fish, meats and vegetables.

The typical adult diet might contain BCAA at a level of 50 to 150 mg/kg of weight, with the level and type of protein consumed being the primary determinant. Meanwhile, the recommneded dietary allowances (RDA) for the BCAA are 12, 14 and 16 mg/kg for isoleucine, valine and leucine. At this time very little is known regarding the potential toxicity of BCAA and the upper limits (UL) have not been set. HMB is found in the diet in small amounts in some protein-rich foods such as fish and milk. Depending on total protein and leucine intake, HMB production in the body may average about .25 to 1 gram per day.

IV. Absorption of Branched-Chain Amino Acids

The BCAA derived from food and supplements must move through the cell layer lining the small intestine and then enter the hepatic portal blood before reaching the systemic circulation. Therefore, BCAA must first pass the small intestine and liver before reaching skeletal muscle. A significant portion of the amino acids will be retained in the wall of the small intestine and liver, where they are metabolized for energy or used to make proteins. More specifically, it has been estimated that the small intestine and liver extract 20–90% of absorbed amino acids.[3] The reason for this broad range is largely the preferential uptake of different amino acids. For instance, only about 20% of absorbed leucine might be extracted, while greater than 50% of the tryptophan is removed.[4,5] In addition, the liver is particularly aggressive at removing glutamate (glutamic acid) during the first pass. Relative to the other amino acids, the BCAA seem to be spared during the first pass, with skeletal muscle being the primary destination tissue.

V. Tissue Distribution and Metabolism of Branched-Chain Amino Acids

The amino acids (BCAA) are found throughout the body as part of proteins (99%) and the free amino acid pool (1%) dissolved in intracellular and extracellular fluids. For instance, an adult male weighing about 154 lb (70 kg) might contain roughly 22 lb (11 kg) of protein. Approximately 14 lb (7 kg) of that would be found in skeletal muscle, which might contain only 120 g of free amino acids.[1] Only about 5 g of amino acids would be present in circulation. The amino acids in the pool are available for protein synthesis, conversion to other molecules (e.g., neurotransmitters, creatine, carnitine, etc.) or energy metabolism.

Certain key tissues throughout the body demonstrate preferential amino acid metabolism. Among the most outstanding examples of this are BCAA and skeletal muscle. The liver is the primary organ for amino acid metabolism and oxidation as it is able to oxidize most amino acids. In contrast, skeletal muscle is able to completely oxidize six amino acids, among which are the BCAA (Figure 4.3). Actually, skeletal muscle is more efficient than the liver at metabolizing BCAA. The carbon skeletons resulting from the

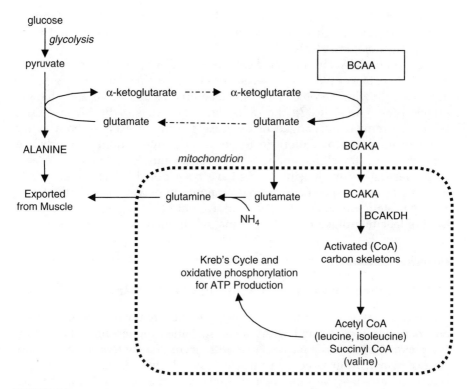

FIGURE 4.3
The BCAA are broken down in skeletal muscle in a series of steps that begin in the cytosol and forms glutamate via transamination of α-ketaglutarate. Glutamate becomes a reactant in a transamination reaction that forms alanine (used for gluconeogenesis in the liver) from pyruvate. The oxidation of BCAA concludes in mitochondria forming acetyl CoA and succinyl CoA, which can be used for energy.

breakdown of BCAA are acetyl CoA and succinyl CoA, which are valuable energy sources.

Physiological stress such as strenuous exercise (prolonged higher submaximal intensity exercise) results in an increased release of amino acids by skeletal muscle into circulation.[1] Yet many amino acids are released from muscle tissue in a disproportionate level to muscle protein concentration. All three BCAA are released into circulation in lower proportion than their presence in muscle protein, and alanine and glutamine in higher proportion than their presence. This strongly suggests that BCAA are broken down at the same time that alanine and glutamine are produced. In fact, much of the alanine and glutamine produced is attributable to the breakdown of BCAA and the transfer of their amine group to pyruvate and glutamate, creating alanine and glutamine (Figure 4.3). Glutamine and alanine account for approximately 15% of the amino acids in muscle protein; however, they can account for as much as 50–80% of the amino acids released from muscle during such stress. In contrast, BCAA, which constitute about 19% of muscle protein, are released from skeletal muscle in lower amounts or not at all during physiological stress.

The level of BCAA in the blood can decrease during and after an exercise bout. For instance, plasma or serum levels of leucine can be reported to decrease following aerobic (11–33%), anaerobic lactic (5–8%) and strength exercise (30%) sessions. At the same time, the levels of BCAA in muscle tissue can remain unchanged or decrease during an exercise bout as the result of increased oxidation. In response to exercise training, the BCAA content in the intracellular amino acid in skeletal muscle can increase.[6] Meanwhile, fasting serum leucine levels have been noted to decrease by 20% during 5 weeks of speed and strength training in power-trained athletes on a daily protein intake of 1.26 g/kg bodyweight.[6] This suggests an increased level of leucine uptake from the plasma into trained skeletal muscle.

VI. Branched-Chain Amino Acids and Exercise-Induced Fatigue

The BCAA have been touted to be ergogenic by reducing fatigue (central and peripheral) associated with longer sustained higher-intensity activity. Peripheral fatigue is often associated with decreased content of energy substrates, primarily glycogen, in muscle tissue during exercise. Meanwhile, central fatigue may be more related to decreased energy substrates to the brain during prolonged exercise and/or fluctuation in neurotransmitter (such as serotonin) levels in the brain.

As mentioned above, exercise training can enhance the level of BCAA in resting muscle tissue.[6] Also, during exercise, the level of BCAA availability in the amino acid pool of active muscle fibers increases. This is the result of decreased protein synthesis and increased protein breakdown, and/or increased BCAA uptake during exercise. This allows the BCAA amino acids to be available to help fuel contraction by providing intermediates of the energy pathways as shown in Figure 4.3. This has led to the notion that BCAA supplementation during exercise might enhance the level of these amino acids in muscle, while at the same time limiting the reduction of BCAA in the plasma during exercise. Thus latter effect would in turn reduce the potential for central fatigue as explained shortly.

The rate-determining reaction involved in utilizing the BCAA is the enzyme system Branch Chain α-ketoacid Dehydrogenase (BCKADH) (Figure 4.3).[1] At rest only about 4% of this enzyme complex is active in human skeletal muscle, however its activity increased several fold during exercise following an overnight fast.[1,7,8] This allows for an increased BCAA oxidation during exercise. However, the activity of BCKADH is influenced by metabolic state and is inversely activated by glycogen levels.[7,9–12] That is to say, if glycogen stores are higher, such as after glycogen loading, the degree of activation of BCKADH would be lower. Also, the ingestion of carbohydrate

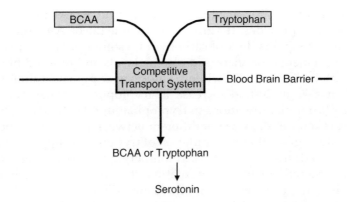

FIGURE 4.4
The BCAA and tryptophan are believed to compete for the same transport system to enter the brain.

(e.g., sport drink) during endurance activity may also limit BCAKDH activity. At this time, little is known as to the effects of BCAA supplementation before or during exercise on the activity level of this enzyme. If the level of BCAA in muscle tissue is enhanced during exercise, could this allow for greater activity of BCAKHD due to greater substrate availability? If so, could this slow glycogen breakdown? While human studies await, one study in rats demonstrated that BCAA supplementation increases the activity of BCA-KHD at rest.[13] In addition, liver and muscle glycogen levels were higher in BCAA-supplemented rats following exhaustive exercise, suggesting a glycogen-sparing effect of greater BCAA dependency.

During sustained higher-intensity exercise, working skeletal muscle will increase its uptake of circulating BCAA, which then can become a fuel source. While this seems beneficial to muscle performance, the reduction in the BCAA content of the plasma has led some researchers to ponder whether this is partly, if not the primary, cause of central fatigue during prolonged exercise. Here the mechanism would be indirect, as BCAA compete with tryptophan for the same transporter to cross the blood–brain barrier (Figure 4.4).[1,14] When the ratio of BCAA to tryptophan (BCAA:tryptophan) decreases, the transport of tryptophan would increase and this amino acid serves as a precursor for serotonin. It seems that the rate-limiting step in the production of serotonin is tryptophan uptake into the brain.[1,15,16] As serotonin is associated with sedation, increased production via increased tryptophan availability will cause central fatigue.

While the central fatigue caused by lowered BCAA-to -tryptophan ratio theory seems plausible, researchers have not been able to demonstrate that it is true nor that BCAA supplementation just prior to and or during exercise extends performance.[1,15] In fact, they have looked at the issue by providing BCAA in an attempt to extend performance as well as tryptophan to bring on fatigue earlier. For instance, researchers provided trained males a sucrose solution (6%) with or without BCAA and tryptophan during a cycling bout of 70–75% maximal power output to exhaustion.[17] Here, tryptophan was provided at 3 grams/liter while the BCAA were provided at either 6 or 18 grams/liter of carbohydrate solution. They reported that BCAA supplementation increased plasma levels and reduced brain tryptophan uptake by 8 to 12% while tryptophan supplementation by itself resulted in a 7- to 20-fold increase in tryptophan uptake. Yet, despite these significant manipulations in tryptophan uptake, time to exhaustion during the exercise trial was not different between treatments. The results of this study suggest that manipulation of tryptophan supply to the brain either has no additional effect on serotoninergic activity during prolonged exhaustive exercise or that manipulation of serotoninergic activity functionally does not contribute to mechanisms of fatigue.

Similarly, BCAA provision did not influence time till exhaustion of endurance-trained males cycling at 75% VO_2max.[18] Here the cyclists began the bout with reduced muscle glycogen stores and were provided either flavored water or a 6% carbohydrate solution with or without BCAA (7 grams/liter). While the carbohydrate solution definitely extended the performance, the addition of the BCAA did not have an additional effect. Other research teams have failed to demonstrate that BCAA has ergogenic potential by pushing back central fatigue during higher-intensity endurance activity.[19] In this study, moderately trained volunteers performed a graded incremental exercise (increases of 35 W every 4 min) to exhaustion. They either received an infusion of a BCAA solution (260 mg x kg-1 x h-1 for 70 min) or saline. No significant differences were observed in total work performed and maximal oxygen in response to the BCAA infusion.

Elsewhere, BCAA supplementation did not improve the endurance trial completion time during 100 km cycling bout.[20] In this study, well-trained cyclists with a maximal oxygen uptake of 63.1 +/− 1.5 ml O_2 min/kg performed three laboratory trials consisting of 100 km of cycling, each separated by 7 days. The cyclists were encouraged to complete the 100 km as fast as possible and they received either a glucose, glucose plus BCAA or placebo solution during the exercise trials. The researchers reported that the BCAA added to one solution failed to improve performance.

On the contrary, there are two reports that suggest that BCAA supplementation may improve endurance performance.[21,22] Aspects of the experimental designs of these studies has led researchers to consider the practical application of the performance results. For instance, the level of carbohydrate ingested was not controlled in one study.[21] Meanwhile, carbohydrate was not provided in the second study.[22] Without question, carbohydrate intake should be a foundational concept when it comes to improving endurance performance. Furthermore, as mentioned above, when BCAA was added to a carbohydrate solution there was no additional benefit over carbohydrate alone.

VII. Branched-Chain Amino Acid Supplementation and Exercise-Induced Changes in Immune Status

While BCAA supplementation may not prolong endurance exercise efforts, other rationales for supplementation among athletes have emerged. Intense long-duration exercise has been associated with immunosuppression, which affects natural killer cells, lymphokine-activated killer (LAK) cells and lymphocytes. For instance, in a study involving male triathletes who swam 2500 m, bicycled 81 km and ran 19 km, the concentration of total serum amino acids decreased (32% reduction in glutamine) during the race, with the lowest values occurring at 2 hours after exercise.[23] Natural killer and LAK cell activities were also suppressed 2 hours after completeing exercise, as was the blood mononuclear cell proliferation rate. The researchers also determined that the plasma concentration of glutamine was statistically related to some of the changes, namely the lymphokine-activated killer cells. Glutamine does seem the critical factor involved in LAK cell activity, at least *in vitro*.[24] However, when researchers provided athletes glutamine supplementation prior to endurance exercise to inhibit the postexercise decline in glutamine levels, the activity of of LAK cells still decreased.[24] The results of another study also suggest that glutamine supplementation prior to strenuous active may not have a favorable impact on lymphocyte distribution and activity.[25] In a recent review it was concluded that plasma glutamine concentration is not likely to play a mechanistic role in exercise-induced immunodepression.[26] The researchers also noted that intracellular glutamine concentration may not be compromised when plasma levels are decreased postexercise.

The mechanisms involved in the reduction in immunocompetence following strenuous exercise are complex and multifactorial, and, at least to some degree, associated with reductions in plasma glutamine levels. However, while reduced plasma glutamine is not the direct cause, it may be a factor. This has led researchers to seek alternative supplements to lessen the postexercise immunosuppression, perhaps ones that also inhibit the reduction in glutamine levels. There is some experimental evidence that BCAA supplementation limits the degree of change in postexercise glutamine levels.[27,28] This has led researchers to speculate whether BCAA supplementation could inhibit the immunocompromising influence of a strenuous exercise by (1) reducing the postexercise glutamine reduction, and (2) having an undetermined role in immune function and exercise.

A couple of research efforts have tested whether BCAA supplementation could limit the degree of immunosuppresion following strenuous exercise and, in turn, lower the incidence of sickness.[27,28] In one such study, BCAA supplementation led to the same levels of plasma glutamine, before and after an endurance trial (Olympic distance triathlon), whereas athletes not receiving BCAA supplementation presented a 22.8% reduction in plasma glutamine following an endurance bout. [27] The placebo group also demonstrated a reduction in IL-1 production after exercise (22.2%), while the BCAA-supplemented group demonstrated a 20% increase in IL-1 as well as a more normalized lymphocyte response to the presence of mitogens. The results of this study also suggested that the inhibition of changes in glutamine levels as well as a normalized immune response were linked to decreased symptoms of infection (33%) in the BCAA-supplemented athletes. In a follow-up study, athletes were provided BCAA supplements or a placebo.[28] Following an Olympic Triathlon or a 30-k run, the nonsupplemented athletes presented a decreased plasma glutamine concentration, which was not observed in the BCAA-supplemented athletes. In addition, the BCAA-supplemented athletes demonstrated a seemingly more favorable immune response in response to mitogens. The researchers concluded that BCAA supplementation recovers the ability of peripheral blood mononuclear cells to proliferate in response to mitogens after long-distance intense exercise, as well as stabilizes plasma glutamine concentration. The results of these studies are suggestive that BCAA supplementation may have a desirable impact on immunocompetence related to strenuous exercise. Further study is clearly warranted along this avenue of BCAA supplementation.

VIII. Branched-Chain Amino Acid Supplementation and Muscle Fiber Hypertrophy and Pathology in Response to Exercise

Leucine and probably the other BCAA are involved in protein production. While the information at present is limited, researchers recently described how leucine is involved in protein synthesis in skeletal muscle by enhancing both the synthesis and activity of proteins involved in mRNA translation.[29] Leucine appears to stimulate mTOR (mammalian target of rapamycin), a protein kinase, which is one of the components of a signal transduction pathway used by insulin.[30] When the cellular energy state is low, stimulation of mTOR by amino acids is prevented by activation of AMP-dependent protein kinase. Whether this information can be applied to BCAA supplementation in an otherwise well nourished individual remains to be determined. Future research will need to focus on the potential effect of BCAA supplementation to stimulate protein synthesis in various nutrition states and in relation to different types of exercise.

Favorable changes in muscle protein turnover potentially resulting from BCAA supplementation could lead to appreciable adaptations in muscle anatomy and measures of strength. However, there is limited research available regarding the impact of chronic BCAA on strength measures. It has been reported that BCAA supplementation reduced levels of plasma markers for muscle tissue damage in the days that followed a bout of endurance exercise.[31] Here BCAA were provided at 12 grams/day for 2 weeks prior to a 2-hour cycling bout at 70% VO_2max, and creatine kinase and lactate dehydrogenase were used as plasma markers of muscle tissue damage. Elsewhere, the results of a study involving a 6-week endurance training program and elderly men suggested that BCAA supplementation did not result in any histological changes in muscle tissue in addition to that induced by training alone.[32] A single treatment of leucine (100 mg x kg/body) provided prior to a strength test failed to lead to increased performance.[33] Further research along these lines is needed to determine the cellular impact of BCAA supplementation with regard to protein turnover, tissue damage and recovery and indicator of hypertrophic adaptation.

IX. Conclusions

The BCAA have a unique relationship with skeletal muscle, which is the primary site of of BCAA oxidation, and these amino acids may be involved in protein synthesis at the translational level. Also, BCAA compete with tryptophan for transport across the blood–brain barrier, which is used for serotonin production. Furthermore, BCAA provided prior to and/or during prolonged strenuous exercise might inhibit the decrease in glutamine, which in turn is linked to immunosuppression and increased incidence of sickness of athletes. This said, BCAA have been speculated to have ergogenic potential by inhibiting fatigue, enhancing protein synthesis and lessening reductions in immune system capabilities following strenuous exercise. While some of the preliminary research efforts related to immunity following strenuous exercise is encouraging, the ability of BCAA supplementation to enhance performance has not been consistently revealed by researchers.

Acknowledgments

Thanks to Lewis Curtright, Director of Corporate Development, Bally Total Fitness Corporation, for supplement industry information.

References

1. Wildman, R.E.C. and Miller, B.S., *Sport Foods, Supplements and Ergogenic Aids, in Sport and Fitness Nutrition*, Wadsworth Publishing, Atlanta, GA, 2004.
2. Supplementwatch Inc. www.supplementwatch.com, July, 2003.
3. Rennie, M.J. and Tipton, K.D., Protein and amino acid metabolism during and after exercise and the effects of nutrition. *Ann. Rev. Nutr.* 20, 457, 2000.

4. Fouillet, H., Bos, C., Gaudichon, C. and Tome D. Approaches to quantifying protein metabolism in response to nutrient ingestion. *J. Nutr.* 132(10), 3208S, 2002.
5. Morens, C., Gaudichon, C., Metges, C.C., Fromentin, G., Baglieri, A., Even, P.C. Huneau, J.F. and Tome, D., A high-protein meal exceeds anabolic and catabolic capacities in rats adapted to a normal protein diet. *J. Nutr.* 130(9), 2312, 2000.
6. Mero, A., Leucine supplementation and intensive training. *Sports Med.* 27(6), 347, 1997.
7. Wagenmakers, A.J., Nutritional Supplements: Effects on Exercise Performance and Metabolism, in *Perspectives in Exercise and Sports Medicine (Volume 12), The Metabolic Basis of Performance and Sport* (Lamb D.R. and Murray R., Eds.). Cooper Publishing Group, 1999.
8. Wagenmakers, A.J., Brookes, J.H., Coakley, J.H., Reilly, T. and Edwards, R.H., Exercise-induced activation of the branched-chain 2-oxo acid dehydrogenase in human muscle, *Eur. J. Appl. Physiol. Occup. Physiol.*, 59, 159, 1989.
9. Wagenmakers, A.J., Beckers, E.J., Brouns, F., Kuipers, H., Soeters, P.B., van der Vusse, G.J. and Saris, W.H., Carbohydrate supplementation, glycogen depletion and amino acid metabolism during exercise, *Am. J. Physiol.*, 260, E883, 1991.
10. Knapik, J., Meredith, C., Jones, B., Fielding, R., Young, V. and Evans, W., Leucine metabolism during fasting and exercise, *J. Appl. Physiol.*, 70, 43, 1991.
11. Wolfe, R.R., Goodenough, R.D., Wolfe, M.H, Royle, G.T and Nadel, E.R., Isotopic analysis of leucine and urea metabolism in exercising humans, *J. Appl. Physiol.*, 52, 458, 1982.
12. van Hall, G., MacLean, D.A., Saltin, B. and Wagenmakers, AJ., Mechanisms of activation of muscle branched-chain alpha-keto acid dehydrogenase during exercise in man, *J. Physiol.*, 494, 899, 1996.
13. Shimomura, Y., Murakami, T., Nakai, N., Nagasaki, M., Obayashi, M., Li, Z., Xu, M., Sato, Y., Kato, T., Shimomura, N., Fujitsuka, N., Tanaka, K. and Sato, M. Suppression of glycogen consumption during acute exercise by dietary branched-chain amino acids in rats. *J. Nutr. Sci. Vitaminol.* (Tokyo). 46, 71, 2000.
14. Pardridge, W.M., Blood-brain barrier transport of nutrients, *Nutr. Rev.*, 44, 15, 1986.
15. Paul, G.L., Gautsch T.A. and Layman, D.K., Amino acid and protein metabolism during exercise and recovery, in Wolinsky, I. (Ed.), *Nutrition in Exercise and Sport.*, CRC Press, Boca Raton, FL, 1998.
16. Lovenberg, W.M., Biochemical regulation of brain function, *Nutr. Rev.*, 44, 6, 1986.
17. van Hall, G., Raaymakers, J.S., Saris, W.H. and Wagenmakers, A.J., Ingestion of branched-chain amino acids and tryptophan during sustained exercise in man: Failure to affect performance, *J. Physiol.*, 486, 789, 1995
18. Blomstrand, E., Andersson, S., Hassmen, P., Ekblom, B. and Newsholme, E.A., Effect of branched-chain amino acid and carbohydrate supplementation on the exercise-induced change in plasma and muscle concentration of amino acids in human subjects, *Acta Physiol. Scand.*, 153, 87, 1995.
19. Varnier, M., Sarto, P., Martines, D., Lora, L., Carmignoto, F., Leese, G.P. and Naccarato, R., Effect of infusing branched-chain amino acid during incremental exercise with reduced muscle glycogen content, *Eur. J. Appl. Physiol. Occup. Physiol.*, 69, 26, 1994

20. Madsen, K., MacLean, D.A., Kiens, B. and Christensen, D., Effects of glucose, glucose plus branched-chain amino acids, or placebo on bike performance over 100 km, *J. Appl. Physiol., 81,* 2644, *1996.*
21. Blomstrand, E., Hassmen, P., Ekblom, B. and Newsholme, E.A., Administration of branched-chain amino acids during sustained exercise — effects on performance and on plasma concentration of some amino acids, *Eur. J. Appl. Physiol. Occup., 63m 83, 1991.*
22. Mittleman, K.D., Ricci, M.R. and Bailey, S.P., Branched-chain amino acids prolong exercise during heat stress in men and women, *Med. Sci. Sports Exerc.,* 30, 83, 1998.
23. Rohde, T., MacLean, D.A., Hartkopp, A. and Pedersen, B.K., The immune system and serum glutamine during a triathlon. *Eur. J. Appl. Physiol. Occup. Physiol.* 74(5), 428, 1996.
24. Rohde, T., MacLean, D.A. and Pedersen, B.K., Effect of glutamine supplementation on changes in the immune system induced by repeated exercise. *Med. Sci. Sports Exerc.* 30(6), 856, 1998.
25. Castell, L.M., Poortmans, J.R., Leclercq, R., Brasseur, M., Duchateau, J. and Newsholme, E.A. Some aspects of the acute phase response after a marathon race and the effects of glutamine supplementation. *Eur. J. Appl. Physiol. Occup.* 75(1), 47, 1997.
26. Hiscock, N. and Pedersen, B.K., Exercise-induced immunodepression — plasma glutamine is not the link. *J. Appl. Physiol.* 93(3), 813, 2002.
27. Bassit, R.A., Sawada, L.A., Bacurau, R.F., Navarro, F., Martins, E. Jr., Santos, R.V., Caperuto, E.C., Rogeri, P. and Costa Rosa, L.F., Branched-chain amino acid supplementation and the immune response of long-distance athletes. *Nutrition,* 18(5), 376, 2002.
28. Bassit, R.A., Sawada, L.A, Bacurau R.F., Navarro, F. and Costa Rosa, L.F., The effect of BCAA supplementation upon the immune response of triathletes, *Med. Sci. Sports. Exerc.,* 32, 1214, 2000.
29. Anthony, J.C., Anthony, T.G., Kimball, S.R. and Jefferson, L.S., Signaling pathways involved in translational control of protein synthesis in skeletal muscle by leucine, *J. Nutr.,* 131, 856S, 2001.
30. Meijer, A.J. Amino acids as regulators and components of nonproteinogenic pathways. *J. Nutr.* 133(6 Suppl 1), 2057S, 2003.
31. Coombes, J.S. and McNaughton, L.R., Effects of branched-chain amino acid supplementation on serum creatine kinase and lactate dehydrogenase after prolonged exercise, *J. Sports Med. Phys. Fitness,* 40, 240, 2000.
32. Freyssenet, D., Berthon, P., Denis, C., Barthelemy, J.C., Guezennec, C.Y. and Chatard, J.C. Effect of a 6-week endurance training programme and branched-chain amino acid supplementation on histomorphometric characteristics of aged human muscle. *Arch. Physiol. Biochem.* 104(2), 157, 1996.
33. Pitkanen, H.T., Oja, S.S., Rusko, H., Nummela, A., Komi, P.V., Saransaari, P., Takala, T. and Mero, A.A., Leucine supplementation does not enhance acute strength or running performance but affects serum amino acid concentration. *Amino Acids,* 25(1), 85, 2003.

5
Carnitine

Robert J. Moffatt and Sara A. Chelland

CONTENTS
I. Introduction 48
   A. History of Carnitine 48
   B. Metabolic Actions of Carnitine 48
II. Carnitine Metabolism 50
   A. Synthesis of Carnitine 50
   B. Distribution and Excretion of Carnitine 51
III. Carnitine Interactions during Exercise 52
   A. Pyruvate Dehydrogenase Complex 52
   B. Acylcarnitines 52
IV. Rationale for Carnitine Supplementation 52
   A. Why use Carnitine as a Supplement? 52
   B. When Would Carnitine Supplementation Be Effective? 53
V. Carnitine Supplementation and Performance 53
   A. Body Carnitine Levels with Supplementation 53
   B. Carnitine Levels and Acute Exercise 54
   C. The Effect of Chronic Training on Carnitine Status 56
   D. Summary of Carnitine and Training Effects 58
   E. Carnitine Supplementation and Performance 58
VI. Recommendations and Conclusions regarding Carnitine Supplementation 61
   A. Recommendations regarding Supplementation of Carnitine 61
VII. Summary 61
References 62

I. Introduction

A. History of Carnitine

The word carnitine is derived from the Latin word *carno* or *carnis*, which means flesh or meat. Carnitine was discovered in muscle extracts by Gulewitsch and Krimberg[1] as well as Kutscher[2] in 1905 and was first thought to be involved with muscle function. Gulewitsch and Krimberg[1] identified the structure of carnitine as 3-hydroxy-4-N-trimethyl-aminobutyric acid ($C_7H_{15}NO_3$), which was later confirmed in 1927 by Tomita and Sendju.[3] Following its discovery, its exact configuration was proposed in 1962[4] and became definite in 1997 as L- or R- 3-hydroxy-4-N,N,N-trimethylamino-butyrate.[5] Initially, carnitine was called vitamin B_T because of its necessity for the growth in the yellow mealworm known as *Tenebrio molitor*. Subsequently, the distribution of carnitine in the organs of mammals, lower animals, plants and microorganisms became well established[6] (Figure 5.1).

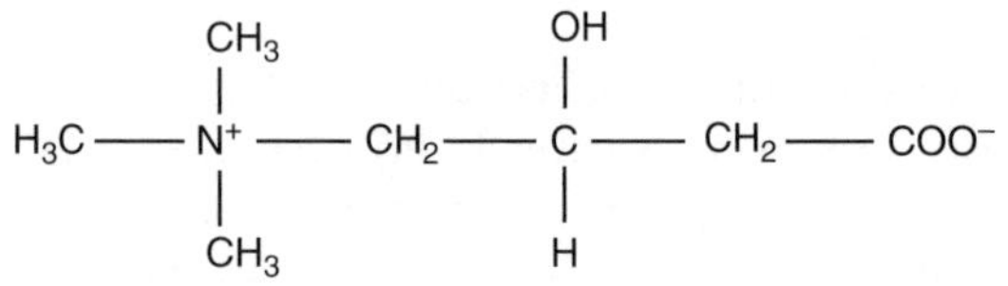

FIGURE 5.1
Chemical structure of L-carnitine.

B. Metabolic Actions of Carnitine

Carnitine serves as a cofactor for several enzymes, including carnitine translocase and acylcarnitine transferases I and II, which are essential for the movement of activated long-chain fatty acids from the cytoplasm into the mitochondria (Figure 5.2). The translocation of fatty acids is critical for the generation of adenosine triphosphate (ATP) within skeletal muscle, via β-oxidation. These activated FA become esterified to acylcarnitines with carnitine via carnitine-acyl-transferase I (CAT I) in the outer mitochondrial membrane. Acylcarnitines can easily permeate the membrane of the mitochondria and are translocated across the membrane by carnitine translocase. Carnitine's actions are not yet complete because the mitochondria has two membranes to cross, thus, through the action of CAT II, the acylcarnitines are converted back to acyl-CoA and carnitine. The acyl-CoA can be used to generate ATP via β-oxidation, Krebs Cycle and the electron transport chain. The carnitine is recycled to the cytoplasm for future use.

Carnitine thus has a unique interaction with acyl-CoA in the mitochondria, and is an important modulator of the acyl-CoA:free CoA ratio. This is demonstrated when acylcarnitines are formed. This relationship is defined by

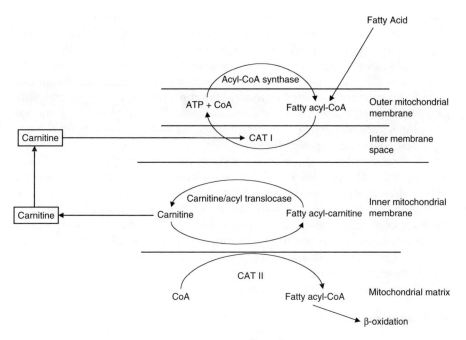

FIGURE 5.2
Roles of carnitine and the movement of long chain fatty acids into the mitochondrial matrix.

the rate of acyl-CoA production — if the acyl-CoAs are produced more rapidly than they are used, then the acyl CoA within the intramitochondrial space is high compared with the free CoA concentration.[7,8] This imbalance can then be corrected because carnitine can bind the acyl-CoAs and the once-elevated ratio can return to normal.

It is important to differentiate the two forms of carnitine, which are L-carnitine and D-carnitine (Dextro form). L-carnitine, the physiologically active form, is endogenously produced within the human, whereas D-carnitine is not physiologically active and is a synthetic.[9] Research has shown D-carnitine given to subjects resulted in depletion of their endogenous stores of L-carnitine[10,11] and may manifest itself as carnitine deficiency, especially during intense bouts of exercise.[12] Supplementation of D-carnitine is therefore not recommended, since it may have a deleterious effect on the body as well as performance.

Carnitine plays a major role in substrate metabolism in healthy individuals, but there are individuals who have carnitine deficiencies and, in them, carnitine supplementation can be utilized as a therapeutic agent. Typically, the daily average American diet has about 100 to 300 mg of carnitine.[13] The primary sources of carnitine are found in red meat and dairy products, whereas vegetables have very little L-carnitine, so vegetarian diets may contain minuscule amounts. Despite adequate intake of carnitine by the majority of the population, a fraction of the population still has carnitine disturbances. This may be due to several metabolic abnormalities that include defective carnitine synthesis, enhanced carnitine degradation, impaired transport of carnitine — whether it is in or out of cells — and finally, abnormal renal handling of carnitine.[14,15] These abnormalities manifest in a syndrome known as primary carnitine deficiency, the myopathic form having symptoms such as muscle fatigue, cramps, hypotona and atrophy of the musculature and the systemic form. The more severe form has symptoms such as nausea, vomiting and coma due to excessive fat storage because of reduced hepatic efficiency.[16-18]

Carnitine supplementation in these states has proven effective; for example, in cardiac diseases, the myocardium uses fatty acids as its primary source of fuel, therefore a deficiency in carnitine may have a serious impact on heart rate and stroke volume and thus cardiac output. In fact, most research on these types of cardiac patients supports the use of carnitine as a supplement and furthermore has found that abnormal fatty acid metabolism is normalized.[19-22] The focus of this chapter, however, will be the effects and efficacy of carnitine supplementation in healthy populations, specifically as a potential ergogenic aid to athletic performance.

II. Carnitine Metabolism

A. Synthesis of Carnitine

In humans, carnitine is synthesized from the essential amino acids lysine and methionine.[9,23-25] Methionine contributes its methyl groups[26-27] and the carbon and nitrogen moieties come from lysine.[28-29] In addition, ascorbic acid, iron, niacin and vitamin B_6 are all requirements for the biosynthesis of carnitine.[30] (Figure 5.3). The liver, kidney, heart and skeletal muscle can convert the trimethyl lysine, which originates from digestion of proteins, to γ-butyrobetaine, but studies have shown that in humans only the liver and kidney can convert γ-butyrobetaine to carnitine.[8,9,31] This has several implications because the heart and skeletal muscles need carnitine but do not produce it, thus carnitine's transport efficiency is critical, especially during activities such as exercise. Carnitine biosynthesis occurs at a rate of two mol/kg of body weight^{-1}/day^{-1}, does not experience significant daily fluctuations, and seems to be related to the availability of N-trimethyllysine.[23]

As mentioned previously, the diet can provide a significant amount of carnitine, approximately 50% (100–300 mg/daily) in the form of either free carnitine or short- and long-chain fatty acids.[32,33] This is true for individuals who consume large amounts of beef, pork and lamb. In addition, this intake is sufficient to maintain normal carnitine homeostasis. Vegetarians who consume less than 0.5 µmol/kg of body weight^{-1}/day^{-1} must rely on endogenous production of carnitine to maintain homeostasis.[34]

Carnitine

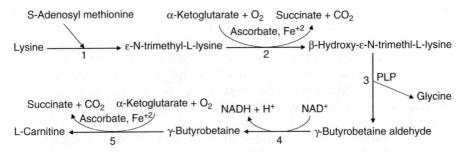

FIGURE 5.3
Biosynthesis of carnitine.

B. Distribution and Excretion of Carnitine

In general, carnitine homeostasis is maintained several ways, including absorption from dietary sources, modest rates of biosynthesis and reabsorption, which is very efficient.[23] Carnitine in its esterified forms as short- and long-chain acylcarnitines is found in several tissues and cellular fluid. In a healthy 70 kg adult, the pool of carnitine is estimated to be about 100 mmol and is distributed between the skeletal and cardiac muscle (98%), liver and kidney (1.6%) and the extracellular fluid (0.4%).[35]

Plasma carnitine concentrations in a normal U.S. population can range from 37-89 µM, with approximate concentrations for males ranging from 59.3±11.9 µM and females 51.5±11.6 µM. Studies have revealed that 54–87% of dietary carnitine is absorbed, which contributes to this plasma pool[9,23] and furthermore, during its metabolism and excretion, it is highly conserved within the kidneys. Approximately 90–98% of ingested carnitine is reabsorbed in the renal tubules.[36] That which is not reabsorbed can be excreted via the feces (about 1–2%) or its typical elimination pathway through the urinary system (about 0.1–0.3%).[9]

III. Carnitine Interactions during Exercise

A. Pyruvate Dehydrogenase Complex (PDC)

Research has postulated that a few interrelated pathways can utilize carnitine to enhance the use of long-chain fatty acids while manipulating acetyl-CoA and pyruvate dehydrogenase (PDH). This would allow for the increased burning of fat as a fuel and the sparing of muscle glycogen during exercise. The rationale behind these assumptions is that *in vitro* experiments have shown that the activation of the pyruvate dehydrogenase complex is inhibited by high ratios of acetyl-CoA:free CoA and NADH:NAD$^+$.[37] Since carnitine is a known regulator of the acetyl-CoA:free CoA via the formation of acylcarnitines it is plausible to assume that carnitine may control the activity of the PDC.

This interaction can be illustrated during exercise when acetyl-CoA produced via the PDC and β-oxidation can be delivered to the Krebs Cycle or accumulate in an acetyl CoA and acylcarnitine pool. Carnitine can then accept the acetyl-CoA groups that continue to accumulate, directly affecting acetyl-CoA and carnitine status during exercise.

B. Acylcarnitines

It is important to exploit the role of the acylcarnitines because an increase in their formation is directly linked to an increase in acetyl CoA, while free carnitine levels decline in the muscle.[32] This yields no change in the overall carnitine level and maintenance of the acetyl CoA:free CoA ratio. This will occur during all types and intensities of exercise. However, it is very important to also recognize follow-up *in vivo* studies examining muscle. They demonstrate that the full potential of PDC activity could be quickly reached and maintained.[38-40] Furthermore, PDC activity was still fully active with low lactate accumulation, which was to be expected. These results suggest that *in vivo* examinations give a more accurate portrayal of the intricate system and role of carnitine in metabolism and can gives us hints into the effectiveness on carnitine as an ergogenic aid.

IV. Rationale for Carnitine Supplementation

A. Why use Carnitine as a Supplement?

There are several rationales behind why carnitine might be a promising ergogenic aid. These are related specifically to its proposed metabolic actions of being capable of transporting long-chain fatty acids into the mitochondria for utilization. The assumption is that "more is better," and the more carnitine available for biological work, the more fatty acids there will be that can be utilized for energy. To be more specific, this rationale is used to argue that, as there will be more fatty acid oxidation, and, since this oxidation is carnitine dependent, muscle glycogen will be spared. In addition, since carnitine promotes fat utilization, supplementation will then be successful in altering body composition to favor decreases in fat mass and aid in weight loss. Other rationales include increasing maximal oxygen consumption (VO_2 max), activating the PDC via the acetyl Co-A:free CoA ratio, replacement of carnitine lost during exercise or redistributed to acylcarnitines and finally to allow muscles to become more resistant to fatigue. All of these scenarios seem plausible and have been extensively researched as to their efficacy.

B. When Would Carnitine Supplementation Be Effective?

When interpreting studies on carnitine, it is important to remember to focus on the changes that occur within the muscle, since that is the major reservoir for carnitine. This will govern whether it is appropriate to supplement and how effective supplementation will be. The majority of research shows that carnitine is typically effective only during a primary or secondary carnitine deficiency, when the body cannot maintain adequate levels of the metabolite,[13-21] although it is important to recognize that there are conflicting results regarding performance.

The remainder of this chapter will focus on the exercise- and performance-related literature that has been completed. This body of literature will examine the possible role of carnitine as an ergogenic aid.

V. Carnitine Supplementation and Performance

A. Body Carnitine Levels with Supplementation

Research has provided many conflicting results when analyzing changes in body carnitine levels. This is primarily because the methodology is inconsistent. Many of the studies have focused on blood or plasma levels of carnitine, but the primary action of carnitine occurs in the muscle. Therefore, those studies that reflect changes in muscle carnitine will most accurately reflect whether supplementation is necessary. Unfortunately, only a few studies have reported muscle carnitine levels.[9,41-43] Those that have measured muscle levels find that supplementation with normal or low doses (approx. 1–6 g orally or intravenously) do not increase levels normally present in the muscle tissue. It seems as though there are only a couple of conditions where carnitine supplementation may actually act to increase concentrations within the muscle. These circumstances include a supraphysiological dose (100 mg/kg body weight). In addition, those individuals who are carnitine deficient will benefit from supplementation. Specifically, this is because those individuals already have low carnitine concentrations and supplementation will allow for the accumulation of the necessary carnitine content within the muscle to perform. These studies[9,41-43] emphasize that there is adequate carnitine within the muscle mitochondria to oxidize lipids and furthermore, that a carnitine deficiency will not occur as a result of aerobic exercise. In addition, when muscle carnitine levels were compared with plasma levels, it was found that the carnitine metabolic state, associated with exercise, was very poorly reflected by changes in both the plasma and urine.[42]

The consensus is that individuals who abide by a standard diet will take in the necessary nutrients to maintain adequate levels of carnitine in the muscle. Furthermore, it is important to emphasize again that caution be exerted when analyzing the carnitine literature because individuals need to recognize the nature of the carnitine pools that were examined, whether in blood or muscle.

B. Carnitine Levels and Acute Exercise

Acute exercise is characterized by a single exercise session that may be repeated for several days, not usually longer than a week. This type of exercise has been widely used in the scientific literature on carnitine to study the changes in carnitine concentrations that occur prior to and after a single exercise session. These types of studies are effective in being able to quantify carnitine concentrations within the body as well as provide an insight into its metabolism during exercise. Acute studies can set the framework for long-term research because the acute data is used to predict how carnitine will act with larger volumes of training.

Lennon et al.[44] examined subjects during submaximal cycle ergometry at 55% of VO_2max and found a significant decrease (about 20%) in muscle carnitine levels. They reported that, as the muscle carnitine dropped, there was a significant and concurrent increase in plasma carnitine from rest to the termination of 40 minutes of exercise. These results led the authors to suggest that some acylcarnitines are lost to the plasma, from the muscle during acute exercise bouts. They further suggested that more intense exercise could severely deplete muscle carnitine levels.

Conflicting results were seen by Carlin et al.[45] who found that with 90 minutes of cycling (50% VO_2max) there were no decreases in total muscle carnitine levels, but increased levels of acylcarnitines and decreased free carnitine within the muscle were observed. Plasma levels of acylcarnitine were shown to increase progressively with exercise as a result of the decline in free carnitine and not the transfer of carnitine from the muscle to the plasma. Again, these studies help to illustrate the importance of measuring and analyzing carnitine pools separately. These results are supported by Harris et al.,[46] who found that both intermittent electrical stimulation and cycling resulted in no change in total muscle carnitine but a significant fall in free carnitine and a concurrent increase in acylcarnitine levels. In addition, Harris' group was able to quantify that resting mean carnitine concentration in muscle is approximately 20.0 mmol/kg dry muscle and that 77% of this was in the form of free carnitine and 19% was acylcarnitines. Their analysis helped make quantifying the concentrations more accurate as well, to improve the validity of results.

To further examine the acute effects, Soop et al.[47] used femoral artery and vein catheters in seven healthy, moderately trained male cyclists who exercised for 2 hours at 50% VO_2 max. The purpose was two-fold, first to examine fatty acid utilization during oral carnitine supplementation (5 g/day) during exercise and second to quantify and examine the changes that occur in plasma and muscle carnitine levels. Results from this study showed that despite a twofold increase in plasma carnitine levels with supplementation there was no change in fatty acid turnover and thus no influence over substrate metabolism.[47] In addition, there were no differences between supplemented and nonsupplemented individuals with regard to plasma carnitine levels. In addition, free carnitine was observed to fall, but there was a release from the leg muscles during exercise. Furthermore, there was an increase in acylcarnitines in the plasma with no evidence of release from the leg muscles. Soop et al.[47] further concluded that there is an alternate site for acylation of carnitine, which they suggested was the liver. Keep in mind that these results do coincide with those of Lennon[44] and Carlin[45] both of whom showed that there were increases in acylcarnitines and decreases in free carnitine. Soop et al. were able to shed more light on the intricate biochemical pathways that are involved in carnitine homeostasis during exercise. These results[44,45,47] support the notion that, within the muscle, adequate carnitine levels can be maintained during exercise and further deficiencies are not likely to occur.

It now seems evident that there are definite changes that occur within the plasma and muscle with regard to free carnitine and acylcarnitine, but not necessarily total carnitine. The next logical step is to analyze the effect of intensity on carnitine concentrations. Hiatt et al.[42] set out to characterize carnitine at two different exercise intensities; 60 minutes at 50% lactate threshold (LT) and 30 minutes at a workload between LT and maximal work capacity for each individual. This was intended to reflect exercise intensities that primarily utilize fatty acids (50% LT) and carbohydrates (above LT to maximum) for energy production. Findings revealed that the lower-intensity exercise was not associated with changes in muscle carnitine metabolism as reflected by alterations in free carnitine and acylcarnitine.[42] In contrast, within 10 minutes of the high-intensity exercise, muscle acylcarnitine increased by 5.5 fold and free carnitine decreased by 66%. These changes remained over the duration of the high-intensity exercise and persisted for 60 minutes into recovery. The changes were seen in both long- and short-chain acylcarnitines. In addition, plasma acylcarnitine levels were also increased, suggesting that there is a redistribution of the carnitine pool that may persist even after recovery.[42] It is important to note that the authors did report that neither exercise bout was associated with changes in carnitine urinary excretion rates or plasma concentrations.

In support of this idea, Sahlin et al.[48] examined carnitine concentrations during several exercise intensities (40, 75 and 100% VO_2 max). No changes were seen with cycling in total muscle carnitine concentration, acylcarnitine or free carnitine during the low-intensity exercise (40% VO_2 max), but the high intensity (75 and 100% VO_2 max) produced a significant increase in acylcarnitine from rest (6.9 ± 1.9 mmol/kg) to exercise (18.1 ± 1.0 mmol/kg). In addition, there was a concurrent decrease in free carnitine levels.[48] Decombaz et al.[43] took the next step and, to determine whether a carnitine deficiency developed, studied cross-country skiers who underwent prolonged strenuous exercise. The exercise consisted of a ski race in the Swiss Alps with an average completion time of 13 hours and 26 minutes. Carnitine intake was evaluated for 2 weeks prior to the race (average 50 ± 4 mg/day) in these highly trained individuals. Again, there was no difference in total carnitine content from rest (17.9 ± 1.0 mmol/kg) to postexercise (18.3 ± 0.8 mmol/kg). In addition there was a 20% decline in free carnitine concentration that was offset by a 108% increase in acylcarnitine. The authors[43] concluded that carnitine deficiency would not develop in trained athletes with a moderate carnitine dietary intake.

It seems evident from the numerous studies that there is no drastic effect on total muscle carnitine concentrations during acute exercise because of the redistribution that occurs; specifically, there is an increase in acylcarnitine and decrease in free carnitine concentrations. Thus, it seems likely that supplementation of carnitine for acute bouts of exercise will not be effective and is consequently unnecessary.

C. The Effect of Chronic Training on Carnitine Status

It is widely accepted that chronic endurance aerobic training can induce changes within the skeletal muscle to enhance performance and increase endurance. Some of these changes include increased capillary density, increased enzymatic concentration and an increase in the size and number of mitochondria. This is associated with an increased ability of the athlete to oxidize fatty acids in the form of intramuscular triglycerides for fuel. This phenomenon has led researchers to analyze the potential changes that might occur with carnitine concentrations in all pools including urinary excretion, plasma and muscle.

Several studies have documented that there are no differences in the resting plasma concentrations of carnitine between trained individuals and their sedentary counterparts.[25,49-51] This is practical because carnitine concentrations are governed by dietary intake and, as stated previously, the majority of the population has a diet containing sufficient precursors to promote adequate carnitine synthesis.[33] Similarly, it seems rather unlikely to lose carnitine in the urine, as it only contributes to approximately 1% of carnitine loss daily. The chance that a carnitine deficiency would develop with training due to changes that occur within the plasma or urinary compartments is low because of the small fraction that they contribute to both exercise and daily excretion.

Since the muscle is the major pool for carnitine, it is important to pay particular attention to the results that have analyzed training and carnitine status. It is important to emphasize here that research findings regarding this topic are divergent because there are researchers who support both sides of the question that asks whether training causes a change in carnitine concentration. There is literature both to support that changes occur as a result of training and to not support changes. The discrepancy seems to be within the methodology of the studies, as there is little uniformity in measurements of carnitine with regard to what pool is being studied or the techniques utilized to measure concentrations. In addition, there are differences in carnitine levels for each person within each muscle and levels vary between genders. Finally, the type of exercise does affect the extent of carnitine status, as shown in the previous section. All these possibilities present sources of error and must be thoroughly evaluated.

Lennon et al.[44] reported that males have significantly higher levels of muscle carnitine than women in both a high- and moderate-training group, which is likely due to the increased muscle mass associated with the male. In addition, they report that training status does not affect muscle carnitine concentrations in either males or females.

Janssen et al.[52] further supported these findings with sedentary individuals who completed an 18- to 20-month marathon training program. Muscle carnitine was evaluated prior to, during and after the completion of the marathon and was not affected during training, nor were sex-related differences observed.[58] In addition, running the marathon did not cause a significant decrease in muscle carnitine levels, suggesting that neither the training nor the strenuous exercise alters carnitine muscle levels. Finally, Decombaz et al.[43] measured the total muscle carnitine level of skiers for 2 years prior to an Alpine ski race and found that there were no significant changes over the course of their 2-year training regiment (mean 17 vs. 16 µmol/g dry wt). It is important to point out that there were consistent individual variations (range 12–22 µmol/g dry wt) that were stable for the course of the training. These results show that, with training, there is little variation in carnitine levels and that deficiencies will not occur as a result of this training, leading to the suggestion that the diet may provide a sufficient amount of precursors to synthesize adequate amounts of carnitine.

It is important to point out that there are studies that found differences in carnitine concentrations after training. One such study by Arenas et al.[53] found that endurance athletes showed a significant decrease in free and total muscle carnitine content after 4 months of training and that these changes were not as severe as those seen in sprinters. The authors suggest these differences were because the endurance athletes had a higher concentration of Type I muscle fibers, thus a higher mitochondrial content compared with the Type II fibers of the sprinters. This led to the conclusion that there was a "wasting" of the short-chain acylcarnitines, a result of chronic training that thus would deplete the carnitine stores for subsequent training sessions. A limitation in this study was that the researchers did not separate the subjects by gender in their analysis and when reporting their findings. Furthermore, there was not an equal number of males and females in both groups. The endurance group consisted of all males, whereas in the sprint group, 5 out of 11 participants were females. It has been shown that there are significant differences between muscle carnitine levels in males and females.[44,54,55]

D. Summary of Carnitine and Training Effects

Exercise, whether acute or chronic, does not seem to influence the total muscle carnitine, but there are definite changes that occur with regard to the accumulation of acylcarnitines and reduction in free carnitine. These changes are well documented,[42,43,45-48,52,58] providing a basis for the subsequent literature analyzing performance. The rationale is that, despite the fact that exercise does not alter muscle carnitine concentration, supplementation may contribute to the already existing pool and furthermore influence performance. Carnitine supplementation has been postulated to influence performance by increasing the delivery of FA into the mitochondria to prolong endurance exercise as well as delay fatigue by sparing muscle glycogen. Improving performance is a central theme in nutrition and sport and provides a basis for the implementation of a supplement to an individual's diet. It is therefore pertinent to examine the effects of supplementation on performance to validate the efficacy of an ergogenic aid.

E. Carnitine Supplementation and Performance

Up to this point, we have examined the effects of carnitine levels in the muscle, plasma and urine during acute and chronic training. More specifically, we examined whether there are changes that occur as a result of a single exercise session or multiple exercise sessions over time. The literature is fairly consistent in that there does not seem to be a depletion of total muscle carnitine, and that an acute bout or chronic training does not change these levels. It is important to now focus on the performance aspect of these and other studies. The intent is to ascertain whether carnitine supplementation enhances performance as measured by such variables as VO_2 max, respiratory quotient (RQ), exercise duration, blood lactate concentrations, substrate metabolism and glycogen sparing. Several studies that have measured all or some combination of these markers will be presented in chronological order to help follow the progression of carnitine research.

An early study by Marconi et al.,[50] using six competitive racewalkers, measured VO_2 max, blood lactate and RQ. Subjects were supplemented with 4 g of oral carnitine for 2 weeks, after which VO_2 max was found to be significantly increased from 54.5 ± 3.7 ml.kg.min^{-1} to 57.8 ± 4.7 ml.kg.min^{-1}. In addition, there were no significant changes in blood lactate accumulation or RQ at a fixed workload. The authors speculated that this slight but significant increase in VO_2 max was due to the activation of substrate flow through the Krebs Cycle. This simply means that, with an increase in substrate flow through the Krebs Cycle, there is a postulated increase in ATP production that allowed the subjects to increase VO_2 max. Subsequent studies[56] could not reproduce these results and found that 2 g of oral carnitine administered to separate groups for either 14 or 28 days produced no significant effects on either VO_2 max or lactate. It is important to note that the training status of the two groups were different, one being trained[50] and the other using two untrained groups.[56] However, neither study was able to alter blood lactate, RQ, heart rate (HR) or ventilation (V_E) suggesting that the increases in VO_2 max found in the Marconi study may not be physiologically significant.

Angelini et al.,[57] using a double-blind protocol, examined untrained subjects who were given either a placebo or carnitine supplementation (50mg/kg/day) for 1 month during an exercise regimen. VO_2 max increased after the supplementation period, but it was not possible to determine whether the affect was due to the carnitine supplementation or the training regimen since untrained individuals can show significant improvements in VO_2 max within 30 days of aerobic training. Furthermore, Cooper et al.[58] showed no significant improvement in marathon race time despite a daily supplementation of 4 g of oral carnitine for 10 days.

Studies by Oyono-Enguelle[59] and Soop[47] found no effect of carnitine supplementation for VO_2, volume of carbon dioxide (VCO_2), lactate, blood glucose at a fixed workload or fatty acid turnover. These studies used untrained[59] and moderately trained[47] individuals and supplemented 5 g orally for 10 days[59] and 2 g orally for 28 days,[47] respectively. Both of these studies concluded that, in healthy subjects, carnitine supplementation does not influence fatty acid utilization, suggesting that the endogenous production of carnitine is sufficient to support exercise.

Gorostiaga[60] utilized ten endurance-trained subjects (eight marathoners, one cyclist, one jogger) and supplemented 2 g of oral carnitine for 28 days. Subjects exercised for 45 minutes at 66% of VO_2 max. Significant differences were observed between supplemented and nonsupplemented groups with respect to RQ during the 38–45-minute interval (0.95 ± 0.01 versus 0.98 ± 0.02 respectively) but not any of the earlier time intervals. There were no other changes reported in any of the other variables measured, which included VO_2 max, HR, blood glycerol and resting blood fatty acid concentrations. These results seem insignificant, since other physiological parameters do not help to substantiate the reduction in RQ at one time interval.

In 1990, Silliprandi[61] and Vecchiet[62] examined the effects of 2 g of oral carnitine in a single dose approximately 1 hour prior to cycle ergometer exercise. Carnitine supplementation was reported to reduce blood lactate and increase VO_2 max postexercise. The authors claim that, during this high-intensity exercise, the PDC is stimulated, thereby reducing lactate production due to the alteration of the acetyl-CoA:free CoA ratio. These findings, however, were not supported, as Constantin-Teodosiu[38–40] showed that full activity of the PDC was reached within a minute of activation and is independent of carnitine supplementation. Subsequent studies [41, 63–68] have tried to extend carnitine research to cover varying lengths of time between 7–14 days, utilizing different exercise modalities including swimming, and altering administration via either oral or intravenous doses. The variables measured included performance times, VO_2 max, VCO_2, substrate utilization, glycogen storage, blood lactate and fatty acid turnover. Findings from these experiments find no support for the ergogenic benefits of carnitine on performance (Table 5.1). These physiological parameters are what govern our performance, and if a supplement is not successful in consistently altering these parameters, the efficacy of carnitine as an ergogenic aid is diminished. It is important to note here that there have also been studies trying to elucidate carnitine as a weight loss supplement due to its association with fatty acid metabolism. The fact remains that these studies[69,70] show that carnitine is not a contributing factor to weight loss since many supplements are given with calorie-restricted diets and/or exercise programs that are in themselves effective tools in reducing weight and increasing muscle mass. Supplement companies then mask their results behind the truly effective tools in weight management.

VI. Recommendations and Conclusions regarding Carnitine Supplementation

A. Recommendations regarding Supplementation of Carnitine

Despite no nutritional deficiency of carnitine, several of the co-factors and precursors necessary for its production are essential. Thus, it should be emphasized that for normal carnitine function a well-balanced diet is vital. Carnitine supplementation, however, is another issue. L-carnitine is very well tolerated and there seems to be no toxicity and very few side effects, even when taken in doses as large as 15g per day.[25,71,72] The problem remains that there is not enough unequivocal evidence to support a necessity for supplementation. The vast majority of the research reveals no benefit to enhancing performance as a result of increased carnitine supplementation. In general, proper training and genetic endowment lead to athletic success, not supplementation, whether it be carnitine or any other supplement.[70] Despite the vast amount of literature on supplementation, it is still prominent to "prescribe" or recommend supplements as a way of enhancing performance.

In conclusion, the majority of studies reveal that carnitine supplementation does not seem to provide an ergogenic benefit to human performance. The body and diet are sufficient to provide enough carnitine to allow an indi-

TABLE 5.1

The Effects of Carnitine Supplementation on Performance

Ref.	Subjects	Carnitine Dosage	Duration of Treatment	Dependent Variables	Effects of Carnitine
50	6 racewalkers	4 g/orally	14 days	VO_2 max, BLa-, RER	↑ VO_2 max, no change BLa-, RER
56	3 M/6 F	2 g/orally	14 days	VO_2 max, BLa-	No changes
59	10 M	2 g/orally	28 days	VO_2, VCO_2, RER, BLa-, plasma glucose	No effect of carnitine
47	7 M	5 g/orally	5 days	FFA, VO_2	No effects of carnitine
60	9 M/1 F	2 g/orally	28 days	RER, VO_2, HR, BLa-, plasma glucose	↓ RER, no other changes
61-62	10 M	2 g/orally	1 hr pre-exercise	VO_2 max, BLa-	↑ VO_2 max, ↓ BLa-
63	9 M	3 g/orally	7 days	RER, RPE, BLa-, HR, fat oxidation	No effect of carnitine
64	12 M	3 g/IV	40 min pre-exercise	VO_2, VCO_2, Substrate oxidation	No change during exercise
65	20 M	2 g/2 times/day oral	7 days	Swim performance, BLa-	No effect of carnitine
41	8 M	6 g/orally	7–14 days	RER, FFA, VO_2	No effect of carnitine

vidual to effectively regulate lipid metabolism both at rest and during various types of exercise.

VII. Summary

The role of carnitine in metabolism is very critical, it is a key component of the enzymes responsible for transporting long-chain fatty acids across the mitochondrial matrix where they can be used to produce energy. Carnitine is endogenously produced from the essential amino acids lysine and methionine in amounts that are sufficient to maintain homeostasis when an individuals' dietary intake includes a low to moderate amount of meat products. If there is a deficiency of carnitine, it is typically a result of impaired synthesis, increased degradation, inefficient transport or abnormal renal handling. This can result in glucose dependency and perhaps even hypoglycemia. Skeletal musculature is weakened and may atrophy in addition to decreased myoglobin concentrations. Finally, cardiac muscle, which utilizes primarily fatty acids for fuel, may experience failure and frequent arrhythmias.

References

1. Guleswitsch, W., Zur Kenntnis der Extraktivstoffe der Muskeln. II. Mitteilung. Über das Carnitin. *Hoppe Seylers Z. Physiol. Chem.*, 45, 326–330, 1905.
2. Kutscher, F., Über Liebig's Fleischextrakt. Mitteilung I. *Z. Untersuch Nahr Genussem.*, 10, 528–537, 1905.
3. Tomita, M. and Sendju, Y., Über die Oxyaminoverbindungen, welche die Biuretreaktion weigen. III. Spaltung der γ-Amino-β-oxybuttersaüre in die optisch-aktiven Komponenten. *Hoppe Seylers Z. Physiol. Chem.*, 169, 263–277, 1927.
4. Kaneko, T. and Yoshida, R., On the absolute configuration of L-carnitine (vitamin B_T). *Bull Chem. Soc. Jpn.*, 35, 1153–1155, 1962.
5. Bau, R., Schreiber, A., Metzenthin, T., Lu, R.S., Lutz, F., Klooster, W.T., Koetzle, T.F., Siem, H., Kleber, H.P., Brewer, F. and Englard, S., Neutron diffraction structure of (2R,3R)-L-(-)-[2-D] carnitine tetrachloroaurate, $[(CH_3)_3N\text{-}CH_2\text{-}CHOH\text{-}CHD\text{-}COOH]^+[AuCl_4]^-$: Determination of the absolute stereochemistry of the crotonobetaine-to-carnitine transformation catalyzed by L-carnitine dehydratase from *Escherichia coli*. *J. Am. Chem. Soc.*, 119, 12055–12060, 1997.
6. Fraenkel, G., Blewett, M. and Coles, M., BT, a new vitamin of the B-group and its relation to the folic acid group and other anti-anemia factors. *Nature*, 161, 981–983, 1948.
7. Siliprandi, N., Carnitine in physical exercise. In *Biochemical Aspects of Physical Exercise*, Benzi, G., Packer, L. and Siliprandi, N. Eds., Elsevier Science Publishers, Amsterdam, 1986, pp. 197–206.
8. Rebouche, C.J., Carnitine function and requirements during the life cycle. *FASEB J.*, 6, 3379–3386, 1992.
9. Heinonen, O.J., Carnitine and physical exercise. *Sports Med.*, 22, 109–132, 1996.
10. Paulson, D.J. and Shug, A.L., Tissue specific depletion of L-carnitine in rat heart and skeletal muscle by D-carnitine. *Life Sci.*, 28, 2931–2938, 1981.
11. Negrao, C.E., Ji, L.L., Schauer, J.E., Nagel, F.J. and Lardy, H.A., Carnitine supplementation and depletion: Tissue carnitines and enzymes in fatty acid oxidation. *J. Appl. Physiol.*, 63, 315–321, 1987.
12. Keith, R.E., Symptoms of carnitine-like deficiency in a trained runner taking DL-carnitine supplements. *JAMA.*, 255, 1137, 1986.
13. DiPalma, J.R., L-Carnitine: Its Therapeutic Potential. *Am. Family Physician.*, 34, 127–130, 1986.
14. Engel, A.G., ReBouche, C.J., Wilson, D.M., Glasgow, A.M., Romshe, C.A. and Cruse, R.P., Primary systemic carnitine deficiency. II. Renal handling of carnitine. *Neurology.*, 31, 819–825, 1981.
15. Borum, P.R., Carnitine function. In *Clinical Aspects of Human Carnitine Deficiency*, Borum, P.R., Ed., Pergamon, New York, 1986, pp. 16–27.
16. Engel, A.G. and Angelini, C., Carnitine deficiency of human skeletal muscle with associated lipid storage myopathy: A new syndrome. *Science*, 179, 899–902, 1973.
17. Pelligrini, G., Scarlato, G. and Moggio, M., A hereditary case in lipid storage myopathy with carnitine deficiency. *J. Neurology.*, 223, 73–84, 1980.
18. Ware, A.J., Burton, W.C., McGarry, J.D., Marks, J.F. and Weinberg, A.G., Systemic carnitine deficiency. Report of a fatal case with multisystemic manifestations. *J. Pediat.*, 93, 959–962, 1978.

19. Bohles, H., The effects of preoperative L-C supplementation on myocardial metabolism during aorto-coronary bypass surgery. *Curr. Ther. Res.*, 39, 429–435, 1986.
20. Brooks, H., Goldberg, L., Holland, R., Klein, M., Sanzari, N. and DeFelice, S., Carnitine-induced effects on cardiac and peripheral hemodynamics. *J. Clin. Pharmacol.*, 17, 561–568, 1977.
21. DiPalma, J.R., Ritchie, D.M. and McMichael, R.F., Cardiovascular and antiarrhythmic effects of carnitine. *Arch. Int. Pharmacody. Ther.*, 217, 246–250, 1975.
22. Silverman, N.A., Schmitt, G., Vishwanath, M., Feinburg, H. and Levitsky, S., Effect of carnitine on myocardial function and metabolism following global ischemia. *Ann. Thorac. Surg.*, 40, 20–24, 1985.
23. Rebouche, C.J. and Seim, H., Carnitine metabolism and its regulation in microorganisms and mammals. *Ann. Rev. Nutr.*, 18, 39–61, 1998.
24. Beiber, L.L., Carnitine. *Ann. Rev. Biochem.*, 57, 261–283, 1988.
25. Ceretelli, P. and Marconi, C., L-carnitine supplementation in humans: The effects on physical performance. *Int. J. Sports Med.*, 11, 1–14, 1990.
26. Wolf, G. and Berger, C.R.A., *Arch. Biochem. Biophys.*, 92, 360–365, 1961.
27. Bremer, J., Carnitine – metabolism and functions. *Physiol. Review.*, 63, 1420–1480, 1983.
28. Tanphaichitr, V. and Broquist, H.P., Role of lysine and trimethyllysine in carnitine biosynthesis. II. Studies in rat. *J. Biol. Chem.*, 248, 2176–2181, 1973.
29. Cox, R.A. and Hoppel, C.L., Biosynthesis of carnitine and 4-N-trimethylaminobutyrate from lysine. *Biochem. J.*, 136, 1083–1090, 1973.
30. Broquist, H.P., Carnitine biosynthesis and function: Introductory remarks. *Fed. Proc.*, 41, 2840–2842, 1982.
31. ReBouche, C.J. and Engel, A.G., Tissue distribution of carnitine biosynthetic enzymes in man. *Biochim. Biophys. Acta.*, 630, 22–29, 1980.
32. Heinonen, O.J., Carnitine: Effect on palmitate oxidation, exercise capacity and nitrogen balance. An experimental study with special reference to carnitine depletion and supplementation. Ph.D. Dissertation, University of Turku, Finland, 1992.
33. Feller, A.G. and Rudman, D., Role of carnitine in human nutrition. *J. Nutr.*, 118, 541-547, 1988.
34. Lombard, K.A., Olson, A.L., Nelson, S.E. and ReBouche, C.J., Carnitine status of lactoovovegetarians and strict vegetarian adults and children. *Am. J. Clin. Nutr.*, 50, 301–306, 1989.
35. Engel, A.G. and ReBouche, C.J., Carnitine metabolism and inborn errors. *J. Inherit. Metab. Dis.*, 7, 38–43, 1984.
36. Frolich, J., Seccombe, D.W. and Hahn, P., Effect of fasting on free and esterified carnitine levels in human serum and urine: correlation with serum levels of free fatty acids and β-hydroxybutyrate. *Metabolism*, 27, 555–561, 1978.
37. Constantin-Teodosiu, D., Regulation of pyruvate dehydrogenase complex activity and acetyl group formation in skeletal muscle during exercise. Ph.D. Dissertation, Huddinge University, Sweden, 1992.
38. Constantin-Teodosiu, D., Carlin, J.I., Cederblad, G., Hariss, R.C. and Hultman, E., Acetyl group accumulation and pyruvate dehydrogenase activity in human muscle during incremental exercise. *Acta. Physiol. Scand.*, 143, 367–372, 1991.
39. Constantin-Teodosiu, D., Cederblad, G. and Hultman, E., PDC activity and acetyl group accumulation in skeletal muscle during prolonged exercise. *J. Appl. Physiol.*, 73, 2403–2407, 1992.

40. Constantin-Teodosiu, D., Cederblad, G. and Hultman, E., PDC activity and acetyl group accumulation in skeletal muscle during isometric contraction. *J. Appl. Physiol.*, 74, 1712–1718, 1993.
41. Vukovich, M.D., Costill, D.L. and Fink, W.J., Carnitine supplementation: Effect on muscle carnitine and glycogen content during exercise. *Med. Sci. Sports Exerc.*, 26, 1122–1129, 1994.
42. Hiatt, W.R., Regensteiner, J.G., Wolfel, E.E., Ruff, L. and Brass, E.P., Carnitine and Acylcarnitine metabolism during exercise in humans. *J. Clin. Investig.*, 84, 1167–1173, 1989.
43. Decombaz, J., Gmuender, B., Sierro, G. and Ceretelli, P., Muscle carnitine after strenuous endurance exercise. *J. Appl. Physiol.*, 72, 423–427, 1992.
44. Lennon, D.L., Stratman, F.W., Shrago, E., Nagle, F.J., Madden, M., Hanson, P. and Carter, A.L., Effects of acute moderate-intensity exercise on carnitine metabolism in men and women. *J. Appl. Physiol.*, 55, 489–495, 1983.
45. Carlin, J.I., Reddan, W.G., Sanjak, M. and Hodach R., Carnitine metabolism during prolonged exercise and recovery in humans. *J. Appl. Physiol.*, 61, 1275–1278, 1983.
46. Harris, R.C., Louise Foster, C.V. and Hultman, E., Acetylcarnitine formation during intense muscular contraction. *J. Appl. Physiol.*, 63, 440–442, 1987.
47. Soop M., Bjorkman, O., Cederblad, G., Hagenfeldt, L. and Wahren, J., Influence of carnitine supplementation on muscle substrate and carnitine metabolism during exercise. *J. Appl. Physiol.*, 64, 2394–2399, 1988.
48. Sahlin, K., Muscle carnitine metabolism during incremental dynamic exercise in humans. *Acta. Physiol. Scand.*, 138, 259–262, 1990.
49. Wagenmakers, A.J.M., L-Carnitine supplementation and performance in man. *Med. Sci. Sports Exer.*, 32, 110–127, 1991.
50. Marconi C., Sassi, G., Carpenelli, A. and Ceretelli, P., Effects of L-carnitine loading on the aerobic and anaerobic performance of endurance athletes. *J. Sports Sci.*, 54, 131–135, 1985.
51. Borum, P.R., Plasma carnitine compartment and red blood cell carnitine compartment of healthy adults. *Am. J. Clin. Nutr.*, 46, 437–441, 1987.
52. Janssen, G.M.E., Scholte, H.R., Vaandrager, M.H.M. and Ross, J.D., Muscle carnitine level in endurance training and running a marathon. *Int. J. Sports Med.*, 10, S153–S155, 1989.
53. Arenas, J., Ricoy, J.R., Encinas, A.R., Pola, P., D'Iddio, S., Zeviani, M., Didonato, S. and Corsi, M., Carnitine in muscle, serum and urine on non-professional athletes: Effects of physical exercise, training and L-carnitine administration. *Muscle and Nerve.*, 14, 598–604, 1991.
54. Cederblad, G., Lindstedt, S. and Lundholm, K., Concentration of carnitine in human muscle tissue. *Clin. Chim. Acta.*, 53, 311–321, 1974.
55. Harper, P., Wadstrom, C. and Cederblad, G., Carnitine measurements in liver, muscle tissue and blood in normal subjects. *Clin. Chem.*, 39, 592–599, 1993.
56. Greig, C., Finch, K.M., Jones, D.A., Cooper, M., Sargeant, A.J. and Forte, C.A., The effect of oral supplementation with L-carnitine on maximum and submaximum exercise capacity. *Eur. J. Appl. Physiol.*, 56, 457–460, 1987.
57. Angelini, C., Vergani, L. and Costa, L., Use of carnitine in exercise physiology. *Adv. Clin. Enzymol.*, 4, 103–110, 1986.

58. Cooper, M.B., Jones, D.A., Edwards, R.H.T., Corbucci, G.C., Montanari, G. and Trevisani, C., The effect of marathon running on carnitine metabolism and on some aspects of mitochondrial activities and antioxidant mechanisms. *J. Sport Sci.*, 4, 79–87, 1986.
59. Oyono-Enguelle, S., Freund, H., Ott, C., Gartner, M., Heitz, A., Marbach, J., Maccari, F., Frey, A., Bigot, H. and Bach, A.C., Prolonged submaximal exercise and L-carnitine in humans. *Eur. J. Appl. Physiol.*, 58, 53–61, 1988.
60. Gorostiaga, E.M., Maurer, C.A. and Eclache, J.P., Decrease in RD during exercise following L-carnitine supplementation. *Int. J. Sports Med.*, 10, 71–80, 1989.
61. Siliprandi, N., DiLisa, F., Peiralisi, G., Ripari, P., Maccari, F., Menabo, R., Giamberardino, M.A. and Vecchiet, L., Metabolic changes induced by maximal exercise in humans following L-carnitine administration. *Biochem. Biophys. Acta.*, 1034, 17–21, 1990.
62. Vecchiet, L., DiLisa, F., Peiralisi, G., Ripari, R., Menabo, R., Giamberardino, M.A. and Siliprandi, N., Influence of L-carnitine administration on maximal physical exercise. *Eur. J. Appl. Physiol.*, 61, 486–490,1990.
63. DeCombaz, J., Deriaz, O., Acheson, K., Gmuender, B and Jequier, E., Effect of L-carnitine on submaximal exercise metabolism after glycogen depletion. *Med. Sci. Sports Exerc.*, 25, 733–740, 1993.
64. Natali, A., Santoro, D., Brandi, L.S., Faraggiana, D., Ciociaro, D., Pecori, N., Buzzigoil, G. and Ferrannini, E., Effects of hypercarnitinemia during increased fatty substrate oxidation in man. *Metabolism.*, 42, 594–600, 1993.
65. Trappe, S.W., Costill, D.L., Goodpaster, B., Vukovich, M.D. and Fink, W.J., The effects of L-carnitine supplementation in performance during interval swimming. *Int. J. Sports Med.*, 15, 181–185, 1994.
66. Brass, E.P., Hoppel, C.L. and Hiatt, W.R., Effects of intravenous L-carnitine on carnitine homeostasis and fuel metabolism during exercise in humans. *Clin. Pharmacol. Ther.*, 55, 681–692, 1994.
67. Barnett, C., Costill, D.L., Vukovich, M.D. Cole, K.J., Goodpaster, B.H., Trappe, S.W. and Fink, W.J., Effect of L-carnitine supplementation on muscle and blood carnitine content and lactate accumulation during high-intensity spring cycling. *Int. J. Sports Nutr.*, 4, 280–288, 1994.
68. Colombani, P., Wenk, C., Kunz, I., Krahenbuhl, S., Kuhnt, M., Arnold, M., Frey-Rindova, P., Frey, W. and Langhans, W., Effects of L-carnitine supplementation on physical performance and energy metabolism of endurance trained athletes: A double-blind crossover study. *Eur. J. Appl. Physiol.*, 73, 434–439, 1996.
69. Gruneweld, K.K. and Bailey, R.S., Commercially marketed supplements for bodybuilding athletes. *Sports Med.*, 15, 90–103, 1993.
70. Williams, M.H., Ergogenic and ergolytic substances. *Med. Sci. Sports Exerc.*, 24, S344–S348, 1992.
71. Snyder, T.M., Little, B.W., Roman-Campos, G. and McQuillen, J.B., Successful treatment of familial idiopathic lipid storage myopathy with L-carnitine and modified lipid diet. *Neurology.*, 32, 1106–1115, 1982.
72. Waber, L.J., Valle, D., Neill, C., DiMauro, S. and Shug, A., Carnitine deficiency presenting as familial cardiomyopathy: a treatable defect in carnitine transport. *J. Pediat.*, 101, 700–705, 1982.

6

Creatine

Richard B. Kreider, Brian C. Leutholtz and Mike Greenwood

CONTENTS

I. Introduction 81
II. Muscle Creatine Content and Phosphocreatine Resynthesis 82
III. Effects Of Short and Long-Term Creatine Supplementation.........85
IV. Safety of Creatine Supplementation..91
V. Potential Medical Uses of Creatine 92
VI. Summary and Conclusions 93
References..93

I. Introduction

During brief explosive exercise lasting several seconds, the energy supplied to rephosphorylate adenosine diphosphate (ADP) to adenosine triphosphate (ATP) is dependent to a large degree on the amount of phosphocreatine (PCr) stored in the muscle.[1,2] As PCr stores become depleted during explosive exercise, energy availability deteriorates due to the inability to resynthesize ATP at the rate required.[1,2] Consequently, the ability to maintain maximal effort exercise declines. Since the availability of PCr in the muscle may significantly influence the amount of energy generated during brief periods of high-intensity exercise, it has been hypothesized that increasing muscle creatine content via creatine supplementation may increase the availability of PCr and allow for an accelerated rate of resynthesis of ATP during and following high intensity, short duration exercises.[1-7]

Studies investigating this hypothesis have demonstrated that short term creatine supplementation (e.g., 20–25 g/d for 4 to 7 d) increases total creatine (TCr) content by 15 to 30% and phosphocreatine (PCr) stores by 10 to 40%.[2,7-16] The increased availability of creatine and phosphocreatine have been reported to maintain ATP levels during high-intensity exercise and facilitate ATP resynthesis following intense exercise.[9,15,17–20] Short-term creatine supplementation has been reported to improve maximal power/strength, work performed during sets of maximal effort muscle contractions, single-effort sprint performance, and work performed during repetitive sprint performance. Moreover, long-term supplementation of creatine or creatine containing supplements (5 to 25 g/d for 5 to 7 d and 2 to 25 g/d thereafter for 7-d to 21 months) have been reported to promote significantly greater gains in strength, sprint performance, and fat-free mass during training in comparison to matched-paired controls.[21–28] While not all studies report ergogenic benefits, most indicate that creatine is an effective nutritional supplement. Consequently, creatine has become one of the most popular nutritional supplements marketed to athletes in recent times. The following chapter overviews the available literature regarding the effects of creatine supplementation on muscle bioenergetics, performance and body composition. In addition, the medical safety and potential medical uses are discussed. The chapter concludes with a summary of findings and suggested areas for additional research.

II. Muscle Creatine Content and Phosphocreatine Resynthesis

The TCr content in the body in free and phosphorylated forms is about 120 g for a 70 kg person.[3,13,29,30] Approximately 95% of creatine is stored in skeletal muscle as PCr (~66%) or free creatine (~33%). The remaining amount of creatine is found in the heart, brain and testes.[3,13,29,30] The normal daily requirement for creatine is approximately 1.6% of the TCr pool (about 2 g for a 70 kg individual). Of this, about half of the daily needs of creatine are obtained from the diet primarily from meat, fish and animal products. For example, there is approximately 1 g of creatine in 250 g of raw red meat. The remaining daily need of creatine is synthesized primarily in the liver, kidney and pancreas from the amino acids glycine, arginine and methionine (Figure 6.1).

The normal muscle TCr concentration ranges between 120 and 125 mmol/kg dry mass muscle.[2,7,8,13–16] Creatine supplementation typically increases TCr content by TCr content by 15 to 30% and phosphocreatine (PCr) stores by 10 to 40%.[2,7–16] However, there is some evidence that some individuals do not appear to respond as well to creatine supplementation (i.e., observe less than a 20 mmol/kg dry mass change in TCr levels).[5,15,31,32] However, more recent studies indicated that ingesting creatine (20 g/d) with glucose,[33–35] a combination of carbohydrate and protein,[36] and D-Pinitol[37] increased muscle creatine content to a greater degree than when creatine was ingested alone. As a result, all subjects experienced significant increases in muscle creatine content. Moreover, there is evidence that co-ingesting creatine with glucose or increasing muscle creatine content prior to carbohydrate loading enhances muscle glycogen retention.[11,38–40] These data suggest that the most effective

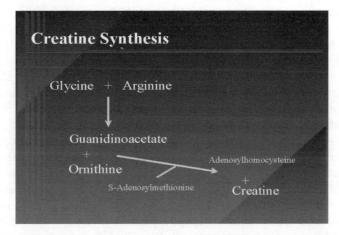

FIGURE 6.1
Biochemical pathway for creatine synthesis.

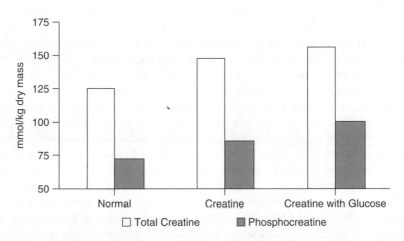

FIGURE 6.2
Average change in concentrations of muscle creatine and phosphocreatine after creatine supplementation with or without large amounts of added glucose.

way to enhance creatine and possibly glycogen retention appears to be to co-ingest creatine and simple carbohydrates or protein. Figure 6.2 presents the average changes in TCr and PCr reported in the literature in response to creatine supplementation with and without glucose.

Since creatine supplementation has been shown to increase intramuscular PCr, a number of studies have evaluated the effects of creatine supplementation on ATP and PCr resynthesis following high-intensity exercise.[9,15,17,18,20,41–43] These studies indicate that creatine supplementation does not appear to alter pre-exercise ATP concentrations.[15,17,44,45] However, the elevated PCr concentrations allow for a greater maintenance of ATP concentrations and power output during a maximal-effort sprint performance.[18,20,42-44,46,47] There is also some evidence that the rate of ATP and PCr resynthesis may be improved after intense exercise, which would allow an individual to perform to a greater degree in subsequent exercise bouts.[15,41,44,48] However, other studies show that, although creatine supplementation increases muscle PCr and improves work output during high-intensity exercise, it does not appear to affect the rate of PCr resynthesis or the rate of creatine kinase (CK) activity.[18,42-44]

Most research on creatine has evaluated the effects of creatine supplementation on the content of high-energy phosphates within muscle and its role on exercise capacity and recovery during sprint exercise, dependent on the phosphagen energy system. However, recent attention has also been paid to the role of creatine supplementation on creatine kinases (CK) in muscle as well as the transfer of high-energy phosphates within the cell via what has been called the creatine phosphate shuttle.[49-54] CK is an important cellular enzyme that facilitates energy transduction in muscle cells by catalyzing the reversible transfer of a phosphate moiety between ATP and PCr.[54] Several isoforms of CK work simultaneously to form a rapid interconversion of PCr and ATP. In this regard, CK is composed of two subunit types in muscle (M) and brain (B) with three isoenzymes (MM-CK, MB-CK, and BB-CK). In addition, a fourth CK isoenzyme (Mi-CK) is located on the outer side of the inner mitochondrial membrane.[52,54,55] CK activity is greatest in skeletal muscle, and CK in the muscle exists almost exclusively in the MM form. MM-CK (also referred to as myofibrillar CK) is bound to the myofibrils and localized to the A-bands as well as being distributed across the entire filament. MM-CK generates ATP from ADP. Mi-CK is found on the outer surface of the inner mitochondrial membrane and is functionally coupled to oxidative phosphorylation. Mi-CK, at the site of oxidative mitochondrial ATP generation, catalyses the phosphorylation of creatine to PCr.[52,54,55] Fast-twitch fibers have greater CK activity because they contain a larger amount of MM-CK than slow-twitch oxidative fibers. However, slow-twitch fibers have a higher percentage of Mi-CK.[52,54-56] Recent studies indicate that dietary and muscle availability of creatine can influence CK activity, particularly in various patient populations.[51,57-63] Creatine supplementation has been reported to influence metabolism or adaptations to training in fast- and slow-twitch muscle fiber.[17,40,64-68]

In addition to its role as an energy buffer, it has been proposed that the CK-PCr system functions in energy transport on the basis of the functional and physical association of CK isoenzymes with subcellular sites of ATP production and hydrolysis. In the creatine phosphate shuttle concept, PCr and Cr act as shuttle molecules between these sites.[56] One proposed shuttle is believed to be functionally coupled to glycolysis,[56] but others believe that the rapid resynthesis of PCr is likely to be oxidative in origin.[54,69,70] Mi-CK promotes the formation of PCr from creatine and from ATP formed via oxidative metabolism in the mitochondria,[52] van Deursen and others (1993) note that PCr is presumed to diffuse from the mitochondria to the myofibrillar M-band, where it locally serves to replenish ATP with MM-CK as catalyzing agent. Finally, Cr diffuses back to sites of ATP synthesis for rephosphorylation. The potential role of modulating CK activity and the shuttling of adenine nucleotides for synthesis and use has prompted a significant amount of research evaluating the potential clinical uses of creatine, as described below. Additionally, since creatine appears to play a role in the shuttling of ATP from the mitochondria to the cytosol, there has been interest in determining whether creatine supplementation may influence high-intensity aerobic exercise capacity or recovery from intense exercise that is facilitated by oxidative metabolism.

III. Effects Of Short and Long-Term Creatine Supplementation

Numerous studies have examined the effects of short-term creatine supplementation (e.g., 20 g/d for 4 to 7 d) on exercise performance. The majority of initial studies suggested that creatine supplementation can significantly increases strength, power, sprint performance or work performed during multiple sets of maximal effort muscle contractions.[5,13,68,71–73] More recent studies have supported these initial observations. For example, Volek and colleagues[66] reported that creatine supplementation (25 g/d for 7 d) resulted in significant increases in the amount of work performed during five sets of bench press and jump squats in comparison with a placebo group. Urbanski and associates[74] reported that creatine supplementation (20 g/d × 5 d) increased maximal isometric knee extension strength and time to fatigue. Tarnopolsky and co-workers[75] reported creatine supplementation (20 g/d × 4 d) increased peak cycling power, dorsi-flexion maximal voluntary contractions (MVC) torque and lactate in men and women with no apparent gender effects. Moreover, Wiroth and colleagues[76] reported that creatine supplementation (15 g/d × 5 d) significantly improved maximal power and work performed during 5 × 10 sec cycling sprints with 60 sec rest recovery in younger and older subjects. These findings and many others support prior reports indicating that creatine supplementation can improve performance when evaluated in controlled laboratory and testing settings.[8,9,23,25,28,77–81]

Some have criticized this type of early creatine research, suggesting that although performance gains have been observed in controlled laboratory settings, it was less clear whether these changes would improve athletic performance on the field.[82,83] Since then, a number of studies have attempted to evaluate the effects of creatine supplementation on field performance. These studies have generally indicated that short-term creatine supplementation may improve high-intensity, short-duration performance in various athletic tasks. For example, Skare and associates[84] reported that creatine supplementation (20 g/d) decreased 100 m sprint times and reduced the total time of 6 × 60 m sprints in a group of well-trained adolescent competitive runners. Mujika and colleagues[85] reported that creatine supplementation (20 g/d × 6 d) improved repeated sprint performance (6 × 15m sprints with 30 sec recovery) and limited the decay in jumping ability in 17 highly trained soccer players. Similarly, Theodorou et al.[86] reported that creatine supplementation (25 g/d × 4 d) significantly improved mean interval performance times in 22 elite swimmers.

These recent findings and many others suggest that creatine supplementation can improve performance of athletes in a variety of sport-related field activities.[27,79–81,87–105]

Since creatine supplementation may affect shuttling of high-energy phosphates between the cytosol and mitochondria, some have suggested that creatine supplementation may affect performance during more prolonged exercise bouts. Recent studies also provide some support this contention. For example, Earnest and associates[106] reported that creatine supplementation (20 g/d × 4 d and 10 g/d × 6 d) improved cumulative run time to exhaustion in two runs lasting approximately 90 sec each. Smith and collaborators[107] reported that creatine supplementation (20 g/d × 5 d) increased work rate during exercise bouts lasting between 90 and 600 s primarily at the shorter, more intense exercise bouts. Nelson and co-workers[108] found that creatine supplementation (20 g/d × 7 d) decreased submaximal heart rate and oxygen uptake (VO_2) while increasing ventilatory anaerobic threshold (VANT) and total time to exhaustion during a maximal exercise test in 36 trained adults. Rico-Sanz et al.[109] reported that creatine supplementation (20 g/d × 5 d) increased time to exhaustion (29.9 ± 3.8 to 36.5 ± 5.7 min) while reducing ammonia levels (a marker of adenine nucleotide degradation) when cycling at 30 and 90% of maximum until exhaustion. Preen and associates[110] evaluated the effects of ingesting creatine (20 g/d × 5 d) on resting and postexercise TCr and PCr as well as performance of an 80-min intermittent sprint test (10 sets of 5–6 × 6 sec sprints with varying recovery intervals). The authors reported that creatine increased resting and postexercise TCR and PCr, mean work performed, and total work performed during 6 × 6 sec sets with 54 sec and 84 sec recovery. In addition, work performed during 5 × 6 sec sprints with 24 sec recovery tended to be greater (p = 0.056). Kambis and colleagues[111] reported that short-term creatine supplementation improved muscle performance in women without significant gains in muscle volume or body weight. Finally, Chwalbinska-Moneta and co-workers[104] reported that creatine supplementation (20 g/d for 5 d) improved endurance (expressed by the individual lactate threshold) and anaerobic performance, independent of the effect of intensive endurance training. Collectively, these findings support contentions that creatine supplementation may provide ergogenic benefit for more prolonged exercise bouts involving both anaerobic and aerobic energy systems.

However, as with previous creatine research, not all of the recent studies have found that creatine supplementation enhances exercise performance. For example, McKenna et al.[112] reported that creatine supplementation (30 g/d × 5 d) did not affect 5 × 10 sec sprints with rest intervals of 180, 50,

20, and 20 sec in 14 untrained subjects. Gilliam and colleagues[113] found that creatine supplementation (20 g/d × 5 d) did not affect isokinetic knee extension performance during 5 × 30 MVC in 23 untrained subjects. Deutekom and co-workers[114] reported that creatine (20 g/d × 6 d) increased body mass but did not affect muscle activation, fatigue or recovery from electrical stimulation of the quadriceps or maximal exercise performance during sprint cycling in 23 well-trained rowers. Similarly, Edwards and associates[115] reported that creatine (20 g/d × 6 d) did not affect running fatigue to exhaustion following performing 4 × 15 sec sprints in 21 moderately active subjects. However, ammonia levels were lower following creatine supplementation, suggesting that may have lessened the degree of adenine nucleotide degradation and improved metabolic efficiency. In another study, Op 't Eijnde et al.[116] reported that creatine (20 g/d × 5 d) did not enhance stroke performance or 70 m agility sprint performance in well-trained tennis players. Finn and colleagues[117] reported that, although creatine supplementation (20 g/d × 5 d) increased TCr content and 1 sec relative peak power in 16 triathletes, no significant effects were observed in repetitive cycling sprint performance (4 × 20 sec sprints with 20 sec rest recovery). Van Schuylenbergh et al.[118] found that 1 week of creatine-pyruvate supplementation at a rate of 7 g/d did not significantly improve endurance capacity or intermittent sprint performance in cyclists. Delecluse et al.[119] reported that no ergogenic effect of creatine supplementation (0.35 g/kg/d for 7 d) on single or repeated 40 m sprint times with varying rest periods was observed in highly trained athletes. Finally, Biwer and associates[120] reported that short-term creatine supplementation (0.3 g/kg/d for 6 d) did not improve performance in submaximal running interspersed with high-intensity intervals.

In our view, when one examines all of the available literature on creatine supplementation, the following conclusions can be drawn. First, although some intrasubject variability has been reported, the vast majority of studies (> 90%) indicate that short-term creatine supplementation significantly increases TCr and PC_r content as determined by assessing muscle biopsies, urinary whole-body creatine retention or magnetic resonance spectroscopy (MRS). Consequently, it is clear that creatine supplementation enhances the potential to perform high-intensity exercise, much like carbohydrate loading enhances the potential to perform endurance exercise to exhaustion. Overall, approximately 70% of short-term studies on creatine supplementation report some ergogenic benefit particularly during high-intensity repetitive exercise. These benefits have been primarily found when performing laboratory tests that have good test-to-test reliability. However, as described above, a number of recent studies have indicated that creatine supplementation can also improve performance in field-type events like running, soccer and swimming. It is also interesting to note that, over the last few years, the percentage of studies reporting ergogenic benefit from creatine supplementation have risen to about 80–85%. presumably due to a greater understanding of how to properly design studies to assess the ergogenic value of creatine supplementation. Benefits have been reported in untrained, trained and diseased children, adolescents, adults and elderly populations.[28] Studies reporting no significant effects of creatine supplementation have generally observed small but nonsignificant improvements in performance (i.e., 1–7%). It should be noted that no study has reported a statistically significant ergolytic (negative) effect from creatine supplementation. Studies that have reported no significant benefit of creatine supplementation often have low statistical power, have evaluated performance tests with large test-to-test reliability or have not incorporated appropriate experimental controls.[28,75,121] Consequently, it is our view that the preponderance of evidence indicates that short-term creatine supplementation enhances performance in a variety of laboratory and on-field exercise tasks.

Theoretically, increasing the ability to perform high-intensity exercise may lead to greater training adaptations over time. Consequently, a number of studies have evaluated the effects of long-term creatine supplementation on training adaptations. For example, Vandenberghe et al.[16] reported that, in comparison with a placebo group, creatine supplementation (20 g/d × 4 d; 5 g/d × 65 d) during 10 weeks of training in women increased TCr and PCr and promoted a 20–25% greater increase in maximal strength, a 10–25% greater increase in maximal intermittent exercise capacity of the arm flexors and a 60% greater increase in fat-free mass (FFM). In addition, the researchers reported that creatine supplementation during 10 weeks of detraining helped maintain training adaptations to a greater degree. Kelly and associates[122] reported that 26 d of creatine supplementation (20 g/d × 4 d; 5 g/d × 22 d) significantly increased body mass, FFM, three-repetition maximum (RM) on the bench press and the number of repetitions performed in the bench press over a series of sets in 18 power lifters. Noonan and collaborators[123] reported that creatine supplementation (20 g/d × 5 d; 100 or 300 mg/kg/d of FFM × 51 d) in conjunction with resistance and speed/agility training significantly improved 40-yd dash time and bench-press strength in 39 college athletes. Kreider and associates[124] reported that creatine supplementation (15.75 g/d × 28 d) during off-season college football training promoted greater gains in FFM and repetitive sprint performance in comparison with subjects ingesting a placebo. Likewise, Stone et al.[27] reported that 5 weeks of creatine ingestion (~10 or 20 g/d with and without pyruvate) promoted significantly greater increases in body mass, FFM, 1 RM bench press, combined 1 RM squat and bench press, vertical jump power output and peak rate of force development during in-season training in 42 Division IAA college football players.

Volek and co-workers[66] reported that 12 weeks of creatine supplementation (25 g/d × 7 d; 5 g/d × 77 d) during periodized resistance training increased muscle TCr and PC, FFM, type I, IIa, and IIb muscle fiber diameter, bench press and squat 1 RM and lifting volume (weeks 5–8) in 19 resistance-trained athletes. Peeters and associates[125] reported that creatine monohydrate and creatine phosphate supplementation (20 g/d × 3 d; 10 g/d × 39 d) during training significantly increased body mass, FFM and 1-RM bench press in 35 resistance-trained males. Kirksey and colleagues[126] found that creatine supplementation (0.3 g/kg/d × 42 d) during off-season training promoted greater gains in vertical jump height and power, sprint cycling performance and FFM in 36 Division IAA male and female track and field athletes. Pearson et al.[127] reported that creatine supplementation (5 g/d × 10 weeks) during resistance training promoted greater gains in strength, power and body mass with no change in percent body fat in 16 Division IA college football players during summer conditioning. Moreover, Jones and collaborators[91] reported that creatine (20 g/d × 5 d; 5 g/d × 10 weeks) promoted greater gains in sprint performance (5 × 15 sec with 15 sec recovery) and average on-ice sprint performance (6 × 80 m sprints) in 16 elite ice-hockey players. Becque and associates[128] found that that creatine supplementation (20 g/d × 5 d; 2 g/d × 37 d) during strength training led to greater gains in arm flexor muscular strength, upper-arm muscle area and FFM than strength training alone in 23 resistance-trained athletes.

Other recent studies also indicate that creatine supplementation improves training adaptations. For example, Burke et al.[129] reported that low-dose creatine supplementation (7.7 g/d × 21 d) during training promoted greater gains in total work until fatigue, peak force, peak power, and fatigue resistance in 41 college athletes. In a follow-up study, Burke and associates[130] found that males supplemented with a combination of whey protein and creatine during resistance training had greater increases in lean tissue mass and bench press than those who supplemented with only whey protein or placebo. Brenner and colleagues[131] reported that creatine supplementation (20 g/d × 7 d; 2 g/d × 28 d) significantly improved upper-body strength gain and decreased percent body fat in 16 female college lacrosse players during preseason training. Larson-Meyer and co-workers[93] reported that creatine supplementation (15 g/d × 7 d; 5 g/d × 84 d) promoted greater gains in bench press and squat maximal strength with no differences in FFM during off-season training in 14 female college soccer players. Interestingly, Jowko et al.[132] reported that creatine supplementation (20 g/d × 7 d; 10 g/d × 14 d) significantly increased FFM and cumulative strength gains during training in 40 subjects initiating training. Additional gains were observed when 3 g/d of calcium beta-hydroxy-beta-methylbutyrate (HMB) was co-ingested with creatine. Finally, Brose et al.[133] found that the addition of creatine to a supervised resistance-training program in older adults enhanced the increase in total and FFM and gains in several indices of isometric muscle strength.

The gains in FFM observed in response to creatine supplementation during training have been suggested to be a result of fluid retention, increased protein synthesis or an enhanced quality of training. For this reason, recent research has attempted to gain insight into the potential mechanisms behind observed changes in strength and muscle mass during training. This research has indicated that creatine supplementation does not appear to promote a disproportionate increase in total body water.[10,124,134-137] These findings suggest that the weight gain observed in response to creatine supplementation cannot simply be attributed to fluid retention. Several studies indicate that creatine supplementation during training promotes muscle hypertrophy, as determined by assessing changes in muscle fiber diameter, muscle volume or advanced methods of assessing body composition.[66,77,137–139] Additionally, recent studies suggest that creatine supplementation may slow rates of muscle atrophy and improve physical therapy outcomes following immobilization.[39,67,140] Although there was some initial evidence that creatine may directly influence markers of acute protein synthesis or turnover,[141–143] recent studies do not support this hypothesis.[144–146] To date, the best evidence to indicate that creatine supplementation during training may influence protein synthesis and muscle hypertrophy comes from an interesting series of experiments from Willougby and associates.[65] These researchers reported that, in comparison with controls, creatine supplementation (6 g/d × 12 weeks) during resistance training (6–8 repetitions at 85–90%; 3 × week) significantly increased total body mass, FFM, and thigh volume, 1 RM strength, myofibrillar protein content, Type I, IIa, and IIx myosin heavy chain (MHC) mRNA expression and MHC protein expression. More recently, Willoughby and colleagues[58] reported that creatine supplementation increases M-CK mRNA expression, likely due to concomitant increases in the expression of myogenin and myogenic regulatory factor-4 (MRF-4). The researchers concluded that the increases in myogenin and MRF-4 mRNA and protein may play a role in increasing myosin heavy-chain expression as they previously observed. This series of studies provides strong evidence that creatine supplementation during intense resistance training leads to greater gains in strength and muscle mass.

It is our view that, after evaluating the available data on the effects of creatine supplementation on training adaptations, the following conclusions can be drawn. Nearly all studies indicate that creatine loading increases TCr and PCr. Creatine supplementation during training is typically associated with a 0.5–2 kg greater increase in body mass or FFM. Although it has been hypothesized that the initial weight gain associated with creatine supplementation may be due to fluid retention, a number of studies indicate that long-term creatine supplementation increases FFM or muscle fiber diameter with no disproportional increase in total body water. These findings suggest that the weight gain observed during training appears to be muscle mass, although it is unclear whether this is simply a result of an improved quality of training or direct effects on modulators of muscle hypertrophy. About 90% of long-term training studies report some ergogenic benefit with gains typically 10–100% greater than controls. Improvements have been reported in untrained and trained adolescents, adults and elderly populations. No clinically significant side effects have been reported in these studies even though many of them involved intense training in a variety of exercise conditions. These findings suggest that creatine supplementation during training serves to enhance training adaptations. Moreover, these beneficial changes may offer some therapeutic benefit for a variety of pathologies involving muscle weakness or muscle wasting.

IV. Safety of Creatine Supplementation

Creatine has become a very popular nutritional supplement among athletes. The only documented side effect from creatine supplementation reported in the scientific or medical literature has been weight gain,[123,125,147,148] and possibly an increase in anterior compartment pressure that some believe may predispose athletes to anterior compartment syndrome,[149,150] although no safety study has reported such a side effect. However, there have also been anecdotal reports and concerns that creatine supplementation may promote muscle cramping, dehydration, increased susceptibility to muscle strains, long-term suppression of creatine synthesis or unknown long-term side effects. Over the last several years, a significant amount of research has been conducted to examine the validity of these concerns. This research has indicated that short- and long-term creatine supplementation does not appear to negatively affect endogenous creatine synthesis,[16,151] renal and liver function,[152–161] muscle and liver enzyme efflux,[152,154,155] blood volume[135,162–164] and electrolyte status,[124,155,165] blood pressure[166] or general markers of health.[125,152,154,159,162,164,166,167] Additionally, recent research does not support contentions that creatine supplementation during intense training in hot or humid environments predisposes athletes to an increased incidence of muscle cramping,[135,168–171] dehydration[168,172,173] or musculoskeletal injuries such as muscle strains.[96,154,163,168,169,174,175–177]

For example, Kreider and colleagues[155] evaluated the effects of short- and long-term creatine supplementation (15.75 g of creatine for 5 d followed by ingesting an average of 5 g/d thereafter administered in 5–10 g doses) over a 21-month period on metabolic markers, muscle and liver enzymes, electrolytes, lipid profiles, hematological markers, lymphocytes and renal in 98 Division IA college football players who did or did not take creatine during training. Multivariate analysis revealed no significant differences among groups in the 54-item panel of quantitative blood and urine markers assessed. Further, univariate analysis revealed no clinically significant interactions among groups in markers of clinical status. In addition, no apparent differences were observed among groups in the 15-item panel of qualitative urine markers. The researchers concluded that long-term creatine supplementation does not appear to adversely affect markers of health status in athletes undergoing intense training in comparison with athletes who do not take creatine.

As a part of this study, the incidence of injury observed during 3 years of NCAA Division IA college football training and competition were also monitored.[178] These athletes practiced or played in environmental conditions ranging from 8 to 40°C (mean 24.7 ± 5°C) and 19 to 98% relative humidity (49.3 ± 17%). It was found that creatine users had fewer incidence of cramping (37 incidences by creatine users to 96 incidences by nonusers, 39%), heat or dehydration (8 to 28, 36%), muscle tightness (18 to 42, 43%), muscle pulls and strains (25 to 51, 49%), noncontact joint injuries (44 to 32, 33%), contact injuries (39 to 104, 44%), illness (12 to 27, 44%), number of missed practices due to injury (19 to 41, 46%), players lost for the season (3 to 8, 38%) and total injuries or missed practices (205 to 529, 39%). Consequently, the researchers concluded that creatine supplementation does not increase the incidence of injury or cramping in college football players. Similar results have been reported by other researchers evaluating the effects of creatine supplementation on injury rates during football,[169] soccer[96] and baseball[179] training and competition. These findings provide additional evidence that short- and long-term creatine supplementation does not appear to cause any short- or long-term detrimental effects. Further, creatine supplementation during training may actually reduce the incidence of injuries in some athletes.

V. Potential Medical Uses of Creatine

There has been significant interest in the potential medical uses of creatine for some time. Initial research evaluated the effects of oral creatine or intravenous creatine phosphate administration on heart function in heart failure patients.[180–184] Additionally, there has been considerable interest in use of creatine in treating patients with creatine synthesis deficiencies.[185–193] Since creatine supplementation has been reported to improve muscle strength and high-intensity exercise performance, and to promote increases in muscle mass, there has been interest in determining the effects of creatine supplementation in patients with a variety of neuromuscular conditions like Duchannes Muscular Dystrophy,[60,62,64,194–196] amyotrophic lateral sclerosis (Lou Gehrig's Disease)[197–202] and Huntington's disease.[59,63,203–208] There is also evidence that creatine may have neuroprotective effects on brain and in spinal cord injuries.[63,206,207,209–212] Finally, creatine supplementation has been studied as a potential therapeutic agent in patients with arthritis, [213] glucose intolerance or diabetes[39] and as a means to reduce cholesterol[124,214,215] and homocysteine levels.[216–219] Finally, as described above, there is evidence that creatine supplementation may lessen atrophy as a result of immobilization, and enhance physical therapy outcomes.[39,67,140] Although creatine has not been shown to be efficacious in all of these conditions and additional research is necessary, it is clear that creatine may have a number of potential therapeutic attributes.

VI. Summary and Conclusions

Based on available research, short-term creatine supplementation appears to improve maximal strength and power by 5 to 15%, work performed during sets of maximal effort muscle contractions by 5 to 15%, single-effort sprint performance by 1 to 5% and work performed during repetitive sprint performance by 5 to 15%. Moreover, long-term supplementation of creatine or creatine-containing supplements (15 to 25 g/d for 5 to 7 d and 2 to 25 g/d thereafter for 7 to 140 d) may promote significantly greater gains in strength, sprint performance and FFM during training in comparison with matched-paired controls. However, not all studies have reported ergogenic benefit, possibly due to differences in subject response to creatine supplementation, length of supplementation, exercise criterion evaluated or the amount of recovery observed during repeated bouts of exercise. The only clinically significant side effect reported in the scientific literature from creatine supplementation has been weight gain. Consequently, creatine supplementation appears to be a safe and effective nutritional strategy to enhance exercise performance. Additional research involving creatine supplementation should continue to examine the possible ergogenic value of creatine supplementation on exercise and training adaptations in a variety of sports. Additionally, research should focus on mechanisms of action and potential therapeutic benefits of creatine.

References

1. Chanutin, A., The fate of creatine when administered to man. *J Biol Chem*, 67, 29, 1926.
2. Hultman, E., et al. Energy metabolism and fatigue, in *Biochemistry of Exercise VII*, Taylor, A., Gollnick, P.D. and Green, H., Eds. 1990, Human Kinetics: Champaign, IL. p. 73.
3. Balsom, P.D., Soderlund, K. and Ekblom, B., Creatine in humans with special reference to creatine supplementation. *Sports Med*, 18(4), 268, 1994.
4. Greenhaff, P., The nutritional biochemistry of creatine. *J Nutr Biochem*, 11, 610, 1997.
5. Greenhaff, P.L. Muscle creatine loading in humans: Procedures and functional and metabolic effects. in *6th Internationl Conference on Guanidino Compounds in Biology and Medicine*, Cincinnati, OH, 2001,
6. Greenhaff, P., Casey, A. and Green, A.L., Creatine supplementation revisited: An update. *Insider*, 4(3), 1, 1996.
7. Harris, R.C., Soderlund, K. and Hultman, E., Elevation of creatine in resting and exercised muscle of normal subjects by creatine supplementation. *Clin Sci (Colch)*, 83, 367, 1992.
8. Zange, J. et al., Creatine supplementation results in elevated phosphocreatine/adenosine triphosphate (ATP) ratios in the calf muscle of athletes but not in patients with myopathies. *Ann Neurol*, 52(1), 126; discussion 126, 2002.
9. Yquel, R.J. et al., Effect of creatine supplementation on phosphocreatine resynthesis, inorganic phosphate accumulation and pH during intermittent maximal exercise. *J Sports Sci*, 20(5), 427, 2002.
10. Francaux, M. et al., Effect of exogenous creatine supplementation on muscle PCr metabolism. *Int J Sports Med*, 21(2), 139, 2000.

11. Zehnder, M. et al., Muscle phosphocreatine and glycogen concentration in humans after creatine and glucose polymer supplementation measured noninvasively by 31P and 13C-MRS. *Med Sci Sports Exer*, 30(5), S264, 1998.
12. Lowe, J.A., Murphy, M. and Nash, V., Changes in plasma and muscle creatine concentration after increases in supplementary dietary creatine in dogs. *J Nutr*, 128(12 Suppl), 2691S, 1998.
13. Hultman, E. et al., Muscle creatine loading in men. *J Appl Physiol*, 81(1), 232, 1996.
14. Green, A.L. et al., Carbohydrate ingestion augments skeletal muscle creatine accumulation during creatine supplementation in humans. *Am J Physiol*, 271(5 Pt 1), E821, 1996.
15. Greenhaff, P.L. et al., Effect of oral creatine supplementation on skeletal muscle phosphocreatine resynthesis. *Am J Physiol*, 266(5 Pt 1), E725, 1994.
16. Vandenberghe, K. et al., Long-term creatine intake is beneficial to muscle performance during resistance training. *J Appl Physiol*, 83(6), 2055, 1997.
17. Rico-Sanz, J., Creatine reduces human muscle PCr and pH decrements and P(i) accumulation during low-intensity exercise. *J Appl Physiol*, 88(4), 1181, 2000.
18. Wiedermann, D. et al., Creatine loading and resting skeletal muscle phosphocreatine flux: A saturation-transfer NMR study. *Magma*, 13(2), 118, 2001.
19. Robinson, T.M. et al., Role of submaximal exercise in promoting creatine and glycogen accumulation in human skeletal muscle. *J Appl Physiol*, 87(2), 598, 1999.
20. Carl, D.L. et al., Effect of oral creatine and caffeine on muscle phosphocreatine resynthesis in competitive swimmers. *Med Sci Sports Exer*, 31(5)1999.
21. Tarnopolsky, M.A., Potential benefits of creatine monohydrate supplementation in the elderly. *Curr Opin Clin Nutr Metab Care*, 3(6), 497, 2000.
22. Demant, T.W. and Rhodes, E.C., Effects of creatine supplementation on exercise performance. *Sports Med*, 28(1), 49, 1999.
23. Wilder, N. et al., The effects of a 10-week, periodized, off-season resistance-training program and creatine supplementation among collegiate football players. *J Strength Cond Res*, 16(3), 343, 2002.
24. Campbell, W.W. et al., Effects of resistance training and chromium picolinate on body composition and skeletal muscle in older men. *J Appl Physiol*, 86(1), 29, 1999.
25. McBride, T.A. and Gregory, M.A., Effect of creatine supplementation during high resistance training on mass, strength and fatigue resistance in rat skeletal muscle. *J Strength Cond Res*, 16(3), 335, 2002.
26. Bemben, M.G. et al., Creatine supplementation during resistance training in college football athletes. *Med Sci Sports Exer*, 33(10), 1667, 2001.
27. Stone, M.H. et al., Effects of in-season (5 weeks) creatine and pyruvate supplementation on anaerobic performance and body composition in American football players. *Int J Sport Nutr*, 9(2), 146, 1999.
28. Kreider, R.B., Effects of creatine supplementation on performance and training adaptations. *Mol Cell Biochem*, 244(1-2), 89, 2003.
29. Williams, M.H., Kreider, R. and Branch, J.D., *Creatine: The Power Supplement* Champaign, IL: Human Kinetics Publishers, 1999, 252.
30. Williams, M.H. and Branch, J.D., Creatine supplementation and exercise performance: An update. *J Am Coll Nutr*, 17(3), 216, 1998.
31. Greenhaff, P.L., Creatine supplementation: recent developments. *Br J Sports Med*, 30(4), 276, 1996.

32. Greenhaff, P.L. et al., Influence of oral creatine supplementation of muscle torque during repeated bouts of maximal voluntary exercise in man. *Clin Sci (Colch),* 84(5), 565, 1993.
33. Green, A.L. et al., Carbohydrate feeding augments skeletal muscle creatine accumulation during creatine supplementation in humans. *Am J Physiol,* 271, E821, 1996.
34. Green, A.L. et al., Carbohydrate ingestion augments creatine retention during creatine feeding in humans. *Acta Physiol Scand,* 158, 195, 1996.
35. Greenwood, M. et al., Differences in creatine retention among three nutritional formulations of oral creatine supplements. *J Exer Physiol Online,* 6(2), 37, 2003.
36. Steenge, G.R., Simpson, E.J. and Greenhaff, P.L., Protein- and carbohydrate-induced augmentation of whole body creatine retention in humans. *J Appl Physiol,* 89(3), 1165, 2000.
37. Greenwood, M. et al., D-Pinitol augments whole body creatine retention in man. *J Exerc Physiol Online,* 4(4), 41, 2001. Available: http://www.css.edu/users/tboone2/asep/Greenwo1.doc.
38. Nelson, A.G. et al., Muscle glycogen supercompensation is enhanced by prior creatine supplementation. *Med Sci Sports Exerc,* 33(7), 1096, 2001.
39. Op 't Eijnde, B. et al., Effect of oral creatine supplementation on human muscle GLUT4 protein content after immobilization. *Diabetes,* 50(1), 18, 2001.
40. Op 't Eijnde, B. et al., Effect of creatine supplementation on creatine and glycogen content in rat skeletal muscle. *Acta Physiol Scand,* 171(2), 169, 2001.
41. Greenhaff, P. et al., The influence of oral creatine supplementation on muscle phosphocreatine Resynthesis following intense contraction in man. *J Physiol,* 467, 75P., 1993.
42. Vandenberghe, K. et al., Phosphocreatine resynthesis is not affected by creatine loading. *Med Sci Sports Exerc,* 31(2), 236, 1999.
43. Smith, S.A. et al., Creatine supplementation influences muscle PCr resynthesis rate. *FASEB J,* 14(4)2000.
44. Smith, S.A. et al., Effects of creatine supplementation on the energy cost of muscle contraction: A 31P-MRS study. *J Appl Physiol,* 87(1), 116, 1999.
45. Kreider R.B., Greenwood , M., Parise, G, Payne, E., Tarnopolsky, M.A. Effects of serum creatine supplementation on muscle creatine content. *J Exerc Physiol Online,* 6(4), In press, 2003.
46. Schneider, D.A. et al., Creatine supplementation and the total work performed during 15-s and 1-min bouts of maximal cycling. *Aust J Sci Med Sport,* 29(3), 65, 1997.
47. Kurosawa, Y. et al., Creatine supplementation enhances anaerobic ATP synthesis during a single 10 sec maximal handgrip exercise. *Mol Cell Biochem,* 244(1-2), 105, 2003.
48. Kent-Braun, J.A. et al., Postexercise phosphocreatine resynthesis is slowed in multiple sclerosis. *Muscle Nerve,* 17, 835, 1994.
49. Wallimann, T. et al., Compartmentation, structure and function of creatine kinases: A rationale for creatine action. *Abstr 6th Int Con Guanidino Compounds in Biology and Medicine,* 2001.
50. Bessman, S. and Savabi, F., The role of phosphocreatine energy shuttle in exercise and muscle hypertrophy, in *Creatine and Creatine Phosphate: Scientific and Clinical Perspectives,* Conway, M.A. and Clark, J.F., Eds. 1988, Academic Press: San Diego, CA. p. 185.

51. Wallimann, T. et al., Some new aspects of creatine kinase (CK): Compartmentation, structure, function and regulation for cellular and mitochondrial bioenergetics and physiology. *Biofactors*, 8(3-4), 229, 1998.
52. Ma, T.M., Friedman, D.L. and Roberts, R., Creatine phosphate shuttle pathway in tissues with dynamic energy demand, in *Creatine and Creatine Phosphate: Scientific and Clinical Perspectives*, Conway, M.A. and Clark, J.F., Eds. 1996, Academic Press: San Diego, CA. p. 17.
53. Newsholme, E. and Beis, I., Old and new ideas on the roles of phosphagens and the kinases., in *Creatine and Creatine Phosphate: Scientific and Clinical Perspectives*, Conway, M.A. and Clark, J.F., Eds. 1996, Academic Press: San Diego, CA. p. 3.
54. Clark, J.F., Field, M.L. and Ventura-Clapier, R., An introduction to the cellular creatine kinase system in contractile tissue., in *Creatine and Creatine Phosphate: Scientific and Clinical Perspectives*, Conway, M.A. and Clark, J.F., Eds. 1996, Academic Press: San Diego, CA. p. 51.
55. Clark, J.F., Creatine and phosphocreatine: A review of their use in exercise and sport. *J Athl Training*, 32, 45, 1997.
56. van Deursen, J. et al., Skeletal muscles of mice deficient in muscle creatine kinase lack burst activity. *Cell*, 74, 621, 1993.
57. Askenasy, N. and Koretsky, A.P., Differential effects of creatine kinase isoenzymes and substrates on regeneration in livers of transgenic mice. *Am J Physiol*, 273(2 Pt 1), C741, 1997.
58. Willoughby, D.S. and Rosene, J.M., Effects of oral creatine and resistance training on myogenic regulatory factor expression. *Med Sci Sports Exerc*, 35(6), 923, 2003.
59. Wyss, M. and Schulze, A., Health implications of creatine: Can oral creatine supplementation protect against neurological and atherosclerotic disease? *Neuroscience*, 112(2), 243, 2002.
60. Tarnopolsky, M.A. et al., Creatine transporter and mitochondrial creatine kinase protein content in myopathies. *Muscle Nerve*, 24(5), 682, 2001.
61. Hespel, P. et al., Creatine supplementation: Exploring the role of the creatine kinase/phosphocreatine system in human muscle. *Can J Appl Physiol*, 26 (Suppl), S79, 2001.
62. Felber, S. et al., Oral creatine supplementation in Duchenne muscular dystrophy: A clinical and 31P magnetic resonance spectroscopy study. *Neurol Res*, 22(2), 145, 2000.
63. Matthews, R.T. et al., Neuroprotective effects of creatine and cyclocreatine in animal models of Huntington's disease. *J Neurosci*, 18(1), 156, 1998.
64. Passaquin, A.C. et al., Creatine supplementation reduces skeletal muscle degeneration and enhances mitochondrial function in mdx mice. *Neuromuscul Disord*, 12(2), 174, 2002.
65. Willoughby, D.S. and Rosene, J., Effects of oral creatine and resistance training on myosin heavy chain expression. *Med Sci Sports Exerc*, 33(10), 1674, 2001.
66. Volek, J.S. et al., Performance and muscle fiber adaptations to creatine supplementation and heavy resistance training. *Med Sci Sports Exerc*, 31(8), 1147, 1999.
67. Hespel, P. et al., Oral creatine supplementation facilitates the rehabilitation of disuse atrophy and alters the expression of muscle myogenic factors in humans. *J Physiol*, 536(Pt 2), 625, 2001.
68. Sipila, I. et al., Supplementary creatine as a treatment for gyrate atrophy of the choroid and retina. *New Engl J Med*, 304, 867, 1981.

69. Blei, M.L., Conley, K.E. and Kushmerick, M.J., Separate measures of ATP utilization and recovery in human skeletal muscle. *J Physiol*, 465, 203, 1993.
70. Radda, G.K., Control of energy metabolism during energy metabolism. *Diabetes*, 45, S88, 1996.
71. Kraemer, W.J. and Volek, J.S., Creatine supplementation. Its role in human performance. *Clin Sports Med*, 18(3), 651, 1999.
72. Kreider, R., Effects of creatine supplementation on performance and training adaptations. *Abstr 6th Int Conf Guanidino Compounds in Biology and Medicine*,2001.
73. Volek, J.S., Creatine supplementation and its possible role in improving physical performance. *ACSM Health Fitness J*, 1(4), 23, 1997.
74. Urbanski, R.L., Vincent, W.J. and Yaspelkis, B.B., 3rd, Creatine supplementation differentially affects maximal isometric strength and time to fatigue in large and small muscle groups. *Int J Sport Nutr*, 9(2), 136, 1999.
75. Tarnopolsky, M.A. and MacLennan, D.P., Creatine monohydrate supplementation enhances high-intensity exercise performance in males and females. *Int J Sport Nutr Exerc Metab*, 10(4), 452, 2000.
76. Wiroth, J.B. et al., Effects of oral creatine supplementation on maximal pedalling performance in older adults. *Eur J Appl Physiol*, 84(6), 533, 2001.
77. Ziegenfuss, T.N. et al., Effect of creatine loading on anaerobic performance and skeletal muscle volume in NCAA Division I athletes. *Nutrition*, 18(5), 397, 2002.
78. Jones, A.M. et al., Effect of creatine supplementation on oxygen uptake kinetics during submaximal cycle exercise. *J Appl Physiol*, 92(6), 2571, 2002.
79. Kilduff, L.P. et al., Effects of creatine on isometric bench-press performance in resistance- trained humans. *Med Sci Sports Exerc*, 34(7), 1176, 2002.
80. Gotshalk, L.A. et al., Creatine supplementation improves muscular performance in older men. *Med Sci Sports Exerc*, 34(3), 537, 2002.
81. Dawson, B., Vladich, T. and Blanksby, B.A., Effects of 4 weeks of creatine supplementation in junior swimmers on freestyle sprint and swim bench performance. *J Strength Cond Res*, 16(4), 485, 2002.
82. Juhn, M.S. and Tarnopolsky, M., Oral creatine supplementation and athletic performance: A critical review. *Clin J Sport Med*, 8(4), 286, 1998.
83. Graham, A.S. and Hatton, R.C., Creatine: A review of efficacy and safety. *J Am Phurm Assoc (Wash)*, 39(6), 803, 1999.
84. Skare, O.C., Skadberg and Wisnes, A.R., Creatine supplementation improves sprint performance in male sprinters. *Scand J Med Sci Sports*, 11(2), 96, 2001.
85. Mujika, I. et al., Creatine supplementation and sprint performance in soccer players. *Med Sci Sports Exerc*, 32(2), 518, 2000.
86. Theodorou, A.S. et al., The effect of longer-term creatine supplementation on elite swimming performance after an acute creatine loading. *J Sports Sci*, 17(11), 853, 1999.
87. Aaserud, R. et al., Creatine supplementation delays onset of fatigue during repeated bouts of sprint running. *Scand J Med Sci Sports*, 8(5 Pt 1), 247, 1998.
88. Cornish, S.M. et al., The effect of creatine monohydrate supplementation on sprint skating in hockey players. *Med Sci Sports Exerc*, 32(5), S135, 2000.
89. Cox, G. et al., Acute creatine supplementation and performance during a field test simulating match play in elite female soccer players. *Int J Sport Nutr Exerc Metab*, 12(1), 33, 2002.

90. Crowder, T. et al., Influence of creatine type and diet on strength and body composition of collegiate lightweight football players. *Med Science Sports Exerc*, 30(5), S264, 1998.
91. Jones, A.M., Atter, T. and Georg, K.P., Oral creatine supplementation improves multiple sprint performance in elite ice-hockey players. *J Sports Med Phys Fitness*, 39(3), 189, 1999.
92. Larson, D.E. et al., Creatine supplementation and performance during off-season training in female soccer players. *Med Sci Sports Exerc*, 30(5), S264, 1998.
93. Larson-Meyer, D.E. et al., The effect of creatine supplementation on muscle strength and body composition during off-season training in female soccer players. *J Strength Cond Res*, 14(4), 434, 2000.
94. Lefavi, R.G. et al., Effects of creatine monohydrate on performance of college baseball and basketball players. *J Strength Cond Res*, 12(4), 275, 1998.
95. Noonan, B., French, J. and Street, G., Creatine supplementation and multiple skating task performance in Division I hockey players. *Med Sci Sports Exerc*, 30, S310, 1998.
96. Ortega-Gallo, P.A. et al., Creatine supplementation in soccer players, effects in body composition and incidence of sport-related injuries. *Med Sci Sports Exerc*, 32(5), S134, 2000.
97. Romer, L.M., Barrington, J.P. and Jeukendrup, A.E., Effects of oral creatine supplementation on high intensity, intermittent exercise performance in competitive squash players. *Int J Sports Med*, 22(8), 546, 2001.
98. Sasaki, H., Hiruma, E. and Aoyama, R., Effects of creatine loading on muscular strength and endurance in female softball players. *Med Sci Sports Exerc*, 33(5), S207, 2001.
99. Stout, J. et al., The effects of a supplement designed to augment creatine uptake on exercise performance and fat free mass in football players. *Med Sci Sports Exerc*, 29, S251, 1997.
100. Selsby, J.T. et al., Swim performance following creatine supplementation in Division III athletes. *Med Sci Sports Exerc*, 33(5), S206, 2001.
101. Warber, J.P. et al., The effect of creatine monohydrate supplementation on obstacle course and multiple bench press performance. *J Strength Cond Res*, 16(4), 500, 2002.
102. Oopik, V. et al., Effects of creatine supplementation during recovery from rapid body mass reduction on metabolism and muscle performance capacity in well-trained wrestlers. *J Sports Med Phys Fitness*, 42(3), 330, 2002.
103. Izquierdo, M. et al., Effects of creatine supplementation on muscle power, endurance and sprint performance. *Med Sci Sports Exerc*, 34(2), 332, 2002.
104. Chwalbinska-Moneta, J., Effect of creatine supplementation on aerobic performance and anaerobic capacity in elite rowers in the course of endurance training. *Int J Sport Nutr Exerc Metab*, 13(2), 173, 2003.
105. Lehmkuhl, M. et al., The effects of 8 weeks of creatine monohydrate and glutamine supplementation on body composition and performance measures. *J Strength Cond Res*, 17(3), 425, 2003.
106. Earnest, C.P., Almada, A. and Mitchell, T.L., Effects of creatine monohydrate ingestion on intermediate duration anaerobic treadmill running to exhaustion. *J Strength Cond Res*, 11(4), 234, 1997.
107. Smith, J.C. et al., Effect of oral creatine ingestion on parameters of the work rate-time relationship and time to exhaustion in high-intensity cycling. *Eur J Appl Physiol Occup Physiol*, 77(4), 360, 1998.

108. Nelson, A.G. et al., Creatine supplementation alters the response to a graded cycle ergometer test. *Eur J Appl Physiol*, 83(1), 89, 2000.
109. Rico-Sanz, J. and Mendez Marco, M.T., Creatine enhances oxygen uptake and performance during alternating intensity exercise. *Med Sci Sports Exerc*, 32(2), 379, 2000.
110. Preen, D. et al., Effect of creatine loading on long-term sprint exercise performance and metabolism. *Med Sci Sports Exerc*, 33(5), 814, 2001.
111. Kambis, K.W. and Pizzedaz, S.K., Short-term creatine supplementation improves maximum quadriceps contraction in women. *Int J Sport Nutr Exerc Metab*, 13(1), 87, 2003.
112. McKenna, M.J. et al., Creatine supplementation increases muscle total creatine but not maximal intermittent exercise performance. *J Appl Physiol*, 87(6), 2244, 1999.
113. Gilliam, J.D. et al., Effect of oral creatine supplementation on isokinetic torque production. *Med Sci Sports Exerc*, 32(5), 993, 2000.
114. Deutekom, M. et al., No acute effects of short-term creatine supplementation on muscle properties and sprint performance. *Eur J Appl Physiol*, 82(3), 223, 2000.
115. Edwards, M.R. et al., The effect of creatine supplementation on anaerobic performance in moderately active men. *J Strength Cond Res*, 15(3), 357, 2000.
116. Op 't Eijnde, B., Vergauwen, L. and Hespel, P., Creatine loading does not impact on stroke performance in tennis. *Int J Sports Med*, 22(1), 76, 2001.
117. Finn, J.P. et al., Effect of creatine supplementation on metabolism and performance in humans during intermittent sprint cycling. *Eur J Appl Physiol*, 84(3), 238, 2001.
118. Van Schuylenbergh, R., Van Leemputte, M. and Hespel, P., Effects of oral creatine-pyruvate supplementation in cycling performance. *Int J Sports Med*, 24(2), 144, 2003.
119. Delecluse, C., Diels, R. and Goris, M., Effect of creatine supplementation on intermittent sprint running performance in highly trained athletes. *J Strength Cond Res*, 17(3), 446, 2003.
120. Biwer, C.J. et al., The effect of creatine on treadmill running with high-intensity intervals. *J Strength Cond Res*, 17(3), 439, 2003.
121. Lemon, P.W., Dietary creatine supplementation and exercise performance: Why inconsistent results? *Can J Appl Physiol*, 27(6), 663, 2002.
122. Kelly, V.G. and Jenkins, D.G., Effect of oral creatine supplementation on near-maximal strength and repeated sets of high-intensity bench press exercise. *J Strength Cond Res*, 12(2), 109, 1998.
123. Noonan, D. et al., Effects of varying dosages of oral creatine relative to fat free body mass on strength and body composition. *J Strength Cond Res*, 12(2), 104, 1998.
124. Kreider, R.B. et al., Effects of creatine supplementation on body composition, strength and sprint performance. *Med Sci Sports Exerc*, 30(1), 73, 1998.
125. Peeters, B.M., Lantz, C.D. and Mayhew, J.L., Effect of oral creatine monohydrate and creatine phosphate supplementation on maximal strength indices, body composition and blood pressure. *J Strength Cond Res*, 13(1), 3, 1999.
126. Kirksey, K.B. et al., The effects of 6 weeks of creatine monohydrate supplementation on performance measures and body composition in collegiate track and field athletes. *J Strength Cond Res*, 13(2), 148, 1999.

127. Pearson, D.R. et al., Long-term effects of creatine monohydrate on strength and power. *J Strength Cond Res*, 13(3), 187, 1999.
128. Becque, M.D., Lochmann, J.D. and Melrose, D.R., Effects of oral creatine supplementation on muscular strength and body composition. *Med Sci Sports Exerc*, 32(3), 654, 2000.
129. Burke, D.G. et al., The effect of continuous low dose creatine supplementation on force, power and total work. *Int J Sport Nutr Exerc Metab*, 10(3), 235, 2000.
130. Burke, D.G. et al., The effect of whey protein supplementation with and without creatine monohydrate combined with resistance training on lean tissue mass and muscle strength. *Int J Sport Nutr Exerc Metab*, 11(3), 349, 2001.
131. Brenner, M., Walberg Rankin, J. and Sebolt, D., The effect of creatine supplementation during resistance training in women. *J Strength Cond Res*, 14(2), 207, 2000.
132. Jowko, E. et al., Creatine and beta-hydroxy-beta-methylbutyrate (HMB) additively increase lean body mass and muscle strength during a weight-training program. *Nutrition*, 17(7-8), 558, 2001.
133. Brose, A., Parise, G. and Tarnopolsky, M.A., Creatine supplementation enhances isometric strength and body composition improvements following strength exercise training in older adults. *J Gerontol A Biol Sci Med Sci*, 58(1), 11, 2003.
134. Powers, M.E. et al., Creatine supplementation increases total body water without altering fluid distribution. *J Athl Train*, 38(1), 44, 2003.
135. Volek, J.S. et al., Physiological responses to short-term exercise in the heat after creatine loading. *Med Sci Sports Exerc*, 33(7), 1101, 2001.
136. Ziegenfuss, T.N., Lowery, L.M. and Lemon, P.W.R., Acute fluid volume changes in men during three days of creatine supplementation. *J Exercise Physiol Online*, 1(3), 1, 1998. Available:http://www.css.edu/users/tboone2/asep/jan3.htm.
137. Volek, J.S. et al., Performance and muscle fiber adaptations to 12 weeks of creatine supplementation and heavy resistance training. *Med Sci Sports Exer*, 31(5)1999.
138. Curtin, S.V. et al., Effect of acute oral creatine supplementation on muscle volume in young and older men. *Med Sci Sports Exer*, 32(5)2000.
139. Kamber, M. et al., Creatine supplementation — part I: Performance, clinical chemistry and muscle volume. *Med Sci Sports Exerc*, 31(12), 1763, 1999.
140. Hespel, P. et al., Oral creatine supplementation facilitates the rehabilitation of disuse atrophy and alters the expression of muscle myogenic factors in humans. *J Physiol*, 536(Pt 2), 625, 2001.
141. Ingwall, J.S. et al., Specificity of creatine in the control of muscle protein synthesis. *J Cell Biol*, 63, 145, 1974.
142. Ingwall, J.S., Creatine and the control of muscle-specific protein synthesis in cardiac and skeletal muscle. *Circ Res*, 38, I, 1976.
143. Bessman, S.P. and Mohan, C., Phosphocreatine, exercise, protein synthesis and insulin, in *Guanidino Compounds in Biology and Medicine*, DeDeyn, P.P. et al., Eds. 1992, John Libbey and Company: London. p. 181.
144. Louis, M. et al., Creatine supplementation does not further stimulate human myofibrillar or sarcoplasmic protein synthesis after resistance exercise. *Am J Physiol Endocrinol Metab*,2003.
145. Louis, M. et al., Creatine supplementation has no effect on human muscle protein turnover at rest in the post-absorptive or fed states. *Am J Physiol Endocrinol Metab*,2002.

146. Parise, G. et al., Effects of acute creatine monohydrate supplementation on leucine kinetics and mixed-muscle protein synthesis. *J Appl Physiol*, 91(3), 1041, 2001.
147. Terjung, R.L. et al., American College of Sports Medicine roundtable. The physiological and health effects of oral creatine supplementation. *Med Sci Sports Exerc*, 32(3), 706, 2000.
148. Earnest, C.P. et al., The effect of creatine monohyerate ingestion on anaerobic power indices, muscular strength and body composition. *Acta Physiol Scand*, 153, 207, 1995.
149. Schroeder, C. et al., The effects of creatine dietary supplementation on anterior compartment pressure in the lower leg during rest and following exercise. *Clin J Sport Med*, 11(2), 87, 2001.
150. Potteiger, J.A. et al., Changes in lower leg anterior compartment pressure before, during and after creatine supplementation. *J Athl Train*, 37(2), 157, 2002.
151. Tarnopolsky, M. et al., Acute and moderate-term creatine monohydrate supplementation does not affect creatine transporter mRNA or protein content in either young or elderly humans. *Mol Cell Biochem*, 244(1-2), 159, 2003.
152. Robinson, T.M. et al., Dietary creatine supplementation does not affect some haematological indices, or indices of muscle damage and hepatic and renal function. *Br J Sports Med*, 34(4), 284, 2000.
153. Poortmans, J.R. et al., Effect of short-term creatine supplementation on renal responses in men. *Eur J Appl Physiol Occup Physiol*, 76(6), 566, 1997.
154. Schilling, B.K. et al., Creatine supplementation and health variables: A retrospective study. *Med Sci Sports Exerc*, 33(2), 183, 2001.
155. Kreider, R.B. et al., Long-term creatine supplementation does not significantly affect clinical markers of health in athletes. *Mol Cell Biochem*, 244(1-2), 95, 2003.
156. Taes, Y.E. et al., Creatine supplementation does not affect kidney function in an animal model with pre-existing renal failure. *Nephrol Dial Transplant*, 18(2), 258, 2003.
157. Poortmans, J.R. and Francaux, M., Adverse effects of creatine supplementation: Fact or fiction? *Sports Med*, 30(3), 155, 2000.
158. Kuehl, K. et al., Effects of oral creatine monohydrate supplementation on renal function in adults. *Med Sci Sports Exerc*, 32(5), S168, 2000.
159. Poortmans, J.R. and Francaux, M., Long term oral creatine supplementation does not impair renal function in healthy athletes. *Med Sci Sports Exerc*, 31(8), 1108, 1999.
160. Poortmans, J.R. and Francaux, M., Renal dysfunction accompanying oral creatine supplements. *Lancet*, 352(9123), 234., 1998.
161. Earnest, C.P., Almada, A. and Mitchell, T.L., Influence of chronic creatine supplementation on hepatorenal function. *FASEB J*, 10, A790, 1996.
162. Kreider, R. et al., Long-term creatine supplementation does not significantly affect clinical markers of health in athletes. *Mol Cell Biochem*, 244, 95, 2003.
163. Greenwood, M. et al., Creatine supplementation patterns and perceived effects in select division I collegiate athletes. *Clin J Sport Med*, 10(3), 191, 2000.
164. Oopik, V. et al., Effect of creatine supplementation during rapid body mass reduction on metabolism and isokinetic muscle performance capacity. *Eur J Appl Physiol Occup Physiol*, 78(1), 83, 1998.
165. Volek, J.S. et al., Physiological response to exercise in the heat after loading with creatine monohydrate. *J Strength Cond Res*, 14(3)2000.

166. Mihic, S. et al., Acute creatine loading increases fat-free mass, but does not affect blood pressure, plasma creatinine, or CK activity in men and women. *Med Sci Sports Exerc*, 32(2), 291, 2000.
167. Stone, M.H. et al., A retrospective study of long-term creatine supplementation on blood markers of health. *J Strength Cond Res*, 13(4), 433, 1999.
168. Greenwood, M. et al., Creatine supplementation during college football training does not increase the incidence of cramping or injury. *Mol Cell Biochem*, 244(1-2), 83, 2003.
169. Greenwood, M. et al. Cramping and injury incidence are not increased by creatine supplementation in collegiate football players. *J Athletic Training*, in press, 2003.
170. Greenwood, L. et al. Effects of creatine supplementation on the incidence of cramping/injury during eighteen weeks of division I football training/competition. *Med Sci Sports Exerc*, 34(5), S146, 2002.
171. Hulver, M.W. et al. The effects of creatine supplementation on total body fluids, performance and muscle cramping during exercise. *Med Sci Sports Exerc*, 32(5), S133, 2000.
172. McArthur, P.D. et al. Creatine supplementation and acute dehydration. *Med Sci Sports Exerc*, 31(5), S263, 1999.
173. Webster, M.J. et al. Creatine supplementation: Effect on exercise performance at two levels of acute dehydration. *Med Sci Sports Exerc*, 31(5), S263, 1999.
174. LaBotz, M. and Smith, B.W., Creatine supplement use in an NCAA Division I athletic program. *Clin J Sport Med*, 9(3), 167, 1999.
175. Watsford, M.L. et al. Creatine supplementation and its effect on musculotendinous stiffness and performance. *J Strength Cond Res*, 17(1), 26, 2003.
176. Greenwood, M. et al. Effects of creatine supplementation on the incidence of cramping/injury during eighteen weeks of collegiate baseball training/competition. *Med Sci Sport Exerc*, 34(S146)2002.
177. Greenwood, M. et al. Perceived health status and side effects associated with creatine supplementation of division I-A football players during 3-a-day training. *Res Q Exerc Sport*, 72(1), A, 2001.
178. Greenwood, M. et al. Creatine supplementation during college football training does not increase the incidence of cramping or injury. in *6th Int Conf Guanidino Compounds in Biology and Medicine* Cincinnati, OH, 2001,
179. Greenwood M, E.A., Creatine supplementation does not increase the incidence of injury or cramping in college baseball players. *J Exerc Physiol Online*, in press 2003.
180. Witte, K.K., Clark, A.L. and Cleland, J.G., Chronic heart failure and micronutrients. *J Am Coll Cardiol*, 37(7), 1765, 2001.
181. Andrews, R. et al. The effect of dietary creatine supplementation on skeletal muscle metabolism in congestive heart failure. *Eur Heart J*, 19(4), 617, 1998.
182. Field, M.L., Creatine supplementation in congestive heart failure. *Cardiovasc Res*, 31(1), 174, 1996.
183. Gordon, A. et al. Creatine supplementation in chronic heart failure increases skeletal muscle creatine phosphate and muscle performance. *Cardiovasc Res*, 30(3), 413, 1995.
184. Ferraro, S. et al. Hemodynamic effects of creatine phosphate in patients with congestive heart failure: A double-blind comparison trial versus placebo. *Clin Cardiol*, 19, 699, 1996.

185. Ganesan, V. et al. Guanidinoacetate methyltransferase deficiency: new clinical features. *Pediatr Neurol*, 17(2), 155, 1997.
186. Stockler, S. and Hanefeld, F., Guanidinoacetate methyltransferase deficiency: A newly recognized inborn error of creatine biosynthesis. *Wiener Klinische Wochenschrift*, 109(3), 86, 1997.
187. Stockler, S. et al. Creatine deficiency in the brain: A new, treatable inborn error of metabolism. *Pediatric Research*, 36, 409, 1994.
188. Stockler, S. et al. Guanidino compounds in guanidinoacetate methyltransferase deficiency, a new inborn error of creatine synthesis. *Metabolism*, 46, 1189, 1997.
189. Stockler, S., Hanefeld, F. and Frahm, J., Creatine replacement therapy in guanidinoacetate methyltransferase deficiency, a novel inborn error of metabolism. *Lancet*, 21(348), 789, 1996.
190. Stockler, S. et al. Guanidinoacetate methyltransferase deficiency: The first inborn error of creatine metabolism in man. *Am J Human Genet*, 58, 914, 1996.
191. Schulze, A. et al. Improving treatment of guanidinoacetate methyltransferase deficiency: reduction of guanidinoacetic acid in body fluids by arginine restriction and ornithine supplementation. *Mol Genet Metab*, 74(4), 413, 2001.
192. Battini, R. et al. Creatine depletion in a new case with AGAT deficiency: Clinical and genetic study in a large pedigree. *Mol Genet Metab*, 77(4), 326, 2002.
193. Leuzzi, V., Inborn errors of creatine metabolism and epilepsy: Clinical features, diagnosis and treatment. *J Child Neurol*, 17 Suppl 3, 3S89, 2002.
194. Tarnopolsky, M.A., Potential use of creatine monohydrate in muscular dystrophy and nuerometabolic disorders. *Abstr 6th Int Con Guanidino Compounds in Biology and Medicine*, 2001.
195. Tarnopolsky, M.A. and Parise, G., Direct measurement of high-energy phosphate compounds in patients with neuromuscular disease. *Muscle Nerve*, 22(9), 1228, 1999.
196. Walter, M.C. et al. Creatine monohydrate in myotonic dystrophy: A double-blind, placebo-controlled clinical study. *J Neurol*, 249(12), 1717, 2002.
197. Andreassen, O.A. et al. Increases in cortical glutamate concentrations in transgenic amyotrophic lateral sclerosis mice are attenuated by creatine supplementation. *J Neurochem*, 77(2), 383, 2001.
198. Drory, V.E. and Gross, D., No effect of creatine on respiratory distress in amyotrophic lateral sclerosis. *Amyotroph Lateral Scler Other Motor Neuron Disord*, 3(1), 43, 2002.
199. Vielhaber, S. et al. Effect of creatine supplementation on metabolite levels in ALS motor cortices. *Exp Neurol*, 172(2), 377, 2001.
200. Mazzini, L. et al. Effects of creatine supplementation on exercise performance and muscular strength in amyotrophic lateral sclerosis: preliminary results. *J Neurol Sci*, 191(1-2), 139, 2001.
201. Derave, W. et al. Skeletal muscle properties in a transgenic mouse model for amyotrophic lateral sclerosis: Effects of creatine treatment. *Neurobiol Dis*, 13(3), 264, 2003.
202. Drory, V.E. and Gross, D., No effect of creatine on respiratory distress in amyotrophic lateral sclerosis. *Amyotroph Lateral Scler Other Motor Neuron Disord*, 3(1), 43, 2002.
203. Andreassen, O.A. et al. Creatine increase survival and delays motor symptoms in a transgenic animal model of Huntington's disease. *Neurobiol Dis*, 8(3), 479, 2001.

204. Persky, A.M. and Brazeau, G.A., Clinical pharmacology of the dietary supplement creatine monohydrate. *Pharmacol Rev,* 53(2), 161, 2001.
205. Kaemmerer, W.F. et al. Creatine-supplemented diet extends Purkinje cell survival in spinocerebellar ataxia type 1 transgenic mice but does not prevent the ataxic phenotype. *Neuroscience,* 103(3), 713, 2001.
206. Ferrante, R.J. et al. Neuroprotective effects of creatine in a transgenic mouse model of Huntington's disease. *J Neurosci,* 20(12), 4389, 2000.
207. Matthews, R.T. et al. Neuroprotective effect of creatine and cyclocreatine in animal models of Huntington's disease. *J Neuroscience,* 18, 156, 1998.
208. Dedeoglu, A. et al. Creatine therapy provides neuroprotection after onset of clinical symptoms in Huntington's disease transgenic mice. *J Neurochem,* 85(6), 1359, 2003.
209. Adcock, K.H. et al. Neuroprotection of creatine supplementation in neonatal rats with transient cerebral hypoxia-ischemia. *Dev Neurosci,* 24(5), 382, 2002.
210. Sullivan, P.G. et al. Dietary supplement creatine protects against traumatic brain injury. *Ann Neurol,* 48(5), 723, 2000.
211. Matthews, R.T. et al. Creatine and cyclocreatine attenuate MPTP neurotoxicity. *Exp Neurol,* 157(1), 142, 1999.
212. Hausmann, O.N. et al. Protective effects of oral creatine supplementation on spinal cord injury in rats. *Spinal Cord,* 40(9), 449, 2002.
213. Willer, B. et al. Effects of creatine supplementation on muscle weakness in patients with rheumatoid arthritis. *Rheumatology (Oxford),* 39(3), 293, 2000.
214. Volek, J.S. et al. No effect of heavy resistance training and creatine supplementation on blood lipids. *Int J Sport Nutr Exerc Metab,* 10(2), 144, 2000.
215. Earnest, C.P., Almada, A. and Mitchell, T.L., High-performance capillary electrophoresis-pure creatine monohydrate reduced blood lipids in men and women. *Clin Sci,* 91, 113, 1996.
216. Jacobs, R.L. et al. Regulation of homocysteine metabolism–effects of insulin, glucagon and creatine. *FASEB J,* 15(4)2001.
217. Steenge, G.R., Verhoef, P. and Greenhaff, P.L., The effect of creatine and resistance training on plasma homocysteine concentration in healthy volunteers. *Arch Intern Med,* 161(11), 1455, 2001.
218. Stead, L.M. et al. Methylation demand and homocysteine metabolism: effects of dietary provision of creatine and guanidinoacetate. *Am J Physiol Endocrinol Metab,* 281(5), E1095, 2001.
219. McCarty, M.F., Supplemental creatine may decrease serum homocysteine and abolish the homocysteine "gender gap" by suppressing endogenous creatine synthesis. *Med Hypotheses,* 56(1), 5, 2001.

7
Gelatin

Douglas S. Kalman

CONTENTS
- I. Introduction ... 105
- II. Gelatin Production ... 106
- III. History of Human Consumption ... 107
- IV. Financials of Worldwide Use ... 108
- V. Uses in the Food and Pharmaceutical Industry ... 108
- VI. Clinical Trials in Bone and Joint Disease ... 109
- VII. Potential Anaphylactic Reaction to Gelatin ... 111
- VIII. Potential Uses for the Athlete ... 111
- IX. Delivery Systems for Gelatin ... 112
- X. Summary ... 112
- References ... 112

I. Introduction

Gelatin is a unique foodstuff that has many applications in the food industry, within nutritional supplements and in the pharmaceutical industry. Gelatin is manufactured from skin and bone (pigskin and the hide split) from slaughtered animals that have been approved for human consumption. The collagen contained in the raw materials is the actual starting material used in the manufacturing of gelatin.

Today, we see gelatin in many foods, especially in sport and snack bars. This is the market where athletes are most exposed to gelatin or hydrolyzed collagen as a protein source. The form of gelatin used in these bars is collagen or hydrolyzed collagen.

Gelatin is derived from collagen, an insoluble fibrous protein that is the principal constituent of connective tissue and bone. Collagen is distinctive in that it contains an unusually high level of the cyclic amino acids proline and hydroxyproline. Collagen consists of three helical polypeptide chains wound around each other and connected by intermolecular cross-links.

Gelatin is primarily composed of the amino acid glycine (about 33%), proline/hydroxyproline (22%) with the remaining 45% made up by 17 amino acids (18 in total). Of these 17 remaining acids, alanine is the most abundant in gelatin, followed by glutamic acid, arginine and aspartic acid. It should be noted that gelatin has no cysteine or cystine, but does have trace amounts of methionine (<1% by weight). Thus, gelatin does not contain all of the essential amino acids, as it lacks tryptophan. Gelatin is the soluble form (via hydrolysis) of collagen. Commercially edible gelatins are 84–90% protein, 8–12% water and 2–4% minerals salts. Gelatin in and of itself contains no fat or carbohydrates.[1] The caloric value of gelatin is 3.6 calories/gram.

II. Gelatin Production

The majority of gelatin is made from pigskin, although it can be produced from cattle skin and bones. All of the raw materials used in the production of gelatin come from registered slaughterhouses, which test all animals aged above 30 months for the presence of bovine spongiform encephalopathy (BSE). Under the auspices of the European Commission, multisite studies were carried out in Scotland (Institute for Animal Health in Edinburgh), the Netherlands (ID-Lelystad) and in the United States (Baltimore Research and Education Foundation) in order to determine whether BSE or related pathogens could survive. The studies found that no harmful organisms survived the manufacturing process for gelatin, thus safety should not be a concern.[2]

Gelatin is recovered from collagen by hydrolysis. There are several varieties of gelatin, the composition of which depends on the source of collagen and the hydrolytic treatment used. Gelatin as it is typically made is not kosher; however, if made from cattle slaughtered under Kashruth religious law and supervision, the gelatin will be certified as kosher.

Cattle hides are the least-used gelatin raw material in North America today.[2] Gelatin recovered from bone is used primarily in photographic applications: some is used for pharmaceutical purposes. The so-called green bone from the slaughter of cattle is cleaned, degreased, dried, sorted and crushed to a particle size of about 1–2 cm. The pieces of bone are then treated with dilute hydrochloric acid to remove mineral salts. The resulting sponge-like material is called ossein. From this point on in the manufacture of type B gelatin, both cattle hides and ossein receive similar treatment.

Cattle hides are available from trimming operations in leather production. They are usually dehaired chemically with a lime-sulfide solution followed by mechanical loosening.

For the production of type B gelatin, both ossein and cattle hide pieces are subjected to lengthy treatment with an alkali (usually lime) and water at ambient temperature. Depending on previous treatment, the nature of the material, the size of the pieces and the exact temperature, liming takes 5–20 weeks, usually 8–12. The process is controlled by the degree of alkalinity of the lime liquor as determined by titration with acid, or by making test extractions.

Pork skin is currently the most abundant raw-material source for production of edible gelatin in North America. Supplied as either fresh or frozen, pork skins come from slaughter houses and meat processing plants already trimmed of fat, flesh and hair. They are usually dehaired by scalding with a hot dilute caustic soda solution.

When pork skins are utilized for production of type A gelatin, they are washed with cold water and then soaked in cold dilute mineral acid for several hours until maximum swelling has occurred. Hydrochloric and sulfuric acids are most commonly employed. The remaining acid is then drained off and the material is again washed several times with cold water. The pork skins are then ready for extraction with hot water.

The pH, time, temperature and number of extractions varies from processor to processor depending on product needs, type of equipment, time of operations and economics. Extraction procedures are closely controlled in the manufacture of both type A and type B gelatin, since they influence both quality and quantity. Although continuous extraction is used by some processors, most methods still employ discrete batch fractions. Extraction is normally carried out in stainless steel vessels equipped with provisions for heating and temperature control. The number of extractions varies, but 3–6 is typical. The type of gelatin produced is dependent on the needs of the customer.

It is imperative to note that gelatin is used in confectionary products (jujubes, gums, marshmallows, toffies, lozenges, licorice etc.), dairy products (for its fat-like mouthfeel), desserts (i.e., jello), meat products (as a coating, extender, glazing or emulsifier), bakery products (i.e., icing) and in the pharmaceutical industry (i.e., gel-caps and as a binding agent or emulsifier). Gelatin is also used to fortify reduced-calorie foods. Soups, shakes and fruit drinks are the products with the most uses for gelatin hydrolysates. Thus, it is obvious, that for the past 400 or more years, gelatin and collagen have been a part of the human diet. In each industry that gelatin is used in, there are quality controls for handling and processing, thus providing one or multiple steps of protection from contamination.

III. History of Human Consumption

In 1682, Papin started the first process for making gelatin (slow cooking of a bone mixture). By 1700, the word gelatin (Latin: gelatinus, meaning stiff or frozen) was coined. In 1754, the first patent was granted in England for the use of gelatine as an adhesive (joiner's glue). Dr. Richard Leach Maddox in 1871 utilized gelatin to improve and further the processing of pictures (photography). With the industrial revolution, factory processing of gelatin was enabled. From 1875 to 1950, gelatin processing evolved slowly; however, in 1974, the first professional association dedicated to the art and science of gelatin was formed (Gelatin Manufacturers of Europe; GME).[3]

IV. Financials of Worldwide Use

Gelatin is a commodity. Worldwide in 2001, 110,400 metric tons of pig skin were made into gelatin. There were also 77,200 metric tons of bovine hides used in gelatin production, 80,800 metric tons of bones and 1,000 metric tons of other animal parts utilized in the production of edible and pharmaceutical gelatin. The gelatin business generated $1.2 billion in worldwide sales.

Western Europe consumes 117,000 metric tons, Eastern Europe 5,000 metric tons, North America 60,500 and South America 39,500 metric tons of gelatin per year. In Europe, Germany, France and Belgium are the largest gelatin consumers, with Switzerland utilizing the least (300 metric tons per year). The sales of gelatin worldwide are estimated to grow 2–3% per annum.[2,3]

V. Uses in the Food and Pharmaceutical Industry

The functional properties of gelatin include its ability to act as a gel (desserts, lunch meats, pate, consommé), whipping agent (nougats, mousses, soufflés, marshmallows, chiffons and whipped cream), a protective colloid (meat rolls, canned meat, confectionary, cheeses, dairy), clarifying agent (wine, beer, fruit juices, vinegar), film coater (coating for meats, fruits, deli items), thickener (powdered drinks, bouillon), gravies, sauces, puddings etc.), process aid (microencapsulation of colors, flavors, oils and vitamins), emulsifier (cream soups, fat-replacer, flavorings etc.), stabilizer (cream cheese, chocolate milk, yogurt, icings) and acts as an adhesive (to affix nonpareils, coconut layered confections, frosting etc.).

Gelatin is used by the pharmaceutical industry for two-piece hard capsules, soft elastic gelatin capsules, tablets, tablet coating and as a gelatin emulsifier/filler. Gelatin is also used for microencapsulation (of oil or hydrophobic compounds), as a water-insoluble sponge during surgical procedures (Gelfoam™), as a film (GelFilm™ for laboratory printing of specific types of analysis (i.e., DNA), it is also used in medical troches and pastilles (lozenge like products). In laboratories, gelatin is also used in bacterial culture media.

It is interesting to note that, within the many uses for gelatin, it is also used in the fabrics industry (to coat yarn), in the manufacture of paper (coating and surface sizing), printing (lithography) and as a cleansing agent.

Other uses for gelatin include as a binder for the material in matches, for coating adhesives, as a filter, in cosmetics (creams and lotions) and in the chemicals industry.[2]

VI. Clinical Trials in Bone and Joint Disease

A 14-week randomized, double-blind, placebo-controlled clinical trial was conducted to test the safety and efficacy of a gelatin-based dietary supplement for symptomatic relief of osteoarthritis, while also noting strength changes. The study entered 175 adults with mild symptoms of knee-based osteoarthritis. Eighty subjects received the product (Knox NutraJoint™ made by Nabisco) while 95 were on placebo. To assess joint strength and work performance, a series of isokinetic and isometric leg strength tests were performed using the Biodex Multi-Joint System B2000. Subjects also filled out questionnaires regarding perceived effects on ability to carry out daily living events (pain, stiffness and mobility). Knox NutraJoint contains 10 grams of hydrolyzed gelatin (collagen source) plus amino acids, as well as vitamins C, D and K, plus calcium, copper, manganese and zinc.[4,5]

After 14 weeks, it was determined that the group receiving the gelatin-based product did show an improvement in knee function and strength compared with those in the placebo group. The principal investigator stated in a press release: "These results suggest that gelatine supplementation has the potential to improve knee function during activities that place a high amount of stress on the joint, like walking or jogging. They also encourage further research to evaluate the long-term benefits of this type of treatment." It should be noted that this study was neither published as a full manuscript in a peer-reviewed journal nor replicated; the findings should thus be taken with caution.

A meta-analysis of studies utilizing collagen hydrolysate in the treatment of osteoarthritis (OA) and osteoporosis was recently published.[6] Hydrolyzed gelatin products are generally regarded as safe and have a long history of use in the foods and pharmaceutical industry. Pharmaceutical-grade collagen hydrolystate use is associated with reduced pain in those with OA of the knee or hip. Pharmacokinetic data indicate an associated increase in serum concentrations of hydroxyproline along with preferential uptake by cartilage suggesting a beneficial effect of the collagen hydrolysate. In ranking pain severity relief of those who achieved perceived benefit when using the collagen hydrolysate, those patients with the most severe symptomology appear to obtain the greatest relief. The dose most common for clinical benefit in OA is 10 grams of collagen hydrolysate.

The potential benefit of collagen hydrolystate supplementation on bone metabolism has been investigated. The reason for research regarding potential uses for collagen hydrolysate in osteoporosis is straightforward; collagen matrix is important for bone integrity. Further, since collagen hydrolysate has been shown to exert a positive impact on cartilage uptake and synthesis, it may also have an effect on the osteoclast and osteoblast activity. Providing calcitonin with the collagen hydrolysate appears to slow the destruction of collagen, as evidenced by reduced urinary pyridinoline cross-links (a marker of bone turnover). Thus, there may be a potential use for collagen hydrolysate in osteoporosis prevention. The collagen hydrolysate dose found beneficial for osteoporosis is 10 grams daily.[6]

Osteoarthritis typically presents as joint pain and joint stiffness, impairs activities of daily living and may reduce quality of life. Treatment of OA is multifaceted and can include the adjunctive use of dietary supplements as well as nonpharmacologic or pharmacologic therapy. Previous clinical trials with soluble undenatured type II collagen supplements from bovine and avian sources have yielded equivocal results in rheumatoid arthritis. A recent study examined whether a hydrolyzed type II collagen product (avian source: Biocell™ Collagen II) would have any positive effects in adults with OA when compared with placebo.[7]

In a randomized double blind placebo-controlled manner, 16 subjects who had definitive OA of the knee or hand were enrolled for a 2-month intervention study. Subjects were balanced at baseline for age, race, gender and location of OA. Subjects who were on a cyclo-oxygenase-2 inhibitor (COX-2 inhibitor) or nonsteroidal anti-inflammatory drugs (NSAID) therapy were allowed to continue this therapy during the study. Subjects who were taking or recently took dietary supplements for joint health were excluded, as were subjects who had clinically significant medical conditions. Those with known allergies to dietary collagen sources or who were pregnant, lactating or who had recently participated in a clinical trial were also excluded. Subjects were randomized to active (Biocell™; 1000 mg BID; n = 8) or placebo (PLA, cellulose; n = 8) treatment. Outcome measurements included changes in the Western Ontario and McMaster Universities Osteoarthritis Index (WOMAC), Quality of Life-short form (QoL), sleep quality ratings (via visual analog scale) as well as clinical markers of safety (i.e., liver function tests, renal function, white blood counts, blood pressure, heart rate etc.).

After appropriate statistical analysis, the data revealed that the group receiving the Hydrolyzed Type II Collagen experienced a significant improvement in their WOMAC pain, stiffness and activities of daily-living scores when compared with placebo ($p < 0.05$ for all subsets and for total WOMAC scores). Additionally, the Biocell™ group achieved a significant improvement in QoL scores ($p < 0.05$) compared with baseline values; however, this change was not significant when compared with placebo. Adverse events were not different between the groups and were neither significant nor considered related to the test product or placebo. Compliance was 93–94% as determined by returned capsule counts. Thus, researchers concluded that, in this limited sample size, hydrolyzed type II collagen appeared to be a safe and effective dietary supplement for the adjunctive treatment of OA.[7]

VII. Potential Anaphylactic Reaction to Gelatin

Infants are given a series of immunizations shots throughout the first few years of life. Some develop anaphylaxis after immunization, though it is rare. Anaphylaxis is serious and potentially life threatening. Gelatin is used in the production and delivery of vaccines for measles, mumps, rubella (MMR), varicella, diphtheria-tetanus-acellular pertussis (DTP) and Japanese encephalitis. Some people who receive any of these vaccines develop a sensitization to the gelatin portion, as evidenced by elevated immunoglobin E (IgE) antibodies (antigelatin antibodies). It is potentially possible that someone who is given a DTP vaccine and develops a sensitization to the gelatin portion could have an anaphylactic reaction when given a subsequent vaccine (i.e., MMR).

A retrospective case control study performed at the Mayo Clinic (Minneapolis, Minnesota) found that, in the United States, 27% of the sample tested had antigelatin antibodies (antigelatin IgE), whereas none of the control subjects had.[8] The Centers for Disease Control and Prevention (CDC) also concluded that, since the introduction of DTP, there has not been a substantial increase in the number of allergic responses reported. The CDC further states that anaphylactic reactions to MMR are rare, and the incidence is unchanged. It is estimated that 25% of all people who have a reaction to MMR are hypersensitive to the gelatin in the vaccine. The new recommendation from this study by the CDC is that allergy testing or evaluation should be conducted on infants who exhibit symptoms of a hypersensitivity to gelatin. Thus, gelatin, while appearing to be healthy as a food and generally regarded as safe, when used medicinally may have an adverse effect in an infant or toddler.[8]

VIII. Potential Uses for the Athlete

At this time there are no studies demonstrating any athletic performance enhancement from the use of gelatin or hydrolyzed collagen. However, since preliminary evidence exists that certain types of gelatin exert benefit for the relief of OA, those who have osteoarthritis and engage in athletic endeavors may want to consider the adjunctive use of the protein.

Athletes also tend to eat a diet that is higher in protein when compared to their sedentary counterparts. Gelatin is derived from collagen. Collagen is a protein source in many foods. However, gelatin is an incomplete protein of low biological value when compared to egg, milk, whey, etc. and thus it should not be the mainstay of anyone's dietary protein intake.

IX. Delivery Systems for Gelatin

Gelatin is used in many foodstuffs, the paper industry, photographic processing and in the delivery of pharmaceutical agents. In the nutrition industry, gelatin is used in tablets and capsules (as a binder or the capsules itself). The more exciting area of growth for the gelatin industry is as a pharmaceutical delivery molecule. The greatest area of interest is twofold, oral-muco delivery (gelfilms that melt on the tongue or in the mouth that deliver a nutritional or pharmaceutical agent) and as part of a microsphere used in the parenteral treatment of specific disease states. Microspheres allow for the delivery of novel bioactive proteins to target tissue for the controlled release of a medication. These compounds are particularly suitable for OA, which affects one joint (monarthritis).[9]

X. Summary

Gelatin, a protein derived from collagen, has many uses within the nutrition, chemical and pharmaceutical industries. Athletes who eat protein bars or consume gelcaps are exposed to gelatin. Hydrolyzed gelatin and hydrolyzed collagen have been found to be beneficial for the symptom relief of OA. In addition, some data provide evidence that hydrolyzed collagen can stimulate change and enhancement of cartilage tissue. Those involved with sports who exhibit signs of osteoarthritis might benefit from the prophalactic use of hydrolyzed gelatin. More research is needed to determine whether gelatin has protective or recovery properties for other collagen or cartilage-based disorders. Gelatin or collagen supplementation as a protein source is not recommended.

References

1. www.gelita.com/dgf-english/gelatine/gelatine_was.html, accessed February 20, 2004.
2. www.gelatin-gmia.com/html/rawmaterials_app.html, accessed February 20, 2004.
3. www.gelatine.org, accessed February 20, 2004.
4. http://www.nutrajoint.com/aboutus/article11-22.shtm, accessed November 11, 2002.
5. McCarthy SM, Carpenter MR, Barrell M, Morrissey D, Jacobson E, Kline G, Rowinski M, Freedson P, Gootman JP, O'Brien D, Knipe SJ, Rippe JM. The effectiveness of gelatine supplementation treatment in individuals with symptoms of mild osteoarthritis. Presented at the 2000 Annual Meeting of the American Academy of Family Physicians. Dallas Texas, September 22, 2000.

6. Moskowitz RW. Role of collagen hydrolysate in bone and joint disease. *Semin Arthritis Rheum* 2000;30(2):87–99.
7. Kalman D, Almada AL, Schwartz, Pachon J, Sheldon E. A randomized double blind clinical pilot trial evaluating the safety and efficacy of hydrolyzed collagen type II in adults with osteoarthritis. *FASEB J*. Annual Experimental Biology Conference. Abstract LB35, 2004.
8. Pool V, Braun MM, Kelso JM, Mootrey G, Chen RT, Yunginger JW, Jacobson RM, Gargiullo PM. Prevalence of anti-gelatin IgE antibodies in people with anaphylaxis after measles-mumps rubella vaccine in the United States. *Pediatrics* 2002;110(6):e71.
9. Brown KE, Leong K, Huang CH, Dalal R, Green GD, Haimes HB, Jiminez PA, Bathon J. Gelatin/chondroitin 6-sulfate microspheres for the delivery of therapeutic proteins to the joint. *Arthritis and Rheumatism* 1998;41(12):2185–2194.

8

Glucosamine and Chondroitin Sulfate

Catherine G.R. Jackson

CONTENTS
I. Introduction 115
II. Description of Products 117
   A. Glucosamine 117
   B. Chondroitin Sulfate 117
III. Mechanisms 118
   A. Glucosamine 118
   B. Chondroitin Sulfate 119
IV. Review of Research Studies and Clinical Trials 119
   A. Glucosamine 119
   B. Chondroitin Sulfate 120
   C. Combined Glucosamine and Chondroitin Sulfate 121
V. Side Effects 122
   A. Glucosamine 122
   B. Chondroitin Sulfate 123
VI. Use in Sport and Exercise 123
VII. Summary and Recommendations 124
References 125

I. Introduction

Americans currently spend more money on natural remedies for osteoarthritis than for any other medical condition,[1] thus producing a significant market for the multibillion dollar supplement industry. Osteoarthritis is destined to become one of the most prevalent and costly diseases in our society. It is estimated that over 21 million adults in the U.S. suffer from osteoarthritis, with age, female gender and obesity as risk factors. Athletes of all types frequently live with chronic joint pain often associated with overuse injuries, and present an additional market for the supplements. Glucosamine and chondroitin sulfate have been widely publicized in the popular media as being capable of decelerating the degenerative processes, decreasing pain and maintaining and improving joint function in osteoarthritis and other conditions where joint pain is the result. However, studies have not been able to confirm these statements. There are numerous anecdotal reports to which supplement manufacturers refer. However, the vast majority of clinical trials have small sample sizes, little or no follow-up and are sponsored by the supplement manufacturers.[2] Problems in evaluation begin with even classifying the agent, as it has been called a drug, a nutriceutical, a food supplement, an alternative therapy, a homeopathic therapy and a complementary therapy. Individuals with joint pain now consume very large quantities of glucosamine, primarily based on a great volume of media coverage as to its value. There is currently much controversy and confusion concerning the topic.

In osteoarthritis, the chondrocytes and aqueous matrix decrease with age, which results in poor-quality cartilage. Bones may become exposed and rub together, creating damage and pain. With time, bones chip and fracture, which can lead to bone growth that produces increased pain and lack of mobility. The individual finds that this disrupts daily life and activity makes symptoms worse. Patients feel unwell and depressed. Active individuals may terminate exercise completely, which increases the risk of inactivity-related chronic diseases.

Osteoarthritis affects approximately 12% of the U.S. population and is a common cause of age-related pain and physical disability. The condition itself, however, is poorly understood. The degenerative process is not slowed or reversed with current treatments, which include aspirin, acetaminophen and nonsteroidal anti-inflammatory drugs (NSAIDS). Interestingly, the origin of pain caused by the condition is unclear and, upon investigation, is more often attributed to lesions or referred pain rather than articular problems, as there are no nerves in articular cartilage. The biochemistry of glucosamine has led to the suggestion that its use might stop and possibly reverse the degenerative process. However, evidence is questionable. Chard and Dieppe[3] showed great insight into the problem by commenting that glucosamine may become the first agent about which we have more published systematic reviews, editorials, meta-analyses and comments than primary research papers. They identified only 24 primary research studies but also found nine reviews and numerous comments and editorials. Most primary research studies are poor and positive results are invariably found in supplement-manufacturer-sponsored research. Chard and Dieppe[3] also concluded that there is more hype than magic, rationales for use are unclear, best dose and route of administration are unknown and published work does not allow conclusions about efficacy or effectiveness. However, since it is safe, toxicity concerns cannot be raised. There is a need for regulation, as there could be long-term side effects while the length of treatment is not known. Other uses of the drugs are to treat migraines,[4] gastrointestinal disorders such as Crohn's disease, ulcerative colitis, atherosclerosis and capsular contracture in breast implants.[5]

II. Description of Products

A. Glucosamine

Glucosamine is an amino monosaccharide (amine-sugar) that can be found in chitin, glycoproteins and the glycosaminoglycans (mucopolysaccharides) such as heparan sulfate and hyaluronic acid. Other chemical designations are 2-amino-2-deoxy-beta-D-glucopyranose, 2-amino-2-deoxyglucose and chitosamine (Hendler).[6] It is available over the counter as a nutritional supplement as either glucosamine hydrochloride (glucosamine HCl), glucosamine sulfate or N-acetyl-glucosamine. Research has used primarily the chloride and sulfate salts, which are those most commonly purchased.

The chemical structure of glucosamine is such that at physiologic and neutral pH the molecule has a positive charge. Negative anions are found in the salt forms that neutralize the charge. In glucosamine sulfate, the anion is sulfate, in glucosamine HCl, the anion is chloride and in N-acetylglucosamine the amino group is acetylated, which results in a neutral charge. All forms are water soluble. Nutritional supplements are usually derived from marine exoskeletons with the chitin extracted from seashells. There are also synthetic forms. Since it falls under the 1994 Dietary Supplement Health and Education Act (DSHEA) and is classified as a medicinal product, its manufacture is not regulated. As a result, there is no standardization of active ingredients, no concentrations or reporting requirements for labels. A consumer cannot know what is contained in the product as glucosamine is inherently unstable and must be combined with other ingredients for stability. Analysis of products consistently produces the result that many formulations do not contain ingredients listed on the label.[7]

B. Chondroitin Sulfate

Chondroitin sulfate is a heteropolysaccharide identified as a glycosaminoglycan or GAG. GAGs form the ground substance in connective tissue extracellular matrix. The molecule itself comprises repeating linear units of D-galactosamine and D-glucuronic acid. It is found in human cartilage, cornea, bone arterial walls and skin; this form is called chondroitin sulfate A (chondroitic 4-sulfate). Cartilage of humans, fish and shark contain chondroitin sulfate C (chondroitin 6-sulfate). The two forms differ in the amino group of chondroitin sulfate A and in the sulfate group of chondroitin sulfate C. A B form called dermatan sulfate is found in heart valves, tendons, skin and arterial walls. The molecular weights of all forms range from 5,000 to 50,000 daltons. It is available over the counter as a nutritional supplement, usually in an isomeric mixture of A and C forms. Nutritional supplements are derived from varied sources such as pork byproducts (ears, snout), bovine trachea cartilaginous rings, whale septum and shark cartilage.[6]

III. Mechanisms

A. Glucosamine

Glucosamine is produced within the body in small amounts in the reactions involving glucose and glutamic acid. It is a small molecule (molecular weight = 179.17) that is absorbed easily *in vivo*. Humans may decrease production with aging. It is not found in any common foods and cannot be obtained externally. If the body is not synthesizing the substance, it needs to be taken as a supplement. It is found in abundance in cartilage, with small amounts measured in tendons and ligaments; it is an essential substrate matrix, which is a component of cartilage.

It is still not clear what the actions are of glucosamine taken as a nutritional supplement. Purported effects are the promotion and maintenance of the structure and function of cartilage in the joints of the body. It has also been reported that glucosamine has anti-inflammatory effects. The biochemistry, however, has been known for quite some time. Glucosamine, a sugar and a sulfated amino-monosaccharide, is involved in glycoprotein metabolism where it is found in proteoglycans as polysaccharide groups called glycosaminoglycans (GAGs). All GAGs contain derivatives of glucosamine or glactosamine. These polysaccharides comprise 95% of the ground substance in the intracellular matrix of connective tissue. One of the GAGs, hyaluronic acid, is essential for the function of articular cartilage and is responsible for the shock absorbing and deformability functions.[6] *In vitro* studies show that it can alter chondrocyte metabolism; it is not clear that oral glucosamine can reach chondrocytes *in vivo*.[8]

Over 90% of the studies in glucosamine pharmacokinetics have used animal models. It has been shown that about 90% of the salt is absorbed from the small intestine and transported to the liver. The majority is then catabolized in the first pass; seldom is it detected in serum after oral ingestion. Free glucosamine is not usually detected in plasma.[9,10] How much is taken into joints is not known for humans while some uptake is seen in articular cartilage in animals.

B. Chondroitin Sulfate

It is still not clear what the actions are of chondroitin sulfate when taken as a nutritional supplement. Purported effects are the promotion and maintenance of the structure and function of cartilage in the joints of the body. It has also been reported that chondroitin sulfate has anti-inflammatory and pain relief effects. The biochemistry has been known for some time. Chondroitin sulfate is a GAG previously described in the glucosamine mechanisms. It is essential for the structure and function of articular cartilage and provides the same properties as hyaluronic acid. While intra-articular injections of hyaluronic acid have been shown to relieve joint pain and improve mobility, the same has not yet been demonstrated for chondroitin sulfate. It is speculated that oral ingestion of chondroitin sulfate may lead to an increase in hyaluronic acid. Thus, cartilage breakdown would be inhibited.[6]

It has been shown that absorption is from the stomach and small intestine. High-molecular-weight forms are not significantly absorbed, while low-molecular-weight forms show significant absorption after oral ingestion. How much is taken into joints is not known for humans, while it is known that some does enter the joint space.

IV. Review of Research Studies and Clinical Trials

A. Glucosamine

Glucosamine was looked at for use in reducing the symptoms of osteoarthritis as early as 1969.[11] A number of years ago, early studies showed, in 20 patients, that the use of glucosamine sulfate resulted in patients who experienced lessening or disappearance of symptoms with use over 6–8 weeks[12] with no adverse reactions. Barclay and associates[13] reviewed the pharmacology and pharmacokinetics of glucosamine and evaluated the available literature regarding safety and efficacy. Of the literature published between 1965 and 1997, three critically evaluated studies were found that reported a decrease in the symptoms of osteoarthritis. However, flaws in the research designs precluded making positive recommendations for improvements in the symptoms of osteoarthritis with oral glucosamine use. Intramuscular glucosamine administration, however, is effective.[14] No statistically significant difference between glucosamine sulfate and placebo were found in managing pain, leading to the conclusion by one group that the supplement was no more effective than the placebo.[15]

A 12-week study of 2,000 mg per day doses of glucosamine in subjects with articular cartilage damage and possible osteoarthritis showed self-reported improvement in symptoms. However, while clinical and functional test scores improved over the evaluation period in both the test and placebo group, there were no significant differences between groups at the end of the study.[16] The "trend" reported was that improvement could be seen after 8 weeks. A 3-year prospective placebo-controlled study evaluating the effect of glucosamine sulfate use in its effect on joint-space narrowing in knee osteoarthritis did not find statistically significant effects in the most severe cases. However, patients with less severe radiographic knee osteoarthritis showed a "trend" toward significant reduction in joint space narrowing.[17] It has been shown that a 3-year treatment of osteoarthritis with glucosamine sulfate use retarded the progression of knee osteoarthritis as determined by a lesser joint-space narrowing as compared with the placebo group.[18] The authors suggested that this retardation of narrowing of joint space might modify and slow the disease process; however, joint space narrowing is not associated with pain.

Positive results are difficult to demonstrate (glucosamine hydrochloride). The objective measurement differences between groups are not usually statistically significant. Results are reported as positive "trends"[19] in objective measurements. More often than not, however, patients report that they feel they are better than at the start of the trial.[19] Glucosamine use was shown to preserve joint space in that significant narrowing did not occur. It was suggested that long-term use prevents joint structure changes and improves disease symptoms.[20] However, a change in joint space is not necessarily associated with a change in pain levels. Some have reported overall positive results.[21]

Literature reviews usually conclude that glucosamine may not only provide symptomatic pain relief but may have a role in chondroprotection.[22] Even though no differences were found between the glucosamine and placeo groups, and positive results were modest, it was still concluded that glucosamine sulfate may be a safe and effective symptomatic slow-acting drug for osteoarthritis.[23] Glucosamine can be administered orally, intravenously, intramuscularly and intra-articularly. Reviews of primarily European and Asian literature have suggested that glucosamine sulfate use may provide pain relief, reduce tenderness and improve mobility in patients with osteoarthritis.[24] Studies in the U.S. do not support these conclusions.

B. Chondroitin Sulfate

A number of years ago, based on *in vitro* studies, chondroitin sulfate was identified as a supplement that might provide chondroprotection.[25] A multicenter randomized double-blind controlled study of 143 subjects with osteoarthritis that used three different formulations of chondroitin sufate showed that improvement of subjective symptoms was achieved after 3 months of treatment.[26] A single daily dose of 1200 mg was found just as effective as three 400 mg doses.

A meta-analysis of chondroitin sulfate supplementation found 16 publications that fit criteria for inclusion. Criteria included types of joint involvement studied, study designs, numbers of patients enrolled and pain index variables analyzed.[27] It was concluded that chondroitin sulfate may be useful in osteoarthritis treatment; however, results of the published studies were clouded by concomitant use of analgesics or NSAIDS, thus making conclusions about benefits difficult.[27] Some have suggested that it can be used as an anti-inflammatory without dangerous effects on the stomach, platelets and kidneys.[28]

Conte and co-workers[29] showed that single daily doses of 0.8 g and two daily doses of 0.4g resulted in an increase of plasma concentration of chondroitin sulfate for a 24-hour period, showing that there was bioavailability. In 20 male volunteers, chondroitin sulfate plasma levels increased in all subjects and peaked after 2 hours.[30] It is questionable, however, as to what level of chondroprotection can be achieved by orally administered chondroitin sulfate. Baici and co-workers[31] found no changes in serum concentrations of glycosaminoglycan concentration before and after ingestion of chondroitin sulfate in six patients with rheumatoid arthritis and six patients with osteoarthritis. They suggested that claims for benefits were biologically and pharmacologically unfounded.

Uebelhart and co-workers[32] assessed the clinical, radiological and biological efficacy and tolerance of chondroitin 4- and 6- sulfate with symptomatic knee osteoarthritis in 42 patients over the period of a year. They reported that the combined preparation was an effective and safe symptomatic slow-acting drug for the treatment of knee osteoarthritis in 42 patients. It was claimed that this was the first study to demonstrate that the natural course of the disease could be changed with symptomatic slow-acting drugs in osteoarthritis (SYSADOAs). Others have made the same suggestion.[32]

C. Combined Glucosamine and Chondroitin Sulfate

There is some evidence that, if positive results in mild to moderate symptoms of osteoarthritis are seen, combined preparations of low-molecular-weight chondroitin sulfate and glucosamine may be more effective, with results reported as synergistic.[33] In 93 patients, a combination preparation was found to be effective when a randomized placebo-controlled study design was implemented.[33] Combination therapy relieved symptoms of knee osteoarthritis and was safe when tested in 34 young males with chronic pain and radiographic evidence of degenerative disease;[34] however, this group was not the older population usually seen with osteoarthritis.

Animal studies show that both chondroitin sulfate and glucosamine sulfate stimulate chondrocyte growth *in vitro* and in animal models.[35] However, no direct evidence that they cause regeneration of cartilage in osteoarthris has been produced. In knee osteoarthritis, glucosamine sulfate can be shown to prevent knee-joint space narrowing, and chondroitin polysulfate has been shown to prevent the same in finger osteoarthritis, as seen on radiographs.[35] These effects are not evidence of regeneration of cartilage. Topical creams have been evaluated using glucosamine sulfate, chondroitin sulfate and camphor that show improvement in relieving pain after 4 weeks.[36]

Some literature reviews have shown that glucosamine and chondroitin sulfates offer safe and effective alternatives to NSAIDs, which may have serious and life-threatening adverse effects.[37] When glucosamine and chondroitin preparations were subjected to meta-analysis, 15 studies were found to fit rigorous criteria. These studies showed some degree of efficacy; trials reported moderate to large effects, but the authors reported that most studies had flawed designs.[38] Chondroitin sulfate is a much larger molecule than glucosamine and is poorly absorbed. Some claim that, in combination with glucosamine, there is no added benefit,[11] but admit to the lack of side effects with use. Manufacturers now use a low-molecular-weight chondroitin sulfate in the hopes of increasing absorbability.

Deal and Moskowitz[39] evaluated the reviews of glucosamine, chondroitin sulfate and collagen hydrolysate use in the symptomatic treatment of osteoarthritis. They came to similar conclusions as most researchers in that recommendations are difficult to make with the current status of nonFDA-evaluated supplements, particularly with long-term use. At a cost of $30–45 per month, older adults on limited incomes may have difficulty sustaining treatment. Some believe that current therapies have little benefit and great risk because of the lack of data in humans; chondroprotection is still questionable, but glucosamine and chondroitin sulfate show modest effectiveness when taken together.[40] Most conclude that there is a modest efficacy for glucosamine and chondroitin sulfate use; however, long-term safety is not yet proved.[41] Meta-analysis does, however, show some degree of positive results, with both supplements[42] claiming that they are effective and safe.[43]

V. Side Effects

A. Glucosamine

There are no known or reported contraindications to glucosamine supplementation. Concerns have been expressed for the potential to increase insulin resistance if glucosamine is given intravenously, as it has been shown to do so in both normal and experimentally diabetic animals. However, this effect is not seen with oral preparations. Some researchers, however, do suggest that it is contraindicated in diabetes, with concerns about its effect on insulin secretion.[43,44] Individuals who are diabetic or overweight should err on the side of caution and carefully monitor blood sugar levels if supplements are taken. Because there is no data, children and pregnant or nursing women should avoid consumption.[6]

Side effects are few and are usually mild digestive problems such as upset stomach, nausea, heartburn and diarrhea. These suggest that it is better taken with food. Short-term adverse effects for glucosamine use also include headache, drowsiness and skin reactions. No allergic reactions have been reported.[6] There are no known interactions with any other nutritional supplement, drug, herb or food. There are no reports of overdosage. Biochemical, hemostatic and hematological measurements indicate that it is safe.[45] The usual dose recommended for benefit is 1500 mg.

B. Chondroitin Sulfate

There are no known or reported contraindications to chondroitin sulfate supplementation. Concerns have been expressed for the theoretical possibility that chondroitin sulfate may have antithrombotic activity and should be avoided by those with hemophilia and those taking anticoagulants such as warfarin. It may also be immunosuppressive.[46] Since the most common form sold is a salt, those on salt-restricted diets should use a salt-free supplement. Because there is no data, children and pregnant or nursing women should avoid consumption.[6]

Side effects are few and are usually mild digestive problems such as nausea, heartburn and diarrhea. No allergic reactions have been reported.[6] There are no known interactions with any other nutritional supplement, drug, herb or food. If chitosan is taken it may decrease absorption. There are no reports of overdosage. Biochemical, hemostatic and hematological measurements indicate that it is safe.[45] The usual dose recommended for benefit is 1200 mg.

VI. Use in Sport and Exercise

Much of the use in sport and exercise is based on the possibility that both glucosamine and chondroitin sulfate will be chondroprotective and will reduce inflammation and pain if injury occurs. People who exercise will use these supplements for varied reasons and many use them more for prophylaxis rather than after an injury. There is a belief that these supplements will help avoid injury, will speed up healing if it occurs and will be a useful adjunct if surgery has occurred.[47]

Exercisers by the nature of what they do put stress on chondral surfaces, and wear and injury can occur. Those most interested in supplementation are runners and those involved in contact and cutting sports where ligaments can be injured.[47] Many athletes injure or tear menisci and chondroprotection is desired. However, there is no data in athletes to support any claims of benefit. The research that has been done has used individuals with osteoarthritis and the supplements have been an adjunct to other therapies used at the same time. Whether the effects will be the same in those without joint damage is not known. The fact remains that consumption of these supplements appears to be safe, although long-term studies have yet to be performed. Athletes will have to judge for themselves, but there are no cautions for use. Knowing that the placebo effect is real, the mere consumption of a product purported to alleviate pain may have a positive effect.

VII. Summary and Recommendations

Glucosamine and chondroitin sulfate have been used as nutriceuticals since 1969.[11] They are believed to ameliorate the symptoms of osteoarthritis by reducing inflammation and by aiding in the restoration of normal cartilage.[48,49] While animal studies have shown positive effects, research in humans is still equivocal. However, as yet, no firm conclusions can be made about these homeopathic remedies.[50] Interestingly, a current recommendation for the use of a nonpharmacological treatment for symptomatic osteoarthritis of the hip and knee includes exercise, both aerobic and strength training, and diet. Exercise was found to be just as effective as NSAIDS for improvement in pain and function.[51]

There is no question that human research needs to be done. Although anecdotal evidence suggests that glucosamine sulfate and chondroitin sulfate are widely used to ameliorate the symptoms of osteoarthritis and may be effective in some cases, the American College of Rheumatology Subcommittee on Osteoarthritis believes it is too early to issue recommendations for use.[52] The National Institute of Arthritis and Musculoskeletal and Skin Diseases (NIAMS) in collaboration with the National Center for Complementary and Alternative Medicine (NCCAM) announced in 1999 a nine-center effort to study the effectiveness of glucosamine and chondroitin sulfate use in a database of over 1,000 patients. The study is in progress and results should be available in 2004.[52]

Athletes consistently look for an advantage in their sport and for natural ways to enhance their performance. While glucosamine and condroitin sulfate, with the current lack of human data on exercisers cannot be considered ergogenic aids, their use cannot be precluded because of their safety. Since little harm can be done, athletes can safely consume these supplements if they believe there will be a benefit. They can be found in pills, powders and in beverages ("Joint Juice," "Motion Potion"). The greatest benefit, if it is does indeed occur, seems to be found in preparations that contain both glucosamine and low-molecular-weight condroitin sulfate. Athletes can safely consume these supplements and need to decide if the cost (30–$45/month) is warranted in light of equivocal research and the fact that, if benefits are noted, it takes 1 to several months before they are observed.[11] The supplements need to be regulated, as there could be long-term side effects and the length of treatment is not known.[4] Since athletes are healthy, effects may not be the same as in those with the diseased joints of osteoarthritis. An excellent book that outlines regimens for reducing pain has been published.[52] While athletes take supplements for osteoarthritis to aid their exercise, those with osteoarthritis may find that exercise is the "drug" that will benefit them the most.

References

1. Morelli, V., Naquin, C. and Weaver, V., Alternative therapies for traditional disease states: Osteoarthritis. *Am. Fam. Physician*, 67(2), 339, 2003.
2. Brief, A.A., Maurer, S.G. and Di Cesare, P.E., Use of glucosamine and chondroitin sulfate in the management of osteoarthritis. *J. Am. Acad. Orthop. Surg.* 9(5), 352, 2001.
3. Chard, J. and Dieppe, P., Glucosamine for osteoarthritis: Magic, hype, or confusion? *Br. Med. J.*, 322, 1439, 2001.
4. Sutton, L., Rapport, L. and Lockwood, B., Gucosamine: Con or cure? Part II. *Nutrition*, 18(6), 693, 2002.
5. Skillman, J.M., Ahmed, O.A. and Rowsell, A.R., Incidental improvement of breast capsular contracture following treatment of arthritis with glucosamine and chondroitin. *Br. J. Plast. Surg.*, 55(5), 454, 2002.
6. Hendler, S., PDR for Nutritional Supplements, Medical Economics, Montvale, N.J., 2001.
7. Abimbola, O., Cox, D.S., Liang, Z. and Eddington, N.D., Analysis of glucosamine and chondroitin sulfate content in marketed products and the Caco-2 permeability of chondroitin sulfate raw materials. *J. Am. Nutraceut. Assoc.*, 3(1), 37, 2000.
8. Towheed, T.E. and Anastassiades, T.P., Glucosamine and chondroitin for treating symptoms of osteoarthritis. Evidence is widely touted but incomplete. *J. Am. Med. Assoc.*, 283, 1483, 2000.
9. Setnikar, I., Palumbo, R., Canali, S. and Zanolo, G., Pharmacokinetics of glucosamine in man. *Arzneimittelforschung*, 43(10), 1109, 1993.
10. Setnikar, I. and Rovati, L.C., Absorption, distribution, metabolism and excretion of glucosamine sulfate. A review. *Arzneimittelforschung*, 51(9), 699, 2001.
11. Sutton, L., Rapport, L. and Lockwood, B., Gucosamine: Con or cure? *Nutrition*, 18(6), 534, 2002.
12. Pujalte, J.M., Llavore, E.P. and Ylescupidez, F.R., Double-blind evaluation of oral glucosamine sulfate in the basic treatment of osteoarthritis. *Curr. Med. Res. Opin.*, 7(2), 110, 1980.
13. Barclay, T.S., Tsourounis, C. and McCart, G.M., Glucosamine. *Ann. Pharmacother.* 32(5), 574, 1998.
14. Reichelt, A., Forster, K.K., Fisher, M., Rovati, L.C. and Setnikar, I., Efficacy and safety of intramuscular glucosamine sulfate in osteoarthritis of the knee. A randomised, placebo-controlled, double-blind study. *Arzneimittelforschung*, 44(1), 1994.
15. Hughes, R. and Carr, A., A randomized, double-blind, placebo-controlled trial of glucosamine sulphate as an analgesic in osteoarthritis of the knee. *Rheumatology(Oxford)*, 41(3), 279, 200216.
16. Braham, R., Dawson, B. and Goodman, C., The effect of glucosamine supplementation on people experiencing regular knee pain. *Br. J. Sports Med.* 37(1), 45, 2003.
17. Bruyere, O., Honore, A., Ethgen, O., Rovati, L.C., Giacovelli, G., Henrotin, Y.E., Seidel, L. and Reginster, J.Y., Correlation between radiographic severity of knee osteoarthritis and future disease progression. Results from a 3-year prospective, placebo-controlled study evaluating the effect of glucosamine sulfate. *Osteoarthritis Cartilage*, 11(1), 1, 2003.

18. Pavelka, K., Gatternova, J., Olejarova, M., Machacek, S., Giacovelli, G. and Rovati, L.C., Glucosamine sulfate use and delay of progression of knee osteoarthritis: a 3-year, randomized, placebo-controlled, double-blind study. *Arch. Intern. Med.*, 162(18), 2113, 2002.
19. Houpt, J.B., McMillan, R., Wein, C. and Paget-Dellio, S.D., Effect of glucosamine hydrochloride in the treatment of pain of osteoarthritis of the knee. *J. Rheumatol.*, 26(11), 2294, 1999.
20. Reginster, J.Y., Bruyere, O., Lecart, M.P. and Henrotin, Y., Naturocetic (glucosamine and chondroitin sulfate) compounds as structure-modifying drugs in the treatment of osteoarthritis. *Curr. Opin. Rheumatol.*, 15(5), 651, 2003.
21. Drovanti, A., Bignamini, A.A. and Rovati, A.L., Therapeutic activity of oral glucosamine sulfate in osteoarthritis: A placebo-controlled double-blind investigation. *Clin. Ther.*, 3, 260, 1980.
22. Phoon, S. and Manolios, N., Glucosamine. A neutraceutical in osteoarthritis. *Aust. Fam. Physician*, 31(6), 539, 2002.
23. Noack, W., Fischer, M., Forster, K.K., Rovati, L.C. and Setnikar, I., Glucosamine sulfate in osteoarthritis of the knee. *Osteoarthritis Cartilage*, 2(1), 51, 1994.
24. da Camara, C.C. and Dowless, G.V., Glucosamine sulfate for osteoarthritis. *Ann. Pharmacother.*, 32(5), 602, 1998.
25. Pipitone, V.R., Chondroprotection with chondroitin sulfate. *Drugs Exp. Clin. Res.*, 17(1), 3, 1991.
26. Bourgeois, P., Chales, G., Dehais, J., Delcambre, B., Kuntz, J.L. and Rozenberg, S., Efficacy and tolerability of chondroitin sulfate 1,200 mg/day vs. chondrotin sulfate 3 x 400 mg/day vs. placebo. *Osteoarthritis Cartilage*, 6 SupplA, 25, 1998.
27. Leeb, B.F., Schweitzer, H. Montag, K. and Smolen, J.S., A meta-analysis of chondroitin sulfate in the treatment of osteoarthritis. *J. Rheumatol*, 27(1), 205, 2000.
28. Ronca, F., Palmieri, L., Panicucci, P. and Ronca, G., Anti-inflammatory activity of chondroitin sulfate. *Osteoarthritis Cartilage*, 6 SupplA, 14, 1998.
29. Conte, A., Volpi, N., Palmieri, L., Bahous, I. and Ronca, G., Biochemical and pharmacokinetic aspects of oral treatment with chondroitin sulfate. *Arzneimittelforschung.*, 45(8), 918, 1995.
30. Volpi, N., Oral bioavailability of chondroitin sulfate (Chondrosulf) and its constituents in healthy male volunteers. *Osteoarthritis Cartilage*, 10(10), 768, 2000.
31. Baici, A., Horler, D., Moser, B., Hofer, H.O., Fehr, K. and Wagenhauser, F.J., Analysis of glycosaminoglycans in human serum after oral administration of chondroitin sulfate. *Rheumatol. Int.*, 12(3), 81, 1992.
32. Uebelhart, D., Thonar, E.J., Delmas, P.D., Chantraine, A. and Vignon, E., Effects of oral chondroitin sulfate on the progression of knee osteoarthritis: A pilot study. *Osteoarthritis Cartilage*, 6 SupplA, 39, 1998.
33. Das, A. and Hammad, T.A., Efficacy of a combination of FCHG49 glucosamine hydrochloride, TRH122 low molecular weight sodium chondroitin sulfate and manganese ascorbate in the management of knee osteoarthritis, *Osteoarthritis Cartilage*, 343, 2000.
34. Leffler, C.T., Philippi, A.F., Leffler, S.G., Mosure, J.C. and Kim, P.D. Glucosamine, chondroitin and manganese ascorbate for degenerative joint disease of the knee or low back: A randomized double-blind, placebo-controlled pilot study. *Mil. Med.*, 164, 85, 1999.

35. Priebe, D., McDiarmid, T., Mackler, L. and Tudiver, F., Do glucosamine or chondroitin cause regeneration of cartilage in osteoarthritis? *J. Fam. Pract.*, 52(3), 237, 2003.
36. Cohen, M., Wolfe, R., Mai, T. and Lewis, D., A randomized, double blind, placebo controlled trial of a topical cream containing glucosamine sulfate, chondroitin sulfate and camphor for osteoarthritis of the knee, *J. Rheumatol.*, 30(3), 523, 2003.
37. de los Reyes, G.C., Koda, R.T. and Lien, E.J., Glucosamine and chondroitin sulfates in the treatment of osteoarthritis: A survey. *Prog. Drug Res.*, 55, 81, 2000.
38. McAlindon, T.E., LaValley, M.P., Gulin, J.P. and Felson, D.T., Glucosamine and chondroitin for treatment of osteoarthritis: A systematic quality assessment and meta-analysis. *J. Am. Med. Assoc.*, 283(11), 1469, 2000.
39. Deal, C.L. and Moskowitz, R.W., Nutraceuticals as therapeutic agents in osteoarthritis. The role of glucosamine, chondroitin sulfate and collagen hydrolysate. *Rheum. Dis. Clin. North Am.*, 25(2), 379, 1999.
40. Walker-Bone, K., Natural Remedies in the treatment of osteoarthritis. *Drugs Aging*, 20(7), 517, 2003.
41. McAlindon, T., Glucosamine and chondroitin for osteoarthritis? *Bull. Rheum. Dis.*, 50(7), 1, 2001.
42. Towheed, T.E., Published meta-analyses of pharmacological therapies for osteoarthritis, *Osteoarthritis Cartilage*, 10(11), 836, 2002.
43. McClain, D.A., Hexaosamines as mediators of nutrient sensing and regulation in diabetes. *J. Diabetes Complications*, 16(1), 72, 2002.
44. McClain, D.A. and Crook, E.D., Hexosamines and insulin resistance. *Diabetes*, 45(8), 1003, 1996.
45. Adebeowale, A., Cox, D.S., Liang, Z. and Eddington, N.D., Analysis of glucosamine and chondroitin sulfate content in marketed products and the Caco-2 permeability of chondroitin sulfate raw materials. *J. Am. Nutraceut. Assoc.*, 3(1), 37, 2000.
46. Volpi, N., Inhibition of human leukocyte elastase activity by chondroitin sulfates. *Chem. Biol. Interact.*, 105(3), 157, 1997.
47. Hungerford, D., Navarro, R. and Hammad, T., Use of nutraceuticals in the management of osteoarthritis, *J. Amer. Nutraceut. Assoc.*,3(1), 23, 2000.
48. Morelli, V., Naquin, C. and Weaver, V., Alternative therapies for traditional disease states: osteoarthritis. *Am. Fam. Physician*, 67(2), 339, 2003.
49. Deal, C. L., Osteoporisis: prevention, diagnosis and management. *Am. J. Med.*,102(1A), 35S, 1997.
50. Long, L. and Ernst, E., Homeopathic remedies for the treatment of osteoarthritis: a systematic review. *Br. Homeopath. J.*, 90(3), 37, 2001.
51. Bischoff, H.A. and Roos, E.M. Effectiveness and safety of strengthening, aerobic and coordination exercises for patients with osteoarthritis. *Curr. Opin. Rheumatol.* 15(2), 141, 2003.
52. Bucci, L., Pain Free: The Definitive Guide to Healing Arthritis, Low-Back Pain and Sports Injuries Through Nutrition and supplements. Summit Group, Fort Worth, TX., 1995.
53. O'Rourke, M., Determining the efficacy of glucosamine and chondroitin for osteoarthritis. *Nurse Pract.*, 26(6), 44, 2001.

9
Glutamine

Satya S. Jonnalagadda and Rob Skinner

CONTENTS

I. Introduction 129
   A. Chemical Structure 130
II. Metabolic Functions 130
   A. Protein Synthesis and Degradation 131
   B. Glucose Regulation 132
   C. Immune Regulation 132
   D. Energy Supply 134
III. Body Stores and Regulation 135
   A. Skeletal Muscle 135
   B. Liver and Kidney 135
IV. Dietary Intake 136
   A. Food Sources 136
   B. Supplemental Sources 136
   C. Digestion and Absorption 137
V. Nutritional Status Assessment 138
VI. Ergogenic Effects 139
   A. Endurance Activities 139
   B. Nonendurance Activities 140
VII. Drug–Nutrient Interactions 142
VIII. Safety and Toxicity 142
IX. Summary 142
References 143

I. Introduction

Glutamine was first isolated in 1883 from beet juice. Glutamine is one of the most abundant amino acids in the body, composing approximately 50% of the free amino acid pool in the blood and skeletal muscle.[1,2] It is considered a nonessential amino acid since it can be produced in the body from other amino acids, namely glutamic acid, in the liver and skeletal muscle. However, under certain physiological conditions (trauma, infection, stress), glutamine can become a conditionally essential amino acid, and exogenous sources (diet and supplements) may be required to meet the needs of the body.

A. Chemical Structure

Glutamine, a 5-carbon compound with two nitrogen groups (Figure 9.1), is a relatively ubiquitous amino acid that is unstable in the liquid phase. Glutamine is an amide derivative of the acid amino acid glutamic acid (or glutamate). The conversion of glutamate to glutamine can occur in key tissues such as liver, brain, kidney, skeletal muscle and intestine, and is catalyzed by the enzyme glutamine synthase.[1,2]

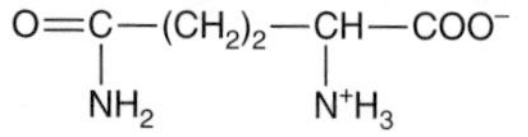

$$O=C-(CH_2)_2-CH-COO^-$$
$$||$$
$$NH_2N^+H_3$$

FIGURE 9.1 Chemical structure of glutamine.

Glutamine is a neutral polar amino acid with an amine functional group, which is highly polar and capable of forming strong hydrogen bonds. This nature of the side chain enables glutamine to contribute to the stability of the protein molecule. The hydrogen bonding capacity of the amine group also enables interaction with water, making glutamine a highly hydrophilic amino acid.[1,2]

Glutamine is degraded to alpha-ketoglutarate, which is an important intermediate in several metabolic pathways, such as the citric acid cycle and the transamination process.[1,2] Glutamine can also be converted to glutamate by the enzyme glutaminase, and thus plays an important role in the transamination process, synthesis of several key proteins, precursors of other nonessential amino acids and synthesis of excitatory neurotransmitters. Figure 9.2 summarizes the metabolism of glutamine

II. Metabolic Functions

Glutamine, the most versatile amino acid, has several key regulatory functions in the body. It plays a central role in acid-base homeostasis, is a precursor for nucleic acids and nucleotide biosynthesis, is used in the synthesis of amino-sugars and participates in inter-organ nitrogen transport. Glutamine is a key anapleuretic energy-yielding substrate under conditions of hypoxia, anoxia and dysoxia, and a key gluconeogenic metabolite under normal postabsorptive conditions.[3] Additionally, glutamine is an important component of the flavor- and taste-enhancing compound monosodium glutamate. Table 9.1 summarizes the metabolic functions of glutamine.

Glutamine

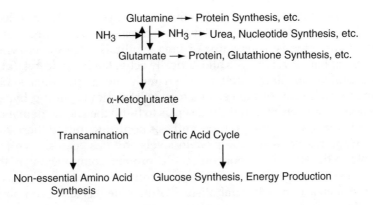

FIGURE 9.2
Brief overview of glutamine metabolism.

TABLE 9.1

Physiological Functions of Glutamine

- Nitrogen carrier
  - Inter-organ transporter
- Metabolic intermediate
  - Energy source
    - Oxidation to lactate (partial) or carbon-dioxide (full)
  - Main energy substrate for
    - Immune system
    - Gastrointestinal system
  - Respiratory substrate
    - Enterocytes
    - Lymphocytes
  - Regulator of protein synthesis
  - Regulator of peptide synthesis
  - Precursor of non-essential amino acid synthesis
  - Purine, pyrimidine, nucleotide synthesis (amine group)
  - Formation of glucosamine (amine group)
  - Fatty acid synthesis (acetyl group)
- Gluconeogenesis (carbon skeleton)
- Glutathione metabolism
- Muscle sparing effect
- Acid-base homeostasis

A. Protein Synthesis and Degradation

Glutamine plays a significant role in the synthesis of proteins and nonessential amino acids. It is referred to as "competence factor" since it stimulates protein synthesis.[4] Glutamine is an amino group donor in various biosynthetic reactions of purine, pyrimidine and amino sugars. It also serves as a storage and transport form of ammonium and plays a significant role in the removal of excess ammonia by playing a critical role in the urea synthesis process. Glutamine is a regulator of urea synthesis since the enzyme carbamyl phosphate synthetase is allosterically activated by N-acetylglutamate for the utilization of the first nitrogen molecule in the urea synthesis. Glutamine utilization of the cells increases dramatically during hypercatabolic conditions such as trauma and sepsis to fulfill the energy demands and to provide nitrogen for the protein synthesis occurring during these states.[1,2]

Glutamine regulates muscle protein levels and has been shown to have antiproteolytic effects on noncontractile protein components in the rat model.[5] Decreasing intramuscluar glutamine concentrations has been shown to increase muscle catabolism.[5] Glutamine has also been observed to regulate the myosin heavy chain synthesis seen in glucocorticoid-induced muscle atrophy.[5] Fatigue due to increased exercise can boost overall protein catabolism and lower plasma and muscle concentration of amino acids; glutamine has been observed to play a key role in restoring these amino acid concentrations.[5]

Glutamine also plays an important role in the regulation of acid–base balance by allowing the kidneys to excrete an acid load, thereby protecting the body against acidosis, and serves as the most important nitrogen shuttle, supplying nitrogen for metabolic purposes.[1,2] Glutamine is a precursor for the synthesis of nucleic acids. It also plays a role in the synthesis of glutathione, a tripeptide molecule synthesized from glycine, cysteine and glutamate. Glutathione plays a role in the transport of certain amino acids, the synthesis of leukotriene (which regulates the inflammation response) and the protection of cells from the harmful effects of free radicals.[6]

B. Glucose Regulation

Glutamine is as important as lactate in contributing to the net gluconeogenesis in the postabsorptive state, contributing to 20–25% of the whole body glucose production. Glutamine gluconeogenesis contributes to approximately 5% of the systemic glucose appearance, and renal production of glucose from glutamine accounts for nearly 75% of all glucose derived from glutamine.[7] However, in trained cyclists, glutamine supplementation did not show any significant impact on the rate of glycogen resynthesis, i.e., muscle glycogen concentrations, despite a twofold increase in plasma glutamine concentrations,[8] suggesting that the impact of glutamine supplementation on glycogen synthesis post-exercise needs to be further evaluated.

C. Immune Regulation

Leukocytes comprise a heterogeneous mix of cells circulating between the lymphoid tissues and organs, and the blood and lymph and play a very important role in maintaining a functional immune response system.[9] Some of the main leukocytes are neutrophils, which make up 60–70% of circulating leukocytes; lymphocytes (T, B and natural killer cells), which compose 20–35% of circulating leukocytes and monocytes, which compose 15% of circulating leukocytes.[9] Additionally, glycoproteins such as immunoglobulins (Ig) and antibodies produced by B cells, are found in the serum and other body secretions, and play an important role in the immune response.

A moderate level of exercise has been suggested to enhance immunity and lower the risk of sickness among active individuals.[4,10] On the other hand, heavy training has been suggested to suppress the immune system of athletes.[4] Among athletes, natural killer cell activity, which protects against viruses and tumor cells, has been observed to be higher than among non-athletes.[10] On the other hand, depending on the severity of the training, neutrophil activity, which is the body's defense against bacteria and viruses, has been observed to be depressed.[10] However, the impact of these changes in the immune cell responses on the risk of infection of athletes is inconclusive. Given the impact of an altered immune system on an athlete's performance, and the role of certain nutrients in the development and maintenance of the immune system, attention has focused on the use of specific nutrients in the prevention of infections among athletes, one such nutrient being glutamine.

Glutamine is considered an immuno-nutrient and is used in medical foods for hypercatabolic situations including burns, cancer, infection, surgery, transplants and trauma.[11,12] Additionally, stressors such as prolonged exercise or overtraining can cause a reduction in skeletal muscle and plasma glutamine concentrations. Muscle glutamine levels have been observed to decrease in a dose-dependent manner based on the degree of stress. Glutamine is also referred to as an anti-catabolic nutrient.[11,12] It has been shown to strengthen immunity under conditions of major trauma, especially in the gastrointestinal tract.[11,12] Glutamine is also needed for wound healing given its role in the regulation of protein synthesis. It serves as a fuel for enterocytes, colonocytes, lymphocytes and proliferating cells. It has been observed to be important for lymphocyte proliferation and generation of lymphokine-activated killer-cell activity *in vitro*. However, Rhode et al.[13,14] did not observe any benefit of glutamine supplementation following prolonged intense exercise on suppressing the decline in lymphokine-activated killer-cell activity. Lymphocyte proliferation, synthesis of interleukin 1 by macrophages and interleukin 2 by lymphocytes have been observed to decrease with glutamine deficiency.

Acute effects of aerobic and anaerobic exercise on the immune system may include increase in leukocyte counts in peripheral circulation with transient immunosuppression in the ability of lymphocytes to respond to immune challenges. The duration and intensity of exercise can also impact the immune response. Exhaustive prolonged exercise can decrease muscle glutamine levels, which is typically magnified after athletic injury. Castell et al.[15] observed that glutamine supplementation (5 g in 330 ml water) immediately after and 2 hours after a marathon decreased the incidence of upper respiratory tract infections up to 7 days following the race. Likewise, Bassit et al.[16] observed that, prior to a triathlon or 30 km run, branched-chain amino acid supplements (6g/day for 15 days), which are precursors for glutamine, prevented a 40% decline in nitrogen-stimulated lymphocyte proliferation observed in the control group post exercise. Additionally, the branched-chain amino acid supplementation prevented the post-exercise fall in plasma glutamine concentrations and was also associated with increase in interleukin-2 and interferon production, suggesting a potential role for these amino acids in the regulation of the immune system. High-intensity exercises have been suggested to suppress neutrophil functions, which play a major role in nonspecific host defense and constitute 50–60% of the blood leukocyte pool. Walsh et al.[17] observed that glutamine supplementation during exercise and recovery helped maintain plasma glutamine concentrations and prevented the reduction in plasma glutamine levels. However, glutamine supplementation had no effect on the post-exercise changes in leukocytosis or whole-blood neutrophil degranulation following prolonged exercise activity. Furthermore, glutamine as a precursor of glutamate plays a major role in cellular synthesis of glutathione in lymphocytes, which is involved in the redox system of the cell by protecting the cells against the harmful effects of oxidants (free radicals) produced during the various metabolic processes.[6]

In athletes, a consequence of overtraining, i.e., excessive training over a prolonged period of time, is immuno-suppression resulting in a higher incidence of infections, slower wound healing, fatigue, impaired immune function and decreased exercise performance. Overtraining has been associated with a decrease in plasma glutamine concentrations.[18–20] Plasma glutamine concentrations have been observed to decrease acutely after intense exercise and during periods of intense training. During periods of stress, glutamine metabolism is increased to promote cell division, antibody production and protein synthesis. During intense training periods, elite Olympic athletes who had lasting fatigue had 33% lower plasma glutamine levels than those who had no long-lasting fatigue; however, the impact on athletic performance was minimal.[21] Additionally, athletes with plasma glutamine levels less than 450 μmol/L were more susceptible to infections, further suggesting that glutamine may have an impact on the synthesis of immune cells.[21]

D. Energy Supply

Glutamine provides nitrogen and carbon molecules for synthesis of macromolecules and production of energy.[1,2] It is a direct respiratory fuel for enterocytes and mucosal epithelial cells of the small intestine, and is also utilized as an energy source by lymphocytes, macrophages, replication cells of the stomach, large intestine, spleen and pancreas.

III. Body Stores and Regulation

Glutamine accounts for more than half the total intramuscular free amino acid pool, making it one of the most abundant and versatile amino acids in the plasma and skeletal muscle.[22] Glutamine is predominantly synthesized and stored in the skeletal muscle by the action of the enzyme glutamine synthetase. Adipose tissue, lungs, liver and brain are also sites of synthesis of glutamine.

A. Skeletal Muscle

In the skeletal muscle, glutamine is needed to maintain protein levels, immune system function and glucose–glycogen metabolism, all of which can be very significant for an active individual such as an athlete.[22] Much of the glutamine is stored in skeletal muscle, and athletes who overuse these muscles may deplete their glutamine stores and increase their susceptibility to infection or slow their rate of recovery from injuries. In the postabsorptive state, skeletal muscle glutamine is a major contributor (48%) to the amino acid nitrogen released into the circulation.[23] Furthermore, 4 hours after a mixed meal, skeletal glutamine accounts for 71% of the amino acid released and 82% of the nitrogen released from the muscle.[23] Plasma glutamine concentrations decreased 10–25% after moderate- to high-intensity exercise (50–80% VO_2 max), which could be due to an increase in muscle protein synthesis during recovery.[23] Glutamine may also be a stimulator of postexercise glycogen synthesis in the muscle, probably due to the direct conversion of glutamine (carbon skeleton) to glycogen- or glutamine-induced cell swelling, which stimulates glycogen synthesis.[24] However, addition of protein to carbohydrate supplements post exercise has not been consistently observed to increase glycogen synthesis beyond that produced with carbohydrate only.[24,25]

B. Liver and Kidney

Glutamine in the liver and kidney is catabolized to glutamate and ammonia with the help of the enzyme glutaminase.[1,2] In the absorptive state or during periods of alkalosis, liver glutaminase activity increases, producing ammonia for the urea cycle.[1,2] Under acidotic state, the use of glutamine in the urea cycle decreases, and instead, glutamine is released from the liver and transported to the kidneys, where it is catabolized by the renal tubular enzyme glutaminase to produce ammonium and glutamate.[1,2] This glutamate may be catabolized by the enzyme glutamate dehydrogenase to produce α-ketoglutarate, which can be used for energy production via the citric acid cycle or nonessential amino acid synthesis via the transamination process and ammonium.[1,2] Renal glutaminase activity and ammonium excretion increase with acidosis and decrease with alkalosis.[2] Thus, given the ubiquitous synthesis of glutamine in the cells and its ability to diffuse in and out of cells, it serves as a major transporter of nitrogen.

IV. Dietary Intake

A. Food Sources

Dietary sources rich in glutamine include all foods that are rich in protein, particularly milk protein and meats. Three ounces (85 g) of meat, chicken or fish contain 3 to 4 g of glutamine. Plant foods such as spinach, parsley and cabbage are also sources of glutamine. Table 9.2 provides some examples of food sources of glutamine and amount per 100 g. Under normal conditions, diet, in addition to what is produced by the body, can meet the glutamine needs of an individual. However, under certain physiological conditions such as injuries, surgery, infections, extreme stress, body glutamine stores can be depleted and supplemental glutamine may be required.

TABLE 9.2

Food Sources of Glutamine*

Food	Amount (g) per 100 g of Cooked, Edible Portion**
Whole wheat bread	2.96
Beef t-bone steak	4.18
Chicken breast	4.64
Pork chops	4.29
Salmon	3.30
Tuna	3.81

* USDA National Nutrient Database for Standard Reference, Available at: http://www.nal.usda.gov/fnic/cgi-bin/nut_search.pl (Accessed May 5th, 2003)
** As glutamic acid

B. Supplemental Sources

Glutamine is a popular supplement consumed by active individuals. Glutamine supplements are available as L-glutamine, either as an individual amino acid supplement or as part of a protein supplement. These supplements are available in powder, capsule, tablet or liquid form, with most claiming to enhance muscular recovery from intense workouts, with the recommended intake being post-workout or with meals. Internet search (February 20th, 2004) using the term "glutamine supplements" produced more than 40,000 (MSN search engine) to more than 70,000 hits (Yahoo and Google search engines). Table 9.3 provides some examples of currently available glutamine supplements. Like all dietary supplements, caution should be used when taking glutamine supplements, which should be taken under the supervision of a health care provider. Individuals with kidney and liver disease should not take glutamine supplements, given the central role of these organs in glutamine metabolism. Glutamine supplements should be taken several hours before or after a meal to prevent interaction with amino acids in the regular diet of the individual.

TABLE 9.3
Summary of Available Glutamine Supplements*

Brand	Type	Amount	Recommended Dose	Cost ($)
EAS	Powder	1000 g	5 g 1–2 times per day	69.99
Pro Performance	Capsule	90 soft gel at 600 mg each	1 capsule 3 times per day	11.99
GNC	Tablet	50 at 1000 mg each	1–2 times per day	11.99
Twinlab	Capsule	120 capsules at 1000 mg each	2–4 daily	31.99
Iron Tech	Powder	1100 g	5g 1–2 times per day	69.95
Optimum Nutrition	Capsules	240 capsules at 1000 mg	1–2 with each meal	20.95
Advantage Rx	Powder	1000 g	5 g 2–3 times per day	38.95
AST	Powder	1200 g	5 g 2 times per day	55.95
MHP	Effervescent powder	400 g	5 g 1–2 times per day	29.95

* Not a comprehensive list

C. Digestion and Absorption

The proteolytic enzymes produced by the pancreas and the brush border facilitate the release of glutamine from dietary protein. Glutamine is absorbed efficiently in the human jejunum and is transported across the intestinal brush border by both sodium-dependent and sodium-independent systems. Following protein digestion and absorption across the brush border and basolateral membranes, glutamine is primarily used by the enterocytes, is the main energy source for these cells and exhibits tropic effects on the gastrointestinal mucosal cells. The gastrointestinal tract utilizes approximately 40% of the glutamine entering the system. Given the dependence of the enterocytes on glutamine, a deficiency of this amino acid can decrease gut function due to a loss of protection against bacterial or endotoxin translocation from the gut lumen into the portal circulation.[1]

V. Nutritional Status Assessment

Plasma glutamine concentrations can be evaluated to determine an individual's glutamine status. Among high-performance athletes (speed skating, swimming, skiing) Smith and Norris[26] observed plasma glutamine concentrations under a variety of training conditions to range from 402 to 741 μmol/L. Under conditions of rest, the average plasma glutamine concentration in these athletes was 585 μmol/L, while, under heavy training conditions, a significant reduction (to 522 μmol/L) was observed. Additionally, under conditions of overtraining, plasma glutamine concentrations decreased to 488 μmol/L. It therefore has been proposed that the glutamine concentrations decrease when the work volume exceeds the athlete's tolerance level; however, low glutamine levels are not necessarily indicative of the training status of an athlete. Hiscock and Mackinnon[18] also observed differences in resting plasma glutamine concentrations based on the athlete's sport, with cyclists having the highest concentration (1395 μM/L), runners and swimmers with intermediate levels (691 and 632 μM/L, respectively) and power lifters having the lowest levels (556 μM/L), suggesting that the physical and metabolic demands of a sport may influence an athlete's glutamine status. However, it is unclear whether this is due to the effect of the sport or a combination of the sport and dietary practices of these athletes. Furthermore, dietary protein intake, when expressed as g/kg body weight, was observed to be negatively associated with plasma glutamine concentrations, which may be attributed to the role of glutamine in maintaining acid–base balance, with high-protein diets increasing the acid load and thereby increasing glutamine needs by the kidneys.

In addition to exercise, an individual's dietary intake can also have an influence on his or her plasma glutamine concentrations. Blanchard et al.[22] observed high-carbohydrate (70%) diets to increase plasma glutamine concentrations compared with low-carbohydrate (45%) (i.e., high-protein) diets in endurance-trained men completing exercise trials. However, muscle glutamine concentrations did not differ between the two groups, and no association was observed between plasma glutamine concentrations and changes in muscle glycogen concentrations. This suggests that the effect of carbohydrate intake on plasma glutamine is not influenced by the muscle glycogen stores. Likewise, Gleeson et al.[23] observed a low-carbohydrate (7%) diet to be associated with a reduction in plasma glutamine concentrations during recovery compared with a high-carbohydrate (75%) diet. Low-carbohydrate and high-protein intakes have been suggested to result in lowering plasma glutamine levels due to a disruption in the acid–base balance, stimulating the kidneys to increase the uptake of glutamine to buffer the hydrogen ion concentration and restore normal pH. Additionally, glutamine may serve as a precursor for gluconeogenesis, with low-carbohydrate intakes, further reducing plasma glutamine concentrations. Branched chain amino acid supplementation was also observed to maintain plasma glutamine concentrations among triathlons compared with placebo, which resulted in a 23% reduction after the triathlon.[16] This ability of branched chain amino acids to maintain glutamine concentrations is attributed to their influence of glutamate metabolism in the skeletal muscle with the subsequent release of NH_3, which is used in the production of glutamine. Therefore, when assessing people's glutamine status, it is important to assess both their dietary intake and exercise patterns.

VI. Ergogenic Effects

A. Endurance Activities

The speculation that glutamine supplementation can enhance performance during endurance activities is largely based on the acute immune system suppression observed during strenuous exercise. Since glutamine is utilized as a fuel source for immune-system cells, the assumption is that supplementary glutamine will attenuate the mobilization of glutamine from the skeletal muscle. Prolonged endurance exercise, such as marathon running, may reduce plasma glutamine concentration. Plasma glutamine concentrations seem to significantly decrease in overtrained athletes compared with control, nonovertrained athletes.[27] Physical stress such as illness or increased physical activity can induce hypercatabolic states, thereby decreasing the body's endogenous rate of glutamine synthesis, making it a conditionally essential amino acid. Early studies suggested that, because of its impact on muscle glycogen synthesis, glutamine may be beneficial to endurance athletes by serving as a substrate for gluconeogensis in the liver, thereby decreasing amino acid release from muscle during extended exercise, and decreasing muscle protein degradation. Glutamine supplementation has also been shown to decrease incidence of infection secondary to overtraining, and to improve the response of cells of the immune system, thereby enabling athletes to maintain training at a greater frequency and intensity.[10] However, other studies have found no beneficial effects of glutamine supplementation in athletes.[13,14,25]

Strenuous exercise can cause significant immuno-suppression along with a reduction in plasma glutamine levels. Following intense exercise of greater than 1 hour, lymphocyte count, natural killer cell activity and the lymphokine-activated killer cell activity were observed to decline.[4,10] Additionally, the lymphocyte proliferative response to T-cell mitogens decreased during exercise. Concomitant decline in plasma glutamine concentrations may play a role in impaired immune function after sustained physical activity. It has been shown that glutaminase activity, which in rapidly dividing cells is a source of nitrogen and carbon that can serve as precursors for macromolecules and energy, is increased during immunologic challenges.[28] Thus, under life stresses and disease states, several dispensable amino acids can become conditionally essential because their utilization within the body exceeds their endogenous production.

While oral or parenteral glutamine supplementation has helped to maintain muscular glutamine concentrations, improve nitrogen balance and increase protein synthesis,[29] the benefits for athletes are not well established.

Castell et al.[15] investigated the effects of consumption of glutamine-containing drinks immediately after heavy exercise and 2 hours after exhaustive exercise in middle-distance, marathon and ultramarathon runners and elite rowers during training and competition. The glutamine-supplemented group reported a reduced instance of infections compared with the placebo group. Castell et al.[30] later observed that glutamine supplementation did not appear to have an effect on immune function (as assessed by lymphocyte distribution) following completion of the Brussels marathon. Similarly, Rhode et al.[13,14] examined the impact of glutamine supplementation on exercise-induced immune change after 30, 45 and 60 minutes at 75% VO_2 max. Arterial glutamine concentration decreased by 20% after the last exercise bout in the placebo trial, whereas the glutamine concentration was maintained at a level above rest at all times in the glutamine-supplemented group (900 mg/kg body weight). However, these differences between the two groups were not significant. The concentration of leukocytes increased during and after each exercise bout, which was attributed to an increase in neutrophils, lymphocytes and monocytes (during) and neutrophils and monocytes (after); however, no differences were observed between the placebo and glutamine groups. Thus, the post-exercise immune changes did not appear to be caused by decreased plasma glutamine concentrations.[13,14] These results further suggest that oral glutamine supplements may prevent post-exercise reduction in plasma glutamine concentrations without influencing the immune system, but the ergogenic benefits of glutamine supplementation need further examination.

B. Nonendurance Activities

Antonio and Street[31] propose that the ergogenic benefits of glutamine observed in certain groups of athletes and not in others may be due to its protective role against protein degradation, potentially enhancing recovery following resistance-training sessions; however, this is yet to be determined among athletes participating in such sports. Since strength and resistance training result in glycogen depletion and increased protein turnover, Antonio and Street have suggested that glutamine supplementation may be beneficial for these athletes. However, the data on nonendurance activities and glutamine supplementation are scarce. The supposition that glutamine may act as a buffering agent for repeated bouts of high-intensity activities originates from the potential ability to maintain the acid–base balance in the body. However, Haub et al.,[32] in a study of the effects of glutamine or placebo supplementation (0.03kg/kg), did not observe any beneficial effects of the supplementation on five repeated bouts of cycling at 100% VO_2 max peak (four bouts lasting 60 seconds and the fifth bout continued to fatigue). The results indicated that acute ingestion of L-glutamine did not enhance buffering potential or performance or delay the onset of fatigue during high-intensity exercise in these athletes.

In another study, Candow et al.[33] examined the effect of oral glutamine supplementation combined with a weight-resistance program on one rep maximum leg press and bench press, peak knee-extension torque, reduction in lean tissue mass and muscle protein degradation. Subjects were supplemented with glutamine (0.9g/kg lean body mass/day) or placebo for 6 weeks of training. Although an increase in strength (31% for squat and 14% for bench press), torque (6%) and lean tissue mass (6%) and 3-methylhistidine (41%) were observed in the glutamine-supplemented group, similar increases were observed in the placebo group. The researchers concluded that glutamine supplementation resulted in no significant effect on muscle performance, body composition or muscle protein degradation during resistance training. Similarly, Van Hall et al.[25] did not observe any beneficial effects of oral glutamine supplementation on glycogen resynthesis following intense interval exercise. Antonio et al.[34] also investigated glutamine supplementation and its effects on weightlifting performance. Resistance-trained individuals (n = 6) performed weightlifting exercises in a double-blind placebo-controlled crossover design supplemented with glutamine or glycine (0.3g/kg, average intake 23 g) mixed with calorie-free fruit juice or placebo. One hour post ingestion of the supplement, subjects performed exercises to muscular failure. Acute ingestion of glutamine did not enhance weightlifting performance and no differences were observed in the average number of maximal repetitions performed in the leg or bench press exercises, suggesting that short-term ingestion of glutamine (1 hour before the event) may not have any ergogenic benefits for individuals participating in resistance-type activities. The long-term ingestion of glutamine is yet to be determined.

This lack of an effect of glutamine supplementation during resistance training may be due to the utilization of glutamine by other tissues before it reaches the peripheral circulation and skeletal muscle. Glutamine serves as a gluconeogenic precursor when muscle glycogen is depleted by approximately 90%; however, resistance training typically produces approximately 40% depletion in muscle glycogen, which may not be severe enough to benefit from glutamine supplementation.[34] Serving as an energy source of the immune cells is another proposed ergogenic effect of glutamine.[15,30] Exhausting endurance exercise has been shown to suppress the immune system in some athletes; however, heavy resistance exercise has not been observed to have any significant impact on the immune system,[35,36] suggesting that some of the ergogenic benefits of glutamine may be dependent on the type of exercise performed.

VII. Drug–Nutrient Interactions

Interactions of glutamine supplements with other nutrients and dietary supplements are not known.[37] Since glutamine is metabolized to glutamate, which can act as an excitatory neurotransmitter, the potential for antagonizing the anticonvulsant effects of certain epilepsy medications exists.[37] Furthermore, since glutamine is metabolized to ammonia, it can potentially antagonize the antiammonia effects of lactulose.[37] Although the aforementioned interactions are theoretically possible, no known adverse effects have been reported with glutamine supplementation. Glutamine supplements, especially the powders, should not be added to hot beverages because heat can destroy the amino acid. Individuals with kidney disease or liver disease should not take glutamine supplements. Individuals undergoing cancer therapy should not take glutamine supplements without consulting their physicians, since preliminary *in vitro* studies suggest that glutamine may stimulate tumor growth. On the other hand, it has been suggested that glutamine may increase the effectiveness and reduce the side effects of chemotherapy treatments, therefore making it essential that individuals check with their healthcare providers before using supplements.

VIII. Safety and Toxicity

Intake of up to 40 grams per day of glutamine supplements did not result in any significant adverse effects other than mild gastrointestinal discomfort in some individuals.[37] Such high intakes can be achieved only through supplement usage and, as such, these amounts should be divided into 2–4 doses throughout the day to result in an increase in total body stores without causing significant competition for absorption with the other amino acids.

IX. Summary

Glutamine has established important physiological functions under both normal and hypercatabolic states. Glutamine has a central role to play in maintaining a healthy immune system and the energy levels of key cells. Adequate dietary protein intake can ensure that these functions of glutamine are fulfilled in the normal healthy individual, with hypercatabolic states requiring additional supplemental sources of glutamine. The immune system-enhancing effects and the antiproteolytic effects of glutamine have implications for athletes involved in intense training activities. However, the ergogenic benefits and the immune system-enhancing effects of glutamine supplementation for the active individual are yet to be realized. Long-term studies examining the effect of glutamine supplementation on protein synthesis, body composition, prevention of infections and increase in muscle glycogen stores are necessary before glutamine supplements can be advocated to athletes.

References

1. Groff, J.L. and Gropper, S.S., Protein, in *Advanced Nutrition and Human Metabolism*, Groff, J.L. and Gropper, S.S., Eds., Wadsworth Thomson Learning, Belmont, CA, 1999, chap. 6.
2. Abcower, S.F. and Souba, W.W., Glutamine and Arginine, in *Modern Nutrition in Health and Disease*, Shils, M.E., Olson, J.A., Shike, M. and Ross, A.C., Lippincott Williams and Wilkins, Media, PA, 1999, chap. 35.
3. Bailey, D.M., Castell, L.M., Newsholme, E.A. and Davies, B., Continuous and intermittent exposure to the hypoxia of altitude: implications for glutamine metabolism and exercise performance, *Br J Sports Med*, 34, 210, 2000.
4. Cynober, L.A., Do we have unrealistic expectations of the potential of immunonutrition?, *Can J Appl Physiol*, 26, S36, 2001.
5. Hargreaves, M. and Snow, R., Amino acids and endurance exercise, *Int J Sports Nutr Exerc Metab*, 11, 133, 2002.
6. Chang, W.K., Yang, K.D. and Shaio, M.F., Lymphocyte proliferation modulated by glutamine: involved in the endogenous redox reaction, *Clin Exp Immunol*, 117, 482, 1999.
7. Gerich, J. and Meyer, C., Hormonal control of renal and systemic glutamine metabolism, *J Nutr*, 130, 995S, 2000.
8. van Hall, G., Saris, W.H.M. and Wagenmakers, A.J.M., Effect of carbohydrate supplementation on plasma glutamine during prolonged exercise and recovery, *Int J Sports Med*, 19, 82, 1998.
9. Mackinnon, L.T., Chronic exercise training effects on immune function, *Med Sci Sports Exerc*, 32, S369, 2000.
10. Nieman, D.C., Exercise immunology: Nutritional countermeasures, *Can J Appl Physiol*, 26, S45, 2001.
11. Gleeson, M. and Bishop, N.C., Modification of immune responses to exercise by carbohydrate, glutamine and antioxidant supplements, *Immunol Cell Biol*, 78, 554, 2000.
12. Field, C.J., Johnson, I. and Pratt, V.C., Glutamine and arginine: Immunonutrients for improved health, *Med Sci Sports Exerc*, 32, S377, 2000.
13. Rohde, T., Asp, S., MacLean, D.A. and Pedersen, B.K., Competitive sustained exercise in humans, lymphokine activated killer cell activity and glutamine — an intervention study, *Eur J Appl Physiol*, 78, 448, 1998.
14. Rohde, R., MacLean, D.A. and Pedersen, B.K., Effect of glutamine supplementation on changes in the immune system induced by repeated exercise, *Med Sci Sports Exerc*, 30, 856, 1998.
15. Castell, L.M., Poortmans, J.R. and Newsholme, E.A. Does glutamine have a role in reducing infections in athletes? *Eur J Appl Physiol*, 1996; 73:488-490.
16. Bassit, R.A., Sawada, L.A., Bacurau, R.F.P., Navarro, R. and Costa Rosa, L.F.B.P., The effect of BCAA supplementation upon the immune response of triathletes, *Med Sci Sports Exerc*, 32, 1214, 2000.
17. Walsh, N.P., Blannin, A.K., Bishop, N.C., Robson, P.J. and Gleeson, M., Effect of oral glutamine supplementation on human neutrophil lipopolysaccharide-stimulated degranulation following prolonged exercise, *Int J Sport Nutr Exerc Metab*, 10, 39, 2000.

18. Hiscock, N. and Mackinnon, L.T., A comparison of plasma glutamine concentration on athletes from different sports, *Med Sci Sports Exerc*, 30, 1693, 1998.
19. Petibios, C., Cazorla, G., Poortmans, J.R. and Deleris, G., Biochemical aspects of overtraining in endurance sports, *Sports Med*, 32, 867, 2002.
20. Walsh, N.P., Blannin, A.K., Clark, A.M., Cook, L., Robson, P.J. and Gleeson, M., The effects of high-intensity intermittent exercise on the plasma concentrations of glutamine and organic acids, *Eur J Appl Physiol*, 77, 434, 1998.
21. Kingsbury, K.J., Kay, L. and Hjelm, M., Contrasting plasma free amino acid patterns in elite athletes: association with fatigue and infection, *Br J Sports Med*, 32, 25, 1998.
22. Blanchard, M.A., Jordan, G., Desbrow, B., Mackinnon, L.T. and Jenkins, D.G., The influence of diet and exercise on muscle and plasma glutamine concentrations, *Med Sci Sports Exerc.*, 33, 69, 2001.
23. Gleeson, M., Blannin, A.K., Walsh, N.P., Bishop, N.C. and Clark, A.M., Effect of low- and high-carbohydrate diets on the plasma glutamine circulating leukocyte responses to exercise, *Int J Sport Nutr*, 8, 49, 1998.
24. Bowtell, J.L., Gelly, K., Jackman, M.L., Patel, A., Simeoni, M. and Rennie, M.J., Effect of oral glutamine on whole body carbohydrate storage during recovery from exhaustive exercise, *J Appl Physiol*, 86, 1770, 1999.
25. van Hall, G., Saris, W.H.M., van de Schoor, P.A.I. and Wagenmakers, A.J.M., The effect of free glutamine and peptide ingestion on the rate of muscle glycogen resynthesis in man, *Int J Sports Med.*, 21, 25, 2000.
26. Smith, D.J. and Norris, S.R., Changes in glutamine and glutamate concentrations for tracking training tolerance, *Med Sci Sports Exerc*, 32, 684, 2000.
27. Williams, M.H., *Nutrition for Health, Fitness and Sport*, 5th ed., WCB/McGraw Hill. Dubuque, IA. 1999.
28. Wilmore, D.W. and Rombeau, J.L., Role of mitochondrial glutaminase in rat renal glutamine metabolism, *J Nutr*, 131, 2491S, 2001.
29. Lacey, J.M. and Wilmore, D.W. Is glutamine a conditionally essential amino acid? *Nutr Rev*, 48, 297, 1990.
30. Castell, L.M., Poortmans, J.R., Leclercq, R., Brasseur, M., Duchateau, J. and Newsholme, E.A., Some aspects of the acute phase response after a marathon race and the effects of glutamine supplementation, *Eur J Appl Physiol*, 75, 47, 1997.
31. Antonio, J. and Street, C., Glutamine: A potentially useful supplement for athletes, *Can J Appl Physiol*, 24, 1, 1999.
32. Haub, M.D., Potteiger, J.A., Nau, K.L., Webster, M.J. and Zebas, C.J., Acute L-glutamine ingestion does not improve maximal effort exercise, *J Sports Med Phys Fitness*, 38, 240, 1998.
33. Candow, D.G., Chilibeck, P.D., Burke, D.G., Davison, K.S. and Smith-Palmer, T., Effect of glutamine supplementation combined with resistance training in young adults, *Eur J Appl Physiol*, 86, 142, 2001.
34. Antonio, J., Sanders, M.S., Kalman, D., Woodgate, D. and Street, C., The effects of high-dose glutamine ingestion on weightlifting performance, *J Strength Cond Res*, 16, 157, 2002.
35. Rall, L.C., Eoubenoff, R., Cannon, J.G., Abad, L.W., Dinarello, C.A. and Meydani, S.N., Effects of progressive resistance training on immune system response in aging and chronic inflammation, *Med Sci Sports Exerc*, 28, 1356, 1996.

36. Flynn, M.G., Fahlman, M., Braun, W.A., Lambert, C.P., Bouillon, L.E., Bronson, P.G. and Armstrong, C.W., Effects of resistance training on selected indexes of immune function in elderly women, *J Appl Physiol*, 86, 1905, 1999.
37. Jellin, J.M., Gregory, P.J., Batz, F. and Hichens, K. Pharmacist's Letter/Prescriber's Letter Natural Medicines Comprehensive Database, 4th ed., Therapeutic Research Faculty, Stockton, CA, pg. 607, 2002.

10

β-Hydroxy-β-Methylbutyrate

Steven L. Nissen

CONTENTS

- I. Introduction 148
  - A. Endogenous Production...148
  - B. Fate...150
  - C. Absorption 150
  - D. Dietary and Supplemental Sources 150
  - E. Mechanism of Action 151
- II. Applications..152
  - A. Resistance Training..152
    1. Increases Strength and Muscle Mass...............................154
    2. Effect of Gender and Training Status 154
    3. Benefit to Older Adults 155
    4. HMB Compared with Other Nutritional Supplements.........157
    5. Combination of HMB and Creatine.................................157
  - B. Endurance 158
  - C. Muscle Damage 158
  - D. Reversing Unwanted Muscle Loss (Non-Exercise)...................159
  - E. Wound Healing..160
- III. Safety 161
  - A. Safety at Recommended Dosages..161
    1. Adverse Events..161
    2. Blood Chemistry and Hematology 161
    3. Blood Lipids..162
    4. Blood Pressure 162
    5. Emotional Profile..162
  - B. Safety at Higher Dosages 162
- IV. Recommendations..163
  1. Well Trained Athletes...163
  2. Casual Athlete and Fitness Enthusiast....................................164
  3. Active Lifestyle...164

V. Future Research 164
VI. Summary 165
References 166

I. Introduction

Since the early 1960s, leucine and its keto acid, α-ketoisocaproate (KIC), have been a subject of research into the regulation of muscle protein synthesis and muscle protein breakdown.[1,2] However, in the early 1990s, a downstream metabolite called β-hydroxy-β-methylbutyrate (HMB) was shown to have a positive effect on muscle protein[3] and was postulated to be responsible for the "leucine effect" on muscle protein metabolism.[4] Since the initial discovery, HMB has been studied extensively as an ergogenic aid in humans especially related to exercise training.[3,5–12]

This chapter will focus on HMB and the benefits related to improving human performance and augmenting the effects of training. The areas that will be examined include: magnification of the strength and fat-free mass gains associated with resistance training,[13] reduction of the muscle damage that occurs during intense exercise,[6,9] the enhancement of indicators of endurance performance,[5] and how HMB combined with other nutrients can restore muscle mass lost from disease[14,15] and aging.[16] Finally, the effects of HMB on cholesterol metabolism[4] and muscle proteolysis[3] will be addressed as a possible mechanism whereby HMB acts to increase lean tissue mass.

A. Endogenous Production

Endogenous production of HMB occurs in muscle and liver[17,18] (Figure 10.1) and possibly other tissues. The first step in HMB formation is the transamination of leucine to KIC, which occurs in both the cytosol and mitochondria of muscle cells.[19] In the mitochondria, KIC is irreversibly oxidized to isovaleryl-CoA by the enzyme branched chain α-ketoacid dehydrogenase. Isovaleryl-CoA then undergoes further metabolic steps within the mitochondria (Figure 10.1) yielding β-hydroxy-β-methylglutaryl-CoA (HMG-CoA). Further metabolism by the enzyme HMG-CoA lyase results in the end products acetoacetate and acetyl-CoA. Approximately 90% of KIC is oxidized to isovaleryl CoA in liver mitochondria and ultimately to acetoacetate and acetyl-CoA.

In the cytosol of cells, the remaining ~10% of the KIC is oxidized to HMB[20–24] via the enzyme KIC-dioxygenase. This enzyme requires molecular oxygen and iron,[24] and may be identical to *p*-phenylpyruvate dioxygenase, which is a key enzyme in the degradation of tyrosine converting 4-hydroxyphenylpyruvate to homogentisate.[25]

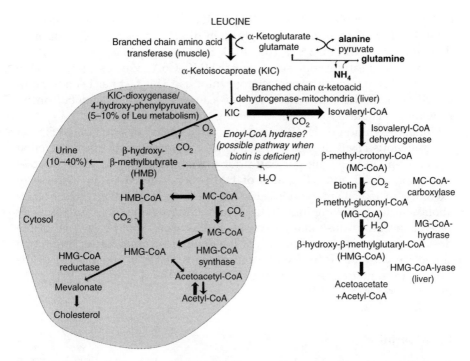

FIGURE 10.1
Overview of leucine, α-ketoisocaproate (KIC) and β-hydroxy-β-methylbutyrate (HMB) metabolism.

A second pathway in the production of HMB has also been postulated through the hydroxylation of methylcrotenoic acid (MCA) but only when biotin is deficient. It is proposed that MCA concentrations become elevated due to the low activity of the biotin-requiring enzyme MC-CoA carboxylase. HMB levels also increase,[26] suggesting MCA may be hydrated to HMB by enol-coA hydrase,[27] an enzyme of the isoleucine pathway. However, it is not clear whether MCA is directly converted to HMB during biotin deficiency, or if the rise in HMB is simply a result of feedback inhibition on the various enzymes along the pathway back to KIC.

Extrapolating from leucine turnover studies in pigs,[28] it is estimated that endogenous HMB production is equal to 0.2 to 0.4 g of HMB per day in a 70 kg man, depending on leucine intake. Furthermore, turnover of HMB is thought to be relatively rapid, as basal plasma concentrations in normal humans range from 1 to 4 nM.[29,30] Plasma levels of HMB increase following ingestion of 1 gram of HMB to approximately 115 nM but are near basal levels 12 hours later, again suggesting rapid metabolism.

B. Fate

HMB has two fates in the body; one being conversion to HMG-CoA[31-33] and the other excretion in the urine.[28-30] The metabolic pathway of HMB metabolism is conversion to HMG-CoA (Figure 10.1). In the cytosol, conversion of HMB to HMG-CoA occurs either through direct carboxylation or through dehydration of MCA-CoA (Figure 10.1). Subsequently, the cytosolic HMG-CoA produced can provide substrate for HMG-CoA reductase, which is the committed step in cellular cholesterol synthesis.[4] This fact has been hypothesized as a mechanism whereby HMB can affect cellular metabolism by providing a cholesterol precursor during times of elevated need.[4]

Urinary excretion of HMB in humans ranges from 10% to almost 50% of an exogenous HMB dosage.[4] Nissen et al.[3] reported urine HMB excretion varied from 10 to 30 mg per day prior to supplementation, while supplementation of 1.5 and 3.0 grams of HMB day resulted in an increase in excretion to 450-500 mg and 950-1200 mg per day, respectively. Recent metabolic studies following ingestion of 1 g of HMB resulted in approximately 14% of the given dose being excreted in the urine. The percentage of the dosage excreted increased to 29% of the given dose after consumption of a single 3 g dose of HMB.[29] With both dosages of HMB given, most of the urinary excretion occurred within six hours of the dosing, paralleling the increases in plasma HMB.[29]

C. Absorption

Absorption of dietary HMB appears to be rapid and complete. Plasma HMB levels have been shown to be elevated as little as 30 minutes following consumption of 1 g of HMB.[29] Following a single 1 g oral dose of HMB, plasma HMB levels peaked at 115 nM at approximately 2 hrs post ingestion and after a single 3 g oral dose of HMB, plasma HMB levels peak at 480 nM at approximately 1 hr post ingestion.[29] Absorption is not increased by concurrent glucose ingestion (75g), rather, HMB absorption is slightly slower when consumed with glucose, but overall HMB absorption is the same either with or without concurrent glucose ingestion.[29] Data from our lab suggest the absorption of supplemental HMB is complete, as supplemental intake of 1.5 or 3 grams of HMB did not appear to affect fecal HMB concentrations.

D. Dietary and Supplemental Sources

Most foods contain trace amounts of HMB. Fruits and vegetables have relatively low HMB concentrations, ranging from 1 to 5 nmol/g, while most meats have higher concentrations, ranging from 15 to 25 nmol/g.[34] However, some foods of plant origin have concentrations of HMB comparable to those seen in products of animal origin. For example, an herbal tea was found to contain 26 nmol HMB/g, aparagus has 22 nmol HMB/g and squash has 22 nmol HMB/g.[35] Although diet is a source of HMB, endogenous production of HMB from leucine generally far exceeds dietary intake of HMB. Therefore, foods containing large concentrations of leucine would probably have a greater influence on the circulating concentrations of HMB in the body than the HMB found in most common foodstuffs. This was demonstrated in pigs when a meal supplemented with 50 grams of leucine was consumed. Plasma HMB concentrations increased 10 fold over that of pigs fed a meal without supplemental leucine.[36] In humans given leucine intravenously, plasma HMB increased from 1.9 to 3.6 µM and plasma HMB rate of appearance increased from 0.19 to 0.27 µmol/kg/hr.[37]

Supplemental HMB is commercially available as calcium β-hydroxy-β-methylbutyrate-monohydrate or $Ca(C_5H_9O_3)_2 \cdot H_2O$ (CaHMB, molecular weight = 292). Supplemental CaHMB is a white powder that is freely soluble in water and has a slightly bitter taste. HMB is chemically synthesized[38-42] and sold commercially under the U.S. patents 5,348,979, 5,360,613 and 6,103,764, which relate to nitrogen sparing, decreasing cholesterol and increasing aerobic capacity, respectively. Based on previously reported data,[3] the recommended dosage is 3 grams (as CaHMB) per day for a 70 kg man; therefore, HMB can easily be supplemented in the form of capsules or tablets and thus individual dosage can be adjusted on a by-weight basis for significantly lighter or heavier individuals (38 mg of HMB per kg of bodyweight per day).

E. Mechanism of Action

Although the mechanism of action for HMB is not known, the primary working theory is that HMB acts by improving cell membrane integrity by supplying adequate substrate for cholesterol synthesis.[4] It is clear that HMB is converted to HMG-CoA in the cytosol, which can be used for cholesterol synthesis in cells.[4] In all cells, cholesterol is needed for the synthesis of new cell membranes as well as the repair of damaged membranes in maintaining proper cell function and growth.[43,44] Certain cells, such as muscle cells, require *de novo* synthesis of cholesterol for cell cholesterol functions. Therefore, during periods of increased stress on cells such as occurs in muscle during intense exercise, the demand for cholesterol for growth or repair of cellular membranes may exceed that which can be made through normal endogenous production from available cellular HMG-CoA. Thus, supplemental HMB may help meet an increased demand for and maintain maximal cell function by supplying intracellular HMG-CoA for cholesterol synthesis. The cholesterol can then be used to build and stabilize muscle cell membranes. This theory is supported by observations on diverse cell functions such as immune function and milk-fat synthesis, which also requires *de novo* synthesis of cholesterol in the cells.[4]

Another possible mechanism of action of supplemental HMB on muscle mass is that HMB somehow directly decreases muscle proteolysis or protein breakdown by having a direct effect on transcriptional or translational control of genes, enzyme activities or other processes involved with proteolysis. When isolated chicken and rat muscles were studied *in vitro*, HMB addition to the muscle strip media decreased muscle proteolysis.[45] Further research *in vivo* has shown that HMB decreases muscle proteolysis through changes in the activity of the proteolytic enzymes such as thiol cathepsins and calpain II.[46] In humans undergoing an intense resistance-exercise program, supplementing HMB resulted in a significant decrease in urinary 3-methylhistidine (3-MH), indicating decreased protein degradation during the first 2 weeks of the exercise program ($p < 0.04$ and $p < 0.001$ for weeks 1 and 2, respectively).[3] Therefore, increasing the circulating levels of HMB through supplementation may decrease the rate of muscle protein breakdown, which would be of benefit in unwanted catabolic conditions such as after heavy exercise or wasting conditions brought about by some diseases.

Both of the mechanisms proposed may contribute to the effects of HMB in helping maintain cellular function. The increase in cell function could be directly through an increase in cellular cholesterol and thus stabilization of the cell membranes, or through a more positive protein balance by decreasing muscle protein breakdown. Whatever mechanism or mechanisms are responsible for the effects of HMB, it appears HMB supplementation has a positive effect on minimizing cell damage and protein breakdown.[3,6,9,45]

II. Applications

Supplemental HMB has been shown to augment the strength and fat-free mass gains associated with resistance training by approximately two fold.[13] Furthermore, HMB has not only been shown to have a positive effect in younger men and women in resistance training,[8] but also in older adult men and women.[10] Other uses of HMB include improving indicators of endurance performance[5] and minimizing muscle damage that occurs during intense exercise.[6,9] HMB in combination with other nutrients has also been shown to have positive effects on muscle mass in nonexercising populations such as those suffering from disease or age-related muscle loss.[14–16]

A. Resistance Training

Nine peer-reviewed studies have so far been published using HMB in exercising humans. These are summarized in Table 10.1. From the original study investigating the effect of HMB on resistance training [3] to a meta-analysis on nutritional supplements,[13] HMB has been consistently shown to increase strength and fat-free mass gains in conjunction with a resistance-training program.[8–11]

1. Increases Strength and Muscle Mass

In the first published study, Nissen et al.[3] reported a linear increase in lean body mass with HMB supplementation. Three weeks of HMB supplementation at 0, 1.5 or 3.0 grams per day resulted in gains of 0.4, 0.8 and 1.2 kg

TABLE 10.1
Summary of Characteristics of All Published Studies Using HMB with Resistance Training That Met the Specific Inclusion Criteria in a Meta-Analysis.

Source	Treatment (n=)	Placebo (n=)	Gender	Dosage/Day*	Age	Training Status[†]	Training (hours/week)	Duration (weeks)	Body Comp.[‡]
Gallagher et al.[9]	12	14	M	38 mg/kg	21.7	U	3	8	SF
Jowko et al.[11]	9	10	M	3 g	19–23	U	3	3	HW
Kreider et al.[7]	13	15	M	3 g	25.1	T	3	4	DEXA
Nissen et al. (short)[3]	15	6	M	3 g	19–22	U	3	3	TOBC
Nissen et al.(long)[3]	13	15	M	3 g	19–29	T	4	7	TOBC
Panton et al. (men)[8]	21	18	M	3 g	24.0	Both	3	4	HW
Panton et al. (women)[8]	18	18	F	3 g	27.0	Both	3	4	HW
Slater et al.[12]	9	9	M	3 g	-	T	3	6	DEXA
Vukovich et al.[10]	14	17	Both	3 g	70.1	U	2	8	DEXA
Average	13.8	13.6			29.2		3.0	5.2	

* Dosages are given in daily dosages.
[†] U = Untrained: No previous resistance training in the last 3 months; T = Trained: Undergoing some form of resistance training prior to study.
[‡] DEXA = Dual energy x-ray absorptiometry, HW = Hydrostatic weighing, SF = Skin fold thickness, TOBC = Total body electrical conductivity.

Source: Nissen SL and Sharp RL. Effect of dietary supplements on lean mass and strength gains with resistance exercise: a meta-analysis. *J Appl Physiol* 2003; 94(2):651–659.

of lean body mass, respectively. Furthermore, similar increases in total body strength were observed with the three levels of HMB supplementation. Total strength increased 338, 529 and 707 kg with the 0, 1.5 and 3.0 grams of HMB per day, respectively. Although positive changes were observed with the lower dose of 1.5 grams of HMB per day, the dose of 3.0 grams per day resulted in the greatest gains with resistance training. Later, in a study by Gallagher et al.,[9] it was concluded that a higher dose of HMB of up to 76 mg per kg of bodyweight per day (6 grams per day) may result in some additional benefit such as attenuating the creatine phosphokinase (CPK) levels during an intense training program. However, the higher dose did not appear to promote additional increases in one-repetition maximum strength or fat-free mass.[7,9]

A meta-analysis by Nissen and Sharp [13] summarized the effects of HMB (and other nutritional supplements) on muscle mass and strength with resistance training. It was found that, in conjunction with resistance training, supplementation of 3 grams of HMB per day resulted in a net increase in lean mass of 0.28% per week and strength of 1.40% per week (Figure 10.2A and Figure 10.3A). When these results are expressed as an effect size, a method of data standardization, HMB resulted in a significant effect size of a net lean mass gain of 0.15 and strength gain of 0.19 (Figure 10.2B and Figure 10.3B). These effect size values indicate a highly positive effect of HMB supplementation on strength and muscle mass ($p < 0.01$). Table 10.1 summarizes the characteristics of all of the HMB studies included in the meta-analysis including dosage used, age and gender of subjects, training status, training duration, and body composition method. The accumulative conclusion of the individual studies,[3,8,9,11] along with the conclusion from the meta-analysis,[13] clearly supports the nutritional supplementation of HMB to augment strength and muscle mass gains from resistance training.

2. Effect of Gender and Training Status

Initially, most of the research performed on HMB supplementation with resistance training was conducted in college-aged men. However, Panton et al.[8] reported that men and women similarly respond to HMB supplementation. When corrected for the initial starting differences and expressed as a percent change, HMB supplementation resulted in similar improvements in strength, body composition and degree of decrease in muscle damage with resistance training for both men and women. Furthermore, it was also reported that the response from HMB supplementation does not appear to be influenced by previous training status.[8] For example, HMB-supplemented trained men had about a 10 kg increase in chest press strength while HMB-supplemented untrained men had approximately a 9 kg increase. The study concluded that regardless of gender or training status, HMB supplementation increases strength gains from resistance training.

Vukovich et al. [10] also reported the effects of HMB in men and women. However, the study subjects were 70 years of age or older. Even older adult

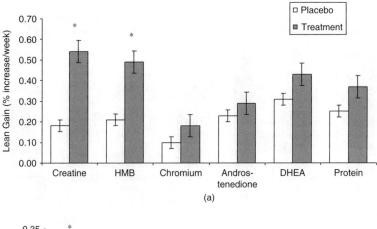
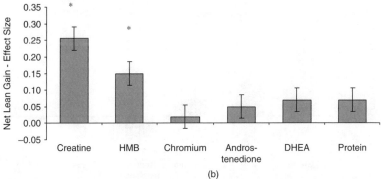

FIGURE 10.2
Comparison of the net lean mass gain of the placebo and treatment groups for each supplement. The upper panel (a) expresses lean gain as percent gained per week, while the lower panel (b) expresses the net effect size for each supplement. * Indicates a significant effect of the treatment versus the placebo ($p < 0.05$). (From Nissen SL and Sharp RL. Effect of dietary supplements on lean mass and strength gains with resistance exercise: A meta-analysis. *J Appl Physiol* 2003; 94(2):651–659.)

men and women were shown to respond equally to HMB supplementation (the study results are presented in the next section).

3. Benefit to Older Adults

Studies have also been conducted in elderly subjects participating in an exercise program, where it was shown that HMB supplementation improved body composition.[10] Seventy-year-old adult men and women undergoing a 5-day-per-week exercise program for 8 weeks were assigned to either 3 grams of HMB or a placebo. The exercise program was a combination of both resistance training (2 days per week) and walking (3 days per week). After 8 weeks, HMB supplementation tended to increase fat-free mass gain ($p < 0.08$) and significantly decreased body fat ($p < 0.05$) compared with the

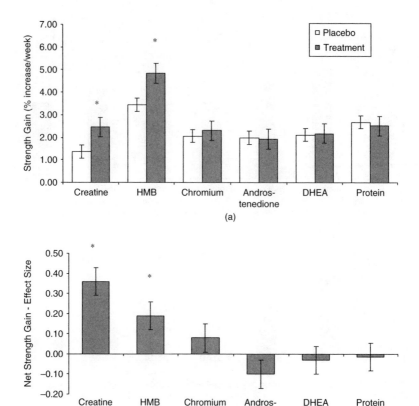

FIGURE 10.3
Comparison of the net strength gain of the placebo and treatment groups for each supplement. The upper panel (a) expresses strength gain as percent increased per week, while the lower panel (b) expresses the net effect size for each supplement. * Indicates a significant effect of the treatment versus the placebo ($p < 0.05$). (Nissen SL and Sharp RL. Effect of dietary supplements on lean mass and strength gains with resistance exercise: A meta-analysis. *J Appl Physiol* 2003; 94(2):651–659.)

placebo-supplemented subjects. In a second study in resistance-training elderly subjects, strength gains during the study were similar between the HMB-supplemented group and the placebo-supplemented group.[47] However, HMB supplementation significantly improved functional ability as measured by a "get-up-and-go" (GUG) test,[47] which measures the time to get up out of a chair, walk a set distance, turn around and go back and sit down again in the chair. Decreasing body fat and improving muscle function are both important findings in the elderly. Body fat and, in particular, visceral fat storage, is linked to the development of insulin resistance (Type II diabetes)[48] and the development of cardiovascular disease.[49] Additionally, improving functional ability in the elderly should improve the quality of life and may result in decreasing falls, a common cause of injury in the elderly population.

4. HMB Compared with Other Nutritional Supplements

The previously mentioned meta-analysis by Nissen and Sharp[13] looked at the effects of common dietary supplements on lean mass and strength gains in conjunction with a resistance-exercise program. A meta-analysis involves the scientific process of gathering and analyzing research results from all previous studies on a related topic to form a final conclusion. Peer reviewed studies during the years 1967 to 2001 were included in the analysis of nutritional supplements and ergogenic aids marketed to improve the results from exercise or athletic performance. Criteria for inclusion in this meta-analysis were studies that were randomized and placebo-controlled, at least 3 weeks in duration, and performed full-body (all major muscle groups) resistance training at least twice per week. Over 250 researched supplements were found in the original search, and only 48 studies met the inclusion criteria for the meta-analysis. Interestingly, only six supplements had more than one published study (at least two studies were needed for inclusion). Creatine had 18 studies, HMB had nine studies (Table 10.1), chromium had 12 studies, dehydroepiandrosterone (DHEA) had two studies, androstenedione had three studies, and protein had four studies. Of the six supplements analyzed, only creatine and HMB were statistically shown to augment strength and muscle mass gains. Furthermore, creatine and HMB gave statistically similar responses.

5. Combination of HMB and Creatine

Most commonly, HMB is available singularly in capsule form. However, HMB is also found combined with other nutritional ingredients. The most popular of these products is a combination of HMB and creatine. As shown in a recent meta-analysis, creatine and HMB are the only two supplements that have been scientifically shown to individually increase strength and muscle mass.[13] One study measured the effect of a combination of creatine and HMB, and this combination was shown to work even better than each of the supplements alone.[11] This suggests creatine and HMB might work by different mechanisms. Lean mass and strength gains were significantly increased in the creatine, HMB and combination of creatine and HMB groups, with the greatest increases reported in the subjects receiving the creatine and HMB combination. Therefore, the interaction between creatine and HMB results in an additive effect on lean body mass gain and strength with resistance training.

B. Endurance

In addition to the studies supporting the use of HMB supplementation during resistance training, there is evidence to suggest that HMB can improve endurance as well. In a study by Vukovich and Dreifort,[5] competitive cyclists were randomly assigned in a double-blinded manner to one of three different supplementation periods (placebo, 3 grams of HMB per day, or 3 grams of leucine per day). Each cyclist completed each 2-week supplementation period with a 2-week washout period in between periods. Maximal oxygen consumption and onset of blood lactate were measured prior to and after the 2 weeks of supplementation for each treatment. The cyclists maintained their current training volume throughout the study period. The results showed that HMB supplementation (3 grams per day) resulted in a 3.6% increase in the time to reach VO_2 peak while time to reach VO_2 peak decreased by 3.6% and 1.2% for the placebo and leucine treatments, respectively ($p < 0.05$). The onset of blood lactate accumulation was also delayed with HMB supplementation. At 2mM blood lactate, VO_2 was significantly increased by 9.1% with HMB supplementation ($p < 0.05$) compared with 2.1% and 0.75% nonsignificant increases during the leucine and placebo supplementation, respectively. These finding suggest that HMB supplementation may improve the ability to exercise at a greater intensity for a longer duration. Similarly, O'Connor and Crowe[50] investigated the effects of HMB supplementation on the aerobic capacity of elite male rugby players. Although not significantly different, HMB supplementation resulted in a 2.3 ml/kg/min increase in aerobic power, while the control group experienced no change in aerobic power.[50] The results from this study need to be cautiously interpreted in that the highly trained and motivated subjects studied were not blinded to the treatments.

Although not showing a significant effect on the indicators of endurance performance (VO_2 max/peak), studies by Knitter et al.[6] and Byrd et al.[51] reported that HMB supplementation can decrease the muscle damage and soreness associated with either prolonged or downhill running. In conclusion, HMB is one of the few, if not only, nutritional supplement shown to improve indicators of endurance performance.

C. Muscle Damage

In studies investigating the effects of HMB on muscle damage following either a single bout of strenuous exercise[6,51] or during an intense resistance-training program,[3,7–9,11] HMB supplementation has been shown to reduce the appearance of creatine phosphokinase (CPK) and lactate dehydrogenase (LDH), both indicators of muscle damage. Both CPK and LDH are muscle enzymes that appear in blood following muscle membrane damage or disruption, and the amount in blood is proportional to severity or magnitude of the muscle damage. Furthermore, supplementation of HMB also results in a significant decrease in plasma 3-MH, which is another marker used as an indicator of muscle breakdown or damage. While undergoing an intense resistance-training program, subjects supplemented with HMB showed a decrease in 3-MH appearance in plasma during the first 2 weeks of the exercise program, which suggests a decrease in protein degradation.[3]

Knitter et al.[6] studied the effects of HMB on muscle damage following a single prolonged run. In this study, runners were recruited and assigned to either a placebo or 3 grams of HMB per day. The subjects were asked to maintain their current running program during the supplementation period. After 6 weeks of supplementation and training, all subjects participated in a 20 km run. Blood was taken and analyzed for CPK and LDH before the run and at several time points after the run. Following the prolonged run, HMB supplementation resulted in a smaller increase in levels of CPK and LDH compared with the placebo-supplemented runners. In addition, Byrd et al.[51] examined the effect of HMB supplementation on muscle soreness and strength following downhill running. Following the downhill running protocol, HMB supplementation resulted in less perceived soreness and less strength loss, which may suggest a protective effect of HMB on muscle and less muscle damage occurring. Therefore, these results suggest HMB appears to minimize the degree of muscle damage that occurs following intense prolonged activity.

HMB supplementation has also been shown to minimize the muscle damage associated with an intense resistance-training program. During the course of a 4-week weight training regimen in both men and women, HMB-supplemented subjects (n = 39) actually showed about a 2% decrease in CPK while the placebo supplemented–subjects (n = 36) had a 26% increase in CPK levels due to the training intensity.[8] Similarly, Gallagher et al.[9] reported results from an 8-week resistance training program where subjects were assigned in a double-blind randomized fashion to receive either a placebo or HMB. This study also found significantly higher blood CPK levels in placebo-supplemented subjects during the start of the strenuous resistance-training program than in those subjects taking HMB.[9] The findings from these studies in addition to several others[3,7,11] suggest HMB supplementation minimizes the muscle damage that occurs from repeated bouts of intense resistance exercise.

D. Reversing Unwanted Muscle Loss (Non-Exercise)

The *in vitro* data and the clinical data in exercise suggest HMB could have a benefit in reducing muscle proteolysis, which should increase muscle mass in individuals suffering from unwanted muscle loss from disease or aging. Although the individual elements were not studied alone, a combination of HMB, arginine and glutamine has been clinically examined in muscle-wasted AIDS[15] and cancer patients.[14] In each of the studies, HMB was supplemented at 3 grams per day while arginine and glutamine were each supplemented at 14 grams per day. The supplement was divided into two equal daily dosages. This combination of HMB and amino acids was shown to result in an increase in lean mass in wasting AIDS and cachexic cancer patients without the need for exercise. Both of the patient populations were in a highly catabolic state and were breaking down muscle tissue.

Supplementing HMB to these patients is thought to minimize the protein breakdown they experience while the amino acids support immune, intestinal and muscle protein synthesis. In the 8-week study in wasted AIDS patients, the HMB, arginine and glutamine-supplemented patients gained on average 2.5 kg of lean mass ($p < 0.01$) while the placebo-supplemented patients continued to lose lean mass and lost on average 0.7 kg during the study period. Similarly, cachexic cancer patients supplemented with the HMB, arginine and glutamine mixture gained 1.4 kg of lean body mass in 8 weeks ($p < 0.05$), while the placebo-supplemented patients continued to lose an additional 1.1 kg of lean body mass. No exercise protocol was used in either of these studies with the nutritional mixture. Therefore, the nutritional mixture alone (HMB, arginine and glutamine) resulted in increased cell function and restoration of muscle mass in catabolic patients losing muscle mass.

A combination of HMB, arginine and lysine was shown to restore protein synthesis in older adult women (71.6 years of age), which resulted in significantly improved functionality, strength and fat-free mass.[16] The HMB, arginine and lysine-supplemented group significantly increased "get-up-and-go" performance time compared with the placebo-supplemented subjects. HMB, arginine and lysine supplementation also resulted in increased leg and grip strength, and tended to increase fat-free mass compared with the placebo-supplemented subjects. The resulting improvement in muscle mass, strength and functionality has the potential to positively impact many aspects of daily activities of the ever-increasing elderly population.

E. Wound Healing

HMB has been tested under circumstances similar to the muscle-wasting conditions with two amino acids to determine whether wound healing is improved.[52] Wound healing requires immune activation, cell proliferation and protein synthesis,[53-55] and the combination of HMB with arginine and glutamine support these cell functions. Williams et al.[52] studied the effects of HMB, arginine and glutamine on wound healing, where a simulated model of wound healing was used. In this model, small subcutaneous catheters were inserted into the deltoid region of the subject's arm. After 2 weeks, the catheters were removed and analyzed for hydroxy-proline (OHP), an indicator of collagen synthesis. The researchers noted that the supplements were well tolerated and that no adverse events were reported during the supplementation period. The results showed that supplementation with HMB, arginine and glutamine increased collagen synthesis by 67% over that of a placebo-supplemented group ($p < 0.03$). It was concluded from the study that the nutritional mixture of HMB, arginine and glutamine would provide a safe nutritional means to increase wound healing in patients.

III. Safety

The safety of any supplement is a function of total dose, dosage schedule and total time of consumption. Most supplements have not been systematically examined relative to safety and toxicity. In some regards, HMB is no different in that long-term, multiyear studies have not been completed. However, in each of the human efficacy studies, extensive safety data were collected.[56]

A. Safety at Recommended Dosages

An analysis of the safety data available from nine published clinical studies on HMB was compiled and published by Nissen et al.[56] The nine studies included both young and elderly males and females ranging in age from 18 to 81 years of age, and the studies were from 3 to 8 weeks in length. Seven of the studies had a resistance-exercise component, one study was in subjects who ran and another study had no exercise component. In each of the studies, data were collected on emotional profile, adverse events and blood chemistry and hematology. The following is a summary of the safety of consuming 3 grams of HMB per day.

1. Adverse Events

Adverse-event questionnaires given during the studies consisted of 32 questions relating to adverse events in major bodily systems. The questionnaires asked the subjects whether they had any adverse symptoms over the past 3 days. Supplementation of HMB resulted in no differences in the occurrence of these events when compared with placebo-supplemented subjects.

2. Blood Chemistry and Hematology

Extensive blood chemistry (about 30 parameters) and hematology (about 20 parameters) data were reported. Supplemental HMB had no negative effects on any of these serum parameters when compared with the placebo-supplemented subjects. Blood potassium was minimally ($< 2\%$) but significantly ($p < 0.01$) decreased in HMB-supplemented subjects. This may have been due to the difference in mineral intakes between the treatment groups as the HMB-supplemented group had a daily intake of 400 mg of calcium, 135 mg of phosphorous and 170 mg of potassium, whereas the placebo group was not balanced for these nutrients. Blood hematology parameters measured during the studies included white blood cells (WBC), red blood cells (RBC), hemoglobin, hematocrit, platelets and WBC sub classes. The only effect of HMB supplementation on any of these parameters was a small ($p < 0.05$)

0.5% decrease in hematocrit; however, no significant difference was noted in initial or ending values for hematocrit for the HMB- and placebo-supplemented groups.

3. Blood Lipids

Blood lipid profiles measured during the studies consisted of total cholesterol, high-density lipoprotein cholesterol (HDL), very-low-density lipoprotein cholesterol (VLDL) and low-density lipoprotein cholesterol (LDL) and triglycerides. In HMB-supplemented subjects, HDL cholesterol showed no change, while in the placebo-supplemented subjects, a 4% increase in HDL cholesterol was seen ($p < 0.04$). Of particular interest is that supplemental HMB significantly ($p < 0.03$) lowered total cholesterol by 3.7% in all subjects and by 5.8% in subjects with cholesterol levels over 200 mg/dL. The decrease in total cholesterol with HMB supplementation was mainly the result of a significant decrease in LDL cholesterol of 5.7% in all subjects ($p < 0.05$) and an even greater decrease of 7.3% ($p < 0.01$) in the high-cholesterol subjects with starting cholesterol levels >200 mg/dL.

4. Blood Pressure

Resting blood pressures were also measured in seven of the nine studies. Supplementation with HMB resulted in a significant decrease in systolic blood pressure of 4.4 mm Hg. In a subset of subjects with systolic blood pressures >130 mm Hg, the decrease was even greater. Diastolic blood pressure was unaffected by HMB supplementation.

5. Emotional Profile

Emotional changes were measured using the Circumplex test of emotion.[57] This consists of a questionnaire of 48 words that describe various emotions. The subjects are instructed to rate each emotion on a scale from 1 (very slightly or not at all) to 5 (extremely) based upon their degree of feeling for that emotion. When compared with the placebo-supplemented group, no negative changes in emotional profile were noted with HMB supplementation during the studies. Supplementation with HMB resulted in a decrease of about 10% in the "unactivated unpleasant effect," which is described by the words tired, drowsy, sluggish, dull, bored and droopy; indicating HMB had a slight but positive effect on emotion.

B. Safety at Higher Dosages

Since the first study was published, two studies have investigated the use of higher dosages of HMB (6 grams per day).[7,9] It was concluded that a higher dose of HMB does not appear to promote additional increases in one-repetition maximum strength or fat-free mass.[7,9] Gallagher et al.[58] studied the effects of HMB consumption at 0, 38, or 76 mg HMB per kg of bodyweight per day for 8 weeks on hematology and hepatic and renal function. The dosages studied corresponded to dosages of approximately 0, 3, or 6 g HMB per day for the average person, with the 6 grams per day level being twice the normally recommended dosage. Thirty-seven healthy male volunteers 18 to 29 years of age took the HMB supplements while undergoing a resistance exercise program. No differences in liver function were seen between any of the treatment groups (0, 38, or 76 mg HMB/kg bodyweight per day). There were also no differences in hematology parameters between the treatment groups, except for an increase ($p < 0.05$) in basophils in the 38 mg HMB/kg body weight per day. However, this increase in basophils was not seen in either the 0 or 76 mg HMB/kg body weight per day supplemented groups. Additionally, no differences were seen in lipid profiles, blood urea nitrogen or hemoglobin between the treatment groups. Renal function was assessed by urine pH and the presence of glucose, protein and ketones. No treatment differences existed for any of the values and all data were within normal limits. Similarly, Kreider et al.[7] also studied the effects of supplementing HMB at 0, 3 and 6 grams per day and showed no adverse effects on hematological or metabolic profiles for the higher-dose group as well as the 3 gram-dose group.

In conclusion, supplementation of 3 grams of HMB per day for several weeks should be considered safe. Additionally, studies of adult males consuming up to 6 g of HMB per day for up to 8 weeks had no adverse effects on measures of hematology or hepatic and renal function. Health-related positive effects of consuming 3 grams of HMB/day include decreasing LDL and total cholesterol, decreasing blood pressure and feeling better (improved mood).

IV. Recommendations

Based on the combined published data, HMB has several benefits, and applications can be recommended for a broad range of people such as athletes, fitness enthusiasts, those simply with an active lifestyle and individuals experiencing unwanted muscle loss. Recommendations are based on the fact that HMB supplementation clearly augments strength and fat-free mass gains from resistance training, minimizes muscle damage and soreness and has been shown to improve endurance.

1. Well Trained Athletes

For a well-trained athlete who is accustomed to a high training volume, it is important to maintain a proper dosage of 38 mg of HMB per kg of body weight (2–5 grams of HMB per day) to minimize the muscle damage that occurs when training intensity or duration changes. For maximal gains, the addition of creatine to the HMB dose should be considered.

2. Casual Athlete and Fitness Enthusiast

A second group that would benefit from HMB supplementation are the fitness enthusiasts or casual athletes who frequently change their training routine through periods of low to moderate to high activity and often participate in strenuous single bouts of competitive activity such as road races, softball games, golf tournaments, etc. Supplementation of 3 grams of HMB per day should improve the gains experienced from either resistance or endurance training, as well as help minimize the muscle damage that occurs during those single bouts of strenuous of activity. Furthermore, HMB supplementation also provides an additive health benefit to exercise.

3. Active Lifestyle

People with an active lifestyle but who rarely participate in a set training regimen will profit mainly through the health benefits of HMB supplementation, but older adults may also benefit by reduced muscle loss and, potentially, muscle gains. Furthermore, supplementation of 3 grams of HMB per day a few weeks prior to and during planned times of strenuous activity (active vacation, golf outing, long hike, etc.) should minimize the muscle damage and soreness that is normally experienced following these activities. Lower doses of 1.5 to 2.0 grams per day may be sufficient, as the physical demands by this group are rather low.

V. Future Research

As recently summarized in a meta-analysis, HMB is one of the few supplements to clearly enhance strength and lean gains with resistance exercise.[13] Further research investigating the effect of HMB on resistance training is not likely based on the current body of evidence. However, research examining the effect of HMB on endurance is not as clear cut. Only three studies on the use of HMB as an endurance supplement have been conducted. While these studies showed positive results on increasing VO_2 max and in decreasing muscle damage, further studies of HMB use for long-distance cyclists, marathon runners or other elite endurance athletes would provide for increased evidence for HMB use by these athletes. Further work should focus on the effect of HMB on VO_2 max or lactic acid accumulation, which could indicate an increase in fatty acid oxidation, therefore benefiting the endurance athlete. Future studies should also concentrate on the muscle damage caused by these types of exercise and the recovery after the endurance event.

Another potential use for HMB is in preserving lean mass in persons losing weight. Many calorie-restrictive diets also limit the amount of protein intake. A combination of protein and calorie restriction causes loss of muscle as well as fat tissue during weight loss. Therefore, it could be hypothesized that HMB could be used to preserve lean muscle during the process of weight loss. Maintenance of muscle while losing body fat would better accomplish the goal of maintaining a permanent weight loss, because maintaining more muscle would result in a greater resting metabolic rate. This in turn would help maintain body weight once the caloric restriction ended.

Although HMB has been shown to decrease muscle protein degradation[3] and CPK leakage from muscle cells,[6] the exact mechanism is still unknown. Evidence points to HMB's having a direct effect on stabilizing muscle cell membranes, however, thus far this is only a hypothesis. While the literature clearly shows that HMB carbon is incorporated into cholesterol,[31,32] and would suggest that through this mechanism supplies more cholesterol for cell membrane synthesis, actual experiments to show that HMB carbon is incorporated into cell membranes have yet to be reported. Additionally, HMB has been shown to decrease muscle protein breakdown.[3] In an animal model to examine protein degradation, it appears HMB may have an effect on decreasing protease activity.[46] Therefore, HMB could have a primary effect on the genes expressing proteases or the proteases themselves, or the decrease in protein degradation may be secondary to minimizing cellular membrane damage.

VI. Summary

HMB is one of only two ergogenic aids that have unequivocal science backing the augmentation of strength and fat-free mass gains associated with resistance training. In general, HMB doubled the effects of resistance training on strength and fat-free mass gains. Furthermore, HMB is the only ergogenic aid that has a positive effect on cardiovascular disease risk profile through lowering blood pressure and cholesterol. HMB also has a strong database of safety showing no harmful effects and, in general, improving emotional profiles. HMB supplementation may also have value in improving endurance and minimizing muscle damage as well as preventing unwanted muscle loss. In conclusion, there is strong evidence that 3 grams of HMB per day can magnify both the strength and lean gains as well as the health benefits of exercise.

References

1. Nair KS, Schwartz RG and Welle S. Leucine as a regulator of whole body and skeletal muscle protein metabolism in humans. *Am J Physiol* 1992; 263:E928–E934.
2. Frexes-Steed M, Lacy DB, Collins J and Abumrad NN. Role of leucine and other amino acids in regulating protein metabolism *in vivo*. *Am J Physiol (Endocrinol Metab)* 1992; 262:E925–E935.

3. Nissen S, Sharp R, Ray M, Rathmacher JA, Rice J, Fuller JC, Jr. Connelly AS and Abumrad N. The effect of the leucine metabolite β-hydroxy β-methylbutyrate on muscle metabolism during resistance-exercise training. *J Appl Physiol* 1996; 81(5):2095–2104.
4. Nissen SL and Abumrad NN. Nutritional role of the leucine metabolite β-hydroxy-β-methylbutyrate (HMB). *J Nutr Biochem* 1997; 8:300–311.
5. Vukovich MD and Dreifort GD. Effect of beta-hydroxy beta-methylbutyrate on the onset of blood lactate accumulation and VO_2 peak in endurance-trained cyclists. *J Strength Cond Res* 2001; 15(4):491–497.
6. Knitter AE, Panton L, Rathmacher JA, Petersen A and Sharp R. Effects of β-hydroxy-β-methylbutyrate on muscle damage following a prolonged run. *J Appl Physiol* 2000; 89(4):1340–1344.
7. Kreider RB, Ferreira M, Wilson M and Almada AL. Effects of calcium beta-hydroxy-beta-methylbutyrate (HMB) supplementation during resistance-training on markers of catabolism, body composition and strength. *Int J Sports Med* 1999; 20(8):503–509.
8. Panton LB, Rathmacher JA, Baier S and Nissen S. Nutritional supplementation of the leucine metabolite β-hydroxy β-methylbutyrate (HMB) during resistance training. *Nutrition* 2000; 16(9):734–739.
9. Gallagher PM, Carrithers JA, Godard MP, Schulze KE and Trappe SW. β-hydroxy-β-methylbutyrate ingestion, Part I: Effects on strength and fat free mass. *Med Sci Sports Exerc* 2000; 32(12):2109–2115.
10. Vukovich MD, Stubbs NB and Bohlken RM. Body composition in 70-year old adults responds to dietary β-hydroxy-β-methylbutyrate (HMB) similar to that of young adults. *J Nutr* 2001; 131(7):2049–2052.
11. Jówko E, Ostaszewski P, Jank M, Sacharuk J, Zieniewicz A, Wilczak and Nissen S. Creatine and β–hydroxy-β-methylbutyrate (HMB) additively increases lean body mass and muscle strength during a weight training program. *Nutrition* 2001; 17:558–566.
12. Slater G, Jenkins D, Logan P, Lee H, Vukovich MD and Rathmacher JA, Hahn A. β-hydroxy β-methylbutyrate (HMB) supplementation does not affect changes in strength or body composition during resistance training in trained men. *Int J Sport Nutr Exerc Metab* 2001; 11:384–396.
13. Nissen SL and Sharp RL. Effect of dietary supplements on lean mass and strength gains with resistance exercise: a meta-analysis. *J Appl Physiol* 2003; 94(2):651–659.
14. Eubanks May P, Barber A, Hourihane A, D'Olimpio JT and Abumrad NN. Reversal of cancer-related wasting using oral supplementation with a combination of β-hydroxy-β-methylbutyrate, arginine and glutamine. *Am J Surg* 2002; 183:471–479.
15. Clark RH, Feleke G, Din M, Yasmin T, Singh G, Khan F and Rathmacher J. Nutritional treatment for acquired immunodeficiency virus-associated wasting using β-hydroxy-β-methylbutyrate, glutamine and arginine: A randomized, double-blind, placebo-controlled study. *J Parenter Enter Nutr* 2000; 24(3):133–139.
16. Levenhagen DK, Vaughan SR, Niedernhofer E, Carr C and Flakoll PJ. Dietary supplementation of arginine, lysine and β-hydroxy-β-methylbutyrate (HMB) to blunt loss of muscle, strength and functionality in elderly females. *FASEB J* 15, A277. 2001 (Abstract).

17. Sabourin PJ and Bieber LL. Formation of β-hydroxyisovalerate from α-ketoisocaproate by a soluble preparation from rat liver. *Dev Biochem* 1981; 18:149–154.
18. Wagenmakers AJM, Salden HJM and Veerkamp JH. The metabolic fate of branched chain amino acids and 2-oxo acids in rat muscle homogenates and diaphragms. *Int J Biochem* 1985; 17:957–965.
19. Krebs HA and Lund P. Aspects of the regulation of the metabolism of branched-chain amino acids. *Advan Enzyme Regul* 1977; 15:375–394.
20. Sabourin PJ and Bieber LL. Subcellular distribution and partial characterization of an α-ketoisocaproate oxidase of rat liver: formation of β-hydroxyisovaleric acid. *Arch Biochem Biophys* 1981; 206:132–144.
21. Sabourin PJ and Bieber LL. Formation of β-hydroxyisovalerate from α-ketoisocaproate by a soluble preparation from rat liver. In: Walser M, Williamson JR, Eds. *Metabolism and Clinical Implications of Branched Chain Amino and Ketoacids*. New York, Elsevier North Holland Inc., 1981: 149–154.
22. Sabourin PJ and Bieber LL. Purification and characterization of an alpha-ketoisocaproate oxygenase of rat liver. *J Biol Chem* 1982; 257:7460–7467.
23. Sabourin PJ and Bieber LL. The mechanism of α-ketoisocaproate oxygenase. Formation of β-hydroxyisovalerate from α-ketoisocaproate. *J Biol Chem* 1982; 257:7468–7471.
24. Sabourin PJ and Bieber LL. Formation of β-hydroxyisovalerate by an α-ketoisocaproate oxygenase in human liver. *Metabolism* 1983; 32:160–164.
25. Lee MH, Zhang ZH, MacKinnon CH, Baldwin JE and Crouch NP. The C-terminal of rat 4-hydroxyphenylpyruvate dioxygenase is indispensable for enzyme activity. *FEBS Lett* 1996; 393(2-3):269–272.
26. Mock DM, Henrich CL, Carnell N and Mock NI. Indicators of marginal biotin deficiency and repletion in humans: validation of 3-hydroxyisovaleric acid excretion and a leucine challenge. *Am J Clin Nutr* 2002; 76(5):1061–1068.
27. Mock DM, Mock NI and Weintraub S. Abnormal organic aciduria in biotin deficiency: the rat is similiar to the human. *J Lab Clin Med* 1988; August:240–247.
28. Van Koevering M and Nissen S. Oxidation of leucine and α-ketoisocaproate to β-hydroxy-β-methylbutyrate *in vivo*. *Am J Physiol (Endocrinol Metab)* 1992; 262:E27–E31.
29. Vukovich MD, Slater G, Macchi MB, turner MJ, Fallon K, Boston T and Rathmacher J. β-Hydroxy-β-methylbutyrate (HMB) kinetics and the influence of glucose ingestion in humans. *J Nutr Biochem* 2001; 12:631–639.
30. Nissen SL and Abumrad NN. Nutritional role of the leucine metabolite β-hydroxy-β-methylbutyrate (HMB). *J Nutr Biochem* 1997; 8:300–311.
31. Bloch K, Clark LC and Haray I. Utilization of branched chain acids in cholesterol synthesis. *J Biol Chem* 1954; 211:687–699.
32. Adamson LF and Greenberg DM. The significance of certain carboxylic acids as intermediates in the biosynthesis of cholesterol. *Biochim Biophys Acta* 1957; 23:472–479.
33. Gey KF, Pletsher A, Isler O, Ruegg R and Wursch J. Influence of iosoprenoid C5 and C6 compounds on the incorporation of acetate in cholesterol. *Helv Chim Acta* 1957; 40:2354–2368.
34. Zhang Z, Rathmacher J, Coates C and Nissen S. Occurrence of β-hydroxy-β-methyl butyrate in foods and feeds. *FASEB J*. 8, A464. 1994 (Abstract).
35. Zhang Z. Distribution of β-hydroxy-β-methylbutyrate in plant and animal tissues. Ames: Iowa State University, M.S. Thesis, 1994.

36. Zhang Z, Talleyrand V, Rathmacher J and Nissen S. Change in plasma β-hydroxy-β-methylbutyrate (HMB) by feeding leucine, alpha-ketoisocaporate (KIC) and isovaleric acid (IVA) to pigs. *FASEB J.* 7, A392. 1993 (Abstract).
37. Zachwieja JJ, Smith SR, Nissen SL and Rathmacher JA. Beta-hydroxy-beta-methylbutyrate (HMB) is produced *in vivo* in humans from leucine. *FASEB J* 14, A747. 2000 (Abstract).
38. Gakhokidze AM. Condensation of ketones with esters of organic acids. I. Condensation of acetone with esters of formic, acetic and propionic acids. *J Gen Chem* (Russian) 1947; 17:1327–1331.
39. Gresham TL, Jansen JE, Shaver FW and Beears WL. Beta-propiolactone. XIV. Beta-isovalerolactone. *J Am Chem Soc* 1954; 76:486–488.
40. Coffman DD, Cramer R and Mochel WE. Synthesis by free-radical reactions. V. A new synthesis of carboxylic acids. *J Am Chem Soc* 1958; 80:2882–2887.
41. Searles S, Ives EK and Nukina S. Base-catalyzed cleavage of 1,3-diols. *J Organ Chem* 1959; 24:1770–1775.
42. Watanabe S, Suga K, Fujita T and Fujiyoshi K. The direct synthesis of beta-hydroxy-acids by lithium naphthalene and acetic acid. *Israel J Chem* 1970; 8:731–736.
43. Chen HW. Role of cholesterol metabolism in cell growth. *Fed Proc* 1984; 43:126–130.
44. Dabrowski MP, Peel WE and Thomson AE. Plasma membrane cholesterol regulates human lymphocyte cytotoxic function. *Eur J Immunol* 1980; 10:821–827.
45. Ostaszewski P, Kostiuk S, Balasinska B, Jank M, Papet I and Glomot F. The leucine metabolite 3-hydroxy-3-methylbutyrate (HMB) modifies protein turnover in muscles of the laboratory rats and domestic chicken *in vitro*. *J Anim Physiol Anim Nutr* (Swiss) 2000; 84:1–8.
46. Jank M, Ostaszewski P, Rosochacki S, Wilczak J and Balasinska B. Effect of 3-hydroxy-3-methylbutyrate (HMB) on muscle cathepsins and calpain activities during the post-dexamethasone recovery period in young rats. *Polish J Vet Sci* 2001; 3(4):213–218.
47. Panton L, Rathmacher J, Fuller J, Gammon J, Cannon L, Stettler S and Nissen S. The effect of β-hydroxy-β-methylbutyrate and resistance training on strength and functional ability in elderly men and women. *Med Sci Sports Exerc* 30, S194. 1998 (Abstract).
48. Fujioka S, Matsuzawa Y, Tokunaga K and Tarui S. Contribution of intra-abdominal fat accumulation to the impairment of glucose and lipid metabolism in human obesity. *Metabolism* 1987; 36(1):54–59.
49. Ernst ND, Obarzanek E, Clark MB, Briefel RR, Brown CD and Donato K. Cardiovascular health risks related to overweight. *J Am Diet Assoc* 1997; 97(7 Suppl):S47–S51.
50. O'Connor DM and Crowe MJ. Effects of β-hydroxy-β-methylbutyrate and creatine monohydrate supplementation on the aerobic and anaerobic capacity of highly trained athletes. *J Sports Med Phys Fitness* 43, 64–68. 2003.
51. Byrd P, Mehta P, DeVita P, Dyck D and Hickner R. Changes in muscle soreness and strength following downhill running: effects of creatine, HMB and Betagen supplementation. *Med Sci Sports Exerc* 31(5), S263. 1999 (Abstract).
52. Williams J, Abumrad N and Barbul A. Effect of a specialized amino acid mixture on human collagen deposition. *Ann Surg* 236, 369–375. 2002 (Abstract).

53. Kirk SJ, Hurson M, Regan MC, Holt DR, Wasserkrug HL and Barbul A. Arginine stimulates wound healing and immune function in elderly human beings. *Surgery* 1993; 114(2):155–159.
54. Karinch AM, Pan M, Lin CM, Strange R and Souba WW. Glutamine metabolism in sepsis and infection. *J Nutr* 2001; 131(9 Suppl):2535S–2538S.
55. Field CJ, Johnson I and Pratt VC. Glutamine and arginine: immunonutrients for improved health. *Med Sci Sports Exerc* 2000; 32(7 Suppl):S377–S388.
56. Nissen S, Panton L, Sharp RL, Vukovich M, Trappe SW and Fuller JC, Jr. β-Hydroxy-β-methylbutyrate (HMB) supplementation in humans is safe and may decrease cardiovascular risk factors. *J Nutr* 2000; 130:1937–1945.
57. Russell JA. A circumplex model of affect. *J Pers Soc Psychol* 1980; 39:1161–1178.
58. Gallagher PM, Carrithers JA, Godard MP, Schutze KE and Trappe SW. β-hydroxy-β-methylbutyrate ingestion, Part II: effects on hematology, hepatic and renal function. *Med Sci Sports Exerc* 2000; 32(12):2116–2119.

11
Lysine

Neal Spruce and C. Alan Titchenal

CONTENTS

I. Introduction 172
   A. Overview of Amino Acid Supplementation in Sports
      and Fitness 172
   B. Lysine's Proposed Sports and Fitness Applications 172
II. Chemical Structure and Metabolism of Lysine 173
   A. Dietary and Supplemental Sources 175
   B. Clinical Applications 176
   C. Nutrient Assessment 177
   D. Toxicity and Precautions 178
   E. Interactions with Other Nutrients and Drugs 178
III. Lysine Use for Sports and Fitness 178
   A. Sports Origin 178
   B. Rationale for Including Lysine in Amino Acid Mixtures
      To Raise Growth Hormone 179
      1. Proposed Mechanism of Action of Lysine for Growth
         Hormone Release 179
      2. Arginine and Lysine in Combination 180
      3. Summary 180
   D. Marketing Claims 181
   E. Current Arginine and Lysine Use in Sports and Fitness ... 182
   F. Lysine Sports- and Fitness-Related Research 182
      1. Other Factors Affecting GH Release 182
      2. Conclusions 187
   G. Future Research Related To Lysine's Use in Sports and
      Fitness 188
IV. Summary and conclusions 188
References 189

I. Introduction

A. Overview of Amino Acid Supplementation in Sports and Fitness

Amino acid and protein intake are generally considered to be synonymous. However, several amino acids with unique characteristics have been proposed as useful dietary supplements with beneficial effects in enhancing athletic performance and exercise-induced muscle hypertrophy, or in treating disease and reducing the effects of the natural aging process.

In the area of sports and fitness, there are three important and reportedly beneficial applications for amino acid supplementation:

1. Accelerating the recovery process following intense exercise with the intent of improving performance during subsequent exercise bouts
2. Enhancing skeletal muscle hypertrophy
3. Boosting endogenous human growth hormone (GH) levels to accomplish the same goals

There is some reason to credit the use of branched-chain amino acid (BCAA) supplementation as an anticatabolic agent. Since supplementation with BCAA may reduce exercise-induced muscle damage, athletes and dietary supplement manufacturers hypothesize that reducing protein catabolism causes the body to spend less time, energy and synthetic substrate on muscle repair, which in turn leads to greater net protein accretion when compared with a nonsupplemented state.[1-4]

However, as subsequent discussion herein will demonstrate, there is little evidence that ingesting designed mixtures of amino acids, specifically lysine, ornithine, arginine or glutamine,[5,6] can raise GH levels and there is no evidence that the putative GH increase could enhance exercise-induced effects on muscle.

B. Lysine's Proposed Sports and Fitness Applications

This chapter discusses the amino acid lysine and its use related to sports and fitness. Although oral and intravenous lysine has been used alone clinically to treat recurrent herpes simplex[7-12] and various metabolic disorders,[13] sports- or fitness-related research on the use of lysine as a solitary supplement is lacking. All relevant studies on the use of lysine to enhance physical training outcomes describe lysine as an ingredient in a mixture containing at least one other amino acid.[5,14-18]

The primary purposes for supplementation with lysine and its mixtures in sport and fitness training are to stimulate GH release and to favorably alter anabolic actions that govern skeletal muscle hypertrophy and regulation of energy substrate utilization. The desired, expected and assumed outcomes are typically greater training-induced muscle hypertrophy, improved performance and reduced body fat.

Injection of recombinant GH (rhGH) to raise GH levels in GH-deficient adults has well known positive effects on lipid metabolism and changes in body composition.[19-22] Additionally, administration of supraphysiological doses to obese women and healthy elderly men has shown a similar outcome, albeit somewhat less dramatic.[23,24]

Recombinant GH injection in healthy strength-training subjects produces no increase in strength compared with strength training alone.[25-29] A review of GH effects by Yarasheski suggests that any increase in fat free mass (FFM) from GH administration in healthy exercisers is primarily due to an increase in tissues other than skeletal muscle and possibly fluids.[29] GH can influence metabolism by causing a shift in substrate oxidation. That shift has in turn been shown to increase basal metabolic rate (BMR) and total energy expenditure (TEE) (possibly by stimulating uncoupling proteins causing futile cycling) in GH-deficient and healthy adults. That effect offers the rationale for elevating GH to decrease body fat.[30-32]

In summary, administering rhGH alone to healthy exercising subjects has offered no ergogenic benefit or increase in skeletal muscle hypertrophy over and above that attributable to training alone, but it may affect energy substrate utilization and TEE in ways that could positively influence body composition.[29,33] Table 11.1, adapted from Zachwieja and Yaresheski's review of studies using rhGH in older exercisers and nonexercisers, summarizes GH effects on body composition.[34] This summary shows that even if oral lysine could stimulate similar increases in GH levels, the only outcome to expect is a shift in substrate utilization and a mild increase in 24-hour energy expenditure.

II. Chemical Structure and Metabolism of Lysine

L-lysine is an indispensable dibasic amino acid (L-2,6-diaminohexanoic acid) required for human growth and for maintaining nitrogen balance in adults. Lysine cannot be synthesized by the body, so it must be extracted from nutrients in the diet.[35] Lysine and threonine are the only essential amino acids (AA) whose amino groups do not contribute to the total body amino pool, because they do not participate in transamination reactions.[36] Metabolites of lysine catabolism enter the Krebs cycle only at the acetyl-CoA site, and they therefore function strictly as ketogenic amino acids,[36] which cannot serve as substrate for glucose synthesis.

Lysine, like most other amino acids, is a building block of body proteins. Among the indispensable amino acids, lysine, at 93.0 mmol/dl and 38 mmol/dl in tissues and serum respectively, represents the greatest concentration in the human body (see Table 11.2).[37]

TABLE 11.1
Studies Investigating the Effects of rhGH Administration on Body Composition and Muscle Force in Older Men and Women

Study	Initial Dosage	Dosage Reduction	Subjects	Study Length	Exercise	Change in LBM	Change in Force	Side Effects
Marcus et al. 1990 (N = 18)[118]	30–120 μ/kg/d	No	Men and women ≥ 60 y	7 d	No	NA	NA	None reported
Kaiser et al. 1991 (N = 5)[114]	100 μ/kg/d	No	Men ≥ 60 y	3 wk	No	NA	NA	None reported
Rudman et al. 1990 (N = 12)[24]	30 μ/kg/d	Yes	Men ≥ 61 y	6 mo	No	+8.8%	NA	Yes
Papadakis et al. 1996 (N = 26)[33]	30 μ/kg/d	Yes	Men ≥ 69 y	6 mo	No	+4.3%	No effect	Yes
Welle et al. 1996 (N = 5)[115]	30 μ/kg/d	No	Men ≥ 60 y	3 mo	No	+5.8%	+14% vs. placebo	Yes
Holloway et al. 1994 (N = 19)[120]	43 μ/kg/d	No	Women ≥ 60 y	6 mo	No	No change	NA	Yes
Thompson et al. 1995 (N = 5)[119]	25 μ/kg/d	No	Women ≥ 65 y	4 wk	No	+3.1%	NA	Yes
Yarasheski and Zachwieja, 1996 (N = 8)[117]	24 μ/kg/d	Yes	Men ≥ 64 y	4 mo	Yes	+8.4%	–3% vs. placebo	Yes
Taafe et al. 1994 (N = 10)[116]	20 μ/kg/d	No	Men ≥ 65 y	10 wk	Yes	+2.3%	No effect	Yes

Source: Zachwieja, J.J. and Yarasheski, K.E. 1999. Does growth hormone therapy in conjunction with resistance exercise increase muscle force production and muscle mass in men and women aged 60 years or older?, *Phys. Ther.* 19, 76–82. American Physical Therapy Association. With permission.

TABLE 11.2

Amino Acid Concentrations in Serum and Muscle (μmol and dl)

Amino Acid	Serum	Muscle
Dispensable		
Alanine	32.6	110
Glutamate	12.0	826
Glutamine	65.2	1038
Glycine	33.6	62
Indispensable		
Lysine	38.0	93.0
Threonine	20.3	33.0
Phenylalanine	6.9	5.9
Leucine	16.6	11.2
Valine	18.8	14.7
Isoleucine	10.8	6.8

Source: Adapted from Morgan, H.E., Earl, D.C., Broadus, A., Wolpert, E.B. Giger, K.E. and Jefferson, L.S. 1971. Regulation of protein synthesis in heart muscle. I. Effect of amino acid levels on protein synthesis. *J. Biol. Chem.* 246(7):2152.

Carnitine, a compound responsible for the transport of long-chain fatty acids into the mitochondria for oxidation, is synthesized in the liver and kidneys from lysine and methionine.[38] Also, lysine is required for collagen synthesis and may be central to bone health.[39,40] Lysine's effects on bone growth also may be related to its ability to enhance calcium absorption and renal conservation, as well as to its participation in the cross-linking process of bone collagen formation.[41–43] In addition, lysine competes with the amino acid arginine for tissue uptake, providing some of the basis for its clinical applications.[44]

A. Dietary and Supplemental Sources

Early in the 20th century, it was found that the protein sources that best support the growth of animals[45] and maintenance of adult humans[46] are high in lysine content. Recently, the Food and Nutrition Board of the Institute of Medicine (FNB, IOM) recommended that diets of infants, children (1 to 3 years of age) and adults (18+ years) contain 69, 51 and 47 mg lysine per gram of protein respectively.[47] For comparison, examples of the lysine content of common food proteins are presented in Table 11.3.[48]

In general, diets that include a variety of protein sources in amounts that provide the RDA for total protein also provide adequate amounts of the indispensable amino acids, including lysine. However, diets that do not include animal protein sources can contain inadequate amounts of lysine unless legumes are a major component of the diet.[47] Isolated L-lysine was

TABLE 11.3

The Lysine Content of Proteins from Various Foods

Food	Lysine (mg/g protein)
Plant foods	
Almonds	29*
Brown rice	38*
Cornmeal	28*
Garbanzo beans	67
Peanut butter	36*
Tofu	66
Wheat bread	28*
Whole wheat flour	30*
Wheat germ	70
Wheat gluten	15*
Animal Foods	
Beef	83
Cheddar cheese	76
Chicken	82
Eggs	70
Gelatin	40*
Milk	79
Navy beans	72
Salmon	92
Tuna, mackerel, sword-fish	86

* Indicates food source of protein lower than FNB recommendation for proper balance of lysine to total protein in total diet of adults.

Source: Data from References 47 and 48 and Food Processor version 7.71, ESHA Research, Salem OR, 2000.

first commercially produced as the L-lysine monohydrochloride salt (LMH) in 1955, and pharmaceutical-grade LMH has been available in dietary supplement form for about 30 years.[49]

B. Clinical Applications

The major clinical application of supplemental L-lysine has been in the treatment and prevention of recurrent herpes simplex infections,[50,51] and it is generally accepted that LMH supplementation is a rational and potentially beneficial treatment for such illnesses,[7,8] although not all studies have reported beneficial treatment or prophylaxis of herpes simplex infections by LMH supplementation. As discussed by Flodin,[49] LMH supplementation may affect herpes simplex by a variety of mechanisms, including the possibility that it simply improves protein nutrition (and thus immune response) in patients with marginal intake of dietary lysine.

TABLE 11.4

Amounts and Energy Contents of Some Common Foods That Provide the Amount of L-lysine Equivalent to the Adult RDA for a 70 kg Person (2.7 g/d)

Food	Amount of food		Energy (kcal)
Beef, round steak, raw	4.8 oz	136 g	190
Chicken breast, skinless, boneless, raw	4.8 oz	136 g	150
Salmon, Atlantic, raw	5.2 oz	147 g	210
Milk, skim	3.6 cups	852 ml	360
Yogurt, low fat, plain	2.6 cups	615 ml	400
Tofu, firm, raw	2 cups	504 g	390
Garbanzo beans	2.8 cups	459 g	750
Almonds, raw	15 oz	425 g	2550
Bread, whole wheat	34 1-oz slices	34 28 g slices	2330

Lysine intake can be deficient in diets low in total energy or protein content. Consequently, it is possible that some athletes consume diets deficient in lysine due to dietary energy restriction for weight control or due to dietary practices limiting high lysine protein sources such as animal proteins and legumes. Deficient lysine intake can compromise protein synthesis and normal adaptation to the stresses of strength and endurance training. Limitations in this adaptive response can compromise any protein-dependent body functions such as muscle-tissue repair, antibody synthesis and maintenance of bone.

The FNB recently established Recommended Dietary Allowance (RDA) values for the indispensable amino acids.[47] The adult RDA for lysine is 38 mg per kg body weight per day. Thus, for a 70 kg adult, the RDA is 2.7 g/day. A listing of the amounts of foods required to provide the RDA for lysine for a 70 kg adult is presented in Table 11.4.

Athletes involved in heavy training may have lysine needs that exceed RDA values. To what extent those needs may exceed the RDA (if they do) is unknown. To put these needs into perspective, mean lysine intake of U.S. adults typically ranges from four to seven g/day,[52] and most athletes are likely to consume even greater amounts.

Although all protein sources contain some lysine, strict vegetarian diets that do not include copious amounts of legumes could impair adequate lysine nutrition. Although lysine supplementation could offset the effects of this problem, it is generally more practical to include good food sources of lysine in the diet.

C. Nutrient Assessment

Assessment of an individual's lysine status is typically accomplished via nutrient analysis of dietary intake. Indications of a potential need for assessment of lysine status include lean-tissue wasting, poor recovery from training bouts and a strict (vegan) vegetarian diet. Diets that include a variety of animal protein sources and meet the energy needs of an athlete are very unlikely to be limited in lysine content.

D. Toxicity and Precautions

There is no evidence of L-lysine toxicity from foods, even with high levels of protein intake. Extremely high levels of supplemental lysine might produce amino acid imbalance effects by interfering with normal utilization of dietary arginine. High lysine intake can compete with L-arginine for absorption in the intestine[44] and for reabsorption at renal tubules.[53] However, clinical trials involving people with herpes infections have utilized doses up to 3 g/day for 3 to 6 months without adverse effects.[47] Lysine is utilized for carnitine synthesis and one study of six healthy adult males found that a single 5 g oral dose increased plasma and urinary levels of carnitine.[54] Due to inadequate data on the adverse effects of L-lysine supplementation, the FNB has been unable to establish a tolerable upper intake level (UL) for L-lysine.[47]

E. Interactions with Other Nutrients and Drugs

Supplemental L-lysine has been reported to increase the fractional absorption of calcium.[41] However, this effect may be apparent only if the diet provides inadequate lysine.[55] High levels of lysine intake can increase the toxicity of amino glycoside antibiotic drugs such as streptomycin and gentamycin. Use of lysine supplementation in the presence of such antibiotics should be avoided without medical supervision.[56]

III. Lysine Use for Sports and Fitness

A. Sports Origin

The most common rationale for lysine supplementation in sport or fitness training is to raise GH levels, which, according to athletic culture theory, in turn improves exercise-induced results. The use of amino acids for stimulating GH release dates back to the 1960s, when infusion of arginine was introduced as a potential diagnostic test for GH secretion.[57] Intravenous infusion of 183 mg of arginine per kg of body weight increased plasma GH 20 fold in females.[58] Infusion of other amino acids including lysine, either singularly or in combination, was also demonstrated to promote between eight- and 22-fold increases in circulating GH levels.[59]

The first noted sport-related use of specific lysine combinations appears to be following research by Isidori that demonstrated a 10-fold increase in plasma levels of GH at 90 minutes following oral ingestion of 1200 mg of lysine in combination with an equal amount of arginine in non-exercising subjects.[5] The prospect of athletes' being able to enhance GH production with oral ingestion of relatively small doses of amino acids was probably the spark that ignited the marketing and use of these supplements as supposed secretagogues for GH release.

B. Rationale for Including Lysine in Amino Acid Mixtures To Raise Growth Hormone

A comprehensive literature search yielded no obvious rationale for the inclusion of lysine in amino acid mixtures to enhance GH release. The literature related to all other amino acid combinations used for increasing GH contained proposed mechanisms of action for each amino acid.[60–64] Even when lysine was part of the mixture, the various authors discussed the rationale for the inclusion of all amino acids except lysine.

There is documentation showing that high protein intakes and or high intakes of amino acids during fasting can elevate GH when compared with the same conditions but with low-protein diets.[65] However, even the Italian research that arguably directed supplement producers and athletes down the lysine path gives the rationale for arginine that is still recognized today — that arginine is believed to stimulate GH release by inhibiting the secretion of somatostatin, an hGH inhibitor[60] — but never discusses lysine's potential mechanism.[5] In the 1993 study by Corpas et al. the authors stated, "Arginine stimulates GH release and lysine may amplify this response." Others have expressed this lack of a clear explanation for lysine's role in stimulating GH.[14]

1. Proposed Mechanism of Action of Lysine for Growth Hormone Release

In our attempt to identify a potential mechanism of action we looked at the major pathways of lysine metabolism. One pathway converts the amino acid to alpha-aminoadipate via pipecolic acid.[66] This is considered an overflow pathway of lysine metabolism and is especially active in the brain. This could possibly link high lysine intake to GH release, since pipecolate is thought to act as an agonist for the gamma-aminobutyric acid (GABA) receptor.[66–68] GABA in turn has been shown to influence the release of GH from the pituitary.[69–72] Therefore, if pipecolic acid levels are increased from lysine metabolism, enhanced GABA receptor activity could potentially increase GH secretion.

Large doses of lysine can enhance calcium absorption and collagen production, leading to the stimulation of osteoblast proliferation and synthetic activity.[41–43] This mechanism has been used in the clinical application of lysine supplementation to treat various human bone pathologies and has demonstrated success in enhancing bone growth factors. Perhaps these effects of lysine on bone have been misconstrued as being related to GH effects on bone, contributing to the inclusion of lysine in AA preparations to be used in increasing GH production.

2. Arginine and Lysine in Combination

One item of rationale for the use of supplemental arginine is the fact that arginine is a substrate for nitric oxide (NO) synthesis.[73] NO is a vasodilator that may enhance blood flow and thereby improve endurance capacity. Patients with vascular and arterial ailments have improved exercise capacity following intravenous infusion with arginine.[74,75] NO is a known GH inhibitor, however.[76] NO reduces the responsiveness of the anterior pituitary to growth hormone releasing hormone (GHRH), but has an insignificant effect on basal levels of GH.[77] The process is catalyzed by NO synthase (NOS). Arginine stimulates NOS,[73] specifically inducible NOS (iNOS).[78] In addition, it has recently been established that a certain lysine compound (N-iminoethyl-L-lysine) may have a specific affinity for the iNOS isoform and can be used to inhibit the production of NO.[77,79] Therefore, if lysine has this effect by itself (which remains questionable), combined supplementation of lysine with arginine could help to increase GH through reduced NO inhibition of GHRH and allow arginine's potential enhancement of GH release by inhibition of somatostatin.[60]

This hypothesis establishes a rationale for the combined use of arginine and lysine for raising GH levels that is supported by the fact that arginine and lysine administered alone demonstrate far less benefit than when applied in combination.[5] The theory also argues against their combined use as an ergogenic aid associated with NO induction of vasodilation, if the assumption is correct that lysine can decrease arginine-induced NO production.

In summary, high doses of arginine can raise human GH levels, purportedly through inhibition of somatostatin with a concurrent rise in NO levels. NO production inhibits the GHRH, limiting supplemental arginine's potential to increase GH release. Lysine may decrease NO production by inhibiting iNOS, thus allowing the full potential of arginine supplementation to enhance GH release, but negate arginine's potential ergogenic benefit according to the "increase in NO" theory. These processes are illustrated in Figure 11.1, in which solid and broken lines represent direct and indirect influences and large and small upward- and downward-pointing arrows represent strong or weak increases or decreases.

3. Summary

a. Lysine

Lysine may exert its putative GH-releasing effects on the anterior pituitary by the production of one of its metabolites, pipecolic acid, which acts as an agonist for the GABA receptors and enhances the GABA influence on GH release.[69-72]

Additionally, it may have been (possibly falsely) assumed that lysine's ability to enhance bone growth is due to a mechanism related to increased GH release. Both of these factors may have led to lysine's inclusion in amino acid preparations promoted for increasing GH.[41-43]

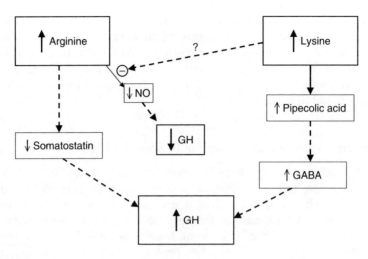

FIGURE 11.1
Possible mechanisms involved in the effect of arginine and lysine on growth hormone release from the anterior pituitary. ⊖ indicates potential inhibition of reaction. Solid and broken lines represent direct and indirect influences respectively. Large and small upward- and downward-pointing arrows represent strong or weak increases or decreases.

b. *Lysine and Arginine*

Arginine's proposed mechanism of action is its inhibitory effect on the secretion of somatostatin, which is a GH inhibitor.[59] Lysine's proposed mechanism of action in combination with arginine has not been previously reported, but we have theorized that lysine may amplify arginine's effects by attenuating NO production (which arginine supplementation enhances), which would otherwise inhibit the GHRH effects on the pituitary release of GH.[77,79]

Finally, if lysine and arginine together truly raise GH levels in combination more than separately, it may simply be the additive effect of stimulating the anterior pituitary (AP) by two different mechanisms: (1) Inhibition of somatostatin by arginine[60] and (2) lysine stimulation of the GABA receptors.[69-72]

We can now focus on lysine as a contributor in increasing GH and no longer view it as a direct ergogenic aid in normal healthy adults. Now that we have proposed a mechanism of action for lysine and discovered no data on the sole use of lysine for enhancing physical training, we will review research that includes lysine as an ingredient in amino acid preparations used to enhance GH release.

D. Marketing Claims

There are marketing claims that specific amino acid combinations containing lysine may increase GH and protein synthesis, leading to increased muscle mass and decreased body fat.[80,81] However, a recent visit by the authors to both GNC and Great Earth health food stores (the country's largest specialty supplement retailers) found no literature or claims regarding lysine or lysine and arginine in combination for the stimulation of GH or improvement of physical performance. This suggests there may be a decreased market demand for the use of lysine for this purpose. Alternatively, these establishments may be avoiding making such claims for these products, since the Dietary Supplement Health and Education Act of 1994 defines such claims as of questionable legality.

E. Current Arginine and Lysine Use in Sport and Fitness

We visited Gold's Gym (perhaps the most famous body-building gym in the United States) in Venice, California and discussed current dietary supplement use for enhancing GH with athletes and bodybuilders. We found that no one is using AA blends for the specific purpose of raising GH, while many stated that they had a non-oral, more guaranteed method of raising levels. Most reported the use of whey protein blends (high concentration of essential amino acids in a liquid delivery) and BCAA supplementation for the goal of reducing exercise-induced catabolism and increasing protein synthesis following exercise bouts. Dietary supplement sales figures from the 24 Hour Fitness chain were made available to us and validated anecdotal evidence gathered at Gold's Gym.[82]

F. Lysine Sports- and Fitness-Related Research

Four published trials associated with exercise and fitness that investigated oral ingestion of AAs including lysine are summarized in Table 11.5.[14–18] For comparison, the table also includes results of the studies by Isidori et al. and Corpas et al. since they are so frequently cited in literature regarding GH and lysine.[5]

1. Other Factors Affecting GH Release

a. Age

Natural GH production declines with age[83–90] and paradoxically, GH secretagogues appear to be more effective on normal subjects in younger age brackets.[5,14,16] Tanaka et al., using AA infusion to stimulate GH release, demonstrated an inverse relationship with age and responsiveness to AA GH stimulus.[91] Corpas et al.[14] reported that healthy male subjects over 60 years of age absorbed arginine as well as young men, but three grams of each AA (arginine and lysine) given twice daily had no effect on GH or insulin-like growth factor-I (IGF-1) secretion.[14] This result was in stark contrast to the 1200 mg administration of the same AAs to 15- to 20-year-old males that elicited up to a 10-fold increase in GH in Isidori's work.[5] Suminski's 16 subjects were 20–25 years of age and, in the absence of exercise, oral ingestion of 1200 mgs each of arginine and lysine caused a 2.4-fold increase. Collectively, these three studies appear to support the theory of age-related decline in GH response to oral administration of arginine and lysine (Arg and Lys).[16]

b. Exercise

Exercise and exercise intensity are strongly related to GH release.[92–99] Research incorporating high-intensity cycling and resistance training demonstrates up to a 10-fold increase in GH concentration when blood is sampled immediately following exercise.[94,100]

The young bodybuilders used in the Lambert's study exercised 5–10 hrs per wk, maintained their normal diet (1.2–2.2 g/kg body weight of protein day) and ingested the Arg and Lys supplement after an 8-hour overnight fast (to take advantage of the anterior pituitary being naturally primed to secrete GH). Blood condition was measured in increments up to 180 minutes following ingestion. Results among subjects were not uniform enough to allow any conclusion, although the mean increase of serum GH in the Arg and Lys group vs. the placebo was approximately two-fold, suggesting the physiological state of individual recipients may affect the Arg and Lys-induced GH response (see Table 11.6). The small group size ($n = 7$) and dramatic individual variations in results allow no firm conclusion to be drawn.[15]

The direct effects of exercise on GH release appear to be significantly greater than the effect of Arg and Lys ingestion. Consequently, exercise may overwhelm and obviate any simultaneous effects of supplementation.

Sumiski et al., using 1500 mgs of both Arg and Lys, measured GH levels in subjects before exercise and in a separate group without exercise. In the exercise group, they measured GH levels before training and in 30- to 90-minute intervals after exercise. GH levels were measured in the nonexercise subjects at the same times.[16] In the two exercise groups (placebo and Arg and Lys), the GH levels were elevated equally (not altered by supplementation) and exercise performances were not improved. In the nonexercise groups (placebo and Arg and Lys), the GH level in the supplemented group was 2.4 times higher than in the placebo users. These results suggest that AA supplementation can raise GH in the absence of exercise, but there is no additive effect to the GH stimulation resulting from exercise, at least when protein intake is adequate. The smaller increase in GH in the nonexercise group as compared with Isidori's results may be partly attributed to the age differences, since Suminski's participants were 20–25 years old, whereas Isidori's 1982 study participants were 15–20 years old.[5,16] In addition, the dosages used by the exercisers may have been too small because of the obligatory tissue uptake of amino acids in response to exercise, which diverts a significant portion of the Arg and Lys to muscle and the splanchnic bed and renders the remaining supplemental AAs ineffective in stimulating the anterior pituitary.[101,102]

Lambert and Fogelholm also studied experienced weight lifters and found no significant increases in GH release using 1.2 gram each of Arg and Lys

TABLE 11.5
Summary of Four Published Trials Associated with Exercise and Fitness That Investigated Oral Ingestion of Amino Acids Including Lysine Compared with Results of Studies by Isidori et al.[5] and Corpas et al.[14]

Study and Reference	No. Treated/Control	Age	Sex	Agent	Dose	No. of Doses	Oral and I.V.	Duration	Time of Ingestion	GH Pretreatment	GH Post Treatment (highest average reported)	IGF-I base and peak	IGF Post treatment	Diet (kcal and protein)	Conclusion
Isidori, A. et al.[5]	15	15–20	M	Arginine + Lysine	1.2 g of each	1	Oral	Single dose	Not given	15.4 ± 5 μg/L	108 ± 7.4 μg/L @ 90 min[4]	0.9 to 3.0 U/ml		Not given	In combination, A/L may increase GH in nonexercising young men.
	8	15–20	M	Arginine	1.2 g	1	Oral	Single dose	Not given	7.3 ± 2.3 μg/L	13.5 ± 7.5 μg/L @ 30 min	ND			Less impact than combining A/L.
				Lysine	1.2 g	1	Oral	Single dose	Not given	4.8 ± 1.9 μg/L	15.8 ± 4.2 μg/L @ 120 min	ND			Less impact than combining A/L.
				Arginine	2.4 g	1	Oral	Single dose	Not given	16.4 ± 4.1 μg/L	6.2 ± 4.1 μg/L @ 120 min				Nitric oxide production from larger arginine dose may have attenuated its potential.
Suminski et al.[16]	16	22–23	M	Arginine + Lysine (C)	1.5 g of each	1	Oral	1 day	9:00 A.M.	2.78 μg/L	7.5 μg/L[3] @ 90 min	Not given	Not given	1.12 g/kg ± 0.06	A/L additive to GH output w/o exercise in young men with normal diets.
				Exercise + Placebo (A)	0	1	Oral	1 day	9:00 A.M.	~3.0 μg/L	~26 μg/L[3]	Not given	Not given	1.11 g/kg ± 0.06	A/L was not additive to exercise effect on GH output.
				Exercise + Arg and Lysine (B)	1.5 g each	1	Oral	1 day	9:00 A.M.	~2.0 μg/L	~22 μg/L[3]	Not given	Not given	1.14 g/kg ± 0.07	A/L was not additive to exercise effect on GH output.

Lysine

Study	n	Age	Sex	Supplement	Dose	Control	Route	Duration	Time	Pretreatment GH	Post-treatment GH			Diet	Comments
Fogelholm et al.[17]	11	19-35	M	6g Arginine, Lysine, Ornithine daily	0	No Ex and No AA (D)	Oral	1 day	9:00 A.M.	~2.5 μg/L	~2.5 μg/L	Not given	Not given		N/A
					6 g		Oral	6 days	1 P.M./9 P.M.	~5.5 μg/L	3.06 ± 1.2 μg/L (no change compared with placebo and pretreatment ($p < 0.55$))		1.11g/kg ± 0.06	3500 kcal and 166 g (2.2g/kg)	A/L had no impact on GH at any time, possibly due to high protein intakes and time of A/L ingestion in relation to meals.
Fry et al.[18]	28		M	Arginine, Ornithine, Lysine, BCAA	2.1 g arg, orn, lys + 2.1 g BCAA + 50 mg gln		Oral	7 days			No change compared with placebo				Exercise, high protein intakes and BCAA inclusion may have eliminated potential GH response to A/L.
Corpas et al.[14]	8 and 8	69+ -5	M	Arginine + Lysine	1.5 g each 2×/d for 3 d then 3 g of each 2×/d for 11 d	2	Oral	14 days	10 P.M./8 A.M.	2.7 ± 2.1 μg/L; 12 h mean; 1.0 ± 0.7 μg/L	2.7 ± 2.5 μg/L; 12 h mean; 1.2 ± 1.1 mcg/L	142 ± 58 mcg/L	147 ± 82 mcg/L	Not given	No change in GH levels may be related to age.
Lambert et al.[15]	7	22-23	M	Arginine + Lysine	2.4 g total	1	Oral	5 1-day visits (1 week between each)	6:00 A.M. after fasting	Placebo ~104 ± 57 ng min and ml[1,2]	283.7 ± 117.2 ng min and ml[2]		1.56 g/kg ± 0.13		Inconsistent results may be related to individual physiological differences that may impact value of A/L administration

[1] Placebo is the same as pretreatment since all subjects did a placebo trial in one of the five trials.
[2] Values in ng min and ml over 180 minutes.
[3] $P<0.05$ compared with pretreatment.
[4] Statistically significant increase from pretreatment value

TABLE 11.6

Integrated Concentrations of Serum Growth Hormone (ng×min×ml-1) for 180 minutes after Ingestion of the Treatment or Infusion of rhGH

Subject	Placebo	A	B	C	rhGH
1	13.5	249.0	54.0	33.0	—
2	64.5	220.5	52.5	667.5	3605.9
3	441.5	357.0	709.5	1725.0	—
4	52.5	937.5	673.5	1597.5	644.3
5	21.0	42.0	30.0	415.5	800.9
6	85.5	144.0	723.0	49.5	654.3
7	51.0	36.0	39.0	36.0	325.5
Mean	104.2	283.7	325.9	646.3	1206.2*
±SE	57.0	117.2	133.1	277.2	511.3

Note: A = arginine and lysine, B = ornithine and tyrosine, C = Bovril[R].
* $p<0.05$, GH-RH vs. placebo, A, B and C.
Source: Lambert, M.I., Hefer, J.A., Millar, R.P. and Macfarlane, P.W. 1993. Failure of commercial oral amino acid supplements to increase serum growth hormone concentrations in male body-builders. *Int. J. Sport. Nutr.* 3(3): 303. With permission.

once daily and 1 gram of arginine, 1 gram ornithine and 1 gram of lysine twice daily respectively. In contrast to Suminski et al., they were unable to demonstrate an acute basal increase in GH in the similar subjects, a result that might be attributed to protein intake differences.

c. Dietary Protein

Exercisers typically have adequate protein intake and the timing of ingestion can play a role in enhancing the anabolic environment following exercise.[103–106] The protein requirements for exercisers may be twice that of their sedentary counterparts.[107] This fact offers a possible explanation for results of the Lampert et al. and Fogelhom et al. studies contrasting with Suminski in relation to basal GH stimulation from Arg and Lys supplementation. The protein intake of Lampert and Fogelholm's subjects during the trials were between 1.5 and two times that of Suminski's participants, and diets high in protein compared with "normal" balanced diets are associated with higher basal GH.[65] These results suggest that the presence of high protein throughout the day may exhaust the GH-releasing potential that Arg and Lys supplementation might otherwise tap.

The only other study showing significant increase in GH levels from Arg and Lys oral supplementation was Isidori. That study did not report protein intake, but the subjects were nonexercisers and we may assume they consumed the typical Italian diet of the times in which protein intake was probably not relatively high.[5]

Finally, in the absence of exercise, high circulating levels of AA may mimic the effects of exercise that lead to an increase in GH production. Increasing the blood levels of amino acids, a condition that takes place during and immediately following exercise,[102,108] may initially give a false signal of catabolism, which then triggers anabolism analogous to cell volumizing with different amino acids.[109]

d. Time of Ingestion

Ingesting Arg and Lys before training should be of no benefit to the GH response, since GH release during exercise does not appear to be amplified by supplementation.[16–18] Pulsatile and continuous intravenous administration of GH have demonstrated similar effects on raising GH, with the latter having a greater effect on raising IGF-1 in GH-deficient subjects.[110,111] These results suggest that it may not be necessary to take advantage of the anterior pituitary (AP) when it is already "primed" for secretion. In all studies reviewed, with the exception of Fogelholm et al., the supplemental AAs were taken in fasting states (or at least post absorptive) presumably to take advantage of a "primed" AP.[112,113]

Fogelholm had subjects taking the Arg and Lys supplement 1.5 hours after the 11:30 A.M. meal and 3 hours following the 6:00 P.M. meal. Considering normal food digestion time, a large flow of AA would normally be entering systemic circulation from the meal at approximately the same time as the 1:00 P.M. supplemented Arg and Lys. The 6:00 P.M. feeding also may have dulled the effects of the 9:00 p.m. Arg and Lys dose and prevented any measurable response. Indeed, the Arg and Lys group did register a small evening peak 90 minutes later than the placebo group, which was 1.5 hours after the 9:00 p.m. ingestion. This latter effect suggests that the Arg and Lys supplement had a minor impact due to less circulating nutrients at the time of ingestion as opposed to the 1:00 p.m. dose. Thus, Arg and Lys supplementation does not appear to be additive to other stimuli affecting GH release.

e. Dosage

All dosages used in the trials were relatively small (see Table 11.5) when compared with infusions of Arg and Lys that show efficacy. However, dosages were based on suggestions by the manufacturers.

2. Conclusions

If amino acid supplementation or high protein intakes stimulate GH secretion, their mechanisms may be the same as exercise.[102,108,109] When supplementation is combined with other stimuli such as exercise, an additive capacity for GH release by the anterior pituitary does not occur, especially in older individuals. In other words, for someone older than 25 years, participating in moderate to intense exercise and consuming between 1.8 and 2.2 gm/kg of protein throughout the day with a portion consumed with carbohydrate shortly after exercise, additional GH response may not be possible by any means other than continuous injections of GH. Conversely, for someone under the age of 20, not exercising and consuming a low-protein diet, Arg and Lys supplementation may exert a potential for increasing GH levels.

All in all, supplementation with Arg and Lys to increase GH seems to be an exercise in futility under most normal conditions. Even GH injections have shown little to no effect on skeletal muscle hypertrophy or performance additive to the effect of training.[26,28,29]

G. Future Research Related To Lysine's Use in Sport and Fitness

Future research on lysine supplementation to enhance athletic performance or adaptation to training through its potential effect on GH levels seems to be without merit for the following reasons:

1. It appears that supplementation with lysine and arginine mixtures has no significant effect on training outcomes.
2. Even if lysine and arginine supplementation did raise GH beyond the level attributable to exercise, elevated GH shows no apparent added value.

Of more significant concern, however, the RDA values for lysine and other indispensable amino acids recently proposed by the Food and Nutrition Board indicate that lysine is the amino acid most likely to be deficient for athletes consuming diets limited in total protein, or those with limited food sources of lysine. Good food sources include animal proteins and legumes. Future research evaluating the adequacy of athletes' diets should consider lysine as a potentially limiting factor for some athletes.

IV. Summary and conclusions

Lysine has a number of important functions in the normal healthy human. It is an indispensable amino acid required for general protein synthesis and may be especially important for the cross-linking process of bone collagen formation and thus maintenance of bone health. Lysine also is used to synthesize carnitine and has been proposed to enhance immune function in the treatment of *herpes simplex* virus.

Although lysine supplementation in combination with arginine has been promoted for its potential to increase GH release, claims of GH's enhancing lean body mass and strength performance have not been supported by studies using injected GH administration. Consequently, even if the proposed mechanisms by which lysine may contribute to enhanced GH release prove to be correct, ergogenic benefits appear to be nonexistent or too minor to measure.

Most athletes likely obtain adequate amounts of lysine when they consume normal balanced diets in amounts that meet energy needs. Conditions that may result in inadequate lysine intake include the consumption of low-calorie diets that do not provide adequate total protein and diets low in sources of animal protein or legumes. Consequently, deficient lysine intake may occur in athletes with alternative dietary practices that provide limited amounts of protein or poor quality protein sources. Chronic lysine deficiency could impair normal adaptive responses to strength and endurance training and compromise normal function of the immune system.

References

1. MacLean, D.A., Kiens, B., Rohde, T., Pederson, B.K., Saltin, B. and Richter, E.A. 1996. Branched chain amino acid supplementation reduces muscle amino acid release after eccentric exercise. *Med. Sci. Sports. Exerc.* 28(5):S181.
2. Blomstrand, E. and Newsholme, E.A. 1992. Effect of branched-chain amino acid supplementation on the exercise-induced change in aromatic amino acid concentration in human muscle. *Acta. Physiol. Scand.* 146(3):293–8.
3. Henderson, S.A., Black, A.L. and Brooks, G.A. 1985. Leucine turnover and oxidation in trained rats during exercise. *Am. J. Physiol.* 249(2 Pt 1):E137–44.
4. MacLean, D.A., Graham, T.E. and Saltin, B. 1994. Branched-chain amino acids augment ammonia metabolism while attenuating protein breakdown during exercise. *Am. J. Physiol.* 267(6 Pt 1):E1010–22.
5. Isidori, A., Lo Monaco, A. and Cappa, M. 1981. A study of growth hormone release in man after oral administration of amino acids. *Curr. Med. Res. Opin*, 7(7):475–81.
6. Bucci, L., Hickson, J.F. Jr., Pivarnik, J.M., Wolinsky, I., McMahon, J.C. and Turner, S.D. 1990. Ornithine ingestion and growth hormone release in bodybuilders. *Nutr. Res.* 10:239.
7. Thein, D.J. and Hurt, W.C. 1984. Lysine as a prophylactic agent in the treatment of recurrent herpes simplex labialis. *Oral Surg. Oral Med. Oral Pathol.* 58(6):659–66.
8. McCune, M.A., Perry, H.O., Muller, S.A. and O'Fallon, W.M. 1984. Treatment of recurrent herpes simplex infections with L-lysine monohydrochloride. *Cutis.* 34(4):366–73.
9. DiGiovanna, J.J. and Blank, H. 1984. Failure of lysine in frequently recurrent herpes simplex infection: Treatment and prophylaxis. *Arch. Dermatol.* 120(1):48–51.
10. Milman, N., Scheibel, J. and Jessen, O. 1980. Lysine prophylaxis in recurrent herpes simplex labialis: A double-blind, controlled crossover study. *Acta. Derm. Venereol.* 60(1):85–7.
11. Griffith, R.S., Walsh, D.E., Myrmel, K.H., Thompson, R.W. and Behforooz, A. 1987. Success of L-lysine therapy in frequently recurrent herpes simplex infection: Treatment and prophylaxis. *Dermatologica.* 175(4):183–90.
12. Griffith, R.S., Norins, A.L. and Kagan, C. 1978. A multicentered study of lysine therapy in Herpes simplex infection. *Dermatologica.* 156(5):257–67.

13. Bondoli, A., Abballe, C., de Cosmo, G., Sabato, A.F. and Magalini, S.I. 1980. Lysine hydrochloride for the control of metabolic alkalosis: A clinical report. *Resuscitation.* 8(4):223–31.
14. Corpas, E., Blackman, M.R., Roberson, R., Scholfield, D. and Harman, S.M. 1993. Oral arginine-lysine does not increase growth hormone or insulin-like growth factor-I in old men. *J. Gerontol.* 48(4):M128–33.
15. Lambert, M.I., Hefer, J.A., Millar, R.P.and Macfarlane, P.W. 1993. Failure of commercial oral amino acid supplements to increase serum growth hormone concentrations in male body-builders. *Int. J. Sport. Nutr.* 3(3):298–305.
16. Suminski, R.R., Robertson, R.J., Goss, F.L., Arslanian, S., Kang, J., DaSilva, S., Utter, A.C. and Metz, K.F. 1997. Acute effect of amino acid ingestion and resistance exercise on plasma growth hormone concentration in young men. *Int. J. Sport Nutr.* 7(1):48–60.
17. Fogelholm, G.M., Naveri, H.K., Kiilavuori, K.T. and Harkonen, M.H. 1993. Low-dose amino acid supplementation: No effects on serum human growth hormone and insulin in male weightlifters. *Int. J. Sport Nutr.* 3(3):290–7.
18. Fry, A.C., Kraemer, W.J., Stone, M.H., Warren, B.J., Kearney, J.T., Maresh, C.M, Weseman, C.A. and Fleck, S.J. 1993. Endocrine and performance responses to high volume training and amino acid supplementation in elite junior weight-lifters. *Int. J. Sport Nutr.* 3(3):306–22.
19. Jorgensen, J.O., Pedersen, S.A., Thuesen, L., Jorgensen, J., Ingemann-Hansen, T., Skakkebaek, N.E. and Christiansen, J.S. 1989. Beneficial effects of growth hormone treatment in GH-deficient adults. *Lancet.* 1(8649):1221–5.
20. Degerblad, M., Elgindy, N., Hall, K., Sjoberg, H.E. and Thoren, M. 1992. Potent effect of recombinant growth hormone on bone mineral density and body composition in adults with panhypopituitarism. *Acta. Endocrinol. (Copenh).* 126(5):387–93.
21. de Boer, H., Blok, G.J. and Van der Veen, E.A. 1995. Clinical aspects of growth hormone deficiency in adults. *Endocr. Rev.* 16(1):63–86.
22. Ahmad, A.M., Hopkins, M.T., Thomas, J., Ibrahim, H., Fraser, W.D. and Vora, J.P. 2001. Body composition and quality of life in adults with growth hormone deficiency: Effects of low-dose growth hormone replacement. *Clin. Endocrinol. (Oxf).* 54(6):709–17.
23. Richelsen, B., Pedersen, S.B., Borglum, J.D., Moller-Pedersen, T., Jorgensen, J. and Jorgensen, J.O. 1994. Growth hormone treatment of obese women for 5 weeks: Effect on body composition and adipose tissue LPL activity. *Am. J. Physiol.* 266(2 Pt 1):E211–6.
24. Rudman, D., Feller, A.G., Nagraj, H.S., Gergans, G.A., Lalitha, P.Y., Goldberg, A.F., Schlenker, R.A., Cohn, L., Rudman, I.W. and Mattson, D.E. 1990. Effects of human growth hormone in men over 60 years old. *N. Engl. J. Med.* 323(1):1–6.
25. Frisch, H. 1999. Growth hormone and body composition in athletes. *J. Endocrinol. Invest.* 22(5 Suppl):106–9. Review.
26. Taaffe, D.R., Jin, I.H., Vu, T.H., Hoffman, A.R., Marcus, R. 1996. Lack of effect of recombinant human growth hormone (GH) on muscle morphology and GH-insulin-like growth factor expression in resistance-trained elderly men. *J. Clin. Endocrinol. Metab.* 81(1):421–5.
27. Yarasheski, K.E., Zachwieja, J.J., Campbell, J.A. and Bier, D.M. 1995. Effect of growth hormone and resistance exercise on muscle growth and strength in older men. *Am. J. Physiol. Endocrinol. Metab.* 268:E268–76.

28. Yarasheski, K.E., Campbell, J.A., Smith, K., Rennie, M.J., Holloszy, J.O. and Bier, D.M. 1992. Effect of growth hormone and resistance exercise on muscle growth in young men. *Am. J. Physiol.* 262(3 Pt 1):E261–7.
29. Yarasheski, KE. 1994. Growth hormone effects on metabolism, body composition, muscle mass and strength. *Exerc. Sport Sci. Rev.* 22:285–312.
30. Jorgensen, J.O., Moller, J., Laursen, T., Orskov, H., Christiansen, J.S. and Weeke, J. 1994 Nov. Growth hormone administration stimulates energy expenditure and extrathyroidal conversion of thyroxine to triiodothyronine in a dose-dependent manner and suppresses circadian thyrotrophin levels: studies in GH-deficient adults. *Clin. Endocrinol. (Oxf).* 41(5):609–14.
31. Wolthers, T., Grofte, T., Norrelund, H., Poulsen, P.L. Andreasen, F., Christiansen, J.S. and Jorgensen, J.O. 1998. Differential effects of growth hormone and prednisolone on energy metabolism and leptin levels in humans. *Metabolism.* 47(1):83–8.
32. Lange, K.H., Isaksson, F., Juul, A., Rasmussen, M., Bulow, J. and Kjaer, M. 2000 Nov. Growth hormone enhances effects of endurance training on oxidative muscle metabolism in elderly women. *Am. J. Physiol. Endocrinol. Metab.* 279(5):E989–96.
33. Papadakis, M.A., Grady, D., Black, D., Tierney, M.J., Gooding, G.A., Schambelan M. and Grunfeld, C. 1996. Growth hormone replacement in healthy older men improves body composition but not functional ability. *Ann. Intern. Med.* 124(8):708–16.
34. Zachwieja, J.J. and Yarasheski, K.E. 1999. Does growth hormone therapy in conjunction with resistance exercise increase muscle force production and muscle mass in men and women aged 60 years or older? *Phys. Ther.* 79(1):76–82.
35. Whitney, E.N. and Rolfes, S.R., Eds. *Understanding Nutrition.* 7th ed. Minneapolis and St. Paul: West Publishing; 1996. p. 197–226.
36. Berdanier, C.D., *Advanced Nutrition: Macronutrients,* 2nd ed., Boca Raton, FL: CRC Press; 2000. p. 177.
37. Morgan, H.E., Earl, D.C., Broadus, A., Wolpert, E.B. Giger, K.E. and Jefferson, L.S. 1971. Regulation of protein synthesis in heart muscle. I. Effect of amino acid levels on protein synthesis. *J. Biol. Chem.* 246(7):2152–62.
38. Rebouche, C.J. and Paulson, D.J. 1986. Carnitine metabolism and function in humans. *Annu. Rev. Nutr.* 6:41–66.
39. Flodin, N.W. 1997. The metabolic roles, pharmacology and toxicology of lysine. *J. Am. Coll. Nutr.* 16(1):7–21.
40. Hall, S.L. and Greendale, G.A. 1998. The relation of dietary vitamin C intake to bone mineral density: Results from the PEPI study. *Calcif. Tissue. Int.* 63(3):183–9.
41. Civitelli, R., Villareal, D.T., Agnusdei, D., Nardi, P., Avioli, L.V. and Gennari, C. 1992. Dietary L-lysine and calcium metabolism in humans. *Nutrition.* 8(6):400–5.
42. Oxlund, H., Barckman, M., Ortoft, G. and Andreassen, T.T. 1995. Reduced concentrations of collagen cross-links are associated with reduced strength of bone. *Bone.* 17(4 Suppl):365S–371S.
43. Torricelli, P., Fini, M., Giavaresi, G., Giardino, R., Gnudi, S., Nicolini A. and Carpi A. 2002. L-Arginine and L-Lysine stimulation on cultured human osteoblasts. *Biomed. Pharmacother.* 56(10):492–7.

44. McCarthy, C.F., Borland, J.L., Lynch, H.J., Owen, E.E. and Tyor, M.P. 1964. Defective uptake of basic amino acids and L-cystine by intestinal mucosa of patients with cystinuria. *J. Clin. Invest.* 43: 1518–24.
45. Osborne, T.B. and Mendel, L.B. 1914. Amino acids in nutrition and growth. *J. Biol. Chem.* 17: 325–49.
46. Hawley, E.E., Murlin, J.R., Nasset, E.S. and Szymanski, T.A. 1948. Biological values of six partially purified proteins. *J. Nutr.* 153–69.
47. Food & Nutrition Board, Institute of Medicine, National Academy of Sciences. Dietary Reference Intakes for Energy, Carbohydrate, Fiber, Fat, Fatty Acids, Cholesterol, Protein and Amino Acids. Washington, DC: National Academy Press, 2002.
48. Food and Agriculture Organization of the United Nations. Amino-Acid Content of Foods and Biological Data on Proteins. FAO Nutrition Studies No. 24, Rome: FAO, 1970. 285 p.
49. Flodin, N.W. 1997. The metabolic roles, pharmacology and toxicology of lysine. *J. Am. Coll. Nutr.* 16(1):7–21.
50. Tankersley, R.W. Jr. 1964. Amino acid requirements of herpes simplex virus in human cells. *J. Bacteriol.* 87: 609–13.
51. Kagan, C. 1974. Lysine therapy for herpes simplex (Letter). *Lancet.* 1(7848):137.
52. U.S. Department of Health and Human Services, National Center for Health Statistics. Third National Health and Nutrition Examination Survey (NHANES III, 1988–1994), 2001.
53. Kamin, H. and Handler, P. 1951. Effect of infusion of single amino acids upon excretion of other amino acids. *Am. J. Physiol.* 164: 654–61.
54. Vijayasarathy, C., Khan-Siddiqui, L., Murthy, S.N. and Bamji, M.S. 1987. Rise in plasma trimethyllysine levels in humans after oral lysine load. *Am. J. Clin. Nutr.* 46(5):772–7.
55. Spencer, H. and Samachson, J. 1963. Effect of lysine on calcium metabolism in man. *J. Nutr.* 81:301–6.
56. Skidmore-Roth, Y.L., *Mosby's Handbook of Herbs & Natural Supplements.* St. Louis: Mosby; 2001. 1032 p.
57. Merimee, T.J., Lillicrap, D.A. and Rabinowitz, D. 1965. Effect of arginine on serum-levels of human growth-hormone. *Lancet.* 2(7414):668–70.
58. Merimee, T.J., Rabinowtitz, D. and Fineberg, S.E. 1969. Arginine-initiated release of human growth hormone. Factors modifying the response in normal man. *N. Engl. J. Med.* 280(26):1434–8.
59. Knopf, R.F., Conn, J.W., Falans, S.S., Floyd, J.C., Guntsche, E.M. and Rull, J.A. 1965. Plasma growth hormone responses to intravenous administration of amino acids. *J. Clin. Endocrinol. Metab.* 25:1140.
60. Alba-Roth, J., Muller, O.A., Schopohl, J. and von Werder, K. 1988. Arginine stimulates growth hormone secretion by suppressing endogenous somatostatin secretion. *J. Clin. Endocrinol. Metab.* 67(6):1186–9.
61. Carlson, H.E., Miglietta, J.T., Roginsky, M.S. and Stegink, L.D. 1989. Stimulation of pituitary hormone secretion by neurotransmitter amino acids in humans. *Metabolism.* 38(12):1179–82.
62. Bucci, L.R., Hickson, J.F. Jr, Wolinsky, I. and Pivarnik, J.M. 1992. Ornithine supplementation and insulin release in bodybuilders. *Int. J. Sport Nutr.* 2(3):287–91.

63. Blachier, F., Leclercq-Meyer, V., Marchand, J., Woussen-Colle, M.C., Mathias, P.C., Sener, A. and Malaisse, W.J. 1989. Stimulus-secretion coupling of arginine-induced insulin release: Functional response of islets to L-arginine and L-ornithine. *Biochim. Biophys. Acta.* 1013(2):144–51.
64. Silk, D.B. and Payne-James, J.J. 1990. Novel substrates and nutritional support: possible role of ornithine alpha-ketoglutarate. *Proc. Nutr. Soc.* 49(3):381–7.
65. Sellini, M., Fierro, A., Marchesi, L., Manzo, G. and Giovannini, C. 1981. Behavior of basal values and circadian rhythm of ACTH, cortisol, PRL and GH in a high-protein diet *Boll. Soc. Ital. Biol. Sper.* 57(9):963–9.
66. Amino Acid Metabolism. The Peroxisome Website, http://www.peroxisome.org/Scientist/Biochemistry/aametabolismtext.html. Accessed 05/15/2003.
67. Charles, A.K. 1986. Pipecolic acid receptors in rat cerebral cortex. *Neurochem. Res.* 11(4):521–5.
68. Gutierrez, M.C. and Delgado-Coello, B.A. 1989. Influence of pipecolic acid on the release and uptake of [3H]GABA from brain slices of mouse cerebral cortex. *Neurochem. Res.* 14(5):405–8.
69. Monteleone, P., Maj, M., Iovino, M. and Steardo, L. 1988. Evidence for a sex difference in the basal growth hormone response to GABAergic stimulation in humans. *Acta. Endocrinol. (Copenh).* 119(3):353–7.
70. McCann, S.M., Vijayan, E., Negro-Vilar, A., Mizunuma, H. and Mangat, H. 1984. Gamma aminobutyric acid (GABA), a modulator of anterior pituitary hormone secretion by hypothalamic and pituitary action. *Psychoneuroendocrinology.* 9(2):97–106.
71. McCann, S.M. and Rettori, V. 1986. Gamma amino butyric acid (GABA) controls anterior pituitary hormone secretion. *Adv Biochem. Psychopharmacol.* 42:173–89.
72. Volpi, R., Chiodera, P., Caffarra, P., Scaglioni, A., Saccani, A. and Coiro, V. 1997. Different control mechanisms of growth hormone (GH) secretion between gamma-amino- and gamma-hydroxy-butyric acid: Neuroendocrine evidence in Parkinson's disease. *Psychoneuroendocrinology.* 22(7):531–8.
73. Koizumi, T., Gupta, R., Banerjee, M. and Newman, J.H. 1994. Changes in pulmonary vascular tone during exercise. Effects of nitric oxide (NO) synthase inhibition, L-arginine infusion and NO inhalation. *J. Clin. Invest.* 94(6):2275–82.
74. Ceremuzynski, L., Chamiec, T. and Herbaczynska-Cedro, K. 1997. Effect of supplemental oral L-arginine on exercise capacity in patients with stable angina pectoris. *Am. J. Cardiol.* 80(3):331–3.
75. Slawinski, M., Grodzinska, L., Kostka-Trabka, E., Bieron, K., Goszcz, A. and Gryglewski, R.J. 1996. L-arginine — substrate for no synthesis — its beneficial effects in therapy of patients with peripheral arterial disease: comparison with placebo-preliminary results. *Acta. Physiol. Hung.* 84(4):457–8.
76. Uemura, S., Rothbard, J.B., Matsushita, H., Tsao, P.S., Fathman, C.G. and Cooke, J.P. 2002. Short polymers of arginine rapidly translocate into vascular cells: effects on nitric oxide synthesis. *Circ. J.* 66(12):1155–60.
77. Vankelecom, H., Matthys, P. and Denef, C. 1997. Involvement of nitric oxide in the interferon-gamma-induced inhibition of growth hormone and prolactin secretion in anterior pituitary cell cultures. *Mol. Cell. Endocrinol.* 129(2):157–67.
78. Moore, W.M., Webber, R.K., Jerome, G.M., Tjoeng, F.S., Misko, T.P. and Currie, M.G. 1994. L-N6-(1-iminoethyl)lysine: A selective inhibitor of inducible nitric oxide synthase. *J. Med. Chem.* 37(23):3886–8.

79. Coiro, V., Volpi, R., Capretti, L., Speroni, G., Caffarri, G., Marchesi, C. and Chiodera, P. 1999. Enhancement of the GH responsiveness to GH releasing stimuli by lysine vasopressin in type 1 diabetic subjects. *Clin. Endocrinol. (Oxf).* 51(4):487–95.
80. Antonio, J., The growth factor diet: How to eat for maximal hormone release. *Muscle and Fitness* Website http://www.Muscleandfitness.com/massgain/p/1372.jsp Accessed 5/6/03.
81. Universal Nutrition animal methoxy stack. Spinelli's Nutrition Center Website: http://www.spinellinutrition.com/Universal_Nutrition_Animal_Methoxy_Stack.html. Accessed 5/22/03.
82. Annual supplement sales of 24 Hour Fitness. Personal communication to the Apex Fitness Group. January 2003.
83. Carlson, H.E., Gillin, J.C., Gorden, P. and Snyder, F. 1972. Absence of sleep-related growth hormone peaks in aged normal subjects and in acromegaly. *J. Clin. Endocrinol. Metab.* 34(6):1102–5.
84. Finkelstein, J.W., Roffwarg, H.P., Boyar, R.M., Kream, J. and Hellman, L. 1972. Age-related change in the twenty-four-hour spontaneous secretion of growth hormone. *J. Clin. Endocrinol. Metab.* 35(5):665–70.
85. Rudman, D., Kutner, M.H., Rogers, C.M., Lubin, M.F., Fleming and G.A., Bain. R.P. 1981. Impaired growth hormone secretion in the adult population: Relation to age and adiposity. *J. Clin. Invest.* 67(5):1361–9.
86. Prinz, P.N., Weitzman, E.D., Cunningham, G.R. and Karacan, I. 1983. Plasma growth hormone during sleep in young and aged men. *J. Gerontol.* 38(5):519–24.
87. Zadik, Z., Chalew, S.A., McCarter, R.J. Jr, Meistas, M. and Kowarski, A.A. 1985. The influence of age on the 24-hour integrated concentration of growth hormone in normal individuals. *J. Clin. Endocrinol. Metab.* 60(3):513–6.
88. Florini, J.R., Prinz, P.N., Vitiello, M.V. and Hintz, R.L. 1985. Somatomedin-C levels in healthy young and old men: Relationship to peak and 24-hour integrated levels of growth hormone. *J. Gerontol.* 40(1):2–7.
89. Vermeulen, A. 1987. Nyctohemeral growth hormone profiles in young and aged men: Correlation with somatomedin-C levels. *J. Clin. Endocrinol. Metab.* 64(5):884–8.
90. Ho, K.Y., Evans, W.S., Blizzard, R.M., Veldhuis, J.D., Merriam, G.R., Samojlik, E., Furlanetto, R., Rogol, A.D., Kaiser, D.L. and Thorner, M.O. 1987. Effects of sex and age on the 24-hour profile of growth hormone secretion in man: Importance of endogenous estradiol concentrations. *J. Clin. Endocrinol. Metab.* 64(1):51–8.
91. Tanaka, K., Inoue, S., Shiraki, J., Shishido, T., Saito, M., Numata, K. and akamura, Y. 1991. Age-related decrease in plasma growth hormone: Response to growth hormone-releasing hormone, arginine and L-dopa in obesity. *Metabolism.* 40(12):1257–62.
92. Pritzlaff, C.J., Wideman, L., Weltman, J.Y., Abbott, R.D., Gutgesell, M.E., Hartman, M.L., Veldhuis, J.D. and Weltman, A. 1999. Impact of acute exercise intensity on pulsatile growth hormone release in men. *J. Appl. Physiol.* 87(2):498–504.
93. Sutton, J. and Lazarus, L. 1976. Growth hormone in exercise: Comparison of physiological and pharmacological stimuli. *J. Appl. Physiol.* 41(4):523–7.
94. Sutton, J. and Lazarus, L. 1974. Effect of adrenergic blocking agents on growth hormone responses to physical exercise. *Horm. Metab. Res.* 6(5):428–9.

95. Gotshalk, L.A., Loebel, C.C., Nindl, B.C., Putukian, M., Sebastianelli, W.J., Newton, R., Hakkinen, K. and Kraemer, W.J. 1997. Hormonal responses of multiset versus single-set heavy-resistance exercise protocols. *Can. J. Appl. Physiol.* 22(3):244–55.
96. Kraemer, W.J., Fleck, S.J., Dziados, J.E., Harman, E.A., Marchitelli, L.J., Gordon, S.E., Mello, R., Frykman, P.N., Koziris, L.P. and Triplett, N.T. 1993. Changes in hormonal concentrations after different heavy-resistance exercise protocols in women. *J. Appl. Physiol.* 75(2):594–604.
97. Kraemer, W.J., Gordon, S.E., Fleck, S.J., Marchitelli, L.J., Mello, R., Dziados, J.E., Friedl, K., Harman, E., Maresh, C. and Fry, A.C. 1991. Endogenous anabolic hormonal and growth factor responses to heavy resistance exercise in males and females. *Int. J. Sports Med.* 12(2):228–35.
98. Kraemer, W.J., Marchitelli, L., Gordon, S.E., Harman, E., Dziados, J.E., Mello, R., Frykman, P., McCurry, D. and Fleck, S.J. 1990. Hormonal and growth factor responses to heavy resistance exercise protocols. *J. Appl. Physiol.* 69(4):1442–50.
99. Vanhelder, W.P., Radomski, M.W. and Goode, R.C. 1984. Growth hormone responses during intermittent weight lifting exercise in men. *Eur. J. Appl. Physiol. Occup. Physiol.* 53(1):31–4.
100. Kraemer, R.R., Kilgore, J.L., Kraemer, G.R. and Castracane, V.D. 1992. Growth hormone, IGF-I and testosterone responses to resistive exercise. *Med. Sci. Sports Exerc.* 24(12):1346–52.
101. Ahlborg, G., Felig, P., Hagenfeldt, L., Hendler, R. and Wahren, J. 1974. Substrate turnover during prolonged exercise in man. Splanchnic and leg metabolism of glucose, free fatty acids and amino acids. *J. Clin. Invest.* 53(4):1080–90.
102. Biolo, G., Maggi, S.P., Williams, B.D., Tipton, K.D. and Wolfe, R.R. 1995. Increased rates of muscle protein turnover and amino acid transport after resistance exercise in humans. *Am. J. Physiol.* 268(3 Pt 1):E514–20.
103. Tipton, K.D., Ferrando, A.A., Phillips, S.M., Doyle, D. Jr. and Wolfe, R.R. 1999. Postexercise net protein synthesis in human muscle from orally administered amino acids. *Am. J. Physiol.* 276(4 Pt 1):E628–34.
104. Kraemer, W.J., Volek, J.S., Bush, J.A., Putukian, M. and Sebastianelli, W.J. 1998. Hormonal responses to consecutive days of heavy-resistance exercise with or without nutritional supplementation. *J. Appl. Physiol.* 85(4):1544–55.
105. Tarnopolsky, M.A., MacDougall J.D. and Atkinson, S.A. 1988. Influence of protein intake and training status on nitrogen balance and lean body mass. *J. Appl. Physiol.* 64(1):187–93.
106. Chandler, R.M., Byrne, H.K., Patterson, J.G. and Ivy, J.L. 1994. Dietary supplements affect the anabolic hormones after weight-training exercise. *J. Appl. Physiol.* 76(2):839–45.
107. Spruce, N. and Titchenal, A., *An Evaluation of Popular Fitness-Enhancing Supplements.* Calabasas: Evergreen Communications, 2001. 417p.
108. Phillips, S.M., Tipton, K.D., Aarsland, A., Wolf, S.E. and Wolfe, R.R. 1997. Mixed muscle protein synthesis and breakdown after resistance exercise in humans. *Am. J. Physiol.* 273(1 Pt 1):E99–107.
109. Haussinger D. 1996. The role of cellular hydration in the regulation of cell function. *Biochem. J.* 313 (Pt 3):697–710.
110. Jorgensen, J.O., Moller, N., Lauritzen, T. and Christiansen, J.S. 1990. Pulsatile versus continuous intravenous administration of growth hormone (GH) in GH-deficient patients: Effects on circulating insulin-like growth factor-I and metabolic indices. *J. Clin. Endocrinol. Metab.* 70(6):1616-23.

111. Laursen, T., Jorgensen, J.O., Jakobsen, G., Hansen, B.L. and Christiansen, J.S. 1995. Continuous infusion versus daily injections of growth hormone (GH) for 4 weeks in GH-deficient patients. *J. Clin. Endocrinol. Metab.* 80(8):2410–8.
112. Macintyre, J.G. 1987. Growth hormone and athletes. *Sports Med.* 4(2):129–42.
113. Quirion, A., Brisson, G., De Carufel, D., Laurencelle, L., Therminarias, A. and Vogelaere, P. 1988. Influence of exercise and dietary modifications on plasma human growth hormone, insulin and FFA. *J. Sports Med. Phys. Fitness.* 28(4):352–3.
114. Kaiser, F.E., Silver, A.J. and Morley, J.E. 1991. The effect of recombinant human growth hormone on malnourished older individuals. *J. Am. Geriatr. Soc.* 39(3):235–40.
115. Welle, S., Thornton, C., Statt, M. and McHenry, B. 1996. Growth hormone increases muscle mass and strength but does not rejuvenate myofibrillar protein synthesis in healthy subjects over 60 years old. *J. Clin. Endocrinol. Metab.* 81(9):3239–43.
116. Taaffe, D.R., Pruitt, L., Reim, J., Hintz, R.L., Butterfield, G., Hoffman, A.R. and Marcus, R. 1994. Effect of recombinant human growth hormone on the muscle strength response to resistance exercise in elderly men. *J. Clin. Endocrinol. Metab.* 79(5):1361–6.
117. Yarasheski, K.E. and Zachwiega, J.J. 1996. Effect of growth hormone administration on muscle strength, protein turnover and glucose metabolism in the elderly. *FASEB. J.* 10:A754.
118. Marcus, R., Butterfield, G., Holloway, L., Gilliland, L., Baylink, D.J., Hintz, R.L. and Sherman, B.M. 1990. Effects of short term administration of recombinant human growth hormone to elderly people. *J. Clin. Endocrinol. Metab.* 70(2):519–27.
119. Thompson, J.L., Butterfield, G.E., Marcus, R., Hintz, R.L., Van Loan, M., Ghiron, L. and Hoffman, A.R. 1995. The effects of recombinant human insulin-like growth factor-I and growth hormone on body composition in elderly women. *J. Clin. Endocrinol. Metab.* 80(6):1845–52.
120. Holloway, L., Butterfield, G., Hintz, R.L., Gesundheit, N. and Marcus, R. 1994. Effects of recombinant human growth hormone on metabolic indices, body composition and bone turnover in healthy elderly women. *J. Clin. Endocrinol. Metab.* 79(2):470–9.

12

Ornithine, Ornithine Alpha-Ketoglutarate and Taurine

Tausha D. Robertson

CONTENTS
I. Ornithine and Ornithine Alpha-Ketoglutarate 197
II. Metabolic Function..198
   A. Body Reserves 198
   B. Nutrient Assessment 199
   C. Toxicity...200
   D. Interaction with Other Nutrients or Drugs 200
III. Summary of Research on the General Population 200
IV. Summary of Research on Athletes and Active Individuals 200
V. Recommendations 201
VI. Taurine...201
VII. Metabolic Function..201
   A. Body Reserves 202
   B. Nutrient Assessment 202
   C. Toxicity...202
   D. Interactions with Other Nutrients or Drugs 202
VIII. Summary of Research on the General Population 203
XI. Summary of Research Athletes and Active Individuals..............203
X. Recommendations 203
XI. Summary 204
References...204

I. Ornithine and Ornithine Alpha-Ketoglutarate

Two variations of ornithine of interest to clinicians and athletes are the nonessential amino L-2,5-diaminovaleric acid and ornithine alpha-ketoglutarate (OKG), a salt composed of one molecule of alpha-ketoglutarate and

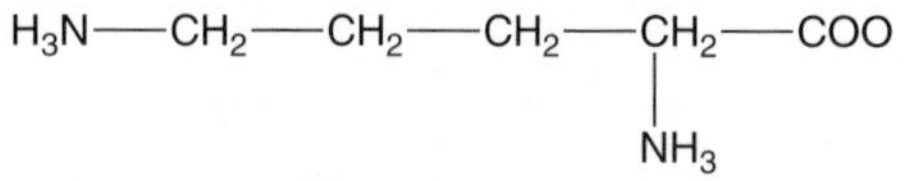

$$H_3N-CH_2-CH_2-CH_2-CH_2-COO$$
$$|$$
$$NH_3$$

FIGURE 12.1
Chemical structure of ornithine (L-5-aminovaline).

two ornithine molecules. Ornithine's role in the urea cycle may contribute to a reduction in ammonia levels.[1] The literature indicates that ornithine in high doses may also stimulate growth hormone secretion.[2] Because the dose used to produced the aforementioned results proved to be impractical, research began to focus on OKG. Investigators have linked OKG to positive changes in nutritional and catabolic status of burn, trauma and surgical patients.[3,4] Current literature related to sports performance focuses on the potential anabolic and anticatabolic properties of OKG.[5] The chemical structure of ornithine is give in Figure 12.1.

II. Metabolic Function

Ornithine is produced in the intestine as a metabolite of arginine in the urea cycle.[1] The catabolism of arginine produces ornithine then glutamate.[6] Ornithine is metabolized differently in various tissues. In the epithelial cells of the small intestine, it is used primarily to synthesize citrulline and arginine. Two different pathways exist in the liver. In live cells surrounding the portal vein, ornithine functions primarily as an intermediate of the urea cycle. Liver cells around the central vein use ornithine to synthesize glutatmate and glutamine. In other peripheral tissues, it is used to synthesize glutamate and proline.

OKG is a salt that is composed of one molecule of alpha-ketoglutarate and two ornithine molecules.[6] Alpha ketolutarate and ornthine share a common metabolic pathway with alpha-ketoglutarate as the end product of ornithine and vice versa (Figure 12.2). These molecules and the direct metabolites enter other pathways that lead to the production of proline, glutamine, polyamines and arginine. Each of these play critical roles in protein homeostasis. [6] OKG increases the secretion of insulin and human growth hormone in both healthy subjects and those with pathologies.[7,8] This may indicate that these hormones are involved in the mechanisms of action of OKG. [8] A review of published research indicates that the combination of ornithine with alpha-ketoglutarate produces greater changes in amino acid metabolism than either compound alone.[9]

A. Body Reserves

Ornithine levels can be measured in various body fluids such as plasma, urine and cerebrospinal fluids. The reference intervals for adults are as follows:

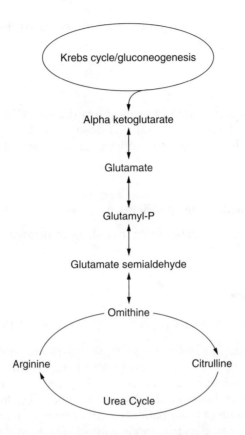

FIGURE 12.2
Ornithine and alpha-ketoglutarate metabolic pathways. Key enzymes are (1) ornthine aminotransferase, (2) ornithine carbamoyltransferase, (3) NO synthase, (4) arginase, (5) ornithine decarboxylase, (6) transaminases, (7) gluatmate deshydrogenase, (8) glutamine synthetase and (9) gludtaminase. From Cynober, L.A., Ornithine alpha-ketoglutarate in *Amino Acid Metabolism and Therapy in Health and Nutritional Disease*, Cynober, L. Ed, CRC Press, Boca Raton, FL, 1995,390. With permission.

plasma 3.0–13.6 micromol/dL, 24 hr urine female 0–80 micromol/24 hr, male 0–30 micro mol/24 hr and cerebrospinal fluid 0.43–0.55 micromol/dL.[10]

B. Nutrient Assessment

Ornithine is a naturally occurring amino acid found in high-protein foods such as meat, fish, chicken, eggs and some vegetables. The USDA National Nutrient Database for Standard Reference does not include values for ornithine content in foods listed.[11] Ornithine can be purchased through various companies in supplement form as L-ornithine. Deficiencies in ornithine can occur if individuals suffer from inherited urea-cycle disorders such as citrullinemia, argininosucinic aciduria and argininemia.[12] OKG is synthetically produced by combining two molecules of ornithine and one molecule of alpha-ketoglutarate.[1] This substance does not occur naturally in foods and is not synthesized by the body.[6]

C. Toxicity

There are no published data indicating ornithine or OKG toxicity.[13] However, Bucci et al. documented that oral doses of 170 mg/kg administered after an overnight fast caused stomach cramps and diarrhea in bodybuilder subjects.[2]

D. Interaction with Other Nutrients or Drugs

There are no published data of negative drug or nutrient interactions with ornithine or OKG.[13]

III. Summary of Research on the General Population

OKG supplementation has improved the nutritional status of of trauma and surgical patients. Burn patients benefit from enteral infusion of OKG. Donati et al.[4] found that burn wound healing time was reduced by OKG treatment. Coudray-Lucas et al . also documented significant decreases in wound healing time in burn patients with enteral administration of OKG as compared with an isonitrogenous control. The study also documented decreases in protein hypercatabolism in less severely burned patients.[14] Studies in rats indicate that the mechanism of OKG action may be to decrease muscle catabolism, the loss of muscle weight and the decreases in muscle glutamine associated with such injuries.[15] OKG also produced anabolic affects on the growth of prepubertal growth-retarded children receiving total parenteral nutrition. At doses of 15g/day, plasma concentrations of insulinlike growth factor 1(IGF1), glutamine and glutamate increased. The IGF1 concentrations correlated positively with height increases.[16] Cynober et al. found that, in healthy subjects, OKG supplementation increased insulin levels.[17]

IV. Summary of Research on Athletes and Active Individuals

Based on their anabolic and anticatabolic properties in trauma patients, ornithine and OKG were investigated for similar effects in athletes. Bucci et al.[2] found that mean serum growth hormone levels rose significantly, ($p<0.05$) at 90 minutes post administration of 170 mg/kg ornithine orally in body builders. The lower doses of 40 mg/kg and 100 mg/kg failed to affect somatotropin levels. The higher but effective dosage caused stomach cramping and diarrhea, thus making its use impractical. This study also found that at each dose, 40, 100 and 170 mg/kg serum ornithine levels were significantly greater at 45 and 90 minutes ($p<0.01$) than at baseline levels.[2] Another ornithine study reported that oral doses of 40, 100 and 170 mg/kg L-ornithine did not increase serum insulin levels in male and female bodybuilders.[18] OKG supplementation of 10g failed to produce anabolic effects in male resistance-trained subjects. After 6 weeks of ingestion, OKG did not increase training intensity or volume, or result in increased muscle mass. The supplement did not alter blood concentrations of insulin or growth hormone during the study.[5]

V. Recommendations

Though research in trauma patients has indicated the anabolic and anticatabolic properties of OKG, oral supplementation in athletes has not produced similar results. Published literature on ornithine and OKG suggest that they have no effect as anabolic agents.[19]

VI. Taurine

Taurine is produced as a metabolite of methionine transsulferation. Classified as a nonessential amino acid, taurine is the second-most abundant free amino acid in muscle after glutamine.[20] Published research has indicated potential roles in cell osmolality, cardiovascular function and antioxidant properties.[21] The chemical structure of taurine is given in Figure 12.3.

VII. Metabolic Function

Taurine, a conditionally essential amino acid, is derived from diet or synthesized primarily in the brain and liver. The body synthesizes taurine from methionine and cysteine (Figure 12.4). Vitamin B_6 is a critical cofactor in this pathway. It is a beta amino acid that is not utilized in protein synthesis.[21]

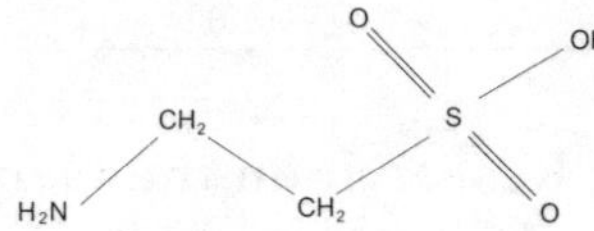

FIGURE 12.3
Chemical structure of taurine (2-aminoethyl sulfonic acid).

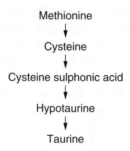

FIGURE 12.4
Taurine biosynthesis.

A. Body Reserves

Laboratory tests can assess taurine levels in plasma, urine and cerebrospinal fluids. Plasma taurine levels in adults range from 3.4–14.5 micromol/dL. 24 hr urine taurine reference values range from 220–1290 micromol/24 hr in females to 350–1850 micromol/24 hr in males. Adult cererbrospinal taurine levels range from 0.059–0.69 micromol/dL.[10]

B. Nutrient Assessment

Taurine is a naturally occurring amino acid in high-protein foods such at meats, dairy products and seafood. Available nutrient analysis databases do not provide information on taurine content in foods.[11] However, Laidlaw et al. published the taurine content of a select sample of foods.[22] The daily dietary intake of nonvegan Americans is estimated at 20–200 mg/day.[23] Although humans are able to synthesize taurine endogenously, the ability may be limited. Inadequate dietary taurine intake, along with low intakes of methionine or cysteine could contribute to low plasma, urine or cerebrospinal fluid levels.[24]

C. Toxicity

There are no published data on taurine toxicity. However, caution must be exercised when taurine supplementation is used for ulcer patients.[13]

D. Interactions with Other Nutrients or Drugs

There are no published data on taurine interactions with other nutrients or drugs.[13]

VIII. Summary of Research on the General Population

Oral supplementation of taurine has produced antihypertensive effects in humans. Six weeks of taurine supplementation (6g/day) reduced systolic, diastolic and mean blood pressure (BP) in sodium-restricted subjects with essential hypertension.[25] Fujita et al. reported that 6 g/day taurine for 7 days reduced systolic, diastolic and mean BP.[26]

Congestive heart failure patients have shown improvements with taurine supplementation. After 6 weeks, 3 g/day taurine improved systolic left ventricular function.[27] Taurine supplementation of 4 g/day improved the clinical symptoms in 19 of 24 congestive heart failure patients.[28] Research by Maturo et al. indicates that the hypoglycemic properties of taurine are linked to its interaction with insulin receptors.[29]

Taurine has been tested *in vitro* and in animal models for other clinical uses. In an animal model, intracerebroventricular taurine supplementation reduced seizures induced by glutamic acid suppression.[30] *In vitro*, taurine acts as an antioxidant and a calcium regulator.[31] This has not yet been shown *in vivo*.

XI. Summary of Research Athletes and Active Individuals

Recent research has drawn attention to the role of taurine supplementation in exercise and sport. Endurance-trained athletes were supplemented with a taurine- and caffeine-containing drink that increased the left atrial contractility. This increase may explain improved maximal performances and lower heart rates at submaximal intensity.[32] Geiss et al. reported that a sports drink (Red Bull) containing taurine and glucuronolactone increases endurance in trained athletes.[33] In other laboratories, Red Bull improved anaerobic and aerobic performance as well as mental performance.[34,35] However, the mechanism of these performances have not been documented. It is unclear if a single ingredient or the combination of ingredients is responsible for the documented changes.

X. Recommendations

Taurine supplementation of 3–6 g/day improved various cardiovascular parameters. The importance of taurine in metabolic functions involved in exercise performance has been well documented.[36] However, there is little research on taurine supplementation in active individuals. These studies on taurine supplementation are conducted using products that contain additional substances. It is difficult to attribute the reported positive gains in performance to taurine alone.

XI. Summary

Ornithine and OKG supplementation have proven effective in improving the nutritional status of trauma patients.[16] In healthy subjects, OKG supplementation increased insulin levels.[17] The anticatabolic and anabolic effects of ornithine and OKG have not been substantiated in active individuals or athletes.[5] Although OKG increased serum growth hormone levels in body builders at doses of 170 mg/kg, the side effects prohibit practical application of such doses.[2]

Taurine supplementation produced antihypertensive effects and improved clinical symptoms in congestive heart failure patients.[25,28] Studies conducted *in vitro* and in animal models report that taurine may be active in other metabolic processes related to calcium regulation, oxidative damage and seizure reduction.[30,31] Published research on taurine supplementation in active individuals and athletes is based on products that contain not only taurine but multiple ingredients.[32-35] Although these mixtures produced favorable results, it is difficult to attribute the results to taurine alone.

References

1. Di Pasquale, M., *Amino Acids and Proteins for the Athlete: The Anabolic Edge*, CRC Press, Boca Raton, FL, 1997, 149.
2. Bucci, L.R., Hickson, J. F., Pivarnik, J.M., Wolinsky, I., McMahon, J.C. and Turner, S.D. 1990. Ornithine ingestion and growth hormone release in body builders. *Nutr. Res.* 10, 239.
3. Cynober,L. 1991. Ornithine alpha-ketoglutarate in nutritional support, *Nutrition.* 7, 313.
4. Donati, L, Signorini, M. and Grappolini, S. 1993. Ornithine alpha-ketoglurate administration in burn injury, *Clin. Nutr.*, 12, 70.
5. Chetlin, R.D., Yeater, R.A., Ullrich, I.H., Hornsby Jr., W.G., Malanga, C.J. and Bryner, R.W. 2000. Effects of ornithine alpha-ketoglutarate on healthy weight trained men. *J. EP,* 3, 37.
6. Cynober, L., Ornithine alpha-ketoglutarate in *Amino Acid Metabolism and Therapy in Health and Nutritional Disease*, Cynober, L. Ed, CRC Press, Boca Raton, FL, 1995, 385.
7. Young, V and El-Khoury, A.E. The notion of the nutritional essentiality of amino acids revisited, with a note on the indispensable amino acid requirements in adults in *Amino Acid Metabolism and Therapy in Health and Nutritional Disease*, Cynober, L. Ed, CRC Press, Boca Raton, FL,1995, 206.
8. Valle, D. and Simell, O. The hyperornithinemias in *The Metabolic and Molecular Bases of Inherited Disease*, 8th ed., Scriver,C.R., Ed., McGraw-Hill, New York, 2001,1857.
9. Cynober, L., Coudray-Lucas, C., de Bandt, J.P., Guechot J., Aussel C., Salvucci M. and Giboudeau J. 1990. Action of ornithine alpha-ketoglutarate, ornithine hydrochloride and calcium alpha-ketoglutarate on plasma amino acid and hormonal patterns in healthy subjects. *J. Am. Coll. Nutr.*, 9, 2.
10. Laboratory Corporation of America Holdings, *Directory of Services and Interpretive Guide*, Lexi-Comp Inc, Hudson, OH, 2003, 414.

11. U.S. Department of Agriculture, Agricultural Research Service. USDA Nutrient Database for Standard Reference, Release 15, 2002, Nutrient Data Laboratory Home Page, http://www.nal.usda.gov/fnic/foodcomp, June 27, 2003.
12. Zieve, L. 1986. Conditional deficiencies of ornithine or arginine. *J. Am. Coll. Nutr.* 5, 167.
13. AltMedDex System. Abt, L. and Hammerly, M., Eds., Micromedex, Greenwood Village, CO, 2003. http://www.micromedex.com/aboutus/legal/cite/
14. Coudray-Lucas, C., Le Bever, H., Cynober, L., De Bandt, J.P., Carsin, H. 2000. Ornithine alpha-ketoglutarate improves wound healing in severe burn patients: A prospective randomized double blind trial versus isonitrogenous controls, *Crit. Care Med.*, 28, 1772.
15. Vaudourdolle, M., Coudray-Lucas, C., Jardel, A., Ziegler F., Ekindjian O.G. and Cynober L. 1991. Action of enternally administered ornithine alpha-ketoglutarate on protein breakdown in skeletal muscle and liver of the burned rat, *JPEN.*, 15, 517.
16. Moukarzel, A.A., Goulet, O., Salas, J.S., Marti-Henneberg, C. Buchman, A.L., Cynober, L., Rappaport, R., Ricour, C. 1994. Growth retardation in children receiving long-term parenteral nutrition: effects of ornithine alpha-ketoglutarate, *Am. J. Clin. Nutr.*, 60, 408.
17. Cynober, L., Vaubourdolle, M., Dore, A. Giboudeau, J. 1984. Kinetics and metabolic effects of orally administered ornithine alpha-ketoglutarate in healthy subjects fed with a standard regimen, *Am. J. Clin. Nutr.*, 39, 514.
18. Bucci, L.R., Hickson,Jr., J. F., Wolinsky, I., Pivarnick, J.M. 1992. Ornithine supplementation and insulin release in bodybuilders, *Int. J. Sport Nutr.*, 2, 287.
19. Robertson, T., Dietary supplements and strength-trained athletes, in *Nutrition and the Strength Athlete*, Jackson, C.G.R., Ed., CRC Press, Boca Raton, 2001, 122.
20. Di Pasquale, M., *Amino Acids and Proteins for the Athlete: The anabolic edge*, CRC Press, Boca Raton, 1997, 144.
21. Lourenco, R. and Camilo, M.E. 2002. Taurine: A conditionally essential amino acid in humans? An overview in health and disease, *Nutricion Hospitalaria*, 17, 262.
22. Laidlaw, S.A., Grosvenor, M., Kopple, J.D. 1990. The taurine content of common foodstuffs, *JPEN*, 14, 183.
23. Kopple, J.D., Vinton, N.E., Laidlaw, S.A. and Ament M.E. 1990. Effects of intravenous taurine supplementation on plasma, blood cell and urine taurine concentrations in adults undergoing long-term parenteral nutrition, *Am. J. Clin. Nutr.*, 52, 846.
24. Struman, J. A., Hepner, G.W. and Hofmann, A.F. 1975. Metabolism of taurine in man, *J. Nutr.*, 105, 1206.
25. Kohashi, N., Katori, R. 1983. Decrease of urinary taurine in essential hypertension. *Prog. Clin. Bio. Med.*, 125, 73.
26. Fujita, T. Ando, K., Noda, H., Ito, Y., Sato, Y.1987. Effects of increased adrenomedullary activity and taurine in young patients with borderline hypertension, *Circulation*, 75, 525.
27. Azuma, J., Sawamura, A., Awata, N. 1992. Usefulness of taurine in chronic congestive heart failure and its prospective application. *Jap. Circ. J.*, 56, 95,
28. Azuma, J., Hasegawa, H., Sawamura, A., Awata, N.,Ogura, K., Harada, H., Yamamura, Y., Kishimoto, S. 1983. Therapy of congestive heart failure with orally administered taurine, *Clin. Ther.* 5, 398.

29. Maturo, J. and Kulakowski, E.C. 1988. Taurine binding to the purified insulin receptor, *Biochem. Pharmacol.*, 37, 3755.
30. Huxtable, R. and Laird., H. 1978. The prolonged anticonvulsant action of taurine on genetically determined seizure susceptibility, *Can. J. Neurol. Sci.*, 5, 215.
31. Schaffer, S.W. and Kocsis, J.J. 1979. Taurine: Research surges after 150 yrs. *Am. Pharm.*, 19, 36.
32. Baum, M. and Weib, M. 2001. The influence of a taurine-containing drink on cardiac parameters before and after exercise measured by echocardiography, *Amino Acids*, 20, 75.
33. Geiss, K.R., Jester, I., Falke, W., Hamm, M., Waag, K.L. 1994. The effect of a taurine-containing drink on performance in 10 endurance athletes, *Amino Acids*, 7, 45.
34. Alford, C., Cox, H., Wescott, R. 2001. The effects of Red Bull energy drink on human performance and mood, *Amino Acids*, 21, 139.
35. Seidl, R., Peyrl, A., Nicham, R., Hauser, E. 2000. A taurine- and caffeine-containing drink stimulates cognitive performance and well-being, *Amino Acids*, 19, 635.
36. Kendler, B. S. 1989. Taurine: An overview of its role in preventive medicine, *Prev. Med.*, 18, 79

Part III

Lipid Derivatives

13

Conjugated Linoleic Acid

Celeste G. Koster and Martha A. Belury

CONTENTS
I. Introduction 209
II. Health Properties of Conjugated Linoleic Acid 210
III. Effects of Conjugated Linoleic Acid on Muscle Mass
 and Strength 210
IV. Effects of Conjugated Linoleic Acid on Body Composition 213
V. Effects of CLA on Energy Intake and Expenditure 215
VI. Summary 216
Acknowledgments 217
References 217

I. Introduction

Conjugated linoleic acid (CLA) refers to a group of stereo and positional isomers of octadecadienoate (18:2) (Figure 13.1).[1] This group of polyunsaturated fatty acids is formed by partial biohydrogenation in the rumen of cattle and lamb and is, therefore, found in the meat and milk products from these ruminant sources (Table 13.1).[2-4] Levels of CLA and of individual isomers in meat and milk result from differences in diet and farming management of cows.[5] CLA can also be prepared synthetically by heating linoleic acid in the presence of alkali or by partial dehydrogenation of linoleic acid; however, synthetic preparation of CLA from precursors such as linoleic acid results in an alteration of the isomeric composition (Table 13.2).[6] In foods, c9t11-CLA, also called rumenic acid,[7] is the most predominant isomer,[4,8] followed by t7c9-CLA, c11t13-CLA, c8t10-CLA and t10c12-CLA.[8] In synthetic preparations of CLA, c9t11-CLA and t10c12-CLA are the dominant isomeric forms followed by t7c9-CLA, c8t10-CLA and c11t13-CLA. Importantly, most

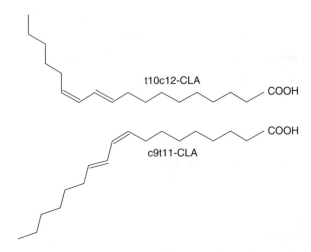

FIGURE 13.1
Structures of t10c12-conjugated linoleic acid and c9t11-conjugated linoleic acid.

research on CLA in health has utilized the synthetic form of CLA oil containing the altered ratio of c9t11-CLA:t10c12-CLA of 0.95.

II. Health Properties of Conjugated Linoleic Acid

Several animal and human studies have found CLA to have various health properties including action as an anticarcinogen[9] and antiatherosclerotic agent,[10,11] as well as an antidiabetic agent.[12] CLA also reduces body mass in growing animals[13–16] and adult humans[17,18] when provided as a synthetic mixture of CLA isomers. The t10c12 isomer appears to have a greater effect on adiposity than the c9t11 isomer in mice.[18,19] Because of favorable effects of CLA on adiposity and lean mass in experimental animals, CLA has received some attention as an ergogenic aid for resistance-trained athletes and bodybuilders. This review will summarize the effects of CLA on energy expenditure, body composition and muscle mass with and without resistance training. We will review and present the current state of knowledge surrounding proposed mechanisms of action and finally, we will speculate about the possibility of CLA to act as an ergogenic aid in people.

III. Effects of Conjugated Linoleic Acid on Muscle Mass and Strength

Along with reducing adiposity in several animal models of obesity, CLA may have positive effects on muscle mass (Table 13.3). A relationship

TABLE 13.1

Content of Conjugated Linoleic Acid in Selected Foods

Food	Total CLA (mg/g fat)	c9t11 Isomer (%)
Dairy Products		
Homogenized Milk	5.5	92
Butter	4.7	88
Sour Cream	4.6	90
Plain Yogurt	4.8	84
Nonfat Yogurt	1.7	83
Ice Cream	3.6	86
Sharp Cheddar Cheese	3.6	93
Mozzarella Cheese	4.9	95
Colby Cheese	6.1	92
Cottage Cheese	4.5	83
American Processed Cheese	5.0	93
Meat (uncooked)		
Fresh Ground Beef	4.3	85
Beef Round	2.9	79
Veal	2.7	84
Lamb	5.6	92
Pork	0.6	82
Poultry (uncooked)		
Chicken	0.9	84
Fresh Ground Turkey	2.5	76
Seafood (uncooked)		
Salmon	0.3	—
Lake Trout	0.5	—
Shrimp	0.6	—
Processed Foods		
Beef Frank	3.3	83
Turkey Frank	1.6	70
Peanut Butter	0.2	—
Canned Foods		
Spam™	1.3	71
Baked Beans	0.7	56
Corned Beef	6.6	85
Vegetable Oils		
Safflower	0.7	44
Sunflower	0.4	38
Canola	0.5	44

Source: From Chin SF, Liu W, Storkson JM, Ha YL and Pariza MW.1992. Dietary sources of conjugated dienoic isomers of linoleic acid, a newly recognized class of anticarcinogens. *J. Food Compos. Anal.*, 5, 185.

TABLE 13.2

Approximate Isomeric Content (%CLA) of Conjugated Linoleic Acid: Selected Foods vs. Synthetic Mixture

	c9t11-CLA	c7t9-CLA, c8t10-CLA	t10c12-CLA, others	c9t11-CLA:t10c12-CLA Ratio
Beef	74.8	15.8	9.0	8.31
Cheese	82.6	8.3	9.0	9.18
CLA Mix	48.7	—	51.3	0.95

TABLE 13.3

Summary of the Role of Conjugated Linoleic Acid to Alter Muscle Mass in Humans

Source	Action with CLA
Lowery et al. 1997	↑ Arm girth in novice bodybuilders ↑ Body mass ↑ Leg press
Ferreira et al. 1997	↑ Bench press in experienced resistance trained men ↑ Leg press
Kreider et al. 1998	↑ Bone mineral content in experienced resistance-trained athletes ↓ Neutrophil/lymphocyte levels
Kreider et al. 2002	↔ No change in experienced resistance-trained athletes

between CLA supplementation and muscle size and strength was first reported in 1997.[20] In a placebo controlled study, 24 male novice bodybuilders were supplemented with 7.2 g/day CLA or a vegetable oil placebo while completing 6 weeks of bodybuilding exercises. Arm circumference and skin-fold measurements were used to determine body mass. Arm girth increased from 7175+/-978 to 7562+/-1000 mm^2 in the CLA group as compared with the placebo (from 7777+/-1532 to 7819+/-1516 mm^2). The greater increase in arm girth in the CLA group suggests that more muscle was built in the CLA group than in the placebo group. Body mass in the CLA group increased from 77.6+/-11.8 to 79.0+/-12 kg, while the control group remained relatively even at 77.8+/-11.9 kg to 77.8+/-11.8 kg. As a measure of strength, leg press increased from 263.6+/-63.0 to 335+/-75 kg in the CLA group, while there was a lesser increase in the placebo group: 271.5+/-62.9 kg to 306.8+/-70.2 kg. Skin-fold measurements, total body fat and total body water measured by bioelectrical impedance analysis (BIA) of both whole body and upper limb were similar between groups.

In a study with similar design by Ferreira et al.,[21] experienced resistance-trained men were supplemented in a randomized, placebo controlled study with CLA or an olive oil placebo for 28 days.[21] In men supplemented with CLA, performance on the 1-RM (repetition maximum) bench press and leg press were slightly improved; however, the improvement was not statistically different from the control group. In addition, blood urea nitrogen levels were lower in men supplemented with CLA, suggesting an anabolic effect. Total body mass measured by dual x-ray absorptiometry (DEXA), fat free mass, fat mass and percent body fat were similar between groups.

In a separate study of CLA in relation to bone mineral content, bone mineral density and immune stress, 23 experienced resistance-trained males were supplemented with 6.0 g/day CLA, with 3.2g/day or an olive oil placebo for 28 days.[22] Bone mineral content increased and neutrophil/lymphocyte ratio decreased in the CLA supplemented group, suggesting less immune stress. In a subsequent study, supplementation with 3.0 g/day of CLA had no effect for 23 experienced resistance-trained males on total body mass, fat-free mass, fat mass, percent body fat, bone mass, strength, serum substrates, general markers of catabolism and immunity during training during a 28-day study.[23] Perhaps the relatively higher doses of 6g/day CLA are more effective than a 3g/day supplement.

Notably, all studies on the role of CLA as an ergogenic aid to enhance strength have involved humans. Unfortunately, the results are mixed, with some studies finding a correlation between supplementation of CLA and strength via leg press[20,21] and others finding no relationship.[22,23] More attention to the prospective relationship between strength and CLA are needed, including using women in addition to men, since it is very possible that the different sexes metabolize and utilize CLA differently.

IV. Effects of Conjugated Linoleic Acid on Body Composition

CLA has been shown to have an inverse relationship with adiposity. A reduction in body fat mass has been found in growing animals such as mice, rats, pigs and cattle as well as adult humans. In fact, when 6-week-old male ICR mice were supplemented with 0.5% CLA plus 5% corn oil to their diet for 32 days, a 57% reduction in body fat and a 5% increase in lean body mass was observed as compared with their respective controls.[13] In the same study, 6-week-old females were fed a 0.5% CLA plus 5% corn oil-supplemented diet for 28 days and showed a 60% reduction in body fat and a 14% increase in lean body mass as compared with their respective controls. The control male and female mice were fed a diet with 5.5% corn oil. It is interesting to note the similarity of body fat reduction, but the nearly threefold difference in lean body mass between the male and female mice, but perhaps male and female mice metabolize and incorporate CLA differently. In a separate study, a 50% reduction in adipose tissue mass was observed when 8-week-old female ICR mice were fed a diet containing 0.5% of a CLA mixture for 4 weeks.[14] The adipose tissue reduction was sustained after the CLA was removed from the diet. When AKR/J mice were fed a semipurified diet supplemented with 2.46 mg/kcal CLA mixture for 6 weeks, a significant reduction in adipose tissue deposition (43–88%) independent of the high-fat (45 kcal%) and low-fat (15 kcal%) diet composition was observed.[15]

To determine whether a metabolite of CLA could explain the reducing effect of CLA on body weight and adiposity, the effect of CLA and CLNA on body fat in male Sprague-Dawley rats was determined.[16] CLNA is a highly unsaturated conjugated fatty acid and is expected to affect lipid metabolism. The 4-week-old male rats were fed a purified diet with either 1% CLA or CLNA for 4 weeks. Peri-renal and epididymal adipose tissue weight was reduced in both the CLA and CLNA groups; however, the effects were heightened within the CLNA group. These results suggest that CLNA and CLA may work differently in reducing adipose tissue weight.

In studies involving human subjects, research findings have been mixed. A study supplementing varying amounts of CLA from 1.7 to 6.8 grams per day to overweight and obese humans for 12 weeks showed a reduction in body fat mass, measured by DEXA, for the CLA group.[17] In people with type 2 diabetes mellitus, supplementation with CLA (mixture of c9+11 CLA and t10c12 CLA isomers) or placebo were provided at 8.0 g per day. Supplementing for 8 weeks with CLA mixture resulted in reduced body weight.[18] Further, it was noted that the t10c12 CLA isomer was more significantly associated with the decreased body weight than the c9+11 CLA isomer, suggesting that c10c12 CLA may be the bioactive isomer responsible for weight loss. Another study, involving weanling ICR rats, also found that the t10c12 CLA isomer was associated with reduced body fat, enhanced body water, enhanced body protein and enhanced body ash, whereas the c9t11 and t9t11 CLA isomers did not affect these parameters.[19]

In addition to reducing adipose tissue mass, the t10c12 CLA isomer has been linked to increased insulin resistance in men who have symptoms of the metabolic syndrome.[26,27] A CLA mixture also appeared to cause hyperinsulinemia in C57BL/6J mice[24] that was accompanied by severe adipose tissue ablation and decreased leptin levels. The effects of the CLA mixture on adipose tissue depletion were reversed by continuous leptin infusion. In a follow-up study, decreasing the amount of a CLA mixture from 1 to 0.1 g/100g diet, while increasing the amount of total fat in the diet from 4 to 34 g fat per 100 g of diet, did not lead to lipodystrophy, while fat mass was modestly reduced.[25] Insulin resistance was present in the group fed the 1g CLA/100 g diet, but not present in the 0.1 g CLA/100g diet group.

In men who were supplemented with 3.4 g/day of a CLA mixture, purified t10c12-CLA isomer, or olive oil placebo,[26] the t10c12 CLA isomer exerted an increase in insulin resistance that correlated with increased urinary isoprostane levels, suggesting an increase in oxidative stress in these same individuals. Previously, in a study from the same group, 60 abdominally obese men supplemented with 3.4 g/day of the t10c12 CLA isomer, a CLA mixture or equal amounts of olive oil, became more insulin resistant when supplemented with t10c12 CLA as compared with people supplemented with the CLA-mix or olive oil control.[27] These results are significant in the clinical usages of various isomers of CLA as a dietary supplement.

In contrast to the above studies, women supplemented with 3 g/day of CLA had no significant difference in body composition as compared with the placebo.[28] Seventeen women were supplemented with either a 3 g/day capsule of a CLA mixture or a sunflower oil placebo while being confined to a metabolic suite for 94 days. Their diet and activity were held constant. It is possible that 3g/day is not enough to elucidate a change in body composition in women.

In 1994, it was proposed that CLA acted as an in-utero anabolic stimulus for rats, since it appeared to enhance weight gain in rats.[29] Eight-week-old female Fisher rats were fed a nonpurified diet and allowed to mate with a male counterpart. Immediately after mating, the females were separated and fed either a mixture of 0.5 gCLA/100g, 0.25 gCLA/100g or a corn oil mixed with a semi-purified diet. Diets with CLA did not affect the weight of dams but increased the weight of the pups in the CLA-fed groups. Conversely, Paulos et al. later performed a similar study in which pregnant Sprague-Dawley rats were provided diets with 6.5g/100g soybean oil and 0.5g/100g CLA or 7g/100g soybean oil.[30] The maternal treatment continued until day 21 of lactation, at which time the pups were weaned. The pups were assigned control or CLA diet until 11 wks of age. No difference was found in the number of pups per litter, weights of whole litters, litter weight gain, litter efficiency, or food intake of the dams fed. Parametrial fat pad weight and retroperitoneal pad weight were less in dams fed CLA, and pups from dams fed CLA were significantly heavier at weaning than the pups from the control dams. Heavier gastrocnemius and soleus muscles and longer tail lengths, which are markers of skeletal growth, were found at 11 weeks of age in male pups from dams fed CLA. These results suggest that CLA treatment in relation to body composition may be dependent on the sex and age of the animal as well as the duration of feeding.

V. Effects of CLA on Energy Intake and Expenditure

The inverse association of CLA with body mass and adiposity prompted research to elucidate the role of CLA to modulate energy intake and expenditure (Table 13.4). Seventeen healthy, non-obese women between the ages of 20 and 41 were supplemented with 3 g/day CLA or a sunflower oil placebo for 64 days.[28] Energy expenditure measured by respiratory gas exchange, energy intake or body composition were associated with CLA provided as a low-dose and in a short-duration protocol. Similarly, no effect of CLA on energy expenditure was found in adult male Syrian hamsters fed diets with the c9t11-CLA isomer to to equate 1.6% of energy or a CLA mixture of 3.2% of energy for 6 to 8 weeks.[31] In contrast, male AKR/J mice supplemented with CLA reduced energy intake and growth rate.[15] The group fed a diet with 1.2% CLA mixture in a high-fat diet and 1.0% CLA mixture in a low-fat diet also had an increased metabolic rate and a decreased nighttime respiratory quotient as compared with the controls fed without CLA. In a separate study, West et al. also found no reduction in energy intake when

TABLE 13.4

Summary of the Role of Conjugated Linoleic Acid in Energy Expenditure and Food Intake

Source	Action with CLA
Animal Models	
West et al. 1998	↓ Energy intake in AKR/J mice
West et al. 2000	↔ No change in energy intake in AKR/J mice ↑ Energy expenditure
Bouthegard et al. 2002	↔ No change in energy expenditure in Syrian hamsters
Human Models	
Zambell et al. 2000	↔ No change in energy expenditure or intake in women

AKR/J mice were fed diets with 1% CLA mixture for 5 weeks.[32] However, energy expenditure was increased and appeared to account for lower body fat stores in the CLA group.

VI. Summary

The effects of CLA as a modulator of muscle mass have been given little attention. While there is evidence that CLA may help to increase muscle mass, there is also some evidence that CLA has no effect as an ergogenic aid. In the studies that presented evidence that CLA did increase muscle mass, the increase was slight and only 20–30 subjects were tested. A larger subject base along with longer supplementation time could possibly provide more significantly impressive data.

CLA has been found to lower adipose tissue, body fat mass and body weight in animal and human models. While lowering adipose tissue in mice, it was also found to increase hyperinsulinemia, which was later found to be lowered with a lower concentration of CLA supplementation. More clinical research varying the amount of CLA and the duration of supplementation is necessary, since the data with human subjects is not conclusive.

The research regarding CLA's effect on energy intake and expenditure is also variant between research groups. In animal models, CLA has been shown to increase or to show no effect on energy intake. Energy expenditure was shown to decrease and to stay at baseline with CLA supplementation. No effect was found in humans. There is good evidence to support that CLA does in fact modulate energy intake and expenditure; however, the mechanisms of this action remain unclear.

Based on ambiguous but suggestive findings, more research examining the effects of CLA on energy expenditure and metabolism are warranted. These studies are clearly required for a better understanding and for making

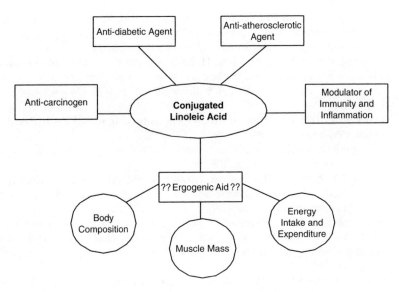

FIGURE 13.2
Health properties that may be responsive to conjugated linoleic acid.

recommendations to the public regarding the practice of using CLA as an ergogenic or performance-enhancing aid. (Figure 13.2).

Acknowledgments

The authors are grateful to Dr. J. Buell, A. Purushotham and A. Wendel for their helpful feedback regarding this manuscript.

References

1. Pariza MW, Park Y and Cook ME. 2000. Mechanisms of action of conjugated linoleic acid: Evidence and speculation. *Exp. Bio. Med.*, 223, 8.
2. Chin SF, Liu W, Storkson JM, Ha YL and Pariza MW. 1992. Dietary sources of conjugated dienoic isomers of linoleic acid, a newly recognized class of anti-carcinogens. *J. Food Compos. Anal.*, 5, 185.
3. Griinari JM, Cori BA, Lacy SH, Chouinard PY, Nurmela KVV and Bauman DE. 2000. Conjugated linoleic acid is synthesized endogenously in lactating dairy cows by delta(9)-desaturase. *J. Nutr.*, 130, 2285.
4. Ma D, Wierzbicki A, Field C and Clandinin MT. 1999. Conjugated linoleic acid in Canadian dairy and beef products. *J. Agric. Food Chem.*, 47, 1956.
5. Parodi PW. 2002. Conjugated linoleic acid. *Food Australia*, 54, 96.
6. Banni S. 2002. Conjugated linoleic acid metabolism. *Current Opinion in Lipidology*, 13, 261.

7. Kramer JKG, Parodi PW, Jensen RG, Mossoba MM, Yurawecz MP and Adlof RO. 1998. Rumenic acid: A proposed common name for the major conjugated linoleic acid isomer found in natural products. *Lipids*, 33, 83.
8. Fritsche J, Rickert R and Steinhart H. 1999. Formation, contents and estimation of daily intake of conjugated linoleic acid isomers and trans-fatty acids in foods. *Advances in Conjugated Linoleic Acid Research*, Vol. 1. AOCS Press, Champaign, IL; 1999:378.
9. Belury MA. 2002. Inhibition of carcinogenesis by conjugated linoleic acid: Potential mechanisms of action. *J. Nutr.*, 132, 2995.
10. Lee KN, Kritchevsky D and Pariza MW. 1994. Conjugated linoleic acid and atherosclerosis. *Atherosclerosis*, 108, 19.
11. Wilson TA, Nicolosi RJ, Chrysam M and Kritchevsky D. 2000. Conjugated linoleic acid reduces early aortic atherosclerosis greater than linoleic acid in hypercholesterolemic hamsters. *Nutr. Res.*, 20, 1795.
12. Houseknect KL, Vanden Heuvel JP, Moya-Camarena SY, Portocarrero CP, Peck LW, Nickel KP and Belury MA. 1998. Dietary conjugated linoleic acid normalizes impaired glucose tolerance in the zucker diabetic fatty rat. *Biochem. Biophys. Res. Comm.*, 244, 678.
13. Park Y, Albright KJ, Liu W, Storkson JM, Cook ME and Pariza MW. 1997. Effect of conjugated linoleic acid on body composition in mice. *Lipids*, 32, 853.
14. Park Y, Albright KJ, Storkson JM, Liu W, Cook ME and Pariza MW. 1999. Changes in body composition in mice during feeding and withdrawal of conjugated linoleic acid. *Lipids*, 34, 243.
15. West DB, Delany JP, Camet PM, Blohm F, Truett AA and Scimeca J. 1998. Effects of conjugated linoleic acid on body fat and energy metabolism in the mouse. *Am. J. Physiol.*, 275, R667.
16. Koba K, Akahoshi A, Masao Y, Tanaka K, Yamada K, Iwata T, Kamegai T, Tsutsumi K and Sugano M. 2002. Dietary conjugated linoleic acid in relation to CLA differently modifies body fat mass and serum and lipid levels in rats. *Lipids*, 37, 243.
17. Blankson H, Stakkestad JA, Fagertun H, Thom E, Wadstein J and Gudmundsen O. 2000. Conjugated linoleic acid reduces body fat mass in overweight and obese humans. *J. Nutr.*, 130, 2943.
18. Belury MA, Mahon A and Banni S. 2003. The conjugated linoleic acid (CLA) isomer, t10c12-CLA, is inversely associated with changes in body weight and serum leptin in subjects with type 2 Diabetes Meillitus. *J. Nutr.*, 133, 257S.
19. Park Y, Storkson JM, Albright KJ, Liu W and Pariza MW. 1999. Evidence that the trans-10, cis-12 isomer of conjugated linoleic acid induces body composition changes in mice. *Lipids*, 34, 235.
20. Lowery LM, Appicelli PA and Lemon PWR. 1998. Conjugated linoleic acid enhances muscle size and strength gains in novice bodybuilders. *Med. Sci. Sports Exerc.*, 30, S182.
21. Ferreira M, Kreider R, Wison M and Almada A. 1997. Effects of conjugated linoleic acid supplementation during resistance training on body composition and strength. *J. Strength Cond. Res.*, 11, 280.
22. Kreider R, Ferreira M, Wilson M and Almada A. 1998. Effects of conjugated linoleic acid (CLA) supplementation during resistance-training on bone mineral content, bone mineral density and markers of immune stress. *FASEB J.*, 12, A244.

23. Kreider R, Ferreira M, Greenwood M, Wilson M and Almada A. 2002. Effects of conjugated linoleic acid supplementation during resistance training on body composition, bone density, strength and selected hematological markers. *J. Strength Cond. Res.*, 16, 325.
24. Tsuboyama-Kasaoka N, Takahashi M, Tanemura K, Kim HJ, Tange T, Okuyama H, Kasai M, Ikemoto S and Ezaki O. 2000. Conjugated linoleic acid supplementation reduced adipose tissue by apoptosis and develops lipodystrophy in mice. *Diabetes.*, 49, 1534.
25. Tsuboyama-Kasaoka N, Miyazaki H, Kasaoka S and Ezaki O. 2003. Increasing the amount of fat in a conjugated linoleic acid-supplemented diet reduces lipodystrophy in Mice. *J. Nutr.*, 133, 1793
26. Riserus U, Arner P, Brismar K and Vessby B. 2002. Treatment with dietary trans 10 cis 12 conjugated linoleic acid causes isomerspecific insulin resistance in obese men with the metabolic syndrome. *Diabetes Care*, 25, 1516.
27. Riserus U, Basu S, Jovinge S, Fredrikson GN, Arnlov J and Vessby B. 2002. Supplementation with conjugated linoleic acid causes isomer-dependent oxidative stress and elevated C-reaction protein. *Circulation*, 106, 1925.
28. Zambell KL, Keim NL, Van Loan MD, Gale B, Benito P, Kelley DS and Nelson GJ. 2000. Conjugated linoleic acid supplementation in humans: Effects on body composition and energy expenditure. *Lipids*, 35, 777.
29. Chin SF, Storkson JM, Albright KJ, Cook ME and Pariza MW. 1994. Conjugated linoleic acid is a growth factor for rats as shown by enhanced weight gain and improved feed efficiency. *J. Nutr.*, 124, 2344.
30. Poulos SP, Sisk M, Hausman DB, Azain MJ and Hausmann GJ. 2001. Pre- and postnatal dietary conjugated linoleic acid alters adipose development, body weight gain and body composition in Sprague-Dawley rats. *J. Nutr.*, 131, 2722.
31. Bouthegourd JC, Even PC, Gripois D, Tiffon B, Blouquit MF, Roseau S, Lutton C, Tome D and Martin JC. 2002. A CLA mixture prevents body triglyceride accumulation without affecting energy energy expenditure in Syrian hamsters. *J. Nutr.*, 132, 2682.
32. West DB, Blohm FY, Truett AA and DeLany JP. 2000. Conjugated linoleic acid persistently increases total energy expenditure in AKR/J mice without increasing uncoupling protein gene expression. *J. Nutr.*, 130, 2471.

14
Medium-Chain Triglycerides and Glycerol

Timothy P. Carr

CONTENTS

I. Introduction 221
II. Overview of Metabolism 222
    A. Chemical and Physical Properties 222
    B. Lipid Digestion and Absorption 223
    C. Fatty Acid and Glycerol Utilization 227
III. Dietary and Supplemental Sources 229
IV. Ergogenic Benefits 231
    A. Medium-Chain Triglycerides 231
    B. Glycerol 233
V. Toxicity 236
VI. Summary and Recommendations 238
References 239

I. Introduction

Dietary medium-chain triglycerides (MCT) have attracted attention as an ergogenic aid because they are quickly absorbed and metabolized by the liver, providing a more immediate energy source than long-chain triglycerides (LCT) commonly found in foods. Proponents of MCT supplementation believe that MCT, when ingested with adequate carbohydrate, spares muscle glycogen during exercise and thus increases performance. MCT are so named because they comprise medium-chain fatty acids (MCFA) having chain lengths of 6 to 12 carbon atoms. Unlike long-chain fatty acids (LCFA) having chain lengths of 14 carbon atoms and greater, MCFA do not require solubilization by bile for digestion and most are absorbed and transported directly to the liver via the portal vein. Rapid oxidization of MCFA in the liver increases the production of ketone bodies that can be used for energy by muscle and other tissues.

Glycerol has also been examined as an ergogenic aid, although its mechanism of action is quite different from MCT. A unique property of glycerol is its ability to retain water in the body; glycerol is therefore used to maintain adequate hydration during exercise, particularly in hot environments. Although glycerol forms the backbone of triglyceride molecules and can be used as a source of energy, its contribution to the body's overall energy utilization is negligible compared with dietary carbohydrate and fatty acids. Consequently, using glycerol supplementation as an energy source is not an effective strategy for sparing muscle glycogen. Current scientific evidence supports the role of glycerol supplementation in increasing fluid retention in the body, but whether hyperhydration improves exercise performance is still uncertain.

This chapter provides a brief overview of fatty acid, triglyceride and glycerol metabolism under normal dietary circumstances. Also included is a discussion on the dietary and supplemental sources of MCT and glycerol. Finally, this chapter reviews human studies that have investigated MCT and glycerol as ergogenic aids. Note that the terms MCT and LCT used throughout the chapter refer to triglyceride molecules composed primarily of MCFA and LCFA, respectively, while recognizing that some LCFA are present in MCT and some MCFA are present in LCT.

II. Overview of Metabolism

A. Chemical and Physical Properties

All triglycerides have the same chemical structure of three fatty acids attached to a glycerol molecule through ester bonds (Figure 14.1). Most fatty acids in nature exist as triglycerides, although free fatty acids are formed during normal metabolism. Triglyceride molecules found in nature may contain the same fatty acid species at all three positions, or they may contain different fatty acids. The unique physical and chemical properties of triglycerides are thus defined by the characteristics of the fatty acids. Table 14.1 shows some common fatty acids found in nature and are categorized as having short, medium or long hydrocarbon chains. The division between lauric (12:0) and myristic (14:0) acid is somewhat arbitrary; some literature sources list lauric acid as a LCFA or myristic acid as a MCFA, while other sources refer to lauric and myristic acid as neither MCFA nor LCFA, but rather "in between." Nevertheless, the classification of fatty acids according to chain length is useful because the groupings reflect certain metabolic attributes unique to each group. Note that increased chain length is generally associated with increased melting point and decreased solubility in water.[1,2] Also note that the presence of double bonds in the LCFA is associated with

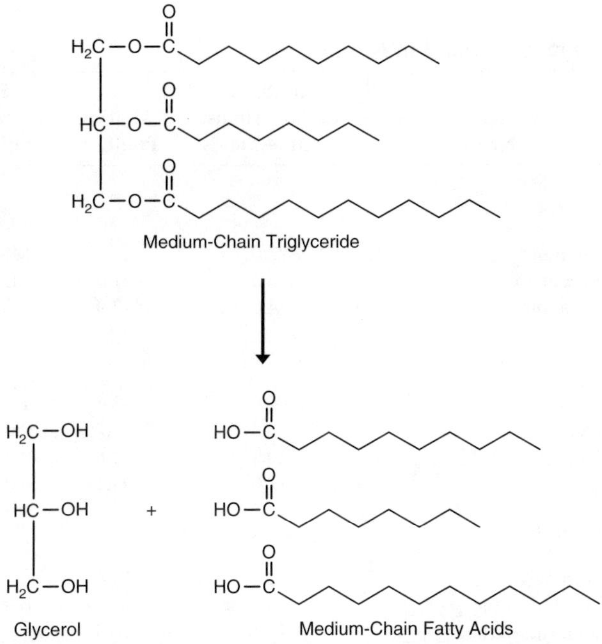

FIGURE 14.1
Chemical structure of a representative medium-chain triglyceride (MCT), medium-chain fatty acids and glycerol. Greater than 90% of fatty acids in commercial MCT preparations are 8 or 10 carbon atoms in length.

decreased melting point. Triglycerides composed mainly of MCFA or unsaturated LCFA are generally liquid at room temperature, whereas triglycerides with a high proportion of saturated LCFA are solid at room temperature. In addition, triglycerides composed of MCFA are more soluble in water than triglycerides composed of LCFA, which influences how MCT and LCT are digested and utilized in the body.

B. Lipid Digestion and Absorption

LCT represent the majority of "fat" consumed by humans, which can exceed 100 grams/day in the U.S.[3] Digestion of LCT begins in the stomach with the action of lingual and gastric lipases.[4] These enzymes preferentially cleave the LCFA at the sn-3 position, leaving 1,2-diglycerides and free fatty acids to pass into the small intestine. The 1,2-diglycerides and triglycerides are further acted upon by pancreatic lipase — stimulated by a protein called colipase — that preferentially cleaves the sn-1 and sn-3 positions.[5] Consequently, the main products of LCT digestion are 2-monoglycerides and free fatty acids, although some free glycerol is also produced. The efficiency of pancreatic lipase is greatly increased by the presence of bile salts released from the gallbladder.

TABLE 14.1

Fatty Acids Commonly Found in Nature

Common Name	Systematic Name	Number of Carbon Atoms: Double Bonds	Melting Point, °C	Solubility in Water,* g/100 g at 20°C
Short-Chain Fatty Acids				
Acetic	Ethanoic	2:0		Miscible
Propionic	Propanoic	3:0	−21.5	Miscible
Butyric	Butanoic	4:0	−7.9	Miscible
Medium-Chain Fatty Acids				
Caproic	Hexanoic	6:0	−3.4	1.08–1.24
Caprylic	Octanoic	8:0	16.7	0.068–0.286
Capric	Decanoic	10:0	31.6	0.015–0.063
Lauric	Dodecanoic	12:0	44.2	0.0055–0.0137
Long-Chain Fatty Acids				
Myristic	Tetradecanoic	14:0	54.4	0.0020–0.0029
Palmitic	Hexadecanoic	16:0	61.3	0.00060–0.00072
Palmitoleic	9-Hexadecenoic	16:1	−0.5	
Stearic	Octadecanoic	18:0	69.6	0.00012–0.00029
Oleic	9-Octadecenoic	18:1	16.3	
Linoleic	9,12-Octadecadienoic	18:2	−6.5	
Linolenic	9,12,15-Octadecatrienoic	18:3	−12.8	

Sources: Ralston, A.W. and Hoerr, C.W., The solubilities of the normal saturated fatty acids. *J. Org. Chem.*, 7, 546, 1942; Westergaard, H. and Dietschy, J.M., The mechanism whereby bile acid micelles increase the rate of fatty acid and cholesterol uptake into the intestinal mucosal cell. *J. Clin. Invest.*, 58, 97, 1976.

Bile salts are biological detergents that help "dissolve" LCT, thus increasing the surface area and exposure of the fat droplets to enzymatic action. Because the main type of fat in human diets is LCT, bile salts and pancreatic lipase are critically important in maintaining proper fat digestion. Metabolic deficiencies in bile salts or pancreatic lipase will result in steatorrhea, a condition of excessive amounts of fat in the feces.[6]

MCT do not require bile salts for digestion.[7] Because MCT are smaller in size and more water-soluble than LCT, digestive lipases have easier access to MCT molecules for fatty acid hydrolysis. Lingual and gastric lipases hydrolyze MCT in the stomach faster and more completely than LCT, thus yielding mostly free fatty acids and free glycerol.[8,9] The products of MCT hydrolysis are absorbed faster than those of LCT,[10] and a significant amount of MCT digestion and absorption can take place even in the absence of pancreatic lipase.[11] For these reasons, MCT have been used therapeutically as an energy source for patients suffering from lipid malabsorption associated with conditions such as pancreatic insufficiency, cystic fibrosis, sprue, short bowel syndrome, HIV and AIDS, Crohn's disease and premature and low-birth-weight infants.[12–17]

LCFA and MCFA are mostly absorbed in the proximal small intestine (duodenum and jejunum),[18,19] although significant amounts of MCFA may be absorbed in the ileum[20] and even the large intestine.[21,22] 2-Monoglycerides remaining from digestion are also absorbed in the proximal small intestine. The mechanisms by which fatty acids cross the cell membrane are not fully understood, but probably involve both diffusion and protein-mediated processes.[23] Once in the intestinal mucosal cell, most of the LCFA — and to a lesser extent some MCFA — are reassembled with 2-monoglycerides to form triglycerides once again.[24–26] The triglycerides are then incorporated into large lipoproteins called chylomicrons and secreted into the lymphatic system, where they eventually enter the blood at the subclavian vein near the heart.[27] The lymphatic routing of chylomicrons has the advantage of bypassing the liver so that dietary LCFA can be delivered directly to muscle and adipose for energy utilization or storage,[28] as illustrated in Figure 14.2. From an evolutionary point of view, humans have benefited from this mechanism of fat absorption because dietary fat when first absorbed is not subject to catabolism by the liver, so the pathway represents an efficient way to capture and conserve dietary energy. In modern times, however, our abundant food supply rich in fat, coupled with the pathway's efficiency, easily contributes to the accumulation of body fat.

In contrast to LCFA, most MCFA are transported directly into the bloodstream via the portal vein, where they bind to serum albumin for delivery to the liver (Figure 14.2). A common misconception is that all MCFA are absorbed into the portal blood and that all dietary LCFA are absorbed into the lymphatic system. In reality, evidence dating back to the 1950s has consistently shown that considerable amounts of LCFA are absorbed directly into the portal vein,[29–34] while MCFA can be reassembled into triglycerides, incorporated into chylomicrons and released into the lymphatic system.[33,35] The solubility and physical characteristics of the fatty acids (see Table 14.1) represent a range of properties that determine the partitioning of fatty acids into lymph or portal blood. Most important is that the extent of triglyceride reassembly and secretion into lymph increases proportionally with increased fatty acid chain length (and decreased solubility), such that lymphatic absorption of capric acid is <20%, lauric acid is about 40%, myristic acid is 60–70% and 18-carbon fatty acids is >80%.[30,35–37] The evidence therefore indicates that the majority — but not all — of the LCFA are absorbed into lymph, whereas the majority of MCFA are quickly absorbed into the portal blood.[38–40]

Free glycerol is also a product of lipid digestion. While LCT hydrolysis yields mostly free fatty acids and 2-monoglycerides, MCT hydrolysis is relatively complete and yields mostly free glycerol and fatty acids. Absorption of glycerol occurs rapidly in the proximal small intestine, although it can also be absorbed in the stomach.[41] Glycerol absorbed in the intestinal mucosal cell is not used for triglyceride synthesis, but enters the portal blood as free glycerol and is quickly taken up by the liver.[42]

Short-chain fatty acids (SCFA) are not major components of the food supply, but are readily synthesized from indigestible carbohydrates by bacteria

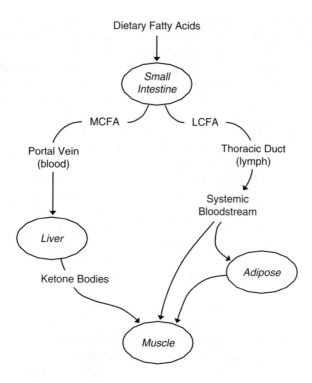

FIGURE 14.2
Routes of transport of dietary medium-chain fatty acids (MCFA) and long-chain fatty acids (LCFA). After digestion of dietary triglycerides, most of the LCFA are resynthesized into triglycerides in the intestinal mucosal cell and incorporated into chylomicrons. The chylomicrons are then secreted into lymph for transport to the bloodstream where the LCT are delivered to adipose for storage or to muscle for energy utilization. In contrast, MCFA are absorbed as free fatty acids and are transported directly to the liver for conversion to ketone bodies. Muscle cells use ketone bodies for energy.

in the large intestine, where they are absorbed and used for energy and other metabolic processes.[43–45] In this way, most SCFA in the gastrointestinal tract are synthesized as free fatty acids and are not produced by enzymatic digestion from food triglycerides. (An exception is milk fat, which contains small amounts of SCFA.) Ruminant animals including cows and sheep consume mostly indigestible carbohydrates as their main fuel source, so that the production of SCFA contributes up to 75% of the animals' total energy. (This also explains why milk and dairy products contain SCFA.) In humans, SCFA are absorbed in the large intestine and can enter the bloodstream or be used for energy by the intestinal mucosal cell. SCFA in the blood can be transported to muscle and used for energy, but their contribution is very small compared with LCFA consumed in food. SCFA have not been widely investigated as ergogenic aids. For centuries, attention has focused on vinegar — diluted acetic acid — as a cure-all for maladies ranging from athlete's foot to yeast infections and for giving Samurai warriors strength and power. But research has shown that acetic acid supplementation does not affect skeletal muscle metabolism.[46] A related compound, dichloroacetate, has been shown to improve exercise performance and delay muscle fatigue,[47-50] but dichloracetate is not considered an SCFA. Another popular dietary supplement, β-hydroxy-β-methylbutyrate (HMB), is similar in chemical structure to butyric acid and has shown promise as an ergogenic aid,[51-53] but HMB is a metabolite of the amino acid leucine and is more similar in chemistry to ketone bodies rather than SCFA.

C. Fatty Acid and Glycerol Utilization

Fatty acids that are absorbed lymphatically appear in the bloodstream as chylomicrons within minutes after consuming a fat-containing meal. The primary function of chylomicrons is to deliver dietary fatty acids (in the form of triglycerides) to peripheral tissues for storage or energy utilization. For this delivery to occur, the chylomicrons must interact with the enzyme lipoprotein lipase anchored on the capillary endothelium of various tissues.[54,55] Lipoprotein lipase converts the triglyceride into free fatty acids and monoglycerides, allowing entry into tissues. Complete hydrolysis of triglyceride is limited, although some free glycerol is produced and can escape into the bloodstream. Those tissues expressing the highest concentration of lipoprotein lipase are muscle and adipose,[56] indicating that these tissues are quantitatively the most important recipients of dietary fat. Lipoprotein lipase associated with cardiac muscle has a particularly high affinity for chylomicrons and other triglyceride-rich lipoproteins, thus ensuring an adequate and constant energy supply for the heart.

Fatty acids (and monoglycerides) taken up by adipose are quickly reassembled into triglycerides for storage, whereas fatty acids taken up by muscle are oxidized and used for energy. The relative proportion of fatty acids taken up by adipose or muscle depends on the energy balance state of the body, which is influenced by the amount and frequency of food consumed and the extent and type of physical activity. Accordingly, triglycerides may continue to accumulate in adipose soon after a meal and when physical activity is low; or adipose triglycerides may be broken down to free fatty acids for transport to muscle between meals or when energy demand is high during exercise. Mobilization of stored triglyceride is initiated by the action of hormone-sensitive lipase, which preferentially cleaves fatty acids at the sn-1 and sn-3 positions.[57,58] The participation of another enzyme, monoglyceride lipase, is required to achieve complete fatty acid hydrolysis.[58,59] The resulting free fatty acids move through the cell membrane, where they bind to serum albumin for transport in the bloodstream. In this way, dietary fatty acids that are absorbed via the lymphatic system (mostly LCFA) ultimately provide energy to muscle, but they may be transiently stored in adipose until needed.

In muscle cells, fatty acids are a rich source of energy, providing more than twice the energy of an equivalent amount of glucose (fatty acids provide 9 kcalories per gram, whereas glucose provides 4 kcalories per gram). The reason is that fatty acids exist in a more reduced state than glucose and therefore undergo a greater extent of oxidation as energy is liberated. While many tissues are capable of oxidizing fatty acids, the large proportion of muscle in the human body makes this tissue a quantitatively significant user of fatty acids. Oxidation of fatty acids occurs in the mitochondria; therefore, the fatty acids must first be transported across the mitochondrial membrane attached to the carrier molecule carnitine.[60,61] Once in the mitochondria, the fatty acids are oxidized via the Krebs cycle, thus producing large amounts of adenosine triphosphate, or ATP. (ATP is the body's main "energy broker" and directly participates in energy-requiring reactions such as muscle contraction and brain function.) Regarding the ergogenic properties of MCT, studies have shown that MCFA quickly enter the mitochondria without the aid of carnitine,[62,63] suggesting that dietary MCT may provide a more immediate source of energy to working muscle.

Fatty acids absorbed into the portal vein are immediately removed from the circulation by the liver. Because most MCFA are absorbed into the portal vein, the majority of dietary MFCA are metabolized in the liver rather than muscle.[7] The liver breaks down MCFA to be used for energy within liver cells. If the liver's energy needs are met and MCFA are abundant, the excess MCFA will be converted to ketone bodies and released into the bloodstream.[64,65] The ketone bodies produced by the liver are acetoacetate and β-hydroxybutyrate (and minor amounts of acetone). The body's cells — including muscle — respond to increased levels of ketone bodies by using them for energy.[66] It is also known that consumption of MCT can significantly increase serum ketone body levels and are frequently used to treat medical conditions where ketogenic diets are desirable.[67,68] The use of MCT as an ergogenic aid is based on the belief that muscle cells will use ketone bodies for energy before using muscle glycogen or body proteins. In theory, MCT supplementation would spare muscle glycogen (important for endurance athletes) and spare muscle protein (important for bodybuilders). However, as discussed below, the effectiveness of MCT has not been firmly established.

Glycerol is also released from adipose during triglyceride mobilization and is transported in the blood to the liver. (Glycerol resulting from the action of lipoprotein lipase and glycerol absorbed during digestion of dietary triglyceride is also transported to the liver.) Here, the glycerol has three possible fates: (1) conversion to glucose, (2) synthesis of triglycerides or (3) oxidized directly for energy. Which pathway is favored depends on the individual's dietary state and level of physical activity. Triglyceride mobilization is greatest when glucose is in short supply, such as during the postabsorptive or fasting states,[69] and when physical activity increases.[70,71] Therefore, most glycerol under these conditions is converted to glucose in the liver to maintain an energy source for the brain, red blood cells and other tissues that cannot use fatty acids for energy. In contrast, excess glycerol may be used for triglyceride synthesis during the well-fed state and when the body is at rest. The amount of glycerol used for energy in the liver is minimal under most conditions.

III. Dietary and Supplemental Sources

Fats and oils found in nature contain a mixture of fatty acids of varying proportions. Table 14.2 shows the fatty acid composition of triglycerides from a variety of plant and animal sources. Coconut and palm kernel oil are particularly rich sources of MCFA, whereas milk fat is one of the few triglycerides of animal origin that contain appreciable amounts of MCFA. This is because the mammary gland is unique in its ability to synthesize MCFA.[72-74] The increased proportion of MCFA in milk fat probably contributes to the efficient digestion of milk fat observed in infants whose pancreatic and biliary function — and hence the digestion of LCT — may not be fully developed. Most infant formulas now contain a mixture of native vegetable oils and MCT to better reflect the fatty acid composition of breast milk.

Table 14.2 clearly shows that, while some fats and oils are relatively rich in MCFA, no single source found in nature contains exclusively MCFA. Therefore, the term "MCT" or "MCT oil" refers to products that are commercially prepared. MCT were first developed in the 1950s by Dr. V.K. Babayan, who was working for Stokely-Van Camp, Inc., a food company that also developed Gatorade®.[75] His procedure for making MCT is fairly straightforward:[76] First, oils rich in MCFA, usually coconut or palm kernel oil, are hydrolyzed to yield free fatty acids and glycerol. The fatty acids are then separated by distillation into fractions containing (1) caproic, caprylic and capric acid, (2) lauric acid and (3) myristic acid and larger. The distillation procedure can be modified so that fraction #2 contains a combination of lauric and myristic acid, but this modification does not affect the MCFA products in fraction #1. The third step in MCT production is to reattach the fatty acids from fraction #1 back to glycerol to create triglycerides once again. Note that lauric acid, although classified as an MCFA, is excluded from fraction #1 and is generally not used in the production of MCT. The reason is partly the high demand for lauric acid in the manufacturing of detergents, shampoos, cosmetics and other household products. Another reason is that lauric acid is the longest MCFA and significant amounts can be absorbed via the lymphatic system,[35,36] a characteristic not desirable for an MCT oil. Thus, the final product is an MCT mixture that contains 1–2% caproic acid, 65–75% caprylic acid, 25–35% capric acid and less than 2% lauric acid.[76]

Development of MCT oils was originally prompted by the desire to create a "fatless" fat that would taste like a fat, that could be eaten and enjoyed like a fat, but would not be metabolized like natural fats and, therefore, not contribute to obesity. Treatment of obesity with MCT has, in fact, proven to be somewhat useful.[77,78] But the major use of MCT has been as a nutritional supplement for medical conditions where normal fat metabolism is compro-

TABLE 14.2
Fatty Acid Composition of Common Fats and Oils

	MCFA						LCFA					
	4:0	6:0	8:0	10:0	12:0	14:0	16:0	16:1	18:0	18:1	18:2	18:3	Other
						% of Total Fatty Acids							
Canola oil						0.1	4.0	0.2	1.8	56.1	20.3	9.3	8.2
Cocoa butter						0.1	25.4	0.2	33.2	32.6	2.8	0.1	5.6
Coconut oil		0.6	7.5	6.0	44.6	16.8	8.2		2.8	5.8	1.8	0.1	5.8
Corn oil						0.1	10.9	0.2	1.8	24.2	58.0	0.7	4.1
Cottonseed oil						0.8	22.7	0.8	2.3	17.0	51.5	0.2	4.7
Olive oil							11.0	0.8	2.2	72.5	7.9	0.6	5.0
Palm oil						1.0	43.5	0.3	4.3	36.6	9.1	0.2	4.9
Palm kernel oil		0.2	3.3	3.7	47.0	16.4	8.1		2.8	11.4	1.6		5.5
Peanut oil						0.1	9.5	0.1	2.2	46.8	32.0		9.3
Safflower oil						0.1	4.3	0.1	1.9	14.4	74.6	0.4	4.2
Soybean oil						0.1	10.3	0.2	3.8	22.8	51.0	6.8	5.0
Beef fat (tallow)					0.9	3.7	24.9	4.2	18.9	36.0	3.1	0.6	7.7
Chicken fat					0.1	0.9	21.6	5.7	6.0	37.3	19.5	1.0	7.9
Milk fat	3.6	2.2	1.2	2.5	2.9	9.8	26.9	2.0	12.1	23.5	3.2	0.4	9.7
Pork fat (lard)				0.1	0.2	1.3	23.8	2.7	13.5	41.2	10.2	1.0	6.0

Source: USDA/ARS Nutrient Data Laboratory (http://www.nal.usda.gov/fnic/foodcomp/ Accessed February 20, 2004)

mised.[79,80] MCT are now routinely included in solutions administered parenterally (intravenous) and enterally (tube feeding). MCT oils can easily replace native fats and oils in a variety of foods products,[81] but they are generally not suitable for frying because they decompose at a lower temperature than most cooking oils containing LCT. Since the U.S. Food and Drug Administration has classified MCT as GRAS (Generally Recognized as Safe), they have become widely available to consumers as oils and powders and as ingredients in a host of food products such as "energy" bars and drinks, desserts and infant formulas. Manufacturers of infant formula usually include MCT in combination with native vegetable oils containing linoleic and linolenic acid to ensure a supply of these essential fatty acids.

Glycerol (also called glycerine) is a colorless, odorless, sweet, syrupy liquid that is commercially produced as a byproduct in the production of soaps and fatty acids. It is available to consumers in its pure form, although it is more widely used as an ingredient in foods and cosmetics and in the manufacturing of printing inks, dyes, lubricants, solvents, glues, toothpaste, antifreeze and dynamite. Glycerol is very hygroscopic; it even absorbs moisture from the air. Because of this strong property, direct consumption of pure glycerol should be avoided and any manner of ingestion should be accompanied by large amounts of water, sports drinks or other liquids.

IV. Ergogenic Benefits

A. Medium-Chain Triglycerides

Depletion of muscle glycogen is the main cause of fatigue during prolonged exercise.[82,83] Therefore, an endurance athlete can benefit from strategies that preserve muscle glycogen and extend its utilization as long as possible. Much attention has focused on carbohydrate (CHO)-feeding strategies before and during exercise as a way to prolong the rate of CHO oxidation and to increase exercise performance. Alternatively, lipids have been examined for their ability to provide energy to working muscle so that glycogen may be spared. Early studies in laboratory rats demonstrated that, when levels of serum free fatty acids are elevated prior to exercise, there is a marked increase in fatty acid oxidation for energy and concomitant sparing of muscle glycogen.[84] In humans, intravenous infusion of LCT during exercise does indeed promote fatty acid oxidation and spares muscle glycogen.[85-87] However, intravenous infusion is obviously not practical and ingesting LCT during exercise is not desirable because: (1) LCT slow gastric emptying, (2) their entry into the bloodstream is relatively slow and, (3) because they enter the circulation as chylomicrons rather than free fatty acids, they do not represent a major fuel source during intense exercise.[88] On the other hand, MCT do not slow gastric emptying[89,90] and are metabolized more similarly to CHO than LCT, suggesting that MCT are quickly used for energy and could thus spare muscle glycogen. This section reviews several human studies that have attempted to determine the impact of MCT ingestion on muscle glycogen, CHO oxidation and/or physical activity and performance (Table 14.3).

In 1980, Ivy et al.[91] conducted the first human study examining the role of MCT in 10 endurance-trained male subjects. The study design allowed for the direct comparison of ingesting 30 g of MCT or 30 g of LCT (both with large amounts of CHO) 1 hour before exercise. Subjects then engaged in cycling or treadmill running for 1 hour at 70% VO_{2max}. The authors found no difference in the rate of CHO oxidation between the MCT or LCT treatments, suggesting that MCT were not preferentially used as energy during exercise. MCT ingestion did, however, cause a significant increase in serum ketone bodies (β-hydroxybutyrate), confirming that MCT are quickly taken up by the liver and converted to ketone bodies. This study raised several questions, including whether MCT were beneficial when consumed without CHO or as a substitute for CHO. In response, Decombaz et al.[92] conducted a study in 1983 that compared the effects of 25 g of MCT or 50 g of CHO given to 12 untrained male subjects 1 hour before exercise. After 1 hour of cycling at 60% VO_{2max}, the results were similar to the Ivy et al.[91] study: no difference in the rate of CHO oxidation, even though serum ketone bodies were higher in the MCT treatment.

These early studies suggest that while MCT ingestion can increase ketone body production, the amount of ketone bodies synthesized may be too small to significantly impact CHO oxidation. Indeed, several investigators have reported that 25–50 g of ingested MCT contributed only 6–14% to total energy expenditure.[92-95] Although three studies have shown a significant reduction in CHO oxidation with MCT treatment,[93,96,97] eight studies have reported no change in CHO oxidation despite increased serum ketone bodies.[90-92,95,98-101]

Another way of quantifying the ergogenic effects of MCT has been to directly analyze muscle glycogen concentrations in tissue biopsies after exercise. Muscle glycogen concentration has been reported in three studies in which subjects consumed MCT before[90,92] and during[100] exercise. In all three studies, MCT ingestion did not alter muscle glycogen concentration despite significantly increased serum ketone body concentrations.[90,92,100] These results indicate that the level of MCT ingested (25–29 g), while high enough to raise serum ketone bodies, was not enough to spare muscle glycogen during exercise.

Surprisingly, only seven studies to date have examined a "performance" outcome as part of their experimental design. In five of these studies, MCT ingestion did not alter exercise performance compared with either CHO[95,98,99] or LCT[95,102,103] ingestion. Van Zyl et al.[97] are the only investigators to report an increase in exercise performance in subjects consuming MCT. Their study consisted of three treatments: 86 g of MCT, 200 g of CHO + 86 g of MCT or 200 g of CHO alone. Subjects consuming MCT + CHO improved their cycling time in a simulated 40-km time trial by 2.5% compared with CHO alone. However, when subjects consumed MCT alone, their cycling time increased

7.9% compared with CHO, indicating that MCT may be detrimental when ingested in the absence of CHO. Jeukendrup et al.[96] repeated the protocol of Van Zyl et al.[97] very closely and found no ergogenic benefit of MCT + CHO compared with CHO alone. Jeukendrup et al.[96] also reported a detrimental effect of MCT, increasing cycling time by 22% compared with CHO. These authors indicated that the high level of MCT (86 g) used in their study was poorly tolerated by the subjects, who experienced severe gastrointestinal side effects.

There may a threshold level at which MCT can be ingested before the onset of gastrointestinal discomfort. Several studies have used 25–29 g of MCT per treatment without adverse side effects,[92-94,100,101] whereas two studies have clearly documented gastrointestinal distress with MCT ingestion.[96,99] One study used 50 g of MCT given 1 hour before exercise, but the authors did not mention the extent to which this high level of MCT was tolerated.[95] Ivy et al.[91] conducted a preliminary experiment in which subjects were given 30, 50 and 60 g of MCT in combination with a variety of solid and liquid foods. They observed gastrointestinal cramping and diarrhea in 100% of the subjects consuming 50 and 60 g of MCT, whereas only 10% of the subjects experienced discomfort at the 30 g level. One approach to minimizing discomfort is to prefeed subjects for several days prior to the exercise event. In one study,[103] subjects were given 34 g of MCT daily for 7 days prior to testing, while in another study,[102] subjects were given 60 g of MCT for 2 weeks prior to testing. All subjects in the latter study[102] complained of gastrointestinal discomfort that lessened in severity or disappeared as the 2-week supplementation period progressed, suggesting that the gastrointestinal tract could be "conditioned" to handle higher amounts of MCT.

The ergogenic benefits of chronic MCT ingestion (greater than 2 weeks) have not been examined in humans, although a study was conducted in mice fed MCT (8% of diet) for 6 weeks.[104] Both trained and untrained mice fed MCT had a significantly greater swimming endurance and also exhibited higher activity of the enzyme, 3-oxo acid CoA-transferase, which plays a role in the uptake of serum ketone bodies by the muscle. Chronic feeding of MCT was also associated with higher serum ketone body levels and higher muscle glycogen levels.[104] The ergogenic benefits of chronic MCT ingestion in humans are not known.

B. Glycerol

Glycerol can be used as a source of energy, but the contribution of glycerol to the body's overall energy use is small compared with carbohydrate and fatty acids, and glycerol supplementation is not a useful strategy for sparing muscle glycogen.[105-107] Nevertheless, glycerol has been touted as an ergogenic aid because of its ability to "absorb" and retain water within the body. This unique property has proven useful in preventing — or at least delaying — the deleterious effects of dehydration during prolonged exercise,

TABLE 14.3
Metabolic Effects of MCT Supplementation in Humans

Ref.	Subjects	Ingestion Protocol	Exercise Protocol	Serum Ketone Bodies	Carbohydrate Oxidation	Muscle Glycogen	Exercise Performance
90	7 trained cyclists	27 g MCT + 54 g CHO or 54 g CHO (given 1 hr before test)	84% VO_{2max} for 30 min	MCT > CHO	No treatment effect	No treatment effect	
91	6 trained cyclists, 4 runners	30 g MCT or 30 g LCT (given 1 hr before test)	70% VO_{2max} for 1 hr	MCT > LCT	No treatment effect		
92	12 untrained cyclists	25 g MCT or 50 g CHO (given 1 hr before test)	60% VO_{2max} for 1 hr	MCT > CHO	No treatment effect	No treatment effect	
93	8 trained cyclists	29 g MCT or 29 g MCT + 149 g CHO or 29 g MCT + 214 g CHO or 214 g CHO (given in partial doses every 20 min during test)	57% VO_{2max} for 3 hr	MCT > MCT+CHO > CHO	MCT < MCT+CHO < CHO		
94	6 untrained cyclists	25 g MCT or 57 g CHO (given 1 hr before test)	65% VO_{2max} for 2 hr	MCT > CHO	No treatment effect		
95	9 trained cyclists	50 g MCT or 44 g LCT or 100 g CHO (given 1 hr before test)	60% VO_{2max} until exhaustion	MCT > CHO, LCT			No treatment effect
96	7 trained cyclists	85 g MCT or 85 g MCT + 170 g CHO or 170 g CHO (given in partial doses every 15 min during test)	60% VO_{2max} for 2 hr, followed by time trial	MCT > MCT+CHO, CHO	MCT < MCT+CHO, CHO		CHO, MCT+CHO > MCT

Medium-Chain Triglycerides and Glycerol

#	Subjects	Treatment	Exercise protocol	Result 1	Result 2	Result 3
97	6 trained cyclists	86 g MCT *or* 86 g MCT + 200 g CHO *or* 200 g CHO (given in partial doses every 10 min during test)	60% VO_{2max} for 2 hr, followed by 40-km time trial	MCT > MCT+CHO > CHO	MCT < MCT+CHO < CHO	MCT+CHO > CHO > MCT
98	8 trained cyclists	11 g MCT + 15 g CHO *or* 15 g CHO (given every 15 min during test)	Time trial to complete 35 kJ/kg body wt	No treatment effect	No treatment effect	No treatment effect
99	9 trained cyclists	13 (or 28) g MCT + 40 g CHO before test, followed by 1.7 (or 3.4) g MCT every 10 min during test	63% VO_{2max} for 2 hr, followed by 40-km time trial	MCT > CHO	No treatment effect	No treatment effect
100	9 trained cyclists	29 g MCT + 149 g CHO *or* 29 g MCT + 214 g CHO *or* 214 g CHO (given in partial doses every 20 min during test)	57% VO_{2max} for 3 hr	MCT+CHO > CHO	No treatment effect	No treatment effect
101	8 trained cyclists	27 g MCT + 87 g CHO *or* 146 g CHO (given in partial doses every 20 min during test)	57% VO_{2max} for 90 min	MCT+CHO > CHO	No treatment effect	
102	12 trained runners	60 g MCT *or* 56 g LCT (given daily for 2 wk before test)	85% VO_{2max} for 30 min, then 75% VO_{2max} until exhaustion	No treatment effect		No treatment effect
103	7 trained runners	34 g MCT *or* 33 g LCT (given daily for 1 wk before test)	80% VO_{2max} until exhaustion	MCT > LCT		No treatment effect

Note: CHO, carbohydrates; MCT, medium-chain triglycerides; LCT, long-chain triglycerides; VO_{2max}, maximum oxygen consumption.

but whether glycerol supplementation improves performance through hyperhydration is still debatable.

There is little doubt that glycerol ingestion effectively promotes hyperhydration, as indicated by the majority of studies in Table 14.4 showing increased fluid retention (as measured by increased plasma volume, decreased urine output or total body water). Riedesel and co-workers[108,109] provided the first evidence that ingestion of glycerol with adequate fluid could significantly reduce urine output over several hours. They further reported that the body's core temperature could be reduced during exercise in a hot dry environment.[109] However, only one other study[110] has observed a reduction in body temperature due to glycerol ingestion, whereas six studies[111-116] have reported no change in body temperature. Thus, the evidence indicates that hyperhydration can easily be achieved with glycerol ingestion around 1 g/kg body weight (accompanied by ample amounts of water), although the relationship between hyperhydration and thermoregulation is less certain and requires further research.

Whether glycerol-induced hyperhydration increases exercise performance is also uncertain. To date, five studies[110,111,115,117,118] have indicated improvements in performance, whereas five studies[113,114,119-121] reported no improvement in performance with glycerol ingestion. Moreover, one study by Gleeson et al.[122] reported that glycerol ingestion (1 g/kg body weight) significantly decreased exercise time to exhaustion compared with a glucose solution. As pointed out by Wagner,[123] the disparity of findings regarding performance outcomes is likely due to methodological differences among studies. There appears to be little consistency in the conditions of exercise — including differences in air temperature, relative humidity, air circulation, duration and intensity of exercise — such that the thermoregulatory system of the body was stressed to different extents. Other confounding factors include differences in the amount of glycerol ingested, the length of time between glycerol ingestion and the start of exercise, and the amount of water (or carbohydrate) co-ingested with glycerol. Despite these differences, the importance of maintaining adequate hydration during exercise, particularly prolonged exercise in hot conditions, cannot be overemphasized.

V. Toxicity

A recent comprehensive review of MCT found no indication of toxicity when consumed at moderate levels in several animal species.[129] There appears to be no acute or subchronic effects of MCT in chicks, mice, rabbits, or rats when MCT are administered in daily doses of up to ~15 g/kg body weight. There is no evidence that MCT adversely affect reproduction in rats fed 12 g/kg per day or in pigs fed 4 g/kg per day. While an LD_{50} has not been established for MCT, it is likely to be > 24 g/kg in mice and > 34 g/kg in rats.[130] In humans, MCT ingestion up to 1 g/kg body weight appears to be

TABLE 14.4

Metabolic Effects of Glycerol Supplementation in Humans

Ref.	Subjects	Ingestion Protocol	Metabolic Effects		
			Fluid Retention	Thermo-Regulation	Exercise Performance
108	22 rested subjects	0.5, 1.0 or 1.5 g/kg bw	↑		
109	6 untrained runners	1 g/kg bw	↑	↓ core temp.	
110	6 trained cyclists	1 g/kg bw	↑	↓ core temp.	↑
111	8 trained cyclists	1 g/kg bw	↑	↔	↑
112	8 trained runners	1.2 g/kg lbm	↔	↔	
113	8 trained runners	1.2 g/kg lbm	↔	↔	↔
114	7 trained cyclists	1.2 g/kg bw		↔	↔
115	11 trained cyclists	1.2 g/kg bw		↔	↑
116	9 untrained cyclists	1.2 g/kg lbm	↑	↔	
117	10 trained triathletes	1.2 g/kg bw	↑		↑
118	8 trained cyclists	1 g/kg bw	↑		↑
119	8 trained triathletes	1 g/kg bw	↔		↔
120	11 untrained runners	1 g/kg bw	↑		↔
121	5 trained cyclists	1 g/kg bw			↔
122	6 untrained cyclists	1 g/kg bw			↓
124	11 rested subjects	1.5 g/kg tbw	↑		
125	8 trained runners	1 g/kg bw	↑		
126	7 rested subjects	~3 g/(kg × day)	↑		
127	8 untrained cyclists	1.1 g/kg bw	↑		
128	9 untrained runners	1% solution ad libitum	↔		

Note: bw, body weight; tbw, total body water; lbm, lean body mass.

safe, and patients receiving MCT parenterally have tolerated doses of 3–9 g/kg body weight per day for periods of several months without adverse effects.[129] Thus, a 70 kg individual could safely consume 70 g per day (1 g/kg body weight) without toxicological consequences. However, doses exceeding 50 g per day clearly produce gastrointestinal disturbances[91] that hinder exercise and decrease performance.

Adverse effects of glycerol ingestion may become more common as athletes seek ways to improve performance through hyperhydration. Lin[41] indicated that an oral dose of 1 g/kg body weight every 6 hours was well tolerated and safe. Humans receiving 1.5 g/kg body weight in a single dose did not report any side effects,[108] although these subjects were rested and did not engage in exercise during the experiment. Bartsch[131] reported the LD_{50} of glycerol in rats was greater than 25 g/kg body weight, suggesting an overdose in humans is possible at higher doses. Some individuals have experienced nausea, vomiting, dizziness and a bloated feeling with glycerol ingestion.[112,116,126] Murray et al.[116] have further cautioned that large doses of glycerol increase the risk of cerebral and intraocular dehydration during exercise because of the slow rates at which glycerol enters the cerebrospinal fluid and aqueous humor.

VI. Summary and Recommendations

Scientific studies have attempted to document the ergogenic benefits of both MCT and glycerol. However, MCT ingestion prior to or during exercise appears to have no effect on muscle glycogen concentration and minimal effects on carbohydrate oxidation. Only one study to date (out of seven reporting performance outcomes) has indicated measurable improvements in exercise performance, but only when MCT were ingested with adequate carbohydrate. Consumption of MCT in the absence of carbohydrate has a negative effect on exercise. This is largely because the dose of MCT must be limited to about 25–30 g per treatment to avoid gastrointestinal disturbances, and because these relatively small doses are negligible compared with the total energy expenditure during exercise. While some individuals can tolerate higher doses, MCT supplementation should be avoided by diabetics (because of the increase in ketone body production) and by people with liver disease. It should also be noted that commercial MCT preparations contain no essential fatty acids (EFA) and could promote EFA deficiency if consumed as the only source of dietary fat. Thus, MCT do not appear to provide a significant advantage over adequate carbohydrate ingestion, indicating that carbohydrate loading (glycogen super-compensation) may still be the best dietary technique for prolonging muscle glycogen utilization.

On the other hand, there is some scientific support for glycerol as an ergogenic aid because of its ability to increase fluid retention in the body. Five (out of 11) studies have reported improvements in performance when subjects "hyperhydrated" with glycerol and water prior to exercise. The greatest benefit of glycerol supplementation is likely to be realized under conditions where the risk of dehydration is increased, such as exercising in hot environments or during prolonged exercise. While glycerol is generally well tolerated at doses of 1 g/kg body weight, side effects have been reported. Individuals ingesting glycerol are advised to drink ample amounts of water; because of its strong hygroscopic properties, glycerol should never be consumed undiluted. It should also be stressed that, as with any dietary supplement, individuals wishing to try glycerol (or MCT) supplementation to improve performance should experiment gradually to establish their own level of tolerance and to avoid serious side effects.

References

1. Ralston, A.W. and Hoerr, C.W., The solubilities of the normal saturated fatty acids. *J. Org. Chem.*, 7, 546, 1942.
2. Westergaard, H. and Dietschy, J.M., The mechanism whereby bile acid micelles increase the rate of fatty acid and cholesterol uptake into the intestinal mucosal cell. *J. Clin. Invest.*, 58, 97, 1976.
3. United States Department of Agriculture, Results from USDA's 1994–96 Continuing Survey of Food Intakes by Individuals. 1997.
4. Hamosh, M., Lingual and gastric lipases. *Nutrition*, 6, 421, 1994.
5. Carey, M.C., Small, D.M. and Bliss, C.M., Lipid digestion and absorption. *Ann. Rev. Physiol.*, 45, 651, 1983.
6. Thomson, A.B., Keelan, M., Thiesen, A., Clandinin, M.T., Ropeleski, M. and Wild, G.E., Small bowel review: Diseases of the small intestine. *Dig. Dis. Sci.*, 46, 2555, 2001.
7. Bach, A.C. and Babayan, V.K., Medium-chain triglycerides: an update. *Am. J. Clin. Nutr.*, 36, 950, 1982.
8. Cohen, M., Morgan, R.G.H. and Hofmann, A.F., Lipolytic activity of human gastric and duodenal juice against medium and long chain triglycerides. *Gastroenterology*, 60, 1, 1971.
9. Hamosh, M., A review. Fat digestion in the newborn: Role of lingual lipase and preduodenal digestion. *Pediatr. Res.*, 13, 615, 1979.
10. Iber, F.L., Relative rates of metabolism MCT, LCT and ethanol in man. *Z. Ernahrungswiss. Suppl.*, 17, 9, 1974.
11. Valdivieso, V.D., Absorption of medium-chain triglycerides in animals with pancreatic atrophy. *Am. J. Dig. Dis.*, 17, 129, 1972.
12. Greenberger, N.J., Ruppert, R.D. and Tzagournis, M., Use of medium chain triglycerides in malabsorption. *Ann. Intern. Med.*, 66, 727, 1967.
13. Roy, C.C., Ste-Marie, M., Chartrand, L., Weber, A., Bard, H. and Doray, B., Correction of the malabsorption of the preterm infant with a medium-chain triglyceride formula. *J. Pediatr.*, 86, 446, 1975.
14. Tantibhedhyangkul, P. and Hashim, S.A., Medium-chain triglyceride feeding in premature infants: effects on fat and nitrogen absorption. *Pediatrics*, 55, 359, 1975.
15. Wanke, C.A., Pleskow, D., Degirolami, P.C., Lambl, B.B., Merkel, K. and Akrabawi, S.S., A medium chain triglyceride-based diet in patients with HIV and chronic diarrhea reduces diarrhea and malabsorption: A prospective, controlled trial. *Nutrition*, 12, 766, 1996.

16. Craig, G.B., Darnell, B.E., Weinsier, R.L., Saag, M.S., Epps, L., Mullins, L., Lapidus, W.I., Ennis, D.M., Akrabawi, S.S., Cornwell, P.E. and Sauberlich, H.E., Decreased fat and nitrogen losses in patients with AIDS receiving medium-chain-triglyceride-enriched formula vs. those receiving long-chain-triglyceride-containing formula. *J. Am. Diet. Assoc.*, 97, 605, 1997.
17. Sakurai, T., Matsui, T., Yao, T., Takagi, Y., Hirai, F., Aoyagi, K. and Okada, M., Short-term efficacy of enteral nutrition in the treatment of active Crohn's disease: A randomized, controlled trial comparing nutrient formulas. *J. Parenter. Enteral Nutr.*, 26, 98, 2002.
18. Sabesin, S.M. and Frase, S., Electron microscopic studies of the assembly, intracellular transport and secretion of chylomicrons by rat intestine. *J. Lipid Res.*, 18, 496, 1975.
19. Wu, A.L., Bennett-Clark, S. and Holt, P.R., Composition of lymph chylomicrons from proximal or distal rat small intestine. *Am. J. Clin. Nutr.*, 33, 582, 1980.
20. Wollaeger, E.E., Role of the ileum in fat absorption. *Mayo Clin. Proc.*, 48, 836, 1973.
21. Jorgensen, J.R., Fitch, M.D., Mortensen, P.B. and Fleming, S.E., In vivo absorption of medium-chain fatty acids by the rat colon exceeds that of short-chain fatty acids. *Gastroenterology*, 120, 1152, 2001.
22. Valdivieso, V.D. and Schwabe, A.D., Absorption medium lipids from the rat cecum. *Am. J. Dig. Dis.*, 11, 474, 1966.
23. Schaffer, J.E., Fatty acid transport: The roads taken. *Am. J. Physiol.*, 282, E239, 2002.
24. Senior, J.R. and Isselbacher, K.J., Activation of long-chain fatty acids by rat gut mucosa. *Biochim. Biophys. Acta*, 44, 399, 1960.
25. Greenberger, N.J., Franks, J.J. and Isselbacher, K.J., Metabolism of 1-C^{14} octanoic and 1-C^{14} palmitic acid by rat intestinal slices. *Proc. Soc. Exp. Biol. Med.*, 120, 468, 1965.
26. Carlier, H. and Bezard, J., Electron microscopy of medium chain fatty acid absorption. *Gastroenterology*, 70, 460, 1976.
27. Hussain, M.M., A proposed model for the assembly of chylomicrons. *Atherosclerosis*, 148, 1, 2000.
28. Cryer, A. and Jones, H.M., The distribution of lipoprotein lipase (clearing factor lipase) activity in the adiposal, muscular and lung tissues of ten animal species. *Comp. Biochem. Physiol. B.*, 63, 501, 1979.
29. Bloom, B., Chaikoff, I.L., Reinhardt, W.O. and Dauben, W.G., Participation of phospholipides in lymphatic transport of absorbed fatty acids. *J. Biol. Chem.*, 189, 261, 1951.
30. Bloom, B., Chaikoff, I.L. and Reinhardt, W.O., Intestinal lymph as pathway for transport of absorbed fatty acids of different chain lengths. *Am. J. Physiol.*, 166, 451, 1951.
31. Borgström, B., Incorporation of saturated fatty acids of different chain lengths in small intestinal and lymph phospholipids of the rat during fat absorption. *Acta Physiol. Scand.*, 25, 315, 1952.
32. Kiyasu, J.Y., The portal transport of absorbed fatty acids. *J. Biol. Chem.*, 199, 415, 1952.
33. Hyun, S.A., Vahouny, G.V. and Treadwell, C.R., Portal absorption of fatty acids in lymph- and portal vein-cannulated rats. *Biochim. Biophys. Acta*, 137, 296, 1967.
34. Lee, D.S., Hashim, S.A. and Van Itallie, T.B., Effect of long chain triglyceride in chylous transport of medium chain fatty acids. *Am. J. Physiol.*, 214, 3294, 1968.

35. Mu, H. and Hoy, C.E., Effects of different medium-chain fatty acids on intestinal absorption of structured triacylglycerols. *Lipids*, 35, 83, 2000.
36. McDonald, G.B., Saunders, D.R., Weidman, M. and Fisher, L., Portal venous transport of long-chain fatty acids absorbed from rat intestine. *Am. J. Physiol.*, 239, G141, 1980.
37. Bernard, A. and Carlier, H., Absorption and intestinal catabolism of fatty acids in the rat: effect of chain length and unsaturation. *Exp. Physiol.*, 76, 445, 1991.
38. Borgström, B., Transport form of ^{14}C decanoic acid in portal and inferior vena cava blood during absorption in the rat. *Acta Physiol. Scand.*, 34, 71, 1955.
39. Bloomstrand, R., Transport form of decanoic acid-1-^{14}C in the lymph during intestinal absorption in the rat. *Acta Physiol. Scand.*, 34, 67, 1955.
40. Vallot, A., Bernard, A. and Carlier, H., Influence of the diet on the portal and lymph transport of decanoic acid in rats. Simultaneous study of its mucosal catabolism. *Comp. Biochem. Physiol. A.*, 82, 693, 1985.
41. Lin, E.C.C., Glycerol utilization and its regulation in mammals. *Ann. Rev. Biochem.*, 46, 765, 1977.
42. Reiser, R., Bryson, M.J., Carr, M.J. and Kuiken, K.A., The intestinal absorption of triglycerides. *J. Biol. Chem.*, 194, 131, 1952.
43. D'Argenio, G. and Mazzacca, G., Short-chain fatty acid in the human colon. Relation to inflammatory bowel diseases and colon cancer. *Adv. Exp. Med. Biol.*, 472, 149, 1999.
44. Scheppach, W., Luehrs, H. and Menzel, T., Beneficial health effects of low-digestible carbohydrate consumption. *Br. J. Nutr.*, 85(Suppl 1), S23, 2001.
45. Topping, D.L. and Clifton, P.M., Short-chain fatty acids and human colonic function: Roles of resistant starch and nonstarch polysaccharides. *Physiol. Rev.*, 81, 1031, 2001.
46. Howlett, R.A., Heigenhauser, G.J. and Spriet, L.L., Skeletal muscle metabolism during high-intensity sprint exercise is unaffected by dichloroacetate or acetate infusion. *J. Appl. Physiol.*, 87, 1747, 1999.
47. Ludvik, B., Mayer, G., Stifter, S., Putz, D., Barnas, U. and Graf, H., Effects of dichloroacetate on exercise performance in healthy volunteers. *Pflugers Arch.*, 423, 251, 1993.
48. Durkot, M.J., De Garavilla, L., Caretti, D. and Francesconi, R., The effects of dichloroacetate on lactate accumulation and endurance in an exercising rat model. *Int. J. Sports Med.*, 16, 167, 1995.
49. Taivassalo, T., Matthews, P.M., De Stefano, N., Sripathi, N., Genge, A., Karpati, G. and Arnold, D.L., Combined aerobic training and dichloroacetate improve exercise capacity and indices of aerobic metabolism in muscle cytochrome oxidase deficiency. *Neurology*, 47, 529, 1996.
50. Howlett, R.A., Heigenhauser, G.J., Hultman, E., Hollidge-Horvat, M.G. and Spriet, L.L., Effects of dichloroacetate infusion on human skeletal muscle metabolism at the onset of exercise. *Am. J. Physiol.*, 277, E18, 1999.
51. Kreider, R.B., Dietary supplements and the promotion of muscle growth with resistance exercise. *Sports Med.*, 27, 97, 1999.
52. Slater, G.J. and Jenkins, D., Beta-hydroxy-beta-methylbutyrate (HMB) supplementation and the promotion of muscle growth and strength. *Sports Med.*, 30, 105, 2000.
53. Schwenk, T.L. and Costley, C.D., When food becomes a drug: Nonanabolic nutritional supplement use in athletes. *Am. J. Sports Med.*, 30, 907, 2002.

54. Goldberg, I.J., Lipoprotein lipase and lipolysis: Central roles in lipoprotein metabolism and atherogenesis. *J. Lipid Res.*, 37, 693, 1996.
55. Goldberg, I.J. and Merkel, M., Lipoprotein lipase: Physiology, biochemistry and molecular biology. *Frontiers Biosci.*, 6, D388, 2001.
56. Camps, L., Reina, M., Llobera, M., Bengtsson-Olivecrona, G., Olivecrona, T. and Vilaro, S., Lipoprotein lipase in lungs, spleen and liver: Synthesis and distribution. *J. Lipid Res.*, 32, 1877, 1991.
57. Fredrikson, G. and Belfrage, P., Postional specificity of hormone-sensitive lipase from rat adipose tissue. *J. Biol. Chem.*, 258, 14253, 1983.
58. Holm, C., Osterlund, T., Laurell, H. and Contreras, J.A., Molecular mechanisms regulating hormone-sensitive lipase and lipolysis. *Ann. Rev. Nutr.*, 20, 365, 2000.
59. Fredrikson, G., Tornqvist, H. and Belfrage, P., Hormone-sensitive lipase and monoacylglycerol lipase are both required for complete degradation of adipocyte triacylglycerol. *Biochim. Biophys. Acta*, 876, 288, 1986.
60. Kerner, J. and Hoppel, C., Fatty acid import into mitochondria. *Biochim. Biophys. Acta*, 1486, 1, 2000.
61. Jeukendrup, A.E., Regulation of fat metabolism in skeletal muscle. *Ann. N. Y. Acad. Sci.*, 967, 217, 2002.
62. Bremer, J., Carnitine and its role in fatty acid metabolism. *Trends Biochem. Sci.*, 2, 207, 1980.
63. McGarry, J.D. and Foster, D.W., Regulation of hepatic fatty acid oxidation and ketone body production. *Ann. Rev. Biochem.*, 49, 395, 1980.
64. Bach, A., Schirardin, H., Weryha, A. and Bauer, M., Ketogenic response to medium-chain triglyceride load in the rat. *J. Nutr.*, 107, 1863, 1977.
65. Guy, D.G. and Tuley, R.J., Jr., Effect of diets high in carbohydrate, soy oil, medium-chain triglycerides or tripelargonin on blood and liver lipid and glucose intermediates in meal-eating rats. *J. Nutr.*, 111, 1437, 1981.
66. Robinson, A.M. and Williamson, D.H., Physiological roles of ketone bodies as substrates and signals in mammalian tissues. *Physiol. Rev.*, 60, 143, 1980.
67. Nebeling, L.C. and Lerner, E., Implementing a ketogenic diet based on medium-chain triglyceride oil in pediatric patients with cancer. *J. Am. Diet. Assoc.*, 95, 693, 1995.
68. Carroll, J. and Koenigsberger, D., The ketogenic diet: A practical guide for caregivers. *J. Am. Diet. Assoc.*, 98, 316, 1998.
69. Randle, P.J., Garland, P.B., Hales, C.N. and Newsholme, E.A., The glucose fatty acid cycle: Its role in insulin sensitivity and the metabolic disturbances of diabetes mellitus. *Lancet*, 1, 785, 1963.
70. Ahlborg, G., Felig, P., Hagenfeldt, S., Hendler, R. and Wahren, J., Substrate turnover during prolonged exercise in man. *J. Clin. Invest.*, 53, 1080, 1974.
71. Ahlborg, G. and Felig, P., Lactate and glucose exchange across the forearm, legs and splanchnic bed during and after prolonged leg exercise. *J. Clin. Invest.*, 69, 45, 1982.
72. Libertini, L.J. and Smith, S., Purification and properties of a thioesterase from lactating rat mammary gland which modifies the product specificity of fatty acid synthetase. *J. Biol. Chem.*, 253, 1393, 1978.
73. Smith, S., Mechanism of chain length determination in biosynthesis of milk fatty acids. *J. Dairy Sci.*, 63, 337, 1980.
74. Thompson, B.J. and Smith, S., Biosynthesis of fatty acids by lactating human breast epithelial cells: An evaluation of the contribution to the overall composition of human milk fat. *Pediatr. Res.*, 19, 139, 1985.

75. Babayan, V.K., Early history and preparation of MCT. *Z. Ernahrungswiss. Suppl.*, 17, 1, 1974.
76. Babayan, V.K., Medium-chain triglycerides: Their composition, preparation and application. *J. Am. Oil Chem. Soc.*, 45, 23, 1968.
77. Bach, A.C., Ingenbleek, Y. and Frey, A., The usefulness of dietary medium-chain triglycerides in body weight control: Fact or fancy? *J. Lipid Res.*, 37, 708, 1996.
78. St-Onge, M.P. and Jones, P.J.H., Physiological effects of medium-chain triglycerides: potential agents in the prevention of obesity. *J. Nutr.*, 132, 329, 2002.
79. Mascioli, E.A., Babayan, V.K., Bistrian, B.R. and Blackburn, G.L., Novel triglycerides for special medical purposes. *J. Parenter. Enteral Nutr.*, 12(6 Suppl), 127S, 1988.
80. Bell, S.J., Bradley, D., Armour Forse, R. and Bistrian, B.R., The new dietary fats in health and disease. *J. Am. Diet. Assoc.*, 97, 280, 1997.
81. Schizas, A.A., Cremen, J.A., Larson, E. and O'Brien, R., Medium-chain triglycerides — use in food preparation. *J. Am. Diet. Assoc.*, 51, 228, 1967.
82. Bergstrom, J., Hermansen, L., Hultman, E. and Saltin, B., Diet, muscle glycogen and physical performance. *Acta Physiol. Scand.*, 71, 140, 1967.
83. Coyle, E.F., Coggan, A.R., Hemmert, M.K. and Ivy, J.L., Muscle glycogen utilization during prolonged strenuous exercise when fed carbohydrate. *J. Appl. Physiol.*, 61, 165, 1986.
84. Rennie, M.J., Winder, W.W. and Holloszy, J.O., A sparing effect of increased plasma fatty acids on muscle and liver glycogen content in the exercising rat. *Biochem. J.*, 156, 647, 1976.
85. Dyck, D.J., Putman, C.T., Heigenhauser, G.J., Hultman, E. and Spriet, L.L., Regulation of fat-carbohydrate interaction in skeletal muscle during intense aerobic cycling. *Am. J. Physiol.*, 265, E852, 1993.
86. Romijn, J.A., Coyle, E.F., Sidossis, L.S., Zhang, X.J. and Wolfe, R.R., Relationship between fatty acid delivery and fatty acid oxidation during strenuous exercise. *J. Appl. Physiol.*, 79, 1939, 1995.
87. Vukovich, M.D., Costill, D.L., Hickey, M.S., Trappe, S.W., Cole, K.J. and Fink, W.J., Effect of fat emulsion infusion and fat feeding on muscle glycogen utilization during cycle exercise. *J. Appl. Physiol.*, 75, 1513, 1993.
88. Jeukendrup, A.E., Saris, W.H.M. and Wagenmakers, A.J.M., Fat metabolism during exercise: A review. Part III: Effects of nutritional interventions. *Int. J. Sports Med.*, 19, 371, 1998.
89. Beckers, E.J., Jeukendrup, A.E., Brouns, F., Wagenmakers, A.J.M. and Saris, W.H.M., Gastric emptying of carbohydrate-medium chain triglyceride suspensions at rest. *Int. J. Sports Med.*, 13, 581, 1992.
90. Horowitz, J.F., Mora-Rodriguez, R., Byerley, L.O. and Coyle, E.F., Preexercise medium-chain triglyceride ingestion does not alter muscle glycogen use during exercise. *J. Appl. Physiol.*, 88, 219, 2000.
91. Ivy, J.L., Costill, D.L., Fink, W.J. and Maglischo, E., Contribution of medium and long chain triglyceride intake to energy metabolism during prolonged exercise. *Int. J. Sports Med.*, 1, 15, 1980.
92. Decombaz, J., Arnaud, M.J., Milon, H., Moesch, H., Philippossian, G., Thelin, A.L. and Howald, H., Energy metabolism of medium-chain triglycerides versus carbohydrates during exercise. *Eur. J. Appl. Physiol.*, 52, 9, 1983.
93. Jeukendrup, A.E., Saris, W.H.M., Schrauwen, P., Brouns, F. and Wagenmakers, A.J.M., Metabolic availability of medium-chain triglycerides coingested with carbohydrates during prolonged exercise. *J. Appl. Physiol.*, 79, 756, 1995.

94. Massicotte, D., Peronnet, F., Brisson, G.R. and Hillaire-Marcel, C., Oxidation of exogenous medium-chain free fatty acids during prolonged exercise: Comparison with glucose. *J. Appl. Physiol.*, 73, 1334, 1992.
95. Satabin, P., Portero, P., Defer, G., Bricout, J. and Guezennec, C.Y., Metabolic and hormonal responses to lipid and carbohydrate diets during exercise in man. *Med. Sci. Sports Exerc.*, 19, 218, 1987.
96. Jeukendrup, A.E., Thielen, J.J.H.C., Wagenmakers, A.J.M., Brouns, F. and Saris, W.H.M., Effect of medium-chain triacylglycerol and carbohydrate ingestion during exercise on substrate utilization and subsequent cycling performance. *Am. J. Clin. Nutr.*, 67, 397, 1998.
97. Van Zyl, C.G., Lambert, E.V., Hawley, J.A., Noakes, T.D. and Dennis, S.C., Effects of medium-chain triglyceride ingestion on fuel metabolism and cycling performance. *J. Appl. Physiol.*, 80, 2217, 1996.
98. Angus, D.J., Hargreaves, M., Dancey, J. and Febbraio, M.A., Effect of carbohydrate or carbohydrate plus medium-chain triglyceride ingestion on cycling time trial performance. *J. Appl. Physiol.*, 88, 113, 2000.
99. Goedecke, J.H., Elmer-English, R., Dennis, S.C., Schloss, I., Noakes, T.D. and Lambert, E.V., Effects of medium-chain triaclyglycerol ingested with carbohydrate on metabolism and exercise performance. *Int. J. Sport Nutr.*, 9, 35, 1999.
100. Jeukendrup, A.E., Saris, W.H.M., Brouns, F., Halliday, D. and Wagenmakers, A.J.M., Effects of carbohydrate (CHO) and fat supplementation on CHO metabolism during prolonged exercise. *Metabolism*, 45, 915, 1996.
101. Jeukendrup, A.E., Saris, W.H.M., Van Diesen, R., Brouns, F. and Wagenmakers, A.J.M., Effect of endogenous carbohydrate availability on oral medium-chain triglyceride oxidation during prolonged exercise. *J. Appl. Physiol.*, 80, 949, 1996.
102. Misell, L.M., Lagomarcino, N.D., Schuster, V. and Kern, M., Chronic medium-chain triacylglycerol consumption and endurance performance in trained runners. *J. Sports Med. Phys. Fitness*, 41, 210, 2001.
103. Oopik, V., Timpmann, S., Medijainen, L. and Lemberg, H., Effects of daily medium-chain triglyceride ingestion on energy metabolism and endurance performance capacity in well-trained runners. *Nutr. Res.*, 21, 1125, 2001.
104. Fushiki, T., Matsumoto, K., Inoue, K., Kawada, T. and Sugimoto, E., Swimming endurance capacity of mice is increased by chronic consumption of medium-chain triglycerides. *J. Nutr.*, 125, 531, 1995.
105. Burelle, Y., Massicotte, D., Lussier, M., Lavoie, C., Hillaire-Marcel, C. and Peronnet, F., Oxidation of [(13)C]glycerol ingested along with glucose during prolonged exercise. *J. Appl. Physiol.*, 90, 1685, 2001.
106. Miller, J.M., Coyle, E.F., Sherman, W.M., Hagberg, J.M., Costill, D.L., Fink, W.J., Terblanche, S.E. and Holloszy, J.O., Effect of glycerol feeding on endurance and metabolism during prolonged exercise in man. *Med. Sci. Sports Exerc.*, 15, 237, 1983.
107. Trimmer, J.K., Casazza, G.A., Horning, M.A. and Brooks, G.A., Autoregulation of glucose production in men with a glycerol load during rest and exercise. *Am. J. Physiol.*, 280, E657, 2001.
108. Riedesel, M.L., Allen, D.Y., Peake, G.T. and Al-Qattan, K., Hyperhydration with glycerol solutions. *J. Appl. Physiol.*, 63, 2262, 1987.
109. Lyons, T.P., Riedesel, M.L., Meuli, L.E. and Chick, T.W., Effects of glycerol-induced hyperhydration prior to exercise in the heat on sweating and core temperature. *Med. Sci. Sports Exerc.*, 22, 477, 1990.

110. Anderson, M.J., Cotter, J.D., Garnham, A.P., Casley, D.J. and Febbraio, M.A., Effect of glycerol-induced hyperhydration on thermoregulation and metabolism during exercise in heat. *Intl. J. Sport Nutr.Exer.Metab.*, 11, 315, 2001.
111. Hitchins, S., Martin, D.T., Burke, L., Yates, K., Fallon, K., Hahn, A. and Dobson, G.P., Glycerol hyperhydration improves cycle time trial performance in hot humid conditions. *Eur. J. Appl. Physiol.*, 80, 494, 1999.
112. Latzka, W.A., Sawka, M.N., Montain, S.J., Skrinar, G.S., Fielding, R.A., Matott, R.P. and Pandolf, K.B., Hyperhydration: Thermoregulatory effects during compensable exercise-heat stress. *J. Appl. Physiol.*, 83, 860, 1997.
113. Latzka, W.A., Sawka, M.N., Montain, S.J., Skrinar, G.S., Fielding, R.A., Matott, R.P. and Pandolf, K.B., Hyperhydration: Tolerance and cardiovascular effects during uncompensable exercise-heat stress. *J. Appl. Physiol.*, 84, 1858, 1998.
114. Marino, F.E., Kay, D. and Cannon, J., Glycerol hyperhydration fails to improve endurance performance and thermoregulation in humans in a warm humid environment. *Pflugers Arch.*, (http://link.springer.de/link/service/journals/00424/contents/03/01058/paper/s00424-003-1058-3ch110.html), 2003.
115. Montner, P., Stark, D.M., Riedesel, M.L., Murata, G., Robergs, R., Timms, M. and Chick, T.W., Pre-exercise glycerol hydration improves cycling endurance time. *Int. J. Sports Med.*, 17, 27, 1996.
116. Murray, R., Eddy, D.E., Paul, G.L., Seifert, J.G. and Halaby, G.A., Physiological responses to glycerol ingestion during exercise. *J. Appl. Physiol.*, 71, 144, 1991.
117. Coutts, A., Reaburn, P., Mummery, K. and Holmes, M., The effect of glycerol hyperhydration on olympic distance triathlon performance in high ambient temperatures. *Intl. J. Sport Nutr.Exer.Metab.*, 12, 105, 2002.
118. Scheett, T.P., Webster, M.J. and Wagoner, K.D., Effectiveness of glycerol as a rehydrating agent. *Intl. J. Sport Nutr.Exer.Metab.*, 11, 63, 2001.
119. Inder, W.J., Swanney, M.P., Donald, R.A., Prickett, T.C. and Hellemans, J., The effect of glycerol and desmopressin on exercise performance and hydration in triathletes. *Med. Sci. Sports Exerc.*, 30, 1263, 1998.
120. Magal, M., Webster, M.J., Sistrunk, L.E., Whitehead, M.T., Evans, R.K. and Boyd, J.C., Comparison of glycerol and water hydration regimens on tennis-related performance. *Med. Sci. Sports Exerc.*, 35, 150, 2003.
121. Maughan, R.J. and Gleeson, M., Influence of a 36 h fast followed by refeeding with glucose, glycerol or placebo on metabolism and performance during prolonged exercise in man. *Eur. J. Appl. Physiol.*, 57, 570, 1988.
122. Gleeson, M., Maughan, R.J. and Greenhaff, P.L., Comparison of the effects of pre-exercise feeding of glucose, glycerol and placebo on endurance and fuel homeostasis in man. *Eur. J. Appl. Physiol.*, 55, 645, 1986.
123. Wagner, D.R., Hyperhydrating with glycerol: Implications for athletic performance. *J. Am. Diet. Assoc.*, 99, 207, 1999.
124. Freund, B.J., Montain, S.J., Young, A.J., Sawka, M.N., DeLuca, J.P., Pandolf, K.B. and Valeri, C.R., Glycerol hyperhydration: Hormonal, renal and vascular fluid responses. *J. Appl. Physiol.*, 79, 2069, 1995.
125. Jimenez, C., Melin, B., Koulmann, N., Allevard, A.M., Launay, J.C. and Savourey, G., Plasma volume changes during and after acute variations of body hydration level in humans. *Eur. J. Appl. Physiol.*, 80, 1, 1999.
126. Koenigsberg, P.S., Martin, K.K., Hlava, H.R. and Riedesel, M.L., Sustained hyperhydration with glycerol ingestion. *Life Sci.*, 57, 645, 1995.

127. Melin, B., Jimenez, C., Koulmann, N., Allevard, A.M. and Gharib, C., Hyperhydration induced by glycerol ingestion: Hormonal and renal responses. *Can. J. Physiol. Pharmacol.*, 80, 526, 2002.
128. Meyer, L.G., Horrigan, D.J., Jr. and Lotz, W.G., Effects of three hydration beverages on exercise performance during 60 hours of heat exposure. *Aviat. Space Environ. Med.*, 66, 1052, 1995.
129. Traul, K.A., Driedger, A., Ingle, D.L. and Nakhasi, D., Review of the toxicologic properties of medium-chain triglycerides. *Food Chem. Toxicol.*, 38, 79, 2000.
130. Elder, R., Cosmetic ingredients — their safety assessment. *J. Environ. Pathol. Toxicol.*, 4, 105, 1980.
131. Bartsch, W., Sponer, G., Dietmann, K. and Fuchs, G., Acute toxicity of various solvents in the mouse and rat. LD50 of ethanol, diethylacetamide, dimethylformamide, dimethylsulfoxide, glycerine, N-methylpyrrolidone, polyethylene glycol 400, 1,2-propanediol and Tween 20. *Arzneimittelforschung.*, 26, 1581, 1976.

15
Wheat Germ Oil and Octacosanol

Susan H. Mitmesser

CONTENTS
I. Introduction 247
II. Mechanism of Action 248
   A. Exercise Performance 248
   B. Antiaggregatory Properties 249
   C. Cholesterol-Lowering Effects 250
III. Biodistribution 251
IV. Summary 252
References 252

I. Introduction

Over the past few years, nutraceuticals have become increasingly popular among the general public.[1] Octacosanol is one such supplement that has been the subject of many research studies. As with other types of supplementation, the benefits must outweigh the risks.

Octacosanol is the main active component of policosanol and wheat germ oil. Octacosanol is a long carbon chain; $CH_3(CH_2)_{26}CH_2O_{14}$. It has a natural mixture of high-molecular-weight alcohols and is primarily isolated from sugar cane (*Saccharum officinarum* L) wax. Small quantities of octacosanol are available in the human diet through plants (mainly as a wax in the superficial layers of fruits), leaves, skins of common plants and whole seeds. Most studies have used wheat germ oil extract or policosanol to elicit an octacosanol response.

Early work by Thomas Cureton examined the physiological effects of wheat germ oil on humans during exercise. He discovered that extracts from wheat germ oil had beneficial effects on the physical performance of athletes.[2] Based on Cureton's work, other researchers began to take notice of the benefits of wheat germ oil. Soon its components were isolated and examined separately. Octacosanol appeared to be the primary active compound in wheat germ oil. With the help of Gonzalez-Bravo et al.,[3] the analytical procedure for determining octacosanol in plasma was refined. This procedure allowed for higher recovery (94.5–98.7%) and precision (1.8–5.8%) than was ever before permitted.

II. Mechanism of Action

While only small amounts of octacosanol are ingested in the diet, many health benefits have surfaced, in part due to researchers developing octacosanol in a supplement form. Most studies to date have used a wheat germ oil extract or policosanol, a natural mixture of primary alcohols isolated from sugar cane wax, of which octacosanol is the primary component. Octacosanol has the potential to benefit human health in many areas, in particular, exercise performance, platelet aggregation and plasma cholesterol levels. Of the potential health benefits associated with octacosanol, its ergogenic properties and cholesterol-lowering effects have been the most studied.

A. Exercise Performance

Ergogenic aids, substances that enhance athletic performance and increase stamina and capability to exercise, are believed to improve performance by either renewing or increasing energy stores in the body, which facilitates biochemical reactions contributing to fatigue or by maintaining optimal body weight.[4] As the U.S. Food and Drug Administration does not require nutritional supplements to be proven safe or effective because they are not classified as "drugs," many athletes take such supplements without realizing potential side effects. Likewise, nutritional supplements are not covered by the United Kingdom Medicines Act, and therefore no regulations for manufacture, quality or usage exist, making them subject to abuse.[5] Additionally, it is often difficult to fully determine whether an improvement in performance or muscle size is due to a nutritional supplement or to other factors, such as improved training techniques and diet.

Octacosanol is one such ergogenic supplement proposed to enhance performance and delay fatigue. Thomas Cureton was the first to report that wheat germ oil had ergogenic effects. During his career, Cureton performed approximately 42 studies that addressed the ergogenic potential of wheat germ oil.[2] In 1963, Cureton published a study involving 30 trainees in the U.S. Navy. The subjects received either 3.7 mL (10 capsules containing 0.37 mL each) of octacosanol in cottonseed oil, cottonseed oil only (used as the placebo), or whole wheat germ oil daily for 6 weeks. While the octacosanol-supplemented group had higher mean exercise performance, they did not differ significantly from that of the placebo group.[6]

In 1986, Saint-John et al. investigated human subjects taking octacosanol supplements in relationship to chest strength, stamina, grip, cardiovascular function and reaction time. Specifically, 1000 μg of octacosanol was administered to healthy human subjects daily. Grip strength and reaction time significantly improved in response to a visual stimulus. The researchers concluded octacosanol was an "active energy releasing factor" and improved performance when compared with a placebo.[7] This research led others to hypothesize that octacosanol might exhibit properties affecting the nervous system because reaction time was able to change according to nerve impulses throughout the body. A study in 1998 on a group of patients with coronary heart disease taking a policosanol supplement improved their response to exercise angina and significantly increased maximum oxygen uptake.[8] However, Bucci reported a failure of other researchers to significantly substantiate the ergogenic benefit of octacosanol.[9]

If octacosanol truly does possess ergogenic properties, the mechanism of action is not fully understood at this point. Kabir and colleagues[10] attempted to address this phenomenon using rats. The researchers set out to investigate the biodistribution of radioactive (^{14}C) octacosanol in response to exercise. They found the amount of voluntary exercise to be significantly higher in octacosanol-fed rats than in the control. Additionally, the amount of radioactive octacosanol in the muscle of exercised animals was significantly higher. The muscle seemed to be able to store a considerable amount of octacosanol in response to exercise. While the exact mechanism behind the increase in physical activity caused by octacosanol is unclear, it is quite possible that octacosanol has the ability to increase the mobilization of free fatty acids from fat cells within the muscle. The results from this study indicate that octacosanol possesses an adipolinetic activity that could potentially affect lipolysis in muscle.[10]

All in all, the studies assessing octacosanol as an ergogenic aid are nonspecific. The trials up to this point appear to have many confounding factors. It is essential in clinical trials, especially when athletic performance is being measured, to eliminate the placebo effect as much as possible. Many of the studies presented here, which are considered the most reliable to date, lack consistency. This lack of congruency makes it difficult to compare studies. Furthermore, completely randomized double-blinded studies comparing athletes participating in similar exercise regimens are warranted.

B. Antiaggregatory Properties

Platelet aggregation occurs when blood is converted from a thin consistency to a thicker form or even a clot, which can cause a deep-vein thrombosis or stroke. Exogenous factors such as arachidonic acid, collagen and adenosine diphosphate can cause aggregation. Platelet aggregation is an important factor in coronary artery disease and vascular arterial disease. Patients at risk for one of these diseases have an increased risk of other diseases occurring as well.[11] Arruzazabala and colleagues[8] found that 5 to 20 mg/kg of policosanol given to rats resulted in antiaggregatory effects. These researchers suggested the action was due to the inhibition of arachidonic acid metabolism. The same group of researchers found that, when 50 to 200 mg/kg of policosanol was administered, a significant inhibition of platelet aggregation was observed.[12] The study emphasized that a large dose was needed for policosanol to exhibit antiaggregatory effects.

In 1997, the same research group investigated the effects of policosanol and aspirin on platelet aggregation.[13] Aspirin is commonly used to prevent thrombosis and cerebral ischemia, but can cause gastric irritation. Policosanol works by inhibiting thromboxane A_2 (TxA_2) synthesis instead of prostaglandin synthesis, therefore fewer side effects such as gastric irritation occur. A daily dose of 20 mg of policosanol caused a significant reduction in platelet aggregation, establishing that policosanol was just as effective as 100 mg/d of aspirin.

C. Cholesterol-Lowering Effects

Researchers noticed that, during exercise endurance experiments on mice, octacosanol had the ability to alter hepatic and serum lipid concentrations.[14] This led them to begin focusing on octacosanol as a cholesterol-lowering agent. In 1992, Hernandez et al. investigated the effects on healthy individuals with normal cholesterol levels given 10 or 20 mg of policosanol or a placebo.[15] The policosanol-supplemented group had a significant decrease in serum cholesterol levels. The group taking 20 mg of policosanol also demonstrated an increase in high-density lipoprotein (HDL) levels. Aside from the cholesterol-lowering effect, this study revealed a good tolerance in all subjects taking the policosanol supplement.

Kato and colleagues set out to test the effect of octacosanol on lipid metabolism through rats fed a high-fat diet and given octacosanol supplements.[16] The addition of 10 g octacosanol/kg to a high-fat diet led to a significant reduction in the perirenal adipose tissue weight without causing a decrease in cell number. This suggests that octacosanol may suppress lipid accumulation in the tissue. Furthermore, octacosanol supplementation decreased the serum triacylglycerol concentration, most likely through the inhibition of hepatic phosphatidate phosphohydrolase. High amounts of lipoprotein lipase in the perirenal adipose tissue and an increase in the total oxidation rate of fatty acid in the muscle were also observed. Lipid absorption, however, was unaffected by octacosanol supplementation. This study suggests that the dietary incorporation of octacosanol affects some aspects of lipid metabolism. The action of octacosanol may depend on specific dietary conditions, such as fat content, as suggested by this study.

While statin drugs (drugs currently used to reduce blood cholesterol levels) inhibit 3-hydroxy-3-methylglutaryl coenzyme A reductase in the cholesterol synthesis pathway, policosanol inhibits the same pathway, but a step earlier.

Policosanol could, therefore, prove to be a useful alternative to statin drugs. In 1999, a study examined just that; the effects of policosanol and pravastatin (a statin drug) on the lipid profile of hypercholesterolemic patients.[17] The researchers demonstrated that policosanol was more effective than pravastatin in lowering low density lipoprotein (LDL), as well as improving the ratios of LDL to HDL and total cholesterol to HDL. As seen in previous studies, policosanol also increased HDL.[15,16] It is important to note that both drugs in this study were well tolerated with little or no side effects reported.

A similar study examined the effects of policosanol and fluvastatin, another commonly used statin drug, on hypercholesterolemic patients.[18] Policosanol exhibited LDL-lowering and HDL-raising effects. Again, policosanol was well tolerated by the hypercholesterolemic patients. As in the studies discussed here, many others have also investigated the effects of policosanol on hypercholesterolemia and in all cases the results were similar; LDL cholesterol decreased and HDL cholesterol increased.[19,20,21,22]

From these studies we can conclude that octacosanol/policosanol seems to be a promising lipid-lowering agent that is well tolerated with fewer reported side effects than its statin counterpart. Additionally, the cost of octacosanol/policosanol supplementation is only a fraction of statin drugs, which would make it more available to lower-income hypercholesterolemic patients. It is important to note that most of the studies examined here are considered short term (<1 year long). Larger trials extending over longer periods of time are needed to assess the longer-term effects of octacosanol.

III. Biodistribution

To evaluate some of the health claims regarding octacosanol, researchers began to focus on the distribution of orally administered octacosanol throughout the body. Kabir and colleagues set out to further understand the mechanism of increased physical exercise and motor endurance by octacosanol.[23] ^{14}C-labeled octacosanol administered to rats was found to primarily accumulate in adipose tissue, especially brown adipose tissue. Additionally, they found excretion through the feces to be very low, but present in measurable quantities in the urine and expired as $^{14}CO_2$. This study suggests that a portion of ^{14}C-octacosanol might be converted into fatty acids, which in turn supplies $^{14}CO_2$ and energy through the process of β-oxidation. There is, however, the possibility that all the free acid produced from octacosanol does not undergo direct oxidation, and part of it is stored in the fat pool of the adipose tissue. Neptune et al. reported similar findings; the conversion of long-chain fatty acids to $^{14}CO_2$ constituted only a small fraction of the total CO_2 in the rat diaphragm.[24]

Kabir and Kimura further investigated the tissue distribution of ^{14}C-octacosanol in liver and muscle of rats after a series of administrations.[25] They found the highest amount of radioactivity in the liver (9.5% of administered dose), followed by the digestive tract (8.2% of administered dose) and the muscle (3.5% of administered dose). Interestingly enough, the radioactivity in the liver disappeared rapidly, while the muscle seemed to be able to store a considerable amount in response to the doses.

IV. Summary

Octacosanol has many potential uses for various health concerns. The most widely studied to date are its cholesterol-lowering properties and exercise performance effects. The studies presented thus far have supported the potential octacosanol has in benefiting human health with regard to cholesterol-lowering and platelet antiaggregation. Additionally, it has the potential to treat these health conditions without the major side effects often associated with currently used therapies. Thus, octacosanol is an important drug of the future, especially in view of the ever-increasing prevalence of obesity and coronary heart disease in the U.S. However, the ability to pinpoint optimal doses for the aforementioned health benefits is lacking in the current literature. Likewise, toxicity of octacosanol has not been observed and remains unknown at this point. More studies are needed to address the long-term effectiveness of octacosanol as well as long-term problems that may arise with supplementation such as this. Furthermore, trials need to be consistent to optimize a comparison of studies that reflect true human function. For example, studies assessing octacosanol as an ergogenic aid need to be completely randomized and double blinded (to eliminate placebo effect), comparing athletes participating in similar exercise regimens. A study containing all of these aspects has not yet been reported. Future research must begin to focus on defining guidelines for octacosanol supplementation, which needs to be specific for each health claim. This is especially important since supplementation of this sort does not fall under the U.S. Food and Drug Administration jurisdiction.

References

1. Wildman, R.E.C., Ed., Handbook of Nutraceuticals and Functional Foods. CRC Press, Boca Raton, FL, 2001.
2. Cureton, T.K, The Physiological Effects of Wheat Germ Oil on Humans in Exercise, Charles C. Thomas Publisher, Springfield, IL, 1972.
3. Gonzalez-Bravo, L., Magraner-Hernandez, J., Acosta-Gonzalex, P.C. and Perez-Souto, N., Analytical procedure for the determination of l-octacosanol ion plasma by solvent extraction and capillary gas chromatography, *J. Chromat. B.*, 682, 359, 1996.
4. Beltz, S.D. and Doering, P.L., Efficacy of nutritional supplements used by athletes, *Clin. Pharm.*, 12, 900, 1993.

5. Cockerill, D.L. and Bucci, L.R., Increases in muscle girth and decreases in body fat associated with a nutritional supplement program, *Chiroprac. Sports Med.*, 1, 73, 1987.
6. Cureton, T.K., Improvements in physical fitness associated with a course of U.S. Navy underwater trainees, with and without dietary supplements, *Res. Q.*, 34, 440, 1963.
7. Saint-John, M. and McNaughton, L., Octacosanol ingestion and its effects on metabolic responses to submaximal cycle ergometry, reaction time and chest and grip strength, *Int. Clin. Nutr. Rev.*, 6, 81, 1986.
8. Arruzazabala, M.L., Mas, R. and Molina, V., Effects of policosanol on platelet aggregation in type II hypercholesterolaemic patients, *Tissue React.*, 20, 119, 1998.
9. Bucci, L.R., Nutritional ergogenic aids. In: Wolinsky, I., Hickson, J.F., Eds., *Nutrition in Exercise and Sport*, Boca Raton, FL, CRC 107–84, 1989.
10. Kabir, Y. and Kimura, S., Distribution of radioactive octacosanol in response to exercise in rats, *Nahrung.*, 38, 373, 1994.
11. Arruzazabala, M.L., Carbajal, D. and Mas, R., Effects of policosanol on platelet aggregation in rats, *Throm. Res.*, 69, 321, 1993.
12. Arruzazabala, M.L., Molina, V. and Carbajal, D., Effect of policosanol on cerebral ischaemia in Mongolian gerbils: Role of prostacyclin and thromboxane A_2, *Prost. Leuk. Essent. Fatty Acids*, 49, 695, 1993.
13. Arruzazabala, M.L., Valdes, S. and Mas, R., Comparative study of policosanol, aspirin and the combination therapy policosanol-aspirin on platelet aggregation in healthy volunteers, *Pharmacol. Res.*, 36, 106, 1997.
14. Shimura, S., Hasegawa, T. and Takano, S., Studies on the effect of octacosanol on motor endurance in mice, *Nutr. Rep. Int.*, 36, 1029, 1987.
15. Hernandez, F., Illait, J. and Mas, R., Effect of policosanol on serum lipids and lipoproteins in healthy volunteers, *Curr. Ther. Res.*, 51, 568, 1992.
16. Kato, S., Karino, K.I., Hasegawa, S., Nagasawa, J., Nagasaki, A., Eguchi, M., Ichinose, T., Tago, K., Okumori, H., Hamatani, K., Takahashi, M., Ogasawara, J., Masushige, S. and Hasegawa, T., Octacosanol affects lipid metabolism in rats fed on a high-fat diet, *Brit. J. Nutr.*, 73, 433, 1995.
17. Castano, G., Mas, R. and Arruzazabala, M., Effects of policosanl and pravastatin on lipid profile, platelet aggregation and endothelemia in older hypercholesterolemic patients, *Int. J. Clin. Pharm. Res.*, 4, 105, 1999.
18. Fernandez, J.C., Mas, R. and Castano, G., Comparison of the efficacy, safety and tolerability of policosanol versus fluvastatin in elderly hypercholesterolaemic women, *Clin. Drug Invest.*, 21, 103, 2001.
19. Pons, P., Rodriguez, M. and Mas, R., One-year efficacy and safety of policosanol in patients with type II hypercholesterolaemia, *Curr. Ther. Res.*, 55, 1084, 1994.
20. Aneiros, E., Mas, R. and Calderon, B., Effect of policosanol in lowering cholesterol levels in patients with type II hypercholesterolaemia, *Curr. Ther. Res.*, 56, 176, 1995.
21. Castano, G., Tula, L. and Canetti, M., Effects of policosanol in hypertensive patients with type II hypercholesterolaemia, *Curr. Ther. Res.*, 57, 691, 1996.
22. Mas, R., Castano, G. and Illnait, J., Effects of policosanol inpatients with type II hypercholesterolaemia and additional coronary risk factors, *Clin. Pharmacol. Ther.*, 65, 439, 1999.
23. Kabir, Y. and Kimura, S., Biodistribution and metabolism of orally administered octacosanol in rats, *Ann. Nutr. Metab.*, 37, 33, 1993.

24. Neptune, E.M., Sudduth, H.C., Foreman, D.R. and Fash, F.J., Phospholipid and triglyceride metabolism of exercised rat diaphragm and the role of these lipids in fatty acid uptake and oxidation, *J. Lipid Res.*, 1, 229, 1960.
25. Kabir, Y. and Kimura, S., Tissue distribution of (8-14C)-octacosanol in liver and muscle of rats after serial administration, *Ann. Nutr. Metab.*, 39, 279, 1995.

Part IV

Other Substances in Foods Not Classified As Essential

16

Buffers: Bicarbonate, Citrate and Phosphate

Mark D. Haub

CONTENTS

I. Introduction 257
II. Hydrogens and Muscle Fatigue 258
III. Lactate Production and Hydrogen Accumulation ... 259
IV. Sodium Bicarbonate Loading on Acid–base Balance and Performance 261
   A. Potential Mechanism 261
   B. Administration and Dose 261
   C. Performance Effects 264
V. Sodium Citrate Loading on Acid–base Balance and Performance 265
   A. Potential Mechanism 265
   B. Administration and Dose 266
   C. Performance Effects 266
   D. Combining Bicarbonate and Citrate 267
VI. Phosphate Loading on Physiology and Performance ... 267
VII. Chronic Buffer Supplementation 268
VIII. Areas of Future Research 268
IX. Recommendations for Coaches and Athletes 269
X. Summary 269
References 270

I. Introduction

With many athletic events placing stress on the acid–base system to maintain a contraction-conducive myocellular environment, the ingestion of buffering compounds has been popular among athletes and scientists. The need for an athlete to maintain a functional pH within the contracting myofibrils places a huge demand on the body's capacity to handle hydrogen (H^+) and other strong ions that alter acid–base homeostasis. The accumulation of metabolic products, such as H^+, lactate (La^-) and inorganic phosphate (P_i), during high-intensity exercise significantly affect pH, excitation-contraction coupling and ultimately power output. If these and other metabolic factors can be maintained homeostatically (e.g., at a steady state) during force production, then the muscle fiber may be able to contract at higher forces or for a longer period of time. Depending on the sport, either of these attributes would be deemed as beneficial. Thus, a legal and safe means to either increase or maintain power output by handling the perturbations in ion concentrations would, theoretically, allow an athlete to enhance performance. Several studies over the past two decades have examined consumable compounds that enhanced the blood buffer capacity and shifted acid–base balance to a state of metabolic alkalosis. However, while the physiologic response to the ingestion of these buffering compounds is undisputed, the effect of buffers on sport performance is anything but consistent. Additionally, while several studies have reported ergogenic benefits with these compounds; the application of those results to sport must be done cautiously. Some of the exercise protocols used in laboratories do not mimic traditional competition (e.g., time trials). This chapter will discuss the physiological basis for the use of ingesting buffering compounds and discuss the results from studies that have assessed the efficacy of these compounds on sport or exercise performance.

II. Hydrogens and Muscle Fatigue

Based on data from muscle biopsies, short-duration high-intensity exercise significantly decreases intramuscular pH (an increase in H^+) from 7.0 at rest to < 6.5 postexercise.[1] At this level of pH, the increased [H^+] seems to inhibit several vital processes (e.g., cross-bridge cycling, enzymatic activity and excitation-contraction coupling) required for muscle contractions to continue at a desired power output. For example, it's been observed that the affinity for Ca^{++} binding to troponin is decreased at low intracellular pH [2], thereby decreasing excitation-contraction (E-C) coupling and decreasing force production in the respective myofibrils. Additionally, Fabiato and Fabiato[3] found that more Ca^{++} was required to maintain a given force output when [H^+] increased. Relative to an athletic event, an athlete would need to increase the action potential going to the respective myofibrils to increase Ca^{++} release from the sarcoplasmic reticulum when the [H^+] increases or the athlete would need to recruit more muscle fibers just to maintain the desired force output.

The accumulation of H^+ has been shown to allosterically inhibit enzymes in the glycolytic pathway. Previous studies have observed that increasing myocellular [H^+] leads to a decreased utilization of glycogen via decreased conversion of phosphorylase "b" (inactive form) to phosphorylase "a" (active form).[4] This decrease in phosphorylase "a" conversion limits the glycolytic flux, which may decrease the rate at which ATP can be generated via glycolysis, although ATP levels are not necessarily affected by decreases in pH.[5] An increase in [H$^+$] has also been shown to decrease the activity of the glycolysis rate-limiting enzyme phosphofructokinase (PFK).[6] A decreased PFK activity would lead to decreased production of downstream glycolytic intermediates (e.g., pyruvate and lactate), in addition to a decreased capacity to regenerate ATP from glycolytic processes. Thus, to maintain force production during high-intensity exercise, an athlete would need to rely more on the one-enzyme systems or recruit more motor units to maintain or increase power output.

Another related aspect of muscle contraction involves the accumulation of P_i and $H_2PO_4^-$, while an increase in [H$^+$] has been suggested to enhance the conversion of HPO_4^{2-} to $H_2PO_4^-$.[7] It was also observed that $H_2PO_4^-$ played a greater role in decreased force production than did [H$^+$].[8] Given these and the previously mentioned results pertaining to E-C coupling, enzymatic activity and cross-bridge cycling, it is apparent the role of [H$^+$] in the onset and progression of fatigue is complex.

III. Lactate Production and Hydrogen Accumulation

With regard to athletes and power output, glycolysis and glycogenolysis are the metabolic pathways that are utilized in most athletic events that require sufficient or sustained force production. Subsequently, a relatively large quantity of lactate is produced. The fate of lactate then lies in the ability of each lactate-producing fiber to either convert La$^-$ back into pyruvate to be oxidized, convert La$^-$ back into a glycolytic intermediate, or transport La$^-$ out of the fiber to be used by surrounding tissue or distributed in circulation for later oxidation or gluconeogenic activity by other peripheral tissues. One benefit of transporting lactate ions out of the cell is that a H$^+$ is also transported with it, thereby decreasing the accumulation of H$^+$ and slowing the drop in pH. The primary means to transport La$^-$ and H$^+$, is by monocarboxylate transporters (MCT, isoforms one and four are primarily found in skeletal muscle) [9] and some lactate may leave the cell via passive diffusion.[10] There is a strong correlation between fiber type and the MCT isoform that is present, with MCT1 being positively associated with Type I fibers (high oxidative capacity), which has led to the suggestion that MCT1 may be better suited for transporting La$^-$ into cells and MCT4 being more responsible for the efflux of lactate out of working cells.

The relevance of this to athletes is that the efflux of lactate and H$^+$ from the working environment helps to remove and delay detrimental acidosis. This theoretical benefit was thought to be one benefit of supplementing compounds that increase the blood-buffering capacity. That is, an increase in the extracellular to intracellular pH gradient was thought to allow for more La⁻ and H⁺ to efflux from the working tissues to delay fatigue. This concept seems to play a role, since many studies have observed increased blood La⁻ levels following buffer ingestion compared with control or placebo conditions even though workloads were identical.[11–14] However, as pointed out by Juel,[9] the increase in blood La⁻ may result from a decreased disposal of lactate and not necessarily the result of increased efflux. A basis for this lies in the fact that the La⁻ to H⁺ transport ratio is 1:1, meaning that one lactate ion is transported with one hydrogen ion. Therefore, if they are effluxed from the working cell on a 1:1 ratio and the H⁺ binds with HCO_3^- via carbonic anhydrase to subsequently yield CO_2 and H_2O, then there is no longer a 1:1 ratio in circulation for lactate to be transported back into cells to be oxidized or used to form glycosyl units.

So, what does all of this mean to an athlete? At rest, blood pH tends to be about 7.3–7.4 and muscle pH around 7.0. However, after high-intensity exercise, the blood pH can drop less than 6.9[15] and muscle pH can decrease to 6.6.[1] Accordingly, power output at the end of a 45–60-second competition decreases significantly. This undesired drop in power output (e.g., fatigue) seems to result, at least partially, from the previously mentioned physiologic processes related to H⁺ and La⁻ production.

Therefore, while H⁺ and La⁻ transport seems to play significant roles in fatigue, our bodies do have buffers to decrease or minimize the force-depleting effects of acidosis. Of these buffers — compounds that minimize changes in pH when acids or bases are added to a solution — intracellular proteins are one of the best defenses against an increase in intracellular acidosis, while blood levels of bicarbonate (HCO_3^-) assist to handle acidosis in circulation. Other extracellular buffers include plasma proteins and hemoglobin. The process of "removing" H⁺ (raising pH) in the blood via HCO_3^- involves the formation of H_2CO_3 by carbonic anhydrase and the subsequent formation of H_2O and CO_2. It is this latter process that buffer loading attempts to increase — and seems to do so quite effectively. Additionally, the use of buffers as potential ergogenic aids seems to be limited to those events that are performed close to or above the lactate threshold and are long enough in duration for acidosis-related fatigue to develop.

The theoretical basis for ingesting buffering compounds (e.g., sodium bicarbonate and sodium citrate) is to increase blood [HCO_3^-] and pH; however, their effectiveness at improving performance is complicated by the fact that myocellular pH needs to be maintained to prevent acidosis-induced muscle fatigue. Just because extracellular or intracellular pH is increased does not necessarily equate to improvements in performance. This is exemplified by the fact that most if not all studies investigating the efficacy of buffer (sodium bicarbonate or citrate) loading report significant changes in blood pH and [HCO_3^-], but improvements in performance are not always observed (Table 16.1). As well, even if intracellular pH is increased, this does not necessarily lead to performance differences.[14]

For more specific information regarding these various aspects of acid–base balance and force production, there are several thorough reviews.[9,16–19] The remainder of this chapter will focus specifically on the efficacy of buffer loading to enhance sport and exercise performance; the results from animal studies have been excluded.

IV. Sodium Bicarbonate Loading on Acid–base Balance and Performance

A. Potential Mechanism

The ingestion of sodium bicarbonate ($NaHCO_3$) leads to an increase in blood pH. What is not fully clear is the means through which this occurs. It has been suggested that the sodium alters the strong ion difference, which leads to the observed rise in pH.[17] This notion is supported by Stewart[20] based on increased Cl^- excretion, given that excess Na^+ is excreted and the fact that bicarbonate is not directly absorbed into circulation.[17] Thus, to handle the increased Na^+ from the supplement, the body consequently increases the excretion of Cl^- concurrently with increased Na^+ excretion because Cl^- remains in the renal tubule as Na^+ is not being returned to circulation.[19] This alteration in the strong ion difference leads to the consequent production of HCO^{3-} and increase in blood pH.

B. Administration and Dose

As with most dietary supplements or ergogenic aids, the administration (oral, intravenous, intramuscular, aerosol inhalation etc.) and dose need to be investigated to determine issues related to safety and efficacy. This is of particular importance to sodium bicarbonate and sodium citrate as the side effects of each may hinder performance or prevent an athlete from competing altogether. Based on work from McNaughton,[21,22] doses were investigated to establish the lowest dose possible that would still elicit an improvement in performance (Table 16.2). One study did show an effect at 0.15 g/kg,[23] which is between the lower doses used by McNaughton.[21,22] A study by Potteiger et al.[13] reported time series data to provide the potentially best times to begin an exercise bout following the ingestion of sodium bicarbonate. It was determined that $[HCO_3^-]$ peaked at 100–120 minutes post-ingestion. This is an important point to note, as some studies that did not observe improvements in performance using the recommended dose had the subjects begin exercise less than 90 minutes or more than 120 minutes after ingesting the buffer.

Even if these doses are ingested at the appropriate time, performance results might be significantly hindered by the side effects of sodium bicarbonate. Based on anecdotal reports from athletes and research volunteers, severe gastrointestinal disturbances including intestinal cramping, diarrhea or flatulence may prevent an athlete from performing as expected, if at all

TABLE 16.1

Results from Investigations Examining the Effect of Buffer Ingestion on Exercise or Sport Performance

Author(s)	Performance	Intensity	Buffer(s)	Dose
No Effect on Performance				
Acute Supplementation				
Ball and Maughan[52]	Cycling	100% VO_{2max}	2	0.3 g/kg
Brien and McKenzie[53]	Rowing	4 min	1	0.3 g/kg
Cox and Jenkins[54]	Cycling	60 s sprints	2	0.5 g/kg
Duffy and Conlee[48]	Running	Exh and leg power	3	1.24 g
Feriche Fernandez-Castanys et al.[55]	Altitude	Maximal	2	0.4 g/kg
Gaitanos et al.[56]	Running	Max Velocity	1	0.3 g/kg
Horswill et al.[57]	Cycling	2 min TT	1	0.1–0.2 g/kg
Ibanez et al.[58]	Running	300 m	2	0.5 g/kg
Johnson and Black[59]	Running	Cross-country	1, 2, 1&2, 3	3.5 g
Katz et al.[60]	Cycling	125% VO_{2max}	1	0.2 g/kg
Kinderman et al.[61]	Running	400m TT	1	190 mmol
Kowalchuk et al.[62]	Cycling	95% VO_{2max}	2	0.3 g/kg
Kozak-Collins et al.[63]	Cycling	95% VO_{2max}	1	0.3 g/kg
Linderman et al.[64]	Cycling	100% VO_{2max}	1	0.2 g/kg
McNaughton and Cedaro[40]	Cycling	10 and 30 sec	2	0.5 g/kg
Parry-Billings and MacLaren[41]	Cycling	3 × 30 sec	1, 1&2	0.3 g/kg
Pierce et al.[65]	Swimming	100 and 200 yd	1	0.2 g/kg
Portington et al.[28]	Leg Press	5 × 12-RM	1	0.3 g/kg
Potteiger et al.[13]	Running	LT + 110%	1, 2	0.3 and 0.5 g/kg
Schabort et al.[43]	Cycling	40 km TT	2	0.2–0.6 g/kg
Stephens et al.[14]	Cycling	~469 kJ TT	1	0.3 g/kg
Tiryaki and Atterbom[66]	Running	600 m TT	1	0.3 g/kg
Webster et al.[29]	Resistance	70% 1-RM	1	0.3 g/kg
Wilkes et al.[67]	800m run	Time trial	1	0.3 g/kg
van Someren et al.[68]	Cycling	45 sec	2	0.3 g/kg
Chronic Supplementation				
Bredle et al.[45]	Running	Exh (70% VO_{2max})	3	176 mmol/d
Duffy and Conlee[48]	Running	Exh and leg power	3	3.73 g/d
Kreider et al.[47]	Running	5-mile TT	3	1 g/d

(continued)

TABLE 16.1 (CONTINUED)

Results from Investigations Examining the Effect of Buffer Ingestion on Exercise or Sport Performance

Author(s)	Performance	Intensity	Buffer(s)	Dose
Performance Effect				
Acute Ingestion				
Bouissou et al.[31]	Cycling	Exh (375 W)	1	0.3 g/kg
Costill et al.[32]	Cycling	1 min sprints	1	0.2 g/kg
Gao et al.[69]	Swimming	5 × 100 yd	1	2.9 mmol/kg
Hausswirth et al.[39]	Knee ext.	Isometric	2	0.4 g/kg
Iwaoka et al.[34]	Cycling	95% VO_{2max}	1	0.2 g/kg
Lavender and Bird[30]	Cycling	10 sec sprints	1	0.3 g/kg
Linossier et al.[37]	Cycling	120% VO_{2max}	2	0.5 g/kg
McKenzie et al.[23]	Cycling	125% VO_{2max}	1	0.15–0.3 g/kg
McNaughton and Thompson[70]	Cycling	90 sec	1	0.5 g/kg
McNaughton[26]	Cycling	120, 240 sec	1	0.3 g/kg
McNaughton[22]	Cycling	60 sec	1	0.1–0.5 g/kg
McNaughton and Cedaro[40]	Cycling	120, 240 sec	2	0.5 g/kg
Parry-Billings and MacLaren[41]	Cycling	3 × 30 sec	2	0.3 g/kg
Potteiger et al.[12]	Cycling	30 km TT	2	0.5 g/kg
Shave et al.[42]	Running	3 km TT	2	0.5 g/kg
Verbitsky et al.[35]	Cycling	117% VO_{2max}	1	0.4 g/kg
Chronic Supplementation				
Kreider et al.[46]	Cycling	40 km TT	3	1 g/d
McNaughton et al.[49]	Cycling	90 sec	1	0.5 g/kg

Note: Acute supplementation = dose was administered and tested within the same day; Chronic supplementation = dose was administered for more than 1 day and testing occurred within a week of starting supplementation; VO_{2max} = maximal amount of O_2 that can be extracted under normal volitional exercise conditions specific to the mode used; knee ext. = knee extension; isometric = isometric contraction endurance at 35% maximal voluntary contraction; TT = time trial; RM = repetition maximum; LT +110% = the run commenced at lactate threshold and was then increased to 110% of VO_{2max} and continued until volitional exhaustion; Exh = exercise to voluntary exhaustion; 1 = sodium bicarbonate; 2 = sodium citrate; 3 = phosphate salts.

The table is divided by those investigations that reported a significant performance difference between treatments and those that did not.

TABLE 16.2

Dosage and Administration Procedures for Sodium Bicarbonate, Sodium Citrate and Phosphate Salts

Buffering Agent	Effective Doses	Peak Levels
Sodium Bicarbonate	0.2–0.5 g/kg[22]	90–120 minutes prior to exercise[13]
Sodium Citrate	0.5 g/kg[21]	90–120 minutes prior to exercise[13]
Phosphate salts	4 g/d	Ingested daily[46]

(depending on the timing or severity of the untoward intestinal events). Therefore, as recommended for all nutritional supplements, it is best to practice this approach prior to a noncompetitive event or a competitive event that is deemed less important. The use of sodium bicarbonate or citrate may lead to other potential health complications such as cardiac arrhythmias, apathy, muscle spasms or gastric ruptures.[17,24] While not explicitly banned by the United States Anti-Doping Agency,[25] the use of these buffers is contrary to the philosophy of fair play because they are ingested for the sole intent of enhancing performance with limited, if any, nutritional justification. A case could be made for ensuring adequate sodium intake, but sodium chloride can achieve that goal without the pH-altering side effects.

A prime advantage of sodium bicarbonate as an ergogenic aid is its low cost. Assuming a performance effect, it is probably the least expensive ingestible supplement (likely because diet-supplement companies cannot compete against the low cost of those companies that sell it for its nonbiological qualities). The downside of purchasing the product in powder form is devising a palatable means to ingest it. Filling gelatin capsules is tedious and time consuming, while ingesting it in solution may require discipline to drink the necessary dose.

C. Performance Effects

Based on the physiological rationale, sodium bicarbonate seems best suited for those events that rely heavily on glycolysis as the primary metabolic pathway. In individual efforts less than 30 seconds, it has been observed that sodium bicarbonate does not seem to enhance performance.[26] Within this short window, the intracellular environment may be able to maintain the desired force output regardless of the extracellular or intracellular pH.[27] Also, while strength or resistance training relies partially on the glycolytic pathway for ATP production, sodium bicarbonate does not appear to elicit a favorable performance response.[28,29] However, once the exercise duration begins to exceed 30 seconds or the shorter events are repeated,[30] then the chance for sodium bicarbonate to be effective begins to increase (Table 16.1).

There seem to be research design or exercise requirements that enhance the potential for sodium bicarbonate to improve performance. The studies that have reported significant improvements in performance seem to be those that place more physiological stress on the acid–base system. For example, these studies[23,30–35] tended to use either repeated high-intensity sprints or had subjects exercise to voluntary exhaustion. A methodological problem with the studies to exhaustion is that few athletic events compete with victory going to the individual who can sustain an activity the longest. Most events are competed on the basis of covering a given distance in the least amount of time. Volitional exhaustion and completion of the event rarely occur at the same time in these events.

That is, if exhaustion occurs prior to completing the event, then that athlete does not finish or finishes only after recovering. The capacity to apply the results of those studies to athletic performance is limited. Based on athletic situations, it appears that sodium bicarbonate is not effective at enhancing performance of single time-trial events (Table 16.1). Therefore, while there is a physiological response with sodium bicarbonate ingestion, that response does not always carry over to sport performance. One interesting point is that sodium bicarbonate seems to be most effective when cycling protocols are utilized, while running protocols infrequently observed improved performance (Table 16.1).

A research design pitfall inherent with buffer research is the issue of placebo effect, or in this case, treatment effect. Given the gastrointestinal distress that tends to occur following buffer ingestion, it can be readily apparent which treatment is received. Therefore, subjects likely know when they have received the treatment versus a placebo. While the use of multiple doses might be able to help alleviate this problem, it does not do so unequivocally, as gastrointestinal distress increases with the dose. A second placebo to mimic the gastrointestinal effects without influencing acid–base status is needed. Therefore, if the subjects improved performance during a trial with this placebo yet did not during use of the typical placebo (e.g., calcium carbonate), then the performance differences may be due to cognitive interference. The presence of a placebo effect in sport nutrition has been illustrated in cyclists.[36]

V. Sodium Citrate Loading on Acid–Base Balance and Performance

A. Potential Mechanism

The ingestion of sodium citrate ($Na_3H_5C_6O_7$) leads to similar increases in blood pH, as observed with sodium bicarbonate ingestion. The strong ion difference theory also explains the acid–base altering effects of this supplement. One potential difference with sodium citrate lies in the fact that, with sodium citrate ingestion, intramuscular citrate levels were reported to increase.[37] This change in myocellular (citrate) may influence glycolytic or oxidative activities. This, in fact, was demonstrated by Hollidge-Horvat et al.[38] They found that, when exercising at 75% VO_{2max}, glycolytic activity was increased and pyruvate oxidation was decreased during alkalosis compared with a control condition. They also observed a decrease in phosphocreatine and subsequent increase in [P_i] at all intensities (30%, 60% and 75% of VO_{2max}). Thus, inducing alkalosis via sodium citrate ingestion indeed alters intracellular metabolism during exercise.

B. Administration and Dose

The administration and dose of sodium citrate is similar to that of sodium bicarbonate. The dose to elicit a response seems to be slightly higher with sodium citrate than the 0.2 g/kg with sodium bicarbonate. The timing prior to exercise is basically the same (Table 16.2).[13]

C. Performance Effects

Much like the responses to sodium bicarbonate ingestion, the performance outcomes following sodium citrate ingestion, compared with a placebo, are mixed. It seems that sodium citrate ingestion does not work to improve performance of single efforts less than 60 seconds (Table 16.1). As the duration increases or the exercise is repeated, performance may be enhanced.[12,37,39-42]

An intriguing study by Potteiger et al.[12] demonstrated for the first time that buffer ingestion may enhance endurance-cycling performance. The rationale for the use of a buffer in these events is that, because athletes perform many of these events at intensities greater than 80% VO_{2max}, glycolysis and lactate production are high enough to lead to significant acidosis even though these events may last up to an hour or more. Subjects in the study were competitive cyclists or triathletes. The subjects and technicians were visually blinded to which treatment (sodium citrate, 0.5 g/kg; or placebo, wheat flour) was being given. The subjects performed a 30 km time trial as fast as possible. Based on this data,[12] it was observed that performance time was significantly lower in the sodium citrate trial; as well, lactate, pH and [HCO_3^-] were significantly elevated at various points throughout the trial with sodium citrate.

More recently, a study by Schabort et al.[43] demonstrated that, while pH and blood-buffering capacity was increased in a group of trained cyclists following sodium citrate administration, no significant performance differences were observed between treatments (sodium citrate at three doses and calcium carbonate). They noted that, in the study by Potteiger et al.,[12] power output was not significantly different between treatments, yet the performance times were different. While there were no differences at any of the individual time points in the study by Potteiger et al.,[12] the average power output was higher throughout the first 25 minutes during the citrate trial. So, even though the power output was not statistically different at any individual time point, the fact that more power was being produced throughout the first 25 minutes led to the significant time difference at the end.

Another aspect that differentiates these two studies is the fact that the protocol by Potteiger et al.[12] was a continuous time trial, and the time trial used by Schabort et al.[43] had the cyclists perform sprints at various times throughout the ride. In actual competition, time trials are performed with athletes pacing themselves to attain the highest power output possible to finish the race without succumbing to fatigue too soon. The protocol used by Schabort et al.[43] better mimics a road race. Another potentially important aspect between these two studies was the ergometer used for each. The cycle ergometer used by Potteiger et al.[12] was a mechanically braked ergometer fitted with clipless pedals and time-trial handlebars, and Schabort et al.[43] used an air-braked ergometer using the cyclists' bicycles. From a physics standpoint, the air-braked ergometer would better mimic a cycling event as, due to aerodynamic forces, there is no linear relationship between velocity and power when bicycle racing. Therefore, the differences in performance observed by Potteiger et al.[12] may have been less in an actual racing situation given the fact that even though power output might have been a little higher in the citrate trial, the subsequent increase in velocity would likely have been less if wind resistance were taken into account. That said, even if the difference between performances would be less in a racing situation, an improvement in performance of 1%, while not statistically significant, might be enough to earn a spot on the podium, which could be athletically significant.

D. Combining Bicarbonate and Citrate

A few studies combined sodium bicarbonate and sodium citrate during the intervention to determine whether the combination elicited similar responses to the treatments alone. Combining the two buffers at half the typical dose of each does not seem to enhance the effects on performance. In fact, Parry-Billings et al.[41] observed that performance was increased only when sodium citrate was ingested. Sodium bicarbonate alone or when combined with sodium citrate did not improve performance.

VI. Phosphate Loading on Physiology and Performance

Relative to the data available for sodium bicarbonate and sodium citrate, fewer studies have investigated the effects of phosphate salts on exercise performance. The rationale for the effectiveness of phosphate salts is somewhat different from that of previously discussed blood buffers (sodium bicarbonate and sodium citrate). Additional effects of phosphate loading may occur due to an increased capacity to store intracellular inorganic phosphate, which may permit a greater quantity of creatine phosphate to be stored for use during short-duration bursts of exercise. With regard to aerobic activity, phosphate loading has been suggested to increase 2,3 diphosphoglycerate (2,3-DPG).[44] Increased levels of 2,3-DPG may yield increased dissociation or release of oxygen from hemoglobin,[45] thereby theoretically creating an opportunity to provide more oxygen to working tissues.

In a study by Kreider et al.,[46] it was observed that the daily ingestion of 4 g of tribasic sodium phosphate improved the subjects' 40 km time-trial performance, aerobic capacity, myocardial ejection fraction and anaerobic threshold. This is contrary to the lack of effect for the same dose to enhance running performance.[47] In a study by Duffy and Conlee,[48] phosphate loading did not elicit a significant acute (1.24 g 1 hour before exercise) or chronic

(3.73 g/day) effect on aerobic capacity, treadmill running performance or isokinetic leg power output. Bredle et al.[45] suggested that oxygen extraction was increased following 4 days of calcium phosphate ingestion (176 mmol/day). However, they did not observe an improvement in time to exhaustion while running at 70% of VO_{2max}.[45]

VII. Chronic Buffer Supplementation

Chronic, or daily, use of these buffer compounds has not been adequately researched. More studies are needed to unequivocally determine whether the daily use of buffering compounds during practice elicits performance on days of competition. Of the few studies that have been performed, most, if not all, have reported significant changes in performance.[46,47,49] It must be noted that those who observed these differences also observed differences with acute treatment. To date, the investigations that did not report an acute treatment effect (Table 16.1) have not investigated the efficacy of chronic supplementation. Due to slight methodological differences between laboratories, it would be interesting to note whether results from those laboratories would support or refute the current data of chronic buffer supplementation.

VIII. Areas of Future Research

Given the limited data regarding chronic supplementation, more research is needed to better establish whether improved performance is consistent. A statistical means of accomplishing this may be through the use of nonparametric procedures. That is, compare or rank how many times the treatment improved performance compared with a placebo and control condition. If the treatment condition always improves performance, then the treatment may elicit a significant performance response, even though the means for performance are not statistically different. Also, the safety of chronic supplementation needs to be verified. It is currently unknown whether the chronic ingestion of buffers throughout a competitive season is safe or effective.

Also, more research is needed to better establish the physiological and performance effects of phosphate loading. Compared with sodium bicarbonate and sodium citrate, there is a dearth of information regarding the use of phosphate salts. With the limited data, it is difficult for coaches and athletes to determine whether it should be used.

Finally, with the plethora of other substances (legal and possibly illegal) being utilized by athletes, studies are needed to investigate whether these different substances interact with one another. These potential interactions may be ergogenic or even ergolytic. Thus, while individually a substance may be observed not to elicit improved performance, performance may improve (or deteriorate) when taken concurrently with another substance. Given the numerous dietary supplements, athletes may ultimately be limiting performance by taking several supplements or purported ergogenic aids that interact in a manner that, in some fashion, hinders force production. An example of this is the combination of caffeine, ephedrine and aspirin and the concomitant health risks.[50]

IX. Recommendations for Coaches and Athletes

Relative to the coach and athlete, issues related to buffer ingestion need to be discussed. First, one concern is the ethical aspect of choosing to use a buffer. While it is not the intent of this review to persuade or dissuade the use of an ergogenic aid, thought or discussions about how the ethical use of these compounds should occur. While these buffer agents are not currently (2003) banned, their use seems to infringe on the philosophical ideal of fair play. On the other side of that philosophical coin, one may question whether competition is "fair" to begin with.

Once the ethical issue is dealt with and if the buffer is deemed appropriate to use, the next concern is deciding whether the buffer of choice has the potential to improve performance. Due to the potential undesirable gastrointestinal events that occur during the digestion of these compounds, it is imperative that the coach or athlete determine dosing and timing strategies in practice. Daily use, as demonstrated by some of the results investigating the chronic effects, may be more advantageous than just using buffers on the day of competition. One note of caution, the chronic effects were only tested over a very short period of time. The long-term (months or years) consequences of chronic use of these buffers at the recommended doses are unknown. Additionally, as with other products, an athlete may need to abstain from using buffers for periods of time (i.e., cycling on or off) for reasons of decreased effectiveness or risks to their health.

X. Summary

It is obvious that buffers elicit a physiological response by increasing the buffering capacity of blood and muscle. Also, there is convincing data that these compounds can enhance performance.[51] However, an athlete's decision to use these compounds should only occur after a careful comparison of his or her sport with studies that mimic that competitive event. Also, it is known that studies that have similar protocols do not always observe similar performance outcomes.[12,43] These discrepancies in research make it difficult for a coach or athlete to discern whether the buffers are really having an effect. With specific reference to phosphate loading, more research is needed to better determine its potential mechanism(s) of action and understand its effects on performance. Given the current data, it is clear that the effects of buffer ingestion on performance are inconsistent and, if administered or dosed inappropriately, may actually hinder performance.

References

1. Costill D.L., Barnett A., Sharp R., Fink W.J., Katz A. Leg muscle pH following sprint running. *Med Sci Sports Exerc*, 15, 325, 1983.
2. Fuchs F., Reddy Y., Briggs F.N. The interaction of cations with the calcium-binding site of troponin. *Biochim Biophys Acta*, 221, 407, 1970.
3. Fabiato A., Fabiato F. Effects of pH on the myofilaments and the sarcoplasmic reticulum of skinned cells from cardiace and skeletal muscles. *J Physiol*, 276, 233, 1978.
4. Chasiotis D. The regulation of glycogen phosphorylase and glycogen breakdown in human skeletal muscle. *Acta Physiol Scand Suppl*, 518, 1, 1983.
5. Spriet L.L., Lindinger M.I., McKelvie R.S., Heigenhauser G.J., Jones N.L. Muscle glycogenolysis and H^+ concentration during maximal intermittent cycling. *J Appl Physiol*, 66, 8, 1989.
6. Trivedi B., Danforth W.H. Effect of pH on the kinetics of frog muscle phosphofructokinase. *J Biol Chem*, 241, 4110, 1966.
7. Nosek T.M., Fender K.Y., Godt R.E. It is diprotonated inorganic phosphate that depresses force in skinned skeletal muscle fibers. *Science*, 236, 191, 1987.
8. Wilson J.R., McCully K.K., Mancini D.M., Boden B., Chance B. Relationship of muscular fatigue to pH and diprotonated Pi in humans: A 31P-NMR study. *J Appl Physiol*, 64, 2333, 1988.
9. Juel C. Current aspects of lactate exchange: Lactate/H^+ transport in human skeletal muscle. *Eur J Appl Physiol*, 86, 12, 2001.
10. Poole R.C., Halestrap A.P. Transport of lactate and other monocarboxylates across mammalian plasma membranes. *Am J Physiol*, 264, C761, 1993.
11. Marx J.O., Gordon S.E., Vos N.H. et al. Effect of alkalosis on plasma epinephrine responses to high intensity cycle exercise in humans. *Eur J Appl Physiol*, 87, 72, 2002.
12. Potteiger J.A., Nickel G.L., Webster M.J., Haub M.D., Palmer R.J. Sodium citrate ingestion enhances 30 km cycling performance. *Int J Sports Med*, 17, 7, 1996.
13. Potteiger J.A., Webster M.J., Nickel G.L., Haub M.D., Palmer R.J. The effects of buffer ingestion on metabolic factors related to distance running performance. *Eur J Appl Physiol Occup Physiol*, 72, 365, 1996.
14. Stephens T.J., McKenna M.J., Canny B.J., Snow R.J., McConell G.K. Effect of sodium bicarbonate on muscle metabolism during intense endurance cycling. *Med Sci Sports Exerc*, 34, 614, 2002.
15. Hermansen L., Osnes J.B. Blood and muscle pH after maximal exercise in man. *J Appl Physiol*, 32, 304, 1972.
16. Fitts R.H. Cellular mechanisms of muscle fatigue. *Physiol Rev*, 74, 49, 1994.

17. Heigenhauser G.J., Jones N.L. Bicarbonate Loading. In: Lamb D, Williams M (Eds.) *Ergogenics: Enhancement of Performance in Exercise and Sport*, Dubuque, IA: Brown & Benchmark, 1991;183.
18. Fitts R.H. Muscle fatigue: The cellular aspects. *Am J Sports Med*, 24, S9, 1996.
19. Heigenhauser G.J. A quantitative approach to acid–base chemistry. *Can J Appl Physiol*, 20, 333, 1995.
20. Stewart P.A. Modern quantitative acid–base chemistry. *Can J Physiol Pharmacol*, 61, 1444, 1983.
21. McNaughton L.R. Sodium citrate and anaerobic performance: implications of dosage. *Eur J Appl Physiol Occup Physiol*, 61, 392, 1990.
22. McNaughton L.R. Bicarbonate ingestion: effects of dosage on 60 s cycle ergometry. *J Sports Sci*, 10, 415, 1992.
23. McKenzie D.C., Coutts K.D., Stirling D.R., Hoeben H.H., Kuzara G. Maximal work production following two levels of artificially induced metabolic alkalosis. *J Sports Sci*, 4, 35, 1986.
24. Downs N.M., Stonebridge P.A. Gastric rupture due to excessive sodium bicarbonate ingestion. *Scott Med J*, 34, 534, 1989.
25. USADA. In: Bowers LD, Wanninger R, Podraza J (Eds.) *Guide to Prohibited Classes of Substances and Prohibited Methods of Doping*, Colorado Springs, CO: United States Anti-Doping Agency 2003.
26. McNaughton L.R. Sodium bicarbonate ingestion and its effects on anaerobic exercise of various durations. *J Sports Sci*, 10, 425, 1992.
27. Horswill C.A. Effects of bicarbonate, citrate and phosphate loading on performance. *Int J Sport Nutr*, 5, S111, 1995.
28. Portington K.J., Pascoe D.D., Webster M.J. anderson L.H., Rutland R.R., Gladden L.B. Effect of induced alkalosis on exhaustive leg press performance. *Med Sci Sports Exerc*, 30, 523, 1998.
29. Webster M.J., Webster M.N., Crawford R.E., Gladden L.B. Effect of sodium bicarbonate ingestion on exhaustive resistance exercise performance. *Med Sci Sports Exerc*, 25, 960, 1993.
30. Lavender G., Bird S.R. Effect of sodium bicarbonate ingestion upon repeated sprints. *Br J Sports Med*, 23, 41, 1989.
31. Bouissou P., Defer G., Guezennec C.Y., Estrade P.Y., Serrurier B. Metabolic and blood catecholamine responses to exercise during alkalosis. *Med Sci Sports Exerc*, 20, 228, 1988.
32. Costill D.L., Verstappen F., Kuipers H., Janssen E., Fink W. Acid–base balance during repeated bouts of exercise: influence of HCO3. *Int J Sports Med*, 5, 228, 1984.
33. Gao J.P., Costill D.L., Horswill C.A., Park S.H. Sodium bicarbonate ingestion improves performance in interval swimming. *Eur J Appl Physiol Occup Physiol*, 58, 171, 1988.
34. Iwaoka K., Okagawa S., Mutoh Y., Miyashita M. Effects of bicarbonate ingestion on the respiratory compensation threshold and maximal exercise performance. *Jpn J Physiol*, 39, 255, 1989.
35. Verbitsky O., Mizrahi J., Levin M., Isakov E. Effect of ingested sodium bicarbonate on muscle force, fatigue and recovery. *J Appl Physiol*, 83, 333, 1997.
36. Clark V.R., Hopkins W.G., Hawley J.A., Burke L.M. Placebo effect of carbohydrate feedings during a 40-km cycling time trial. *Med Sci Sports Exerc*, 32, 1642, 2000.

37. Linossier M.T., Dormois D., Bregere P., Geyssant A., Denis C. Effect of sodium citrate on performance and metabolism of human skeletal muscle during supramaximal cycling exercise. *Eur J Appl Physiol Occup Physiol*, 76, 48, 1997.
38. Hollidge-Horvat M.G., Parolin M.L., Wong D., Jones N.L., Heigenhauser G.J. Effect of induced metabolic alkalosis on human skeletal muscle metabolism during exercise. *Am J Physiol Endocrinol Metab*, 278, E316, 2000.
39. Hausswirth C., Bigard A.X., Lepers R., Berthelot M., Guezennec C.Y. Sodium citrate ingestion and muscle performance in acute hypobaric hypoxia. *Eur J Appl Physiol Occup Physiol*, 71, 362, 1995.
40. McNaughton L., Cedaro R. Sodium citrate ingestion and its effects on maximal anaerobic exercise of different durations. *Eur J Appl Physiol Occup Physiol*, 64, 36, 1992.
41. Parry-Billings M., MacLaren D.P. The effect of sodium bicarbonate and sodium citrate ingestion on anaerobic power during intermittent exercise. *Eur J Appl Physiol Occup Physiol*, 55, 524, 1986.
42. Shave R., Whyte G., Siemann A., Doggart L. The effects of sodium citrate ingestion on 3,000-meter time-trial performance. *J Strength Cond Res*, 15, 230, 2001.
43. Schabort E.J., Wilson G., Noakes T.D. Dose-related elevations in venous pH with citrate ingestion do not alter 40-km cycling time-trial performance. *Eur J Appl Physiol*, 83, 320, 2000.
44. Cade R., Conte M., Zauner C. et al. Effects of phosphate loading on 2,3-diphosphoglycerate and maximal oxygen uptake. *Med Sci Sports Exerc*, 16, 263, 1984.
45. Bredle D.L., Stager J.M., Brechue W.F., Farber M.O. Phosphate supplementation, cardiovascular function and exercise performance in humans. *J Appl Physiol*, 65, 1821, 1988.
46. Kreider R.B., Miller G.W., Schenck D. et al. Effects of phosphate loading on metabolic and myocardial responses to maximal and endurance exercise. *Int J Sport Nutr*, 2, 20, 1992.
47. Kreider R.B., Miller G.W., Williams M.H., Somma C.T., Nasser T.A. Effects of phosphate loading on oxygen uptake, ventilatory anaerobic threshold and run performance. *Med Sci Sports Exerc*, 22, 250, 1990.
48. Duffy D.J., Conlee R.K. Effects of phosphate loading on leg power and high intensity treadmill exercise. *Med Sci Sports Exerc*, 18, 674, 1986.
49. McNaughton L., Backx K., Palmer G., Strange N. Effects of chronic bicarbonate ingestion on the performance of high-intensity work. *Eur J Appl Physiol Occup Physiol*, 80, 333, 1999.
50. Shekelle P.G., Hardy M.L., Morton S.C. et al. Efficacy and safety of ephedra and ephedrine for weight loss and athletic performance: A meta-analysis. *Jama*, 289, 1537, 2003.
51. Matson L.G., Tran Z.V. Effects of sodium bicarbonate ingestion on anaerobic performance: a meta-analytic review. *Int J Sport Nutr*, 3, 2, 1993.
52. Ball D., Maughan R.J. The effect of sodium citrate ingestion on the metabolic response to intense exercise following diet manipulation in man. *Exp Physiol*, 82, 1041, 1997.
53. Brien D.M., McKenzie D.C. The effect of induced alkalosis and acidosis on plasma lactate and work output in elite oarsmen. *Eur J Appl Physiol Occup Physiol*, 58, 797, 1989.
54. Cox G., Jenkins D.G. The physiological and ventilatory responses to repeated 60 s sprints following sodium citrate ingestion. *J Sports Sci*, 12, 469, 1994.

55. Feriche Fernandez-Castanys B., Delgado-Fernandez M., Alvarez Garcia J. The effect of sodium citrate intake on anaerobic performance in normoxia and after sudden ascent to a moderate altitude. *J Sports Med Phys Fitness*, 42, 179, 2002.
56. Gaitanos G.C., Nevill M.E., Brooks S., Williams C. Repeated bouts of sprint running after induced alkalosis. *J Sports Sci*, 9, 355, 1991.
57. Horswill C.A., Costill D.L., Fink W.J. et al. Influence of sodium bicarbonate on sprint performance: relationship to dosage. *Med Sci Sports Exerc*, 20, 566, 1988.
58. Ibanez J., Pullinen T., Gorostiaga E., Postigo A., Mero A. Blood lactate and ammonia in short-term anaerobic work following induced alkalosis. *J Sports Med Phys Fitness*, 35, 187, 1995.
59. Johnson W., Black D. Comparison of effects of certain blood alkalinizers and glucose upon competitive endurance performance. *J Appl Physiol*, 5, 557, 1953.
60. Katz A., Costill D.L., King D.S., Hargreaves M., Fink W.J. Maximal exercise tolerance after induced alkalosis. *Int J Sports Med*, 5, 107, 1984.
61. Kindermann W., Keul J., Huber G. Physical exercise after induced alkalosis (bicarbonate or tris-buffer). *Eur J Appl Physiol Occup Physiol*, 37, 197, 1977.
62. Kowalchuk J.M., Maltais S.A., Yamaji K., Hughson R.L. The effect of citrate loading on exercise performance, acid–base balance and metabolism. *Eur J Appl Physiol Occup Physiol*, 58, 858, 1989.
63. Kozak-Collins K., Burke E.R., Schoene R.B. Sodium bicarbonate ingestion does not improve performance in women cyclists. *Med Sci Sports Exerc*, 26, 1510, 1994.
64. Linderman J., Kirk L., Musselman J., Dolinar B., Fahey T.D. The effects of sodium bicarbonate and pyridoxine-alpha-ketoglutarate on short-term maximal exercise capacity. *J Sports Sci*, 10, 243, 1992.
65. Pierce E.F., Eastman N.W., Hammer W.H., Lynn T.D. Effect of induced alkalosis on swimming time trials. *J Sports Sci*, 10, 255, 1992.
66. Tiryaki G.R., Atterbom H.A. The effects of sodium bicarbonate and sodium citrate on 600 m running time of trained females. *J Sports Med Phys Fitness*, 35, 194, 1995.
67. Wilkes D., Gledhill N., Smyth R. Effect of acute induced metabolic alkalosis on 800 m racing time. *Med Sci Sports Exerc*, 15, 277, 1983.
68. van Someren K., Fulcher K., McCarthy J., Moore J., Horgan G., Langford R. An investigation into the effects of sodium citrate ingestion on high-intensity exercise performance. *Int J Sport Nutr*, 8, 356, 1998.
69. Gaenzer H., Sturm W., Neumayr G. et al. Pronounced postprandial lipemia impairs endothelium-dependent dilation of the brachial artery in men. *Cardiovasc Res*, 52, 509, 2001.
70. Mc Naughton L., Thompson D. Acute versus chronic sodium bicarbonate ingestion and anaerobic work and power output. *J Sports Med Phys Fitness*, 41, 456, 2001.

17

Caffeine

Faidon Magkos and Stavros A. Kavouras

CONTENTS

- I. Introduction 276
- II. Caffeine Use in Sports 276
- III. Pharmacokinetics of Caffeine 277
  - A. Absorption 277
  - B. Distribution 278
  - C. Metabolism 278
  - D. Excretion 279
- IV. Cellular and Molecular Mechanisms of Caffeine Action *in Vitro* 280
  - A. Calcium Release 280
  - B. Inhibition of Phosphodiesterase 281
  - C. Inhibition of Glycogen Phosphorylase 281
  - D. Antagonism of Adenosine Receptors 282
  - E. Stimulation of Na^+/K^+-ATPase 284
  - F. Inhibition of Phosphoinositide Kinases 284
- V. Physiological and Metabolic Effects of Caffeine *in Vivo* 285
  - A. Catecholamine Release 285
  - B. Lipolysis and Fat Oxidation 287
  - C. Glycogen Breakdown 292
  - D. Neuromuscular Function 294
- VI. Caffeine and Exercise Performance 295
  - A. Long-Term Submaximal Exercise 296
  - B. Short-Term Intense Exercise 308
  - C. Incremental Exercise 309
  - D. Caffeine and Ephedrine 310
- VII. Summary 311
- References 313

I. Introduction

Athletes have always sought extrinsic means to improve their performance beyond what systematic training offers. This attempt is as old as competitive sports — ancient Greek Olympians ate mushrooms and consumed protein-rich diets to win. Nowadays, the field of ergogenic aids has expanded tremendously. Ergogenic comes from the Greek words ergo meaning "work" and geno meaning "giving birth" or "producing." Accordingly, ergogenic means "work-producing." Numerous compounds, including potential energy sources, metabolites, recovery aids and drugs, are currently available alone or in combination and promise enhanced endurance, power, strength and speed.[1] Nevertheless, scientific evidence for such claims is scarce, and the efficiency of these products is mainly based on anecdotal reports and personal testimonies. On the other hand, while some agents may indeed improve physical performance under certain circumstances, their use is tightly regulated and they are banned from sports.[2] Additionally, the carries a substantial risk of adverse health effects.[3] In this general scheme, it is not surprising that many athletes, both elite and recreational, use caffeine as an ergogenic aid.

The scope of the present chapter is to provide an overview of caffeine use in sports, its pharmacokinetics in man and the molecular mechanisms of its action at the cellular level, and also to review available evidence regarding some of the hormonal, metabolic and physiological effects of acute caffeine ingestion in humans *in vivo*, both at rest and during exercise. The main focus is targeted on the effects that may be of relevance to the ergogenicity of the drug. The crux of the chapter deals with the effects of caffeine on several parameters of athletic performance; a detailed discussion on the timely combination of caffeine with ephedrine is also included. It should be mentioned that long-term consumption of caffeine or coffee has been implicated in several pathological conditions,[4] but these will not be considered here.

II. Caffeine Use in Sports

Caffeine is the most widely consumed psychoactive drug in the world, with almost 80% of the population being habitual users.[5] After its chemical isolation and synthetic production, caffeine has been used in several prescription and over-the-counter preparations as well.[6] The main sources of caffeine are coffee, tea, cocoa products and several cola beverages. Energy drinks represent a relatively new category of beverages that contain caffeine in amounts that exceed those in soft drinks and approach the low end of the concentration range found in coffee.[7] Caffeine consumption depends on many factors, such as natural source, age, gender, nutritional status, fitness level, peer behavior and habituation.[8,9] The most pertinent factor, however, is probably the source of caffeine, since its content in a number of natural and nonnatural sources varies in several ways.[10]

Athletes use caffeine for the specific purpose of enhancing performance. The most important reason for that is probably its perceived ergogenic efficacy, but caffeine is also inexpensive, can be consumed in generally legal amounts, has little or no acute adverse health effects and finally, is a socially acceptable drug. It has been reported that about 27% of both Canadian high-school students[11] and U.S. male adolescent athletes[12] used caffeine for improving performance, while about 70% of young U.S. athletes consumed the drug, but claimed to do so for social reasons only.[13] In competitive sports, there is only one described incidence of purported use of caffeine suppositories — by the U.S. Olympic Cycling Team in 1984[14] — and disqualification cases due to caffeine abuse are extremely scarce.[15] These figures notwithstanding, use of stimulants in general[16] and caffeine in particular[17,18] is believed to be much more widespread, among both professional and amateur athletes.

III. Pharmacokinetics of Caffeine

In principle, caffeine is a pharmacologically active compound with no essential physiological functions, but with various effects on most body tissues and organs.[19,20] To gain a useful insight into caffeine's action, it is necessary to have some knowledge of its behavior inside the body.

A. Absorption

Ingested caffeine is rapidly and completely absorbed by the gastrointestinal tract; absorption is complete within approximately 30–60 min after ingestion.[21-24] The absorption rate constant of caffeine is influenced by the physicochemical properties of the dose formulation, including pH, volume and composition.[25] Hence, for example, caffeine absorption is faster from a gum than from a capsule,[26] from a capsule than from coffee,[27] cola or chocolate,[24] and from coffee and tea than from cola.[22] These differences, however, are not always observed and it seems that when absolute caffeine dose and administration volume are the same, absorption rates are comparable regardless of formulation type.[28] It should be mentioned than a quicker rate of absorption could result in a quicker onset of effects, but not necessarily of a greater magnitude.

B. Distribution

With moderate caffeine consumption, i.e., about 5 to 6 cups of coffee daily, plasma concentrations remain typically around 50 µmol/l,[19] while it has been estimated that a dose of 1 mg/kg (considered to be equivalent to one cup of coffee) produces plasma levels of approximately 5 to 10 µmol/l.[29] Peak plasma caffeine is usually reached between 15 and 120 min after oral ingestion, with a tendency for slightly faster times when smaller doses are administered.[23,24,26,30] Following ingestion of different doses, plasma caffeine concentrations rise in a dose-dependent manner at rest[26] and during exercise,[31] and exhibit a similar rise-then-fall pattern. A linear relationship between caffeine dose and peak plasma concentration has been demonstrated for oral doses ranging from 1 to 10 mg/kg.[25]

Caffeine is sufficiently hydrophobic to pass through all biological membranes and is readily distributed throughout all tissues of the body.[29] In humans, caffeine dilutes in total body water and achieves a steady-state volume of distribution between 500 and 800 ml/kg.[21,26,30,32] Caffeine concentrations in the extracellular fluid of subcutaneous adipose tissue (and presumably other peripheral tissues) in man are generally not different from those in plasma and follow a similar time course,[33] probably reflecting the ability of the unbound drug to readily penetrate epithelial membranes. Likewise, studies in rats have shown that caffeine concentrations in the extracellular fluid of several tissues (brain, adipose, liver, muscle) are virtually identical to those in blood.[34] With respect to intracellular caffeine concentrations, available data from animal studies indicate that they tend to show some variation between individual organs, e.g., brain, heart, kidneys, liver, lungs, spleen, testes and muscle, especially early after administration; nevertheless, 30 to 60 min later, no significant differences between the concentrations of caffeine in plasma and in intracellular fluids are apparent.[35]

C. Metabolism

Caffeine (1,3,7-trimethylxanthine, molecular weight 194) is a purine alkaloid and, like most xenobiotics, undergoes extensive metabolism in the liver microsomes; at least 25 metabolites have been identified.[36] Formation of the three related dimethylxanthines via demethylation reactions is quantitatively the most important metabolic pathway, accounting for more than 95% of all caffeine biotransformations (Figure 17.1). N3-demethylation to form paraxanthine, N1-demethylation to form theobromine and N7-demethylation to form theophylline account for approximately 80%, 11% and 5%, respectively, of total caffeine elimination in humans *in vivo*.[37] The pathways leading to trimethyluric acid and trimethyluracil, as well as renal elimination of unchanged drug, account for the remainder (less than 5%) of total caffeine elimination.[37] Once formed, paraxanthine, theobromine and theophylline are subjected to further metabolic transformations.[21,26,38]

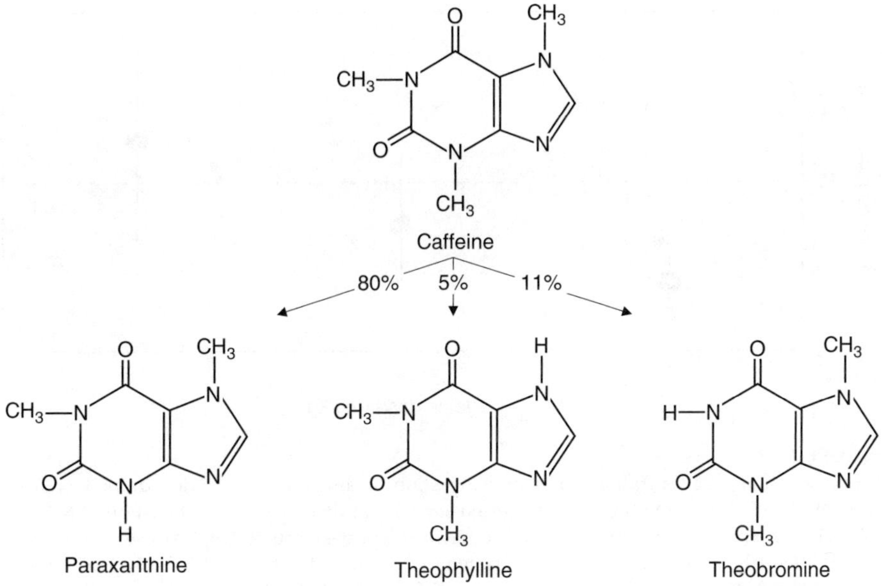

FIGURE 17.1
Major metabolites of caffeine in man.

D. Excretion

The half-life for elimination of caffeine from the plasma compartment for doses lower than 10 mg/kg ranges from 2.5 to 10 h in humans, and clearance rates are approximately 1–3 ml/kg/min.[21,23,26,30,32,39] The large variability in caffeine elimination parameters implies that inter-individual differences in caffeine pharmacokinetics are mainly due to the several-fold variable efficiency to metabolize and eliminate the drug, rather than to absorb it. For instance, urine and steady-state plasma concentrations of caffeine after ingestion of equal doses have been reported to range by 15.9-fold and 8.1-fold, respectively, among healthy volunteers.[40] A large inter-individual variability in urinary caffeine levels for a given dose is a particularly consistent finding.[41]

As noted above, urinary excretion of caffeine accounts for considerably less than 5% of the orally administered dose,[37] and typically, only around 0.5–3% is eliminated unchanged in urine.[38,42] Also, urinary concentrations peak within 1–3 h after ingestion,[42,43] thus the duration of exercise may indirectly impact on urinary caffeine measurements by lengthening or shortening the time interval between ingestion of dose and sampling of urine. This provides a clear advantage for athletes, since caffeine is a restricted compound in sports.*

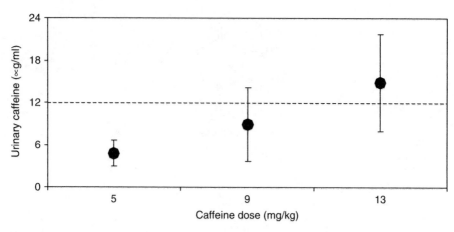

FIGURE 17.2
Caffeine urinary concentrations after oral ingestion of different doses. Nine trained cyclists ingested 0, 5, 9, or 13 mg/kg caffeine, rested for 1 hour, then cycled to exhaustion at 80% of W_{max}. Urine samples were obtained approximately 2 h post-ingestion. For the dose of 0 mg/kg, urine caffeine levels were generally undetected (not shown). The broken line is the IOC cut-off point. Values are means ± SD. (From Pasman, W. J., van Baak, M. A., Jeukendrup, A. E. and de Haan, A., The effect of different dosages of caffeine on endurance performance time, *Int. J. Sports Med.*, 16, 225, 1995.)

To further illustrate this point, doses of 5 to 9 mg/kg taken about 1 h prior to exercise result in post-exercise (approximately 2–2.5 h after the dose) urine levels that are below the acceptable level of 12 µg/ml set by the International Olympic Committee (IOC).[44–46] Only with doses of 13 mg/kg is there an increased likelihood for urine levels to exceed the IOC cut-off point (Figure 17.2).[46] However, caffeine can be of ergogenic benefit at considerably lower amounts than 9 or 13 mg/kg; the recommended dose lies between 3 and 6 mg/kg.[47]

IV. Cellular and Molecular Mechanisms of Caffeine Action *in Vitro*

A number of actions ascribed to caffeine from *in vitro* studies in animals and humans are briefly discussed below. Not surprisingly, due to their similar purine structure (xanthine derivatives), caffeine and dimethylxanthines share some common mechanisms of action, nevertheless with varying potency.[27]

---

* At the time these lines were written, caffeine was a restricted compound in sports. As of January 2004, caffeine and ephedra alkaloids (pseudoephedrine and phenylpropanolamine) have been removed from the list of banned substances drawn up by the World Anti-Doping Agency (WADA). The new list is available at WADA website: the World Anti-Doping Code–The 2004 Prohibited List–International Standard (http://www.wada-ama.org/docs/web/standards_harmonization/code/list_standard_2004.pdf).

A. Calcium Release

Evidence that caffeine can induce the contractility of isolated muscle dates back to 1910.[48] The site and mode of caffeine's action were elucidated several years later, in studies using isolated fragments of the sarcoplasmic reticulum (SR) from frog muscle: these experiments showed an increase in the release of calcium (Ca^{2+}) from the SR after application of the drug.[49,50] Soon, it became clear that caffeine could interfere with excitation-contraction coupling, i.e., the process that couples electrochemical events at the surface membrane with the release of Ca^{2+} from the SR. Caffeine induces SR Ca^{2+} release by activating the SR Ca^{+2} release channel formed by ryanodine receptors. More specifically, by using the planar lipid bilayer vesicle fusion technique, caffeine was found to modify the gating behavior, but not the conductance, of both skeletal[51] and cardiac[52] SR Ca^{2+} release channels. That is, caffeine increased the frequency and duration of open events, i.e., increased the open probability of the channels.[51,52] Reviewing these and several other studies, it was concluded that the regulatory effect of caffeine on excitation-contraction coupling is bimodal: (1) a Ca^{2+}-dependent activation with threshold concentrations greater than 100 μmol/l caffeine that renders ryanodine receptors more sensitive to endogenous activators (e.g., Ca^{2+}, ATP) and (2) a Ca^{2+}-independent activation occurring at higher (mmol/l) caffeine concentrations.[53]

B. Inhibition of Phosphodiesterase

The inhibitory effect of methylxanthines on phosphodiesterase activity, leading to accumulation of intracellular cAMP (and cGMP), has been known for some time. In fact, caffeine and theophylline were the first compounds to be identified as inhibitors of phosphodiesterase.[54] Therefore, they have been widely used to elucidate the physiological roles of cAMP as an integral component of hormone action in various tissues and, most extensively, in the regulation of adipose tissue lipolysis.[55,56] It is of interest to note that the relative potencies of a series of 56 xanthine derivatives to enhance lipolysis correlated well with their ability to inhibit phosphodiesterase in fat cells.[57] Subsequently, sufficient evidence has emerged for methylxanthine-induced suppression of phosphodiesterase activity in several other tissues as well, e.g., liver[58] and skeletal muscle.[59] Enzyme inhibition is of competitive nature and theophylline is generally more potent than caffeine, while both are substantially less potent than several more substituted, chemically synthesized xanthine derivatives. Concentrations for half-maximal inhibition (IC_{50}) typically range from 100 μmol/l to about 6 mmol/l, depending on the specific characteristics of the phosphodiesterase isoform (i.e., substrate specificity, tissue, cellular and subcellular localization, etc.), as well as on the prevailing concentrations of various other enzyme effectors.

C. Inhibition of Glycogen Phosphorylase

Some 50 years ago, caffeine was reported to inhibit glycogen phosphorylase.[60] The drug acts in concert with glucose in a synergistic, competitive and nonexclusive manner to inhibit both liver and skeletal muscle glycogen phosphorylase a[61] and b[62] forms. In other words, the binding of one ligand facilitates the binding of the other and the inhibition caused by the binding of both is not only greater than that caused by either ligand alone, but also greater than if the interaction was additive. Glucose binds at the catalytic site of the enzyme, whereas caffeine binds at the nucleoside-inhibitor site.[63] The synergism between these two compounds is evidenced by the fact that, in the presence of glucose concentrations between 5 and 10 mmol/l, estimated IC_{50} values for caffeine are decreased by approximately 5–10 fold, compared with when glucose is absent.[64] For instance, the IC_{50} estimate for inhibition of human liver glycogen phosphorylase a by caffeine in the absence of glucose is 240 µmol/l, whereas it falls to only 26 µmol/l in the presence of 7.5 mmol/l glucose.[64] Recent studies have also shown that the inhibitory effect of caffeine and the synergism with glucose are preserved even within a complex, yet physiological, mixture of various other effectors.[65,66]

D. Antagonism of Adenosine Receptors

The mounting evidence for a link between methylxanthine action and adenosine metabolism and the increasing understanding of the structure–function relationships of cell surface adenosine receptors, eventually led to the concept of methylxanthine-mediated antagonism of adenosine's actions via blockade of adenosine receptors.[67–70] Initially, theophylline was observed to counteract adenosine-induced cAMP accumulation in guinea pig cerebral cortex slices under several *in vitro* conditions.[71] This was a surprising discovery, since reduced cAMP was exactly the opposite effect of what one would expect from a phosphodiesterase inhibitor. Later on, methylxanthine-induced adenosine antagonism was demonstrated in several peripheral tissues as well, including smooth and striated muscle, adipocytes, heart, platelets, adrenals, pancreas and many others.[72] The most interesting feature of this discovery was that adenosine antagonism was evident at considerably lower concentrations than those needed to elicit other effects.[67–70] In other words, these methylxanthines were shown to be significantly more potent as adenosine receptor antagonists, rather than as phosphodiesterase inhibitors or SR Ca^{2+} release agents.

On the basis of structural and pharmacological criteria, four types of adenosine receptors have been identified: A_1, A_{2A}, A_{2B} and A_3, all of which belong to the family of rhodopsin-like G protein-coupled receptors.[73] Activation of these receptors by adenosine usually results in the generation of a diverse array of second messengers that modulate a variety of cellular processes.[27,73] This feature, in addition to the wide receptor distribution throughout the body tissues, accounts for the multiplicity of adenosine's actions. In

TABLE 17.1
Some Major Characteristics of Adenosine Receptor Subtypes

Receptor Subtype	A_1	A_{2A}	A_{2B}	A_3
	\multicolumn{4}{c}{Agonist and Antagonist Potency (µmol/l)[a]}			
Adenosine	0.31 (0.18–0.53)	0.73 (0.56–0.95)	23.5 (16.2–64.1)	0.29 (0.18–0.53)
Caffeine	33.8 (28–39)	12.3 (9.8–14.8)	15.5 (9.3–21.8)	>100
Theophylline	8.9 (6.5–14)	7.9 (6.3–10.0)	4.8 (3.9–5.8)	>100
Paraxanthine	15.8 (4.5–34)	5.3 (3.8–6.9)	5.5 (5.1–5.8)	>100
	\multicolumn{4}{c}{Physiological Functions[b]}			
	Inhibition of lipolysis; reduced glomerular filtration; tubero-glomerular feedback; antinociception; reduction of sympathetic and parasympathetic activity; presynaptic inhibition; ischemic preconditioning	Regulation of sensorimotor integration in basal ganglia; inhibition of platelet aggregation and polymorphonuclear leukocytes; vasodilatation; protection against ischemic damage; stimulation of sensory nerve activity	Relaxation of smooth muscle in vasculature and intestine; inhibition of monocyte and macrophage function; stimulation of mast cell mediator release (some species)	Enhancement of mediator release from mast cells (some species); preconditioning (some species)

[a] Potencies for agonists (EC_{50}) and antagonists (K_i) are given as means and 95% confidence intervals.[74]
[b] Physiological tissue functions of adenosine A_1, A_{2A} and A_3 receptors have been confirmed in knockout mouse.[73,75]

principle, by blocking these receptors, caffeine and related methylxanthines can antagonize the actions of adenosine, with varying potency at each receptor.[27,74] The binding of xanthines does not generate any kind of response, i.e., these compounds simply occupy the receptor site where adenosine normally binds. Adenosine A_3 receptors are generally insensitive to methylxanthine-induced inhibition, while the other three subtypes are effectively inhibited by both caffeine and theophylline, the latter being more potent. A summary of some major features of these receptors is shown in Table 17.1.[73-75]

E. Stimulation of Na^+/K^+-ATPase

The basic framework of the cell membrane Na^+/K^+-ATPase response to caffeine was delineated in a series of experiments performed some 30 years ago, indicating that the drug was able to increase the activity of the pump.[76,77] Recent studies have shown that paraxanthine can mimic this effect and, in fact, is a more potent stimulator than caffeine: 10-500 µmol/l paraxanthine increased unidirectional K^+ transport into resting rat hind limb muscle by approximately 25%, whereas 50–1,000 µmol/l caffeine had no effect.[78] The exact biochemical events responsible for the methylxanthine-induced stimulation of Na^+/K^+-ATPase activity are not known. Reviewing a number of relevant studies, Lindinger et al.[79] have concluded that there are five main candidate mechanisms, three of which are mediated by Ca^{2+} and two by cAMP:

1. Methylxanthines may stimulate SR Ca^{2+} release, resulting in increased intracellular Ca^{2+} concentration ($[Ca^{2+}]_i$), which in turn stimulates the Na^+/Ca^{2+} exchanger and results in increased $[Na^+]_i$, that subsequently activates the pump.
2. Methylxanthines may stimulate the opening of sarcolemmal L-type Ca^{2+} channels, resulting in Ca^{2+} influx and increased $[Ca^{2+}]_i$ and subsequent activation of Na^+/K^+-ATPase as in mechanism 1.
3. Increased $[Ca^{2+}]_i$ may in some way stimulate the Na^+/H^+ exchanger, resulting in intracellular alkalinization (decreased $[H^+]_i$) and increased $[Na^+]_i$ and subsequent activation of the pump.
4. Methylxanthines may bind to adenosine receptors coupled to G_i proteins and cause an increase in cAMP and protein kinase phosphorylation, which in turn in some way stimulate Na^+/K^+-ATPase activity.
5. Methylxanthines may diffuse into the cell and stimulate SR Ca^{2+} release or Ca^{2+} entry across the sarcolemma, leading to increased $[Ca^{2+}]_i$ and subsequently increased cAMP, which in turn activates the pump.

F. Inhibition of Phosphoinositide Kinases

There is some evidence from earlier studies that methylxanthines like caffeine and theophylline can lead to decreased phosphoinositide synthetic rate.[80,81] The molecular basis of this effect has not yet been established, but it was recently demonstrated that caffeine (and theophylline) can inhibit the lipid kinase activity of class Ia phosphoinositide-3-OH-kinases (PI3K; p110α, β, γ and δ),[82] as well as the protein kinase activity of PI3K-related kinases,[83,84] although with many-fold varying potency. In fact, only the lipid kinase activity of p110δ was inhibited by physiological caffeine concentrations (75 µmol/l),[82] whereas IC_{50} values for the other kinase activities ranged from 200 µmol/l to 10 mmol/l.[82-84] It is well established that activation of the class Ia PI3Ks is an integral component of the insulin signaling network and is necessary to elicit many of insulin's effects on glucose and lipid metabolism.[85] For instance, downstream activation of protein kinase B via phosphorylation is controlled by the lipid products of PI3Ks and is a physiological consequence of PI3K pathway activation by insulin. It was also reported that caffeine and theophylline are able to suppress insulin-stimulated protein kinase B phosphorylation-activation; the inhibitory effect of these methylxanthines and especially of caffeine was not restricted to cell culture models, but was also evident in intact rat soleus muscle, and with even greater potency.[82] In particular, IC_{50} values for caffeine and theophylline in cell culture were in the range of 1,000 and 500 µmol/l, respectively, while they fell to approximately 100 µmol/l in the case of intact muscle.[82] The significance of these findings awaits further research.

V. Physiological and Metabolic Effects of Caffeine *in Vivo*

It is clear from the above discussion that *in vitro* studies provide a wealth of different mechanisms that could potentially account for, or contribute to, the pharmacodynamic actions of caffeine in humans. On this basis, one might anticipate that the effects of caffeine on human metabolism are relatively complex and refractory to accurate investigation, since they are brought about by the integration of a variety of independent or interrelated cellular actions, probably with varying contribution. Nevertheless, not all mechanisms are responsible for the observed effects *in vivo*, since caffeine concentrations needed for each vary several-fold, from micromolar to millimolar levels. Apparently, only these actions elicited at the low end of the micromolar range (0-100 µmol/l) will have any relevance to the *in vivo* condition in man. Therefore, there is a consensus that the effects of caffeine and other methylxanthines on various central and peripheral tissue functions in humans *in vivo* result primarily from antagonising adenosine's actions via blockade of adenosine receptors.[27,68,69]

A. Catecholamine Release

The catecholamine response to the ingestion of caffeine and related methylxanthines has been the subject of extensive study during the second half of the 20th century. In some of the earliest *in vivo* studies in resting humans, administration of moderate amounts of caffeine was consistently shown to increase urinary excretion of adrenaline and noradrenaline; the relative increase was always greater for the former than for the latter.[86,87] This was confirmed in a double-blind, placebo-controlled study by Robertson et al.[88] These investigators were also the first to measure blood catecholamine concentrations in addition to their urinary excretion, following acute ingestion of caffeine (250 mg) in healthy volunteers. Plasma adrenaline and noradrenaline were significantly increased after 1 h of rest in the caffeine trial compared with placebo.[88] Also, caffeine ingestion resulted in greater urinary excretion of both metanephrine (metabolite of adrenaline) and normetanephrine (metabolite of noradrenaline) by approximately 100% and 52% over placebo, respectively.[88]

A number of investigations using various experimental protocols have provided several interesting but sometimes conflicting results. Acute ingestion of moderate caffeine doses (3–6 mg/kg) by caffeine-naive individuals has generally resulted in increased plasma adrenaline concentrations by approximately 30–100% over placebo after 60 min of rest and until 2–4 h post-ingestion.[88-93] Only rarely has this not been observed.[45] Increases of similar magnitude have been reported in studies where no information on subjects' prior caffeine use is disclosed,[94-97] or in those recruiting subjects with variable habitual caffeine consumption,[31,44,98,99] provided that 48 h of caffeine abstinence preceded the tests. The results for noradrenaline are less consistent, with increases in plasma noradrenaline being less frequently observed. In fact, besides the initial studies by Robertson et al.,[88,89] none of the remaining investigations in caffeine-naive individuals have reported any effects of caffeine on blood noradrenaline concentrations at rest.[45,90-92] Studies recruiting subjects of unknown or variable habitual caffeine consumption have also provided mixed results. Graham and colleagues observed a substantial increase (approximately 50–75% over placebo) in plasma noradrenaline after 60 min of rest following ingestion of caffeine (3–9 mg/kg),[31,96] while in another experiment, a 32% increase in venous noradrenaline 60 min post-ingestion of caffeine (9 mg/kg) was noted, but only in one of two trials.[44] Most similar investigations, however, reported no enhancement of noradrenaline levels by caffeine at rest.[94,95,97-100]

The effects of the drug on catecholamines during physical activity are qualitatively similar to those in the resting state. Most frequently, acute caffeine ingestion has been associated with higher adrenaline concentrations during subsequent exercise of various modalities,[31,44,45,94,96-99,101,102] although this has not always been the case.[95,103] A question remains, however, as to whether increases in circulating adrenaline during exercise merely reflect a residual effect of the drug at rest. Some studies observed a progressive increase of the difference between caffeine and placebo above that of rest as exercise continued,[44,45] while most showed no additional increase[94,96-98] and one reported a disappearance of the resting difference in adrenaline once exercise began.[95] Caffeine-induced increases in noradrenaline concentrations during exercise are rarely observed,[96,102] and most studies report either no effect of caffeine on the exercise-induced increase[44,45,94,97-99,101] or just sporadic increases above placebo at some time points during exercise.[31,95,103] Nonetheless, in a recent investigation using muscle arteriovenous catheterizations, it was observed that leg noradrenaline spillover during exercise more than doubled after caffeine ingestion and the difference from placebo progressively increased as exercise continued.[96]

It is important to refer to some variables that may influence the catecholamine response to methylxanthine ingestion and should, therefore, be taken into account when designing or interpreting such studies. Probably the most relevant in this respect is prior caffeine status. In a landmark study, Robertson et al.[89] had nine healthy adults who were regular coffee drinkers abstain from caffeine consumption for 3 weeks. These volunteers then received placebo for 3 days, caffeine (3 × 250 mg with each meal) for 7 days (days 1 to 7) and again placebo for 4 days. Catecholamine responses to acute caffeine (250 mg) ingestion during the subsequent 4 h of rest on day 1 (caffeine-naive) and day 7 (caffeine-habituated) were determined. Caffeine increased both adrenaline and noradrenaline concentrations after 15–30 min from ingestion on day 1; both catecholamines peaked at 3 h post-ingestion and remained elevated for 3–4 h after dosing. In striking contrast, there was no effect on plasma catecholamines on day 7. This experiment clearly showed that catecholamine responses to acute caffeine ingestion are subjected to complete tolerance and a period of caffeine abstinence is therefore needed to reestablish the catecholaminergic sensitivity to the effects of the drug.[89]

Part of the discrepancy in the abovementioned results may also rest on the confounding effects of simultaneous exercise, since exercise by itself is accompanied by significant increases in circulating adrenaline and noradrenaline.[104,105] An additional confounding factor may be the training status of the individuals — well-trained subjects were shown to be more sensitive to caffeine-induced increases in plasma adrenaline than untrained ones; no training effects were observed in noradrenaline responses.[106] Interestingly, the form of ingested caffeine also assumes a role in the catecholamine response to the drug,[99] while it seems that caffeine is more effective than theophylline in increasing catecholamine concentrations, both at rest and during submaximal exercise.[97]

B. Lipolysis and Fat Oxidation

In one of the earliest investigations with human subjects, Bellet et al.[107] reported that caffeine administration increased free fatty acid (FFA) concentrations by 40–70% over placebo; peak levels were reached after 2–3 h of rest.

The same investigators also examined the effects of ingesting placebo, decaffeinated coffee or regular coffee (250 mg caffeine) and showed that regular coffee drinking resulted in almost a doubling of plasma FFA after 3–4 of rest, while decaffeinated coffee and placebo had no effect.[108] Several studies have followed up on these results and, for the most part, confirm that caffeine ingestion increases FFA levels after approximately 1 h of rest; the increase may range from 30% to more than 100% above placebo. This effect seems to be independent of habitual caffeine consumption, since it has been observed in both caffeine-naive[45,90,93,109–113] and caffeine-habituated[45,106,109,111,114–116] subjects, as well as in those with variable[31,98,117] or unknown[96,118–120] habitual caffeine use. There seems to be good evidence, therefore, that caffeine ingestion significantly increases FFA concentrations in resting humans. Only a few investigators have reported no such effect.[44,97,121] On the other hand, data is not as homogenous with respect to resting plasma glycerol concentrations, with about half the studies showing an increase over placebo 60–90 min after caffeine administration,[44,45,90,93,96,98,111,117] and the other half not.[31,45,95,97,99,122]

The results of the studies examining fat metabolism during exercise after caffeine ingestion are summarized in Table 17.2.[31,44,45,96–98,101,103,109,110,112–115,117–119,121–130] It is evident that a large number of such experiments have been carried out, but the vast differences in design result in considerable heterogeneity of available evidence. From a simplistic point of view, caffeine may or may not lead to increased FFA concentrations during exercise. Still, it seems that increments in circulating FFA during exercise are less frequently observed and are of a smaller magnitude than at rest. This may also be due to the lipolytic effect of exercise per se, which may mask that of caffeine. As in the resting condition, glycerol concentrations may be increased after caffeine vs. placebo ingestion, either throughout[117,126] or only at some time points[31,44,95] during exercise, but generally, they remain unaltered regardless of training status, caffeine habituation and withdrawal, dose and time of administration, type, duration and intensity of exercise, etc.[45,96–99,101,122–125]

In spite of the large number of studies measuring FFA and glycerol concentrations, only a few have used more complicated methodologies to investigate FFA and glycerol metabolism after methylxanthine administration. In one of these studies, Graham et al.[96] used arteriovenous balance measurements and reported that, although ingestion of caffeine 60 min pre-exercise increased arterial FFA and glycerol levels above placebo, it had no effect on net uptake or release of these substrates by the active muscles (Figure 17.3). Similarly, Raguso et al.[131] had seven overnight-fasted endurance-trained cyclists receive intravenous infusion of theophylline starting 60 min prior to a 30-min cycling bout at 75% of VO_{2max}. Theophylline concentrations immediately before exercise were 34.1 ± 1 µmol/l. These investigators quantified fat metabolism by stable isotope methodology and non steady-state kinetics and reported no effect of theophylline on FFA and glycerol concentrations, as well as on whole-body FFA and glycerol rates of appearance and disappearance, either at rest or during exercise.[131]

TABLE 17.2.
Effects of Caffeine on Free Fatty Acid Concentrations ([FFA]) During Exercise

Study	Experimental Design	FFA Response
Costill et al. (1978)[126]	9 competitive cyclists (7 M, 2 F) ingested P or C (330 mg) after 6-12 h of fast. Following 60 min of rest, they cycled to exhaustion at 80% VO_{2max}.	No effect.
Ivy et al. (1979)[124]	9 well-trained cyclists (7 M, 2 F) ingested P or C (250 mg) after a 12 h fast. Following 60 min of rest, they performed 120 min of isokinetic cycling at 80 rpm, during which time an additional 250 mg dose was ingested.	At the end of exercise, [FFA] was higher (+30%) in the C trial (1.00 ± 0.16 mmol/l) vs. P (0.77 ± 0.12 mmol/l).
Knapik et al. (1983)[123]	10 M (5 trained and 5 untrained) ingested P or C (5 or 9 mg/kg). After 60 min of rest, they ran for 60 min at 60% of VO_{2max}.	No effect of either C dose in either trained or untrained subjects.
Giles and MacLaren (1984)[119]	6 M well-trained athletes ingested P or C (5 mg/kg) with or without glucose after a 24-h C abstinence. Following 60 min of rest, they ran for 2 h at 65% of VO_{2max}.	[FFA] was higher in the C trials (alone or with glucose) vs. the non-C trials (P alone or with glucose).
Casal and Leon (1985)[118]	9 M well-trained marathon runners ingested P or C (400 mg) after a 4-h fast. Following 60 min of rest, they ran for 45 min at 75% of VO_{2max}.	No effect.
Wells et al. (1985)[128]	10 M recreationally trained marathon runners ingested P or C (5 mg/kg) 60 min prior to and throughout a 20-mile run at self-selected pace.	No effect.
McNaughton (1987)[130]	10 M well-trained athletes (C-naive) ingested P or C (10 or 15 mg/kg) after a 12-h fast and a 24-C abstinence. Following 60 min of rest, they performed incremental cycling exercise (at 50 rpm) to exhaustion with variable resistance.	Both C doses increased [FFA] by approximately 100% over P at all workloads.
Sasaki et al. (1987)[103]	7 M recreational athletes ingested P or C (800 mg) after an 8-h fast, immediately prior to and during 4 × 30 min of running at 62–67% of VO_{2max}. This was followed by running for 10 min at 80% of VO_{2max} and then at 90% of VO_{2max} until exhaustion.	[FFA] was generally higher in the C trial and this was mainly evident after the first hour of exercise.
Sasaki et al. (1987)[121]	5 M distance runners ingested P or C (300 mg) after an 8-h fast. Following 60 min of rest, they ingested an additional 60-mg dose and then ran for 45 min at 80% of VO_{2max}. Then they ingested another 60 mg C, rested for 5 min and ran to exhaustion at 80% VO_{2max}.	C increased [FFA] by approximately 20% over P throughout exercise.
Chad and Quigley (1989)[110]	5 F untrained (C-naive) ingested P or C (5 mg/kg) after a 5-h fast. Following 60 min of rest, they walked vigorously for 90 min at 55% of VO_{2max}.	[FFA] was increased in the C trial vs. P.

(continued)

TABLE 17.2 (CONTINUED)
Effects of Caffeine on Free Fatty Acid Concentrations ([FFA]) During Exercise

Study	Experimental Design	FFA Response
Tarnopolsky et al. (1989)[114]	6 M well-trained (C users) ingested P or C (6 mg/kg) after 15 h of C abstinence and 3 h of fast. Following 60 min of rest, they ran for 90 min at 70% of VO_{2max}.	C increased [FFA] vs. P at 60 min into exercise (0.47 vs. 0.28 mM, respectively), but not before or after.
Flinn et al. (1990)[112]	9 M recreational cyclists (C-naive) ingested P or C (10 mg/kg) after 24 h of C abstinence and 12 h of fast. Following 180 min of rest, they performed an incremental cycling test to exhaustion.	[FFA] was increased in the C trial by approximately 100% over P.
Dodd et al. (1991)[109]	17 M recreationally trained (9 C users and 8 C-naive) ingested P or C (3 or 5 mg/kg) after a 24-C abstinence and a 4-h fast. Following 60 min of rest, they underwent a graded, incremental cycle ergometer test to exhaustion.	[FFA] was higher in both C trials vs. P in C-naive, but not in C-habituated subjects.
Graham et al. (1991)[101]	6 M recreationally trained (C-naive) ingested P or C (5 mg/kg) after 12-14 h of fast. Following 30 min of rest in a thermoneutral environment, they moved to a cold environment and cycled for 2 h at 40% of VO_{2max}.	No effect.
Graham and Spriet (1991)[44]	6 well-trained runners (6 M, 1 F) with variable habitual C consumption ingested P or C (9 mg/kg) after 2-4 h of fast and 48 h of C abstinence. Following 60 min of rest, they ran or cycled to exhaustion at 85% of VO_{2max}.	Generally no effect.
Titlow et al. (1991)[129]	5 M recreationally trained ingested P or C (200 mg) after a 12-h fast. Following 60 min of rest, they ran for 60 min at 60% of VO_{2max}.	No effect.
Donelly and McNaughton (1992)[115]	6 F untrained (light C users) ingested P or C (5 or 10 mg/kg) after a 12-h fast. Following 60 min of rest, they cycled for 90 min at 55% of VO_{2max}.	Both C doses similarly increased [FFA] above P.
Spriet et al. (1992)[98]	8 recreational cyclists (7 M, 1 F) with variable habitual C consumption ingested P or C (9 mg/kg) after 2-4 h of fast and 72 h of C abstinence. Following 60 min of rest, they cycled to exhaustion at 75-85% of VO_{2max}.	Generally no effect.
Van Soeren et al. (1993)[45]	14 M recreationally trained (7 C users and 7 C-naive) ingested P or C (5 mg/kg) after a 6-h fast. Following 60 min of rest, they cycled for 60 min at 50% of VO_{2max}.	No effect in either C users or non-users.

Study	Description	Result
Graham and Spriet (1995)[31]	8 M well-trained runners with variable habitual C consumption ingested P or C (3, 6, or 9 mg/kg) after 2-4 h of fast and 48 h of C abstinence. Following 60 min of rest, they ran to exhaustion at 85% of VO_{2max}.	No effect of any C dose.
Trice and Haymes (1995)[113]	8 M well trained (not regular C users) ingested P or C (5 mg/kg) after a 12-h fast and a 24-h C abstinence. Following 60 min of rest, they performed 3 × 30-min bouts of intermittent cycling at 85-90% of maximal workload.	Post-exercise [FFA] was higher in the C trial (1.71 ± 0.06 mmol/l) vs. P (1.44 ± 0.16 mmol/l).
Cole et al. (1996)[117]	10 M with variable training status and habitual C use ingested P or C (6 mg/kg) after 10-14 h of fast and 48 h of C abstinence. Following 60 min of rest, they performed 30 min of isokinetic, variable resistance cycling at 80 rpm.	[FFA] was higher in the C trial, but the difference from P progressively declined as exercise continued.
Mohr et al. (1998)[125]	6 M spinal cord injured (quadriplegic) patients (heavy C users), ingested P or C (6 mg/kg) in the early postprandial period after a 48-h C abstinence. Following 60 min of rest, they performed electrically stimulated involuntary leg cycling exercise at constant oxygen consumption.	No effect.
Van Soeren and Graham (1998)[122]	6 M recreationally trained (C users) ingested P or C (6 mg/kg) after 2-4 h of fast and 0, 2, or 4 days of C abstinence. Following 60 min of rest, they cycled to exhaustion at 80-85% of VO_{2max}.	No effect, regardless of withdrawal period.
Graham et al. (2000)[96]	10 M ingested P or C (6 mg/kg) after a 48-h C abstinence and an overnight fast. Following 60 min of rest, they cycled for 60 min at 70% of VO_{2max}.	Arterial [FFA] was elevated in the C trial during the first 30 min of exercise, but not thereafter.
Greer et al. (2000)[97]	8 M recreationally trained ingested P or C (6 mg/kg) in the early postprandial period after a 48-h C abstinence. Following 90 min of rest, they cycled to exhaustion at 80-85% of VO_{2max}.	No effect.
Van Baak and Saris (2000)[127]	15 M well-trained endurance athletes ingested P, propranolol, or C (5 mg/kg) plus propranolol after a 2-h fast and a 24-h C abstinence. Following 120 min of rest, they cycled to exhaustion at 70% of maximal aerobic power.	C had no effect on the propranolol-induced decrease in [FFA].

Note: FFA= free fatty acids; P = placebo; C = caffeine; M = males; F = females.

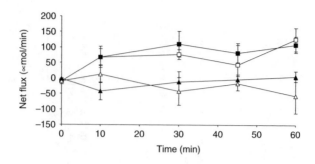

FIGURE 17.3
Free fatty acid (FFA) and glycerol fluxes across the active muscle during steady-state exercise after caffeine ingestion. Ten overnight-fasted young adult males ingested caffeine (closed symbols, 6 mg/kg) or placebo (open symbols) after a 48-h caffeine abstinence. Following 1 h of rest, they cycled for 60 min at 70% of VO_{2max}. Teflon catheters were inserted into a femoral artery and vein of one leg and FFA (squares) and glycerol (triangles) turnover rates were determined. Positive and negative values indicate net uptake and release, respectively. Values are mean ± SE. (From Graham, T. E., Helge, J. W., MacLean, D. A., Kiens, B. and Richter, E. A., Caffeine ingestion does not alter carbohydrate or fat metabolism in human skeletal muscle during exercise, *J. Physiol.*, 529, 837, 2000.)

The observations on the effects of caffeine on fat metabolism can be summarized as follows: Caffeine ingestion will probably lead to increased FFA concentrations at rest and perhaps also during exercise, while glycerol levels may or may not increase. Still, the drug will not enhance active muscle FFA uptake, although its effects on other tissues remain unknown. This casts doubt on the popular notion that caffeine results in accelerated fat oxidation, which in turn reduces muscle glycogen use (see section on glycogen breakdown). In support of this, many studies have examined the effects of the drug on the respiratory exchange ratio during exercise and, while some have indeed reported a reduced value in the caffeine trial compared with placebo,[98,106,110,112,115,119,121,124,126] indicating increased oxidation of fat, most have observed no such effect.[31,44,45,90,95–97,101,103,109,114,117,118,122,123,125,129,132,133] Hence, although caffeine may indeed stimulate lipolysis and increase the concentration and availability of fat fuels in the circulation, it will probably not enhance fat oxidation, either at the whole-body level or in the active muscles. Whether, however, such an effect is manifested within other tissues (e.g. inactive muscles, liver), remains unclear at present.

C. Glycogen Breakdown

The effects of caffeine on skeletal muscle glycogen metabolism have attracted much interest throughout the years, especially in the field of sports physiology, since it is well known that glycogen availability is a performance-limiting factor during prolonged submaximal exercise.[134,135] Although a number of researchers have attributed caffeine's ergogenic efficiency on its putative glycogen-sparing effects, only a few have actually measured glycogen utilization during exercise after caffeine or placebo ingestion. In the first of these investigations, Essig et al.[136] had subjects cycle for 30 min at 65–75% of VO_{2max} and observed that caffeine (5 mg/kg) reduced glycogen use by approximately 42% compared with the placebo trial. A second report has also noted a glycogen-sparing effect of caffeine — ingestion of the drug (5 mg/kg) resulted in approximately 31% less glycogen used during the subsequent 90 min of cycling at 65–70% of VO_{2max}.[137] The study by Spriet et al.[98] made an interesting point in that caffeine (9 mg/kg) caused a 56% sparing of muscle glycogen during the initial 15 min of exercise, but not for the remainder of a cycling bout to exhaustion at approximately 75–85% of VO_{2max}. This could imply that the effects of the drug are manifested in resting muscle, so that individuals commence exercise with an intramuscular metabolic profile favoring fat oxidation rather than carbohydrate. As a result, muscle glycogen use is reduced early during exercise, i.e., at a time period when glycogen utilization rates are the greatest.[134,138]

Although attractive in theory, subsequent studies have provided equivocal findings in support of this mechanism. Jackman et al.[94] reported no effect of caffeine (6 mg/kg) ingestion on skeletal muscle glycogen breakdown during short-term intense exercise at approximately 100% of VO_{2max} and, in fact, the drug resulted in significantly greater muscle lactate concentrations during the first two bouts of cycling. On the other hand, Chesley et al.[139] observed a variable glycogen-sparing effect of caffeine (9 mg/kg) during 15 min of submaximal cycling: a group of six subjects used on average 28 ± 2.6% (range 21–38%) less glycogen during the caffeine trial, while the drug had no effect on glycogenolysis in the remaining six subjects. The two groups did not differ with respect to pre-exercise glycogen content, VO_{2max} or oxygen consumption during exercise, plasma caffeine concentrations and muscle citrate synthase or maximal glycogen phosphorylase activity; muscle lactate levels were also similar and unaffected by caffeine.[139] The three most recent studies investigating the effects of caffeine on glycogen metabolism in human skeletal muscle during steady state (65–70% of VO_{2max}) exercise have failed to observe any kind of glycogen sparing. Graham et al.[96] reported that caffeine (6 mg/kg) ingestion did not affect glycogen utilization in overnight-fasted humans, either during the initial 10 min or throughout a 60-min bout of submaximal cycling. Likewise, no effect of caffeine (6 mg/kg) or theophylline (4.5 mg/kg) on muscle glycogenolysis during the initial 15 min or throughout the remainder of a cycling bout to exhaustion was noted by Greer et al.[97] Finally, Laurent et al.[140] measured glycogen breakdown in the thigh muscle by natural abundance ^{13}C NMR spectroscopy (all other studies used needle biopsies from the vastus lateralis muscle), after having subjects glycogen-depleted or loaded or not: no differences were observed in the glycogenolytic rates during 2 h of exercise after caffeine (6 mg/kg) or placebo ingestion in either group.

Clearly, available studies do not allow reaching consensus with respect to the effects of caffeine on human skeletal muscle glycogen metabolism. Whether the variability in the results rests on differences in study design

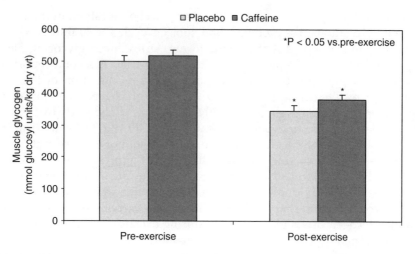

FIGURE 17.4
Effect of caffeine on skeletal muscle glycogen utilization in exercising humans. The data were pooled from four studies showing variable glycogen-sparing effects of caffeine.[96–98,139] In all these studies, subjects (n = 37) ingested placebo or caffeine (6 or 9 mg/kg), rested for 1 h, then cycled for 10 or 15 min at submaximal intensity (65-85% of VO_{2max}). Values are mean ± SE. (From Graham, T. E., Caffeine, coffee and ephedrine: Impact on exercise performance and metabolism, *Can. J. Appl. Physiol.*, 26 Suppl, S103, 2001. With permission.)

(e.g., caffeine dose, mode, intensity and duration of exercise) or subject characteristics (e.g., prior caffeine use, training status, individual sensitivity to the drug) remains to be evaluated. Furthermore, the relatively small number of subjects who can be recruited in studies of such type precludes drawing unequivocal conclusions. It is of interest, however, that, in a recent retrospective analysis of data from four of these studies including a total of 37 subjects, Graham[141] concluded that caffeine ingestion does not alter glycogen utilization in exercising humans (Figure 17.4). From the four studies reanalyzed, two showed no glycogen-sparing effect of caffeine[96,97] and one did,[98] while the fourth observed a variable sparing effect of the drug.[139] A point should also be made that animal studies provide no evidence of reduced glycogen utilization in skeletal muscle of resting or exercising rats after intravenous injection of caffeine, regardless of training status, fasting or fed states, type of muscle (fast, slow, or mixed), caffeine dose and time of administration.[142–144]

D. Neuromuscular Function

Despite the wealth of *in vitro* evidence that caffeine may interfere with Ca^{2+} handling in skeletal muscle (see section on calcium release), only a few studies have been conducted to examine the effects of the drug on neuromuscular function. The latter term may be considered to encompass all processes involved in muscle contraction.[145] Lopes et al.[146] were the first to describe the effects of caffeine ingestion on voluntary and electrically stimulated contractions of the adductor pollicis muscle in man. In that study, 5 healthy adults were given 500 mg of caffeine (approximately 7–9.5 mg/kg) after a 12-h fast. Following 60 min of rest (plasma caffeine concentration was 66.0±25.3 µmol/l), electrical stimulation of the ulnar nerve was performed at the wrist with supramaximal square-wave pulses of 50–100 µs duration. Force-velocity curves were constructed before and after fatigue at 50% of maximal voluntary contraction. Ingestion of caffeine produced an increase in the tension developed for a given frequency of stimulation, except at 100 Hz; this effect was more marked at 20 and 30 Hz and smaller at 10 and 50 Hz.[146] The effects of the drug were qualitatively similar both before and after fatigue.[146] In other words, and in agreement with data from animal studies, caffeine potentiated force development at submaximal but not at maximal frequencies of stimulation.

Importantly, this effect is probably not subjected to tolerance, since ingestion of the drug (6 mg/kg) similarly potentiated torque output in both caffeine-naive and caffeine-habituated subjects at low stimulation frequencies (20 Hz), but not at higher ones (40 Hz).[147] Nevertheless, another study failed to support a potentiating effect of caffeine on neuromuscular function in man. Six well-trained young male runners (moderate caffeine consumers) received 6 mg/kg caffeine after a 15-h abstinence and a 3-h fast; they rested for 60 min and then ran for 90 min at 70% of VO_{2max}.[114] A number of neuromuscular function measurements were performed, including maximal voluntary contraction, peak twitch torque, muscle compound action potentials (M-waves) and percent motor unit activation. No effects of caffeine on any of these parameters were observed, either at rest or during exercise.[114] Still, the authors commented that this could be due to small sample size or inadequacy of the measurements.[145] Perhaps relevant in this respect, an elegant recent study reported that caffeine (6 mg/kg) ingestion increased the occurrence of self-sustained firing in tibialis anterior motor units in caffeine-naive male volunteers; in theory, this could offset the effects of central fatigue normally seen during low-intensity, long-duration activities.[148]

Although several other studies have shown that caffeine may enhance some aspects of the neuromuscular function *in vivo* in man, results are generally isolated and inconclusive.[149,150] Also, the attenuation by caffeine of the exercise-induced increase in plasma [K^+] during submaximal aerobic exercise[151] (but not during maximal anaerobic exercise[95]), probably resulting from the stimulation of Na^+/K^+-ATPase in tissues other than the active muscles,[96] could theoretically preserve resting membrane potential and thus facilitate force development and delay fatigue. All these aspects of study are still in their infancy and additional research is certainly warranted, because results to date imply that the ergogenic effects of caffeine may also reside on direct actions on the muscle itself, in addition to possible metabolic or central mechanisms.

VI. Caffeine and Exercise Performance

The first well-designed and placebo-controlled study examining the effects of caffeine on athletic performance in man was published approximately 100 years ago by Rivers and Webber.[152] These investigators reported that ingestion of 500 mg caffeine increased the work done in pulling a weight with a finger.[152] A number of similar studies over the next 40 years provided considerable evidence that caffeine may enhance exercise capacity and recovery from fatigue; these effects were mainly attributed to putative stimulation of the central nervous system.[153] Of interest are the experiments carried out by Foltz and colleagues: they observed that the drug was effective in increasing performance in trained individuals,[154] but not in untrained ones.[155] Costill and colleagues[124,126,136] renewed interest in caffeine as an ergogenic aid and also provided a metabolic basis for its effects, while Graham and Spriet[44,98] demonstrated beyond doubt that it can result in substantial improvements in athletic performance. Theoretically, some of the effects elicited by caffeine consumption could be expected to impair exercise capacity under some circumstances,[156] but this is only rarely observed. Also, fear that ingestion of caffeine may cause fluid-electrolyte imbalances during exercise (e.g., dehydration) that could hamper performance is unsubstantiated, as comprehensively outlined by Armstrong.[157]

A. Long-Term Submaximal Exercise

The results of the studies examining the effects of caffeine on endurance time in situations where fatigue generally ensues within 30–60 min have demonstrated that the drug is capable of increasing time to exhaustion (Table 17.3),[31,44,46,97–99,113,120–122,125–127] although this may not always be true.[158–160] Nevertheless, the relevance of aerobic endurance to true performance is obscure, since all these studies used experimental protocols where power and speed were kept constant and endurance time could be easily measured. This, however, does not accurately reflect race conditions. In essence, these results suggest that caffeine ingestion will enable one to continue exercising at a given submaximal intensity for longer periods. Other parameters of performance during long-term exercise, such as speed and power output, have seldom been quantified. Still, available studies (Table 17.4) indicate that caffeine ingestion may result in increased cycling[161] and swimming[162] speed and also in increased work production during cycling.[117,124,161] On the other hand, the drug does not seem to enhance performance during ultra-endurance events such as simulated 21-km run[163] or 100-km cycle[164] races.

Some interesting concepts with respect to the effects of caffeine on endurance time during long-term submaximal exercise merit special reference and are therefore highlighted in Figure 17.5. The type of exercise does not seem to influence the ergogenicity of the drug, as shown by the significant per-

TABLE 17.3
Effects of Caffeine on Athletic Performance — Endurance in Long-Term Submaximal Exercise

Subjects	Fitness Level	Caffeine Habits	Caffeine Abstinence	Fasting	Caffeine Administration	Exercise Test	Performance
8 M[31]	Well-trained distance runners	Variable (0–940 mg/d)	48 h	Early postprandial period	0, 3, 6 and 9 mg/kg at 60 min prior to exercise	Running to exhaustion at 85% of VO_{2max}	C (3 and 6 mg/kg) increased time to exhaustion by 22% over P; the high dose caused a non-significant 11% increase
6 M, 1 F[44]	Well-trained distance runners	2 moderate users (450–720 mg/d), 3 low users (120–150 mg/d) and 2 non-users (<20 mg/d)	48 h	Early postprandial period	9 mg/kg at 60 min prior to exercise	Running or cycling to exhaustion at 85% of VO_{2max}	C increased time to exhaustion over P during both running (+44%) and cycling (+51%)
9 M[46]	Well-trained cyclists	5 low users (100–250 mg/d) and 4 moderate or high users (>250 mg/d)	72 h	Variable (from 2h to overnight)	0, 5, 9 and 13 mg/kg at 60 min prior to exercise	Cycling to exhaustion at 80% of W_{max}	All C doses similarly increased time to exhaustion (58 ± 11, 59 ± 12 and 58 ± 12 min for 5, 9 and 13 mg/kg vs. P (47 ± 13 min)
8 M[97]	Recreationally trained	N/A	48 h	Early postprandial period	6 mg/kg at 90 min prior to exercise	Cycling to exhaustion at 80–85% of VO_{2max}	C increased time to exhaustion by 22% over P
7 M, 1 F[98]	Recreational cyclists	5 C-naive (<50 mg/d), 2 low users (180–250 mg/d) and 1 heavy user (600 mg/d)	72 h	2–4 h	9 mg/kg at 60 min prior to exercise	Cycling to exhaustion at 75–85% of VO_{2max}	C increased time to exhaustion (96.2 ± 8.8 min) vs. P (75.8 ± 4.8 min)

(continued)

TABLE 17.3 (CONTINUED)
Effects of Caffeine on Athletic Performance — Endurance in Long-Term Submaximal Exercise

Subjects	Fitness Level	Caffeine Habits	Caffeine Abstinence	Fasting	Caffeine Administration	Exercise Test	Performance
8 M, 1 F[99]	Well-trained endurance runners	1 C-naive, 2 light users (<100 mg/d) and 6 moderate users (<500 mg/d)	48 h	Early postprandial period	4.45 mg/kg in pure form, added in decaffeinated coffee, or as regular coffee at 60 min prior to exercise	Running to exhaustion at 85% of VO_{2max}	Only C in pure form increased time to exhaustion (40-45 min) compared with the other trials (30-35 min)
8 M[113]	Well-trained	Not regular users	24 h	12 h	5 mg/kg in decaffeinated coffee at 60 min prior to exercise	Intermittent exercise: 3 × 30-min bouts of alternating 1-min cycling (85-90% of maximal workload) and rest intervals	C increased time to exhaustion by 29% (77.5 ± 5.26 min) over P (61.25 ± 2.20 min)
4 M, 4 F[120]	Recreationally trained	N/A	48 h	N/A	0, 2.2, 4.4 and 8.8 mg/kg at 60 min prior to exercise	Running to exhaustion at 80% of VO_{2max}	Only the medium C dose increased time to exhaustion (73.4 min) vs. P (53.4 min); the low (67.8 min) and high (57.9 min) C doses had no effect
5 M[121]	Distance runners	N/A	N/A	8 h	300 mg at 60 min prior to exercise, 60 mg immediately before exercise and 60 mg after 45 min of exercise (total: 384 ± 13 mg; approximately 6.2–6.6 mg/kg)	Running for 45 min at 80% of VO_{2max}, resting for 5 min and then running to exhaustion at 80% of VO_{2max}	C increased time to exhaustion by 34.4% (53.02 ± 9.16 min) over P (39.45 ± 11.19 min)

Caffeine

Subjects	Population	Caffeine history	Withdrawal	Timing	Dose	Exercise	Results
6 M[122]	Recreationally trained	Heavy users (761 ± 12 mg/day)	0, 2, or 4 days	2-4 h	6 mg/kg at 60 min prior to exercise	Cycling to exhaustion at 80-85% of VO_{2max}	C increased time to exhaustion by 25-35% over P, regardless of withdrawal period
6 M[125]	Spinal cord injured (SCI) patients (quadriplegic)	Heavy users (898 ± 197 mg/day)	48 h	Early postprandial period	6 mg/kg at 60 min prior to exercise	Involuntary leg cycling exercise through functional electrical stimulation of the quadriceps, hamstrings and gluteal muscles via skin electrodes	Time to exhaustion increased with C by 6% over P (27.24 ± 2 vs. 25.48 ± 1.48 min, respectively)
7 M, 2 F[126]	Competitive cyclists	N/A	N/A	6-12 h	330 mg (4.4-5.8 mg/kg) in decaffeinated coffee at 60 min prior to exercise	Cycling to exhaustion at 80% of VO_{2max}	C increased time to exhaustion by 19.5 ± 2.3% (90.2 ± 7.2 min) over P (75.5 ± 5.1 min)
15 M[127]	Well-trained runners and cyclists	N/A	N/A	2 h	5 mg/kg with or without propranolol (80 mg) at 120 min prior to exercise	Cycling to exhaustion at 70% of W_{max}	C plus propranolol tended to increase endurance time vs. propranolol alone (31.2 ± 17.2 vs. 22.6 ± 10.8 min, respectively, P = 0.056)

Note: P = placebo; C = caffeine; M = males; F = females; N/A = not available.

TABLE 17.4
Effects of Caffeine on Athletic Performance — Speed and Power in Long-Term Submaximal Exercise

Subjects	Fitness Level	Caffeine Habits	Caffeine Abstinence	Fasting	Caffeine Administration	Exercise Test	Performance
10 M[117]	2 recreational runners, 2 competitive runners and 6 triathletes	8 C-naive (<20 mg/day), 1 medium user (180 mg/d) and 1 heavy user (720 mg/day)	48 h	10–14 h	6 mg/kg at 60 min prior to exercise	30 min of isokinetic, variable resistance cycling at 80 rpm (during each trial, subjects cycled at what they perceived to be a rating of perceived exertion of 9 for the first 10 min, a rating of 12 for the next 10 min and a rating of 15 for the last 10 min)	C increased total work output by 12.6% (277.82 ± 26.05 kJ) over P (246.73 ± 21.45 kJ); the increase was independent of the rate of perceived exertion (i.e., improvement at all three ratings)
7 M, 2 F[124]	Trained cyclists	N/A	N/A	12 h	250 mg (3.4–4.4 mg/kg) in artificially sweetened drink at 60 min prior to exercise and 250 mg at 15-min intervals during the 2 h of exercise (7 × 35 mg)	120 min of isokinetic cycling at 80 rpm (speed is fixed, resistance is variable, thus work production is variable)	C increased total work production by 7.4% (117,016 ± 1,437 kpm) vs. P (108,984 ± 1,251 kpm); in essence, C resulted in a slower rate of loss of work produced, especially during the last 30 min of exercise
15 M[161]	Well-trained triathletes and cyclists	Variable (20–410 mg/d)	48 h	1.5–2 h	154, 230 and 328 mg (approximately 2.1, 3.2 and 4.5 mg/kg, respectively) in 7% carbohydrate solution during warm-up and exercise	After the warm-up, subjects rested for 35 min and then completed a set amount of work (cycling) as fast as possible	Subjects completed the time trial significantly faster and with a significantly higher mean work output after ingestion of the moderate (58.9 ± 1.0 min; 308 ± 9 W) and high (58.9 ± 1.2 min; 309 ± 10 W) C doses compared with P (62.5 ± 1.3 min; 292 ± 10 W), carbohydrate solution alone (61.5 ± 1.1 min; 295 ± 9 W) and the low C dose (60.4 ± 1.0 min; 299 ± 10 W)

Subjects	Training status	Caffeine use	Time since last caffeine	Dose	Protocol	Results
7 M, 4 F[162]	Competitive distance swimmers	C-naïve (<300 mg/wk)	48 h	6 mg/kg in drink 150 min prior to exercise	After 120 min of rest, subjects warmed-up for 20 min and rested again for 10 min; then they performed a time trial (1500 m swim)	C resulted in faster swim times (20:58.8 ± 0:36.4 min) vs. P (21:21.8 ± 0:38.2 min)
5 M, 2 F[163]	Competitive distance runners	Low and moderate users (0–300 mg/day)	24 h	0, 5, or 9 mg/kg at 60 min prior to exercise	21-km outdoor road race under high heat stress (race pace)	No effect of either C dose on time to finish the race
8 M[164]	Competitive endurance cyclists	N/A	48 h	6 mg/kg at 60 min prior to exercise and 0.33 mg/kg (maintenance dose) every 15 min throughout exercise (total: 9 mg/kg in 7% carbohydrate solution)	100-km time trial, consisting of 5 × 1-km sprints after 10, 32, 52, 72 and 99 km, as well as 4 × 4-km sprints after 20, 40, 60 and 80 km	No effect of C on average power, time to complete the sprints or total time to complete the 100-km trial vs. carbohydrate solution alone

Note: P = placebo; C = caffeine; M = males; F = females; N/A = not available.

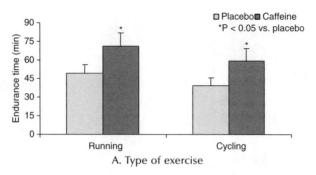

FIGURE 17.5A
Influence of various factors on the caffeine-induced increase in endurance during long-term submaximal exercise. Seven competitive runners ingested placebo or caffeine (9 mg/kg), rested for 1 h, then ran or cycled to exhaustion at approximately 85% of VO_{2max}. (From Graham, T. E. and Spriet, L. L., Performance and metabolic responses to a high caffeine dose during prolonged exercise, *J. Appl. Physiol.*, 71, 2292, 1991.)

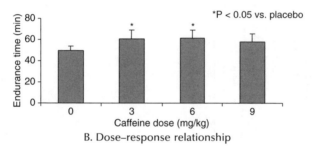

FIGURE 17.5B
Influence of various factors on the caffeine-induced increase in endurance during long-term submaximal exercise. Eight well-trained endurance athletes ingested various doses of caffeine (0, 3, 6, or 9 mg/kg), rested for 1 h, then ran to exhaustion at 85% of VO_{2max}. (From Graham, T. E. and Spriet, L. L., Metabolic, catecholamine and exercise performance responses to various doses of caffeine, *J. Appl. Physiol.*, 78, 867, 1995. With permission.)

formance improvements during both running and cycling (Figure 17.5A).[44] Also, there seems to be no dose-response relationship between the dose of caffeine and the increase in endurance time (Figure 17.5B).[31,46,120] The form of administration, however, may play a key role, since only caffeine in pure form (capsule) increased time to exhaustion, in striking contrast to when it was added in decaffeinated coffee or when ingested as regular coffee (Figure 17.5C).[99] Dimethylxanthines like theophylline are also able to bring about similar improvements in aerobic endurance, when administrated in pharmacologically active doses (Figure 17.5D),[97] which is important in light of the fact that theophylline is currently not regulated by IOC's legislation; hence, it is not considered a doping agent. Habituation to and withdrawal from caffeine do not seem to impair its performance-enhancing efficacy, since acute ingestion of the drug after 0, 2, or 4 days of withdrawal caused substantial and similar increments in exhaustion time (Figure 17.5E).[122] The biochemical basis of caffeine's ergogenicity probably involves some direct

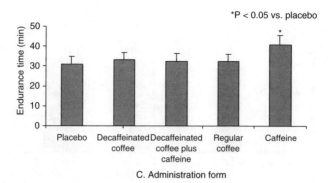

FIGURE 17.5C
Influence of various factors on the caffeine-induced increase in endurance during long-term submaximal exercise. Nine well-trained endurance athletes ingested either water (with caffeine or placebo) or coffee (decaffeinated coffee, decaffeinated coffee with added caffeine, or regular coffee), rested for 1 h, then ran to exhaustion at 85% of VO_{2max}. In all three trials, the dose was 4.45 mg/kg. (From Graham, T.E., Hibbert, E. and Sathasivam, P., Metabolic and exercise endurance effects of coffee and caffeine ingestion, *J. Appl. Physiol.*, 85, 883, 1998. With permission.)

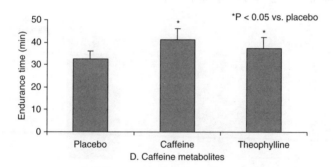

FIGURE 17.5D
Influence of various factors on the caffeine-induced increase in endurance during long-term submaximal exercise. Eight recreationally trained males ingested placebo, caffeine (6 mg/kg), or theophylline (4.5 mg/kg), rested for 1.5 h, then cycled to exhaustion at 80% of VO_{2max}. (From Greer, F., Friars, D. and Graham, T. E., Comparison of caffeine and theophylline ingestion: Exercise metabolism and endurance, *J. Appl. Physiol.*, 89, 1837, 2000. With permission.)

effects on muscle, in addition to central effects, inasmuch as spinal-cord-injured patients were able to increase their endurance by approximately 6% after caffeine consumption (Figure 17.5F).[125] The role of increased catecholamine release, however, is debatable because this improvement was not accompanied by an increase in plasma catecholamines.[125] On the other hand, caffeine ingestion combined with β-blockade did not significantly prolong time to exhaustion vs. β-blockade alone, but still, a strong trend for improvement was apparent (Figure 17.5G).[127] Finally, the drug was shown to be more effective in increasing endurance after acute altitude exposure (+54%), than after chronic altitude exposure (+24%, not significant) or at sea level (Figure 17.5H).[160]

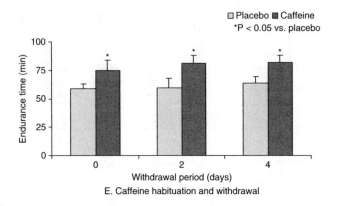

FIGURE 17.5E
Influence of various factors on the caffeine-induced increase in endurance during long-term submaximal exercise. Six recreationally trained, habitual caffeine users ingested placebo or caffeine (6 mg/kg), rested for 1 h, then cycled to exhaustion at 80-85% of VO_{2max}, under conditions of 0, 2 or 4 days of withdrawal from dietary caffeine. (From Van Soeren, M. H. and Graham, T. E., Effect of caffeine on metabolism, exercise endurance and catecholamine responses after withdrawal, *J. Appl. Physiol.*, 85, 1493, 1998.)

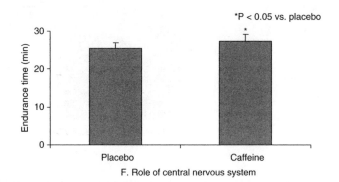

FIGURE 17.5F
Influence of various factors on the caffeine-induced increase in endurance during long-term submaximal exercise. Six quadriplegic patients ingested placebo or caffeine (6 mg/kg), rested for 1 h, then performed involuntary leg cycling exercise through functional electrical stimulation of the quadriceps, hamstrings and gluteal muscles via skin electrodes. (From Mohr, T., Van Soeren, M., Graham, T. E. and Kjaer, M., Caffeine ingestion and metabolic responses of tetraplegic humans during electrical cycling, *J. Appl. Physiol.*, 85, 979, 1998.)

What is important and should be borne in mind is that the ergogenic effects of caffeine have been observed during exercise in the laboratory, or in controlled settings simulating race conditions. These studies do not simulate real competition in sport events and consequently, extrapolations from these results to field conditions may not be valid. Field trials are necessary to confirm caffeine's ergogenic effects, but this kind of study is generally associated with many design problems and interpretative limitations. Probably the only field trial to date is that by Berglund and Hemmingsson.[165] These

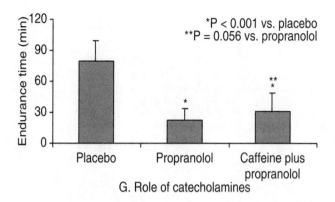

FIGURE 17.5G
Role of catecholamines: 15 male endurance athletes ingested placebo, propranolol (80 mg), or caffeine (5 mg/kg) plus propranolol (80 mg), rested for 2 h, then cycled to exhaustion at 70% of W_{max}. (From Van Baak, M. A. and Saris, W. H., The effect of caffeine on endurance performance after nonselective β-adrenergic blockade, *Med. Sci. Sports Exerc.*, 32, 499, 2000.)

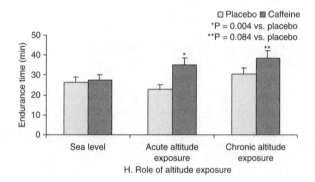

FIGURE 17.5H
Eight young males ingested placebo or caffeine (4 mg/kg), rested for 1 h, then cycled to exhaustion at approximately 80% of their altitude-specific maximal aerobic power during each of three phases: at sea level and after acute (1 h) or chronic (2 wk) exposure at 4,300 m. (From Fulco, C. S., Rock, P. B., Trad, L. A., Rose, M. S., Forte, V. A., Jr., Young, P. M. and Cymerman, A., Effect of caffeine on submaximal exercise performance at altitude, *Aviat. Space Environ. Med.*, 65, 539, 1994.)

investigators examined the effects of caffeine on skiing time over a specified distance (20–23 km) in 14 well-trained cross-country skiers, both at low and high altitudes. Caffeine ingestion (6 mg/kg) resulted in faster halfway and finishing times at low altitude (by 33 and 59 s, respectively) and this effect was even greater at high altitude (by 67 and 101 s, respectively), representing improvements of approximately 1–2% (low altitude) and 2–3% (high altitude).[165] Overall, one ought to point out that a great variability is usually observed in the individual performance responses to caffeine ingestion. For all these reasons, it has been suggested that athletes should experiment with

TABLE 17.5
Effects of Caffeine on Athletic Performance on Speed and Power in Short-Term Intense Exercise

Subjects	Fitness Level	Caffeine Habits	Caffeine Abstinence	Fasting	Caffeine Administration	Exercise Test	Performance
9 M[95]	Recreationally trained	N/A	48 h	2–3 h	6 mg/kg at 60 min prior to exercise	Subjects performed 4 × 30-s Wingate tests, with 4 min of rest in between	No effect of C on peak power, average power or rate of power loss for any of the 4 bouts; in fact, peak power (bout 4) and average power (bouts 3 and 4) were less during the C trial
3 M, 3 F[102]	Recreationally trained	N/A	1 wk	8 h	5 mg/kg at 60 min prior to exercise	Subjects performed a Wingate test (30 s of supramaximal cycling against a resistance load)	No effect of C on anaerobic power, anaerobic capacity and rate of power loss
10 M, 4 F[169]	Recreational athletes	N/A	N/A	N/A	250 mg (3.72 mg/kg) at 30 min prior to exercise	Subjects performed a force-velocity test to determine maximal anaerobic power (W_{max}): 4 × 6-s maximal cycling sprints against a progressively increasing load (2, 4, 6 and 8 kg) with 5 min of rest in between	C increased W_{max} (964 ± 66 W) vs. P (904 ± 53 W), because of an increase in pedaling frequency (since W = force × velocity and force was constant)
8 F[170]	Competitive rowers	N/A	72 h	8–12 h	6 or 9 mg/kg at 60 min prior to exercise	Subjects performed a 2,000-m (lasting approximately 2 min) time trial on a rowing ergometer	The high C dose improved time to complete the trial by 1.3%; the moderate dose only tended (P = 0.06) to do so by 0.7%; mean power output increased after the high C dose by 2.7%, but not after the moderate dose (+1.4%, not significant)

Subjects	Subject type				Dose	Protocol	Results
8 M[171]	Competitive rowers	N/A	72 h	8–12 h	6 or 9 mg/kg at 60 min prior to exercise	Subjects performed a 2,000-m time trial on a rowing ergometer (lasting approximately 2 min)	Both C doses improved time to complete the trial (by 1–1.3%) and mean power output (by 2.7%) vs. P
8 M[172]	Moderately trained	N/A	1 wk	1 h	250 mg (3.56 mg/kg) 60 min prior to exercise	Subjects cycled for 10 min at 90% of VO_{2max}, rested for 10 min and then performed a 10-min mock test (simulated cycle race)	No effect of C vs. P on mean power (240.1 ± 15.3 vs. 234.9 ± 16 W, respectively)
18 M, 10 F[173]	Recreationally trained runners	Variable	N/A	N/A	3 g of regular or decaffeinated coffee (approximately 200 mg C, i.e., 2–2.5 mg/kg) at 60 min prior to exercise	Subjects performed a 1,500-m time trial, or a 1,100-m run at controlled speed, followed by a 1-min "finishing burst"	C resulted in faster time to complete the run vs. P (286.0 vs. 290.2 s, respectively); C also resulted in increased running speed during the last 400 m (23.5 vs. 22.9 km/h, respectively)
5 M, 9 F[174]	7 competitive (3M, 4 F) and 7 recreational (2M, 5 F) swimmers	N/A	1 wk	2 h	250 mg (4.3 ± 0.2 mg/kg) at 60 min prior to exercise	Subjects performed 2 × 100-m freestyle swims at maximal speed with 20 min of passive recovery in between	C improved swimming velocity during both swims in competitive (greater improvement was observed during the second swim), but not in recreational swimmers

Note: P = placebo; C = caffeine; M = males; F = females.

the drug during the training season, before competing in any major event.[18,166]

B. Short-Term Intense Exercise

Although seldom examined, caffeine has also been shown to improve aerobic endurance during short-term intense exercise lasting as little as 5 minutes. Jackman et al.[94] had 14 recreationally trained subjects (11 males and three females) ingest caffeine (6 mg/kg) at 60 min before commencement of exercise. After this period of rest, they performed two 2-min bouts of cycling at approximately 100% of VO_{2max}, followed by a third bout to exhaustion at the same intensity, with 6-min rest intervals between the bouts. Caffeine increased time to exhaustion by 19.7% over placebo (4.93 ± 0.60 vs. 4.12 ± 0.36 min, respectively).[94] However, when the high-intensity test was preceded by long-term submaximal exercise, no effects of caffeine on endurance times were observed. In one study, subjects cycled for 3 h at 60% of VO_{2max} and then performed a high-intensity (85% of VO_{2max}) cycle ergometer test at 50 rpm until exhaustion.[167] In a second study, a similar high-intensity (90% of VO_{2max}) exhaustive test was performed after an 8-h march involving 40 km of marching on level dirt roads while carrying a 20–25-kg backpack and walking at a rate of 1.4–1.9 m/s (45–50% of VO_{2max}).[168] In a third study, subjects performed four 30-min bouts of running at 62–67% of VO_{2max} with 5 min of rest between; thereafter, they ran for 10 min at 80% of VO_{2max} and then at 90% of VO_{2max} until exhaustion.[103] Endurance times in all three studies were comparable for the caffeine (9–13 mg/kg) and the placebo trial.[103,167,168] It may be that the shorter and the more intense the event, the less obvious will be caffeine's effect. Alternatively, this could only be due to the difficulties encountered in quantifying endurance in such types of exercise, as well as due to the minimal potential for improvement.

The effects of the drug on speed and power output during short-term intense exercise are summarized in Table 17.5. The limited number of studies and the heterogeneous experimental designs preclude drawing unequivocal conclusions. For example, caffeine ingestion has been reported to increase maximal anaerobic power during a force-velocity test,[169] and also mean power output during rowing,[170,171] but to have no effects on mean power sustained during a mock test[172] or on anaerobic power, anaerobic capacity and rate of power loss during the 30-s Wingate test.[102] Similarly, no effects on peak power, average power and rate of power loss were observed during a sequence of four repeated Wingate tests and, in fact, performance was impaired during the last two bouts in the caffeine trials.[95] On the other hand, the results of the studies measuring speed in time trials during running,[173] swimming[174] and rowing[170,171] demonstrate a clear ergogenic effect of the drug. Interestingly, an improvement in swimming velocity was noted only in well-trained swimmers and not in recreational ones.[174]

C. Incremental Exercise

The number of studies examining the effects of caffeine ingestion on performance parameters during incremental exercise is rather limited. In one of the few investigations with female participants, ingestion of various doses of caffeine (0, 4, 7 and 10 mg/kg) failed to improve time to exhaustion during a progressive cycling protocol lasting approximately 6 min (endurance times were 299.5, 312.1, 299.8 and 303.2 s for the 0, 4, 7 and 10 mg/kg caffeine trials, respectively).[175] Similarly, in three other studies with male subjects, no effects of caffeine (3–5 mg/kg) consumed 60 min prior to exercise were observed during a subsequent graded, incremental cycle ergometer test to exhaustion lasting 10–20 min.[109,176,177]

By contrast, Flinn et al.[112] and McNaughton[130] reported an ergogenic effect of the drug on endurance during incremental exercise in caffeine-naive male volunteers. In these studies, subjects ingested a high caffeine dose (10–15 mg/kg) after an overnight fast and then rested for 60[130] or 180 min.[112] Caffeine consumption resulted in considerable increases in time to exhaustion during the subsequent incremental cycling tests. In the one lasting 7–8 min, endurance was prolonged by approximately 8.5% irrespective of dose (430.7 ± 5.8 s, 466.8 ± 5.6 s and 467.9 ± 6.6 s for the 0, 10 and 15 mg/kg caffeine trials, respectively).[130] In the one lasting approximately 15–20 min, an improvement of more than 20% was observed (17.5 ± 3.7 min for the caffeine trial vs. approximately 14.5 min for the placebo and control trials).[112] Consequent to that, the total amount of work completed was significantly greater in the caffeine trial (206.4 ± 8.2 kJ) vs. the placebo and control trials (approximately 165 kJ).[112] Similar results have been reported in another study with almost identical experimental design, the only difference being that the incremental cycle test was preceded by submaximal steady-state exercise (45 min of running at 75% of VO_{2max}); the total distance run was significantly increased after caffeine ingestion (12,899 ± 457 m) compared with the placebo and control trials (approximately 12,650 m).[178]

D. Caffeine and Ephedrine

Ephedrine is a sympathomimetic drug acting as an adrenergic agonist on both α- and β-adrenergic receptors in central and peripheral tissues. The recent surge of interest in the simultaneous use of caffeine and ephedrine for improving exercise performance, as well as the extensive media coverage justify a detailed discussion of the topic. Use of ephedra alkaloids in sports is in fact "old news" and investigators have been studying the effects of such formulations for more than 60 years.[154] Earlier studies examining the effects of acute administration of ephedrine and related compounds (pseudoephedrine and phenylpropanolamine) on exercise capacity in man, at doses generally considered to be safe, have shown no enhancements on performance parameters such as time to exhaustion and muscular strength.[179] The regeneration of interest witnessed during the last decade, however, has been fueled by several contemporary reports by Bell and colleagues, indicating that combined use of caffeine and ephedrine may be of potential ergogenic benefit.[141]

In the first of these investigations, eight male volunteers ingested caffeine (5 mg/kg) or ephedrine (1 mg/kg) alone, caffeine plus ephedrine, or placebo, rested for 90 min and then cycled to exhaustion at 85% of VO_{2peak}.[180] The caffeine–ephedrine treatment resulted in significantly prolonged time to exhaustion (17.5 ± 5.8 min) compared with caffeine alone (14.4 ± 4.1 min), ephedrine alone (15.0 ± 5.7 min), or placebo (12.6 ± 3.1 min).[180] In that study, four subjects experienced vomiting and nausea during the exercise trial involving the combined treatment, and were thus excluded from the analysis.[180] In a follow-up experiment, 12 male subjects ingested placebo or various combinations of caffeine plus ephedrine, rested for 90–120 min and then cycled to exhaustion as previously.[181] All caffeine–ephedrine treatments resulted in significant improvements in endurance time (27–28 min) over placebo (17.0 ± 3.0 min), while no incidents of nausea or vomiting occurred with the lowest dosage (4 mg/kg caffeine plus 0.8 mg/kg ephedrine).[181] The same investigators undertook two field trials where they also demonstrated an ergogenic effect of the "combination" treatment. In the first study, nine male recreational runners ran 3.2 km as quickly as possible while wearing 11-kg gear; run times were significantly improved by approximately 5% when caffeine (375 mg) plus ephedrine (75 mg) were ingested 120 min prior to exercise, compared with placebo.[182] In the second study, 10 male and two female recreational runners performed a 10-km race while wearing the same 11-kg gear, approximately 90 min after ingesting placebo, caffeine (4 mg/kg), ephedrine (0.8 mg/kg) or caffeine plus ephedrine.[183] Run times for the ephedrine trials (alone or with caffeine) were significantly reduced by approximately 2% compared with the nonephedrine trials (caffeine alone and placebo), because of an increase in pace over the last 5 km of the run.[183]

Seemingly, the ergogenic effects of the "combination" treatment are not limited to submaximal short- and long-term exercise. In one study, 16 male volunteers ingested placebo, caffeine (5 mg/kg), ephedrine (1 mg/kg), or caffeine plus ephedrine and then performed the 30-s Wingate test after 90 min of rest. Again, administration of ephedrine (alone or with caffeine) produced a small (1–2%) but significant increase in power output early during the ride compared with the nonephedrine trials (caffeine alone and placebo), but this effect disappeared after 15 s.[184] In another experiment presented in the same report, eight male volunteers underwent the maximal oxygen deficit (MAOD) test (supramaximal cycling at 125% of VO_{2peak}); treatments were as above. In this case, administration of caffeine (alone or with ephedrine) resulted in a 7–8% increase in time to exhaustion and oxygen deficit compared with the noncaffeine trials (ephedrine alone and placebo).[184] The most recent study in this area of research examined the effects of caffeine and ephedrine on muscular endurance during weightlifting.[185] Ninety minutes after ingesting caffeine (4 mg/kg), ephedrine (0.8 mg/kg), a combination of both, or placebo, 13 male subjects performed a weight-training circuit consisting of three supersets, each comprising leg press followed by bench press with 2 min of rest in between. Compared with the nonephedrine trials (caffeine alone and placebo), ephedrine ingestion (alone or with caffeine) caused significant increases in the mean number of repetitions completed during the first superset for both the leg-press and bench-press exercises, but no effect was evident for the remaining two supersets.[185]

Taken together, these data lend strong support to the premise that both aerobic and anaerobic performance is improved, albeit modestly, after ingestion of a combination of caffeine and ephedrine. Still, available research is limited and quite heterogeneous in nature, thus conclusions remain tentative. The two drugs act independently and probably additively rather than interactively, i.e., there are no synergistic effects.[180,183,184,186,187] Caffeine tolerance does not seem to impair the ergogenic efficiency of the "combination" treatment, since most subjects participating in the abovementioned experiments were regular coffee drinkers. Before recommending such a formulation to the public, however, one ought to be aware of the health risks associated with the simultaneous use of caffeine and ephedrine.[188] These may involve serious cardiovascular and central nervous system events, including hypertension, palpitations, tachycardia, chest pain, stroke, cerebral vascular accidents, myocardial infarction, seizures and other psychiatric and autonomic symptoms; permanent disabilities and even death have also been documented in several cases.[3,189-191]

VII. Summary

Caffeine is used by both elite and recreational athletes in an attempt to enhance exercise performance and delay fatigue. The drug is completely absorbed within 1 h after ingestion and is readily distributed throughout all tissues of the body, achieving similar steady-state concentrations in plasma and in extracellular and intracellular fluids of various tissues. Caffeine's metabolism is remarkably complex and a number of metabolites have been identified, the major class of which are the dimethylxanthines paraxanthine, theobromine and theophylline. Caffeine elimination from the body varies many-fold among healthy humans, but only a small percentage of the ingested dose is excreted unchanged in urine. Therefore, moderate doses (3–6 mg/kg) of the drug prior to exercise will generally not produce illegal levels in urine.

Evidence from *in vitro* studies provides a wealth of different molecular actions at the cellular level that could potentially contribute to caffeine's pharmacodynamic effects. Among others, caffeine has been shown to potentiate muscle contractility by inducing sarcoplasmic reticulum calcium release, inhibit several phosphodiesterase isoenzymes and increase the concentration of cyclic monophosphates, inhibit glycogen phosphorylase enzymes in liver and muscle, nonselectively antagonize adenosine's actions by blocking adenosine receptors, stimulate the cell membrane sodium-potassium pump and impair phosphoinositide metabolism by inhibiting phosphoinositide-3-OH-kinases. Not all actions, however, seem to account for the observed effects *in vivo*. The most physiologically relevant mechanism of action is probably the blockade of adenosine receptors, although a variable degree of contribution cannot be readily discounted on the basis of experimental data.

Studies in humans have demonstrated that caffeine ingestion leads to increased levels of circulating catecholamines and especially adrenaline, both at rest and during exercise. Also, the drug results in increased plasma free fatty acid and, to a lesser extent, glycerol concentrations. However, it does not seem to enhance muscle free fatty acid uptake or whole-body free fatty acid and glycerol appearance and disappearance rates during exercise. Furthermore, the weight of evidence suggests that caffeine fails to reduce whole-body respiratory quotient. Muscle glycogen catabolism during exercise will probably not be affected by prior caffeine ingestion, although some studies have reported a sparing effect and various, but circumstantial, intramuscular metabolic shifts. On the other hand, the limited data available indicate that the drug may facilitate neuromuscular function in man *in vivo*.

There is mounting evidence showing that ingestion of caffeine improves aerobic endurance, i.e., increases time to exhaustion, during submaximal exercise bouts lasting approximately 30–60 min. Speed and power output during such activities may also improve following caffeine administration, but this does not seem to be the case during ultra-endurance events. Field trials are necessary to confirm the performance-enhancing efficiency of caffeine under race conditions, but these are generally difficult to construct and evaluate. The ergogenic effects of the drug are less frequently observed during short-term intense exercise, although available research indicates an increase in speed after caffeine consumption. The drug may also prolong time to fatigue during intense exercise bouts, but this effect is not evident if the maximal test is preceded by long-term submaximal exercise. The results for power output during short-term intense exercise, as well as for performance during incremental exercise, are quite mixed and thus no generalized statements can be made. The combination of caffeine with ephedrine may be more effective in increasing performance during various exercise modes than each compound alone, but safety and health-related reasons preclude advocating use of such formulations by sportspersons.

More research is needed to elucidate the exact nature and the underlying basis of caffeine's ergogenicity, but future studies should be well-designed and account for confounding factors such as caffeine dose and time of administration, habitual caffeine use and period of abstinence, training status and previous diet, type, duration and intensity of exercise, etc. In addition, the hormonal and metabolic response to the drug should be addressed with contemporary and perhaps more invasive methodologies like stable isotope tracers, arteriovenous balance measurements and tissue biopsies. Merely measuring the concentrations of the various substrates in blood adds nothing but confusion. Studies into the neuromuscular effects of the drug relevant to performance enhancement are also warranted. Caffeine has numerous actions in most tissues of the body and only well-designed and controlled research will help unravel the multiplicity and the basis of its effects.

References

1. Williams, M. H., *The Ergogenics Edge: Pushing the Limits of Sports Performance*, Human Kinetics, Champaign, IL, 1998.
2. Mottram, D. R., Banned drugs in sport. Does the International Olympic Committee (IOC) list need updating?, *Sports Med.*, 27, 1, 1999.
3. Ahrendt, D. M., Ergogenic aids: Counseling the athlete, *Am. Fam. Physician*, 63, 913, 2001.
4. Nawrot, P., Jordan, S., Eastwood, J., Rotstein, J., Hugenholtz, A. and Feeley, M., Effects of caffeine on human health, *Food Addit. Contam.*, 20, 1, 2003.
5. Gilbert, R. M., Caffeine consumption, in *The Methylxanthine Beverages and Foods: Chemistry, Consumption and Health Effects*, Spiller, G. A., Ed., Alan R. Liss, Inc., New York, 1984, 185.
6. Waldvogel, S. R., Caffeine — a drug with a surprise, *Angew. Chem. Int. Ed. Engl.*, 42, 604, 2003.
7. Mandel, H. G., Update on caffeine consumption, disposition and action, *Food Chem. Toxicol.*, 40, 1231, 2002.
8. Brice, C. F. and Smith, A. P., Factors associated with caffeine consumption, *Int. J. Food Sci. Nutr.*, 53, 55, 2002.
9. Harland, B. F., Caffeine and nutrition, *Nutrition*, 16, 522, 2000.
10. Paluska, S. A., Caffeine and exercise, *Curr. Sports Med. Rep.*, 2, 213, 2003.
11. Melia, P., Pipe, A. and Greenberg, L., The use of anabolic-androgenic steroids by Canadian students, *Clin. J. Sport Med.*, 6, 9, 1996.
12. Forman, E. S., Dekker, A. H., Javors, J. R. and Davison, D. T., High-risk behaviors in teenage male athletes, *Clin. J. Sport Med.*, 5, 36, 1995.
13. Wagner, J. C., Enhancement of athletic performance with drugs. An overview, *Sports Med.*, 12, 250, 1991.
14. Rogers, C. C., Cyclists try caffeine suppositories, *Phys. Sportsmed.*, 13, 38, 1985.
15. Nehlig, A. and Debry, G., Caffeine and sports activity: A review, *Int. J. Sports Med.*, 15, 215, 1994.
16. Jones, A. R. and Pichot, J. T., Stimulant use in sports, *Am. J. Addict.*, 7, 243, 1998.
17. Delbeke, F. T. and Debackere, M., Caffeine: Use and abuse in sports, *Int. J. Sports Med.*, 5, 179, 1984.
18. Spriet, L. L., Caffeine and performance, *Int. J. Sport Nutr.*, 5 Suppl, S84, 1995.
19. Benowitz, N. L., Clinical pharmacology of caffeine, *Annu. Rev. Med.*, 41, 277, 1990.
20. Leonard, T. K., Watson, R. R. and Mohs, M. E., The effects of caffeine on various body systems: A review, *J. Am. Diet. Assoc.*, 87, 1048, 1987.
21. Arnaud, M. J., The pharmacology of caffeine, *Prog. Drug Res.*, 31, 273, 1987.
22. Marks, V. and Kelly, J. F., Absorption of caffeine from tea, coffee and coca cola, *Lancet*, 1, 827, 1973.

23. Blanchard, J. and Sawers, S. J., The absolute bioavailability of caffeine in man, *Eur. J. Clin. Pharmacol.*, 24, 93, 1983.
24. Mumford, G. K., Benowitz, N. L., Evans, S. M., Kaminski, B. J., Preston, K. L., Sannerud, C. A., Silverman, K. and Griffiths, R. R., Absorption rate of methylxanthines following capsules, cola and chocolate, *Eur. J. Clin. Pharmacol.*, 51, 319, 1996.
25. Bonati, M., Latini, R., Galletti, F., Young, J. F., Tognoni, G. and Garattini, S., Caffeine disposition after oral doses, *Clin. Pharmacol. Ther.*, 32, 98, 1982.
26. Kamimori, G. H., Karyekar, C. S., Otterstetter, R., Cox, D. S., Balkin, T. J., Belenky, G. L. and Eddington, N. D., The rate of absorption and relative bioavailability of caffeine administered in chewing gum versus capsules to normal healthy volunteers, *Int. J. Pharm.*, 234, 159, 2002.
27. Fredholm, B. B., Battig, K., Holmen, J., Nehlig, A. and Zvartau, E. E., Actions of caffeine in the brain with special reference to factors that contribute to its widespread use, *Pharmacol. Rev.*, 51, 83, 1999.
28. Liguori, A., Hughes, J. R. and Grass, J. A., Absorption and subjective effects of caffeine from coffee, cola and capsules, *Pharmacol. Biochem. Behav.*, 58, 721, 1997.
29. Carrillo, J. A. and Benitez, J., Clinically significant pharmacokinetic interactions between dietary caffeine and medications, *Clin. Pharmacokinet.*, 39, 127, 2000.
30. McLean, C. and Graham, T. E., Effects of exercise and thermal stress on caffeine pharmacokinetics in men and eumenorrheic women, *J. Appl. Physiol.*, 93, 1471, 2002.
31. Graham, T. E. and Spriet, L. L., Metabolic, catecholamine and exercise performance responses to various doses of caffeine, *J. Appl. Physiol.*, 78, 867, 1995.
32. Lelo, A., Birkett, D. J., Robson, R. A. and Miners, J. O., Comparative pharmacokinetics of caffeine and its primary demethylated metabolites paraxanthine, theobromine and theophylline in man, *Br. J. Clin. Pharmacol.*, 22, 177, 1986.
33. Stahle, L., Arner, P. and Ungerstedt, U., Drug distribution studies with microdialysis. III: Extracellular concentration of caffeine in adipose tissue in man, *Life Sci.*, 49, 1853, 1991.
34. Stahle, L., Segersvard, S. and Ungerstedt, U., Drug distribution studies with microdialysis. II: Caffeine and theophylline in blood, brain and other tissues in rats, *Life Sci.*, 49, 1843, 1991.
35. Burg, A. W. and Werner, E., Tissue distribution of caffeine and its metabolites in the mouse, *Biochem. Pharmacol.*, 21, 923, 1972.
36. Somani, S. M. and Gupta, P., Caffeine: A new look at an age-old drug, *Int. J. Clin. Pharmacol. Ther. Toxicol.*, 26, 521, 1988.
37. Lelo, A., Miners, J. O., Robson, R. A. and Birkett, D. J., Quantitative assessment of caffeine partial clearances in man, *Br. J. Clin. Pharmacol.*, 22, 183, 1986.
38. Miners, J. O. and Birkett, D. J., The use of caffeine as a metabolic probe for human drug metabolizing enzymes, *Gen. Pharmacol.*, 27, 245, 1996.
39. Kaplan, G. B., Greenblatt, D. J., Ehrenberg, B. L., Goddard, J. E., Cotreau, M. M., Harmatz, J. S. and Shader, R. I., Dose-dependent pharmacokinetics and psychomotor effects of caffeine in humans, *J. Clin. Pharmacol.*, 37, 693, 1997.
40. Birkett, D. J. and Miners, J. O., Caffeine renal clearance and urine caffeine concentrations during steady state dosing. Implications for monitoring caffeine intake during sports events, *Br. J. Clin. Pharmacol.*, 31, 405, 1991.
41. Van der Merwe, P. J., Luus, H.G. and Barnard, J.G., Caffeine in sport. Influence of endurance exercise on the urinary caffeine concentration, *Int. J. Sports Med.*, 13, 74, 1992.

42. Delbeke, F. T. and Debackere, M., The influence of diuretics on the excretion and metabolism of doping agents. Part IV — Caffeine, *Biopharm. Drug Dispos.*, 9, 137, 1988.
43. Van der Merwe, P. J., Muller, F. R. and Muller, F. O., Caffeine in sport. Urinary excretion of caffeine in healthy volunteers after intake of common caffeine-containing beverages, *S. Afr. Med. J.*, 74, 163, 1988.
44. Graham, T. E. and Spriet, L. L., Performance and metabolic responses to a high caffeine dose during prolonged exercise, *J. Appl. Physiol.*, 71, 2292, 1991.
45. Van Soeren, M. H., Sathasivam, P., Spriet, L. L. and Graham, T. E., Caffeine metabolism and epinephrine responses during exercise in users and nonusers, *J. Appl. Physiol.*, 75, 805, 1993.
46. Pasman, W. J., van Baak, M. A., Jeukendrup, A. E. and de Haan, A., The effect of different dosages of caffeine on endurance performance time, *Int. J. Sports Med.*, 16, 225, 1995.
47. Graham, T. E., Caffeine and exercise: Metabolism, endurance and performance, *Sports Med.*, 31, 785, 2001.
48. Veley, V. H. and Waller, A. D., On the comparative toxicity of theobromine and caffeine, as measured by their direct effect upon the contractility of isolated muscle, *Proc. R. Soc. Lond. B Biol. Sci.*, 82, 568, 1910.
49. Weber, A. and Herz, R., The relationship between caffeine contracture of intact muscle and the effect of caffeine on reticulum, *J. Gen. Physiol.*, 52, 750, 1968.
50. Weber, A., The mechanism of the action of caffeine on sarcoplasmic reticulum, *J. Gen. Physiol.*, 52, 760, 1968.
51. Rousseau, E., Ladine, J., Liu, Q. Y. and Meissner, G., Activation of the Ca^{2+} release channel of skeletal muscle sarcoplasmic reticulum by caffeine and related compounds, *Arch. Biochem. Biophys.*, 267, 75, 1988.
52. Rousseau, E. and Meissner, G., Single cardiac sarcoplasmic reticulum Ca^{2+}-release channel: activation by caffeine, *Am. J. Physiol.*, 256, H328, 1989.
53. Herrmann-Frank, A., Luttgau, H. C. and Stephenson, D. G., Caffeine and excitation-contraction coupling in skeletal muscle: a stimulating story, *J. Muscle Res. Cell Motil.*, 20, 223, 1999.
54. Butcher, R. W. and Sutherland, E. W., Adenosine 3',5'-phosphate in biological materials. I. Purification and properties of cyclic 3',5'-nucleotide phosphodiesterase and use of this enzyme to characterize adenosine 3',5'-phosphate in human urine, *J. Biol. Chem.*, 237, 1244, 1962.
55. Fain, J. N., Pointer, R. H. and Ward, W. F., Effects of adenosine nucleosides on adenylate cyclase, phosphodiesterase, cyclic adenosine monophosphate accumulation and lipolysis in fat cells, *J. Biol. Chem.*, 247, 6866, 1972.
56. Butcher, R. W., The role of cyclic AMP in the actions of some lipolytic and antilipolytic agents, *Horm. Metab. Res.*, 2, 5, 1970.
57. Beavo, J. A., Rogers, N. L., Crofford, O. B., Hardman, J. G., Sutherland, E. W. and Newman, E. V., Effects of xanthine derivatives on lipolysis and on adenosine 3',5'-monophosphate phosphodiesterase activity, *Mol. Pharmacol.*, 6, 597, 1970.
58. Ingebretsen, C., Clark, J. F., Allen, D. O. and Ashmore, J., Effect of glucagon, dibutyryl adenosine 3',5'-cyclic monophosphate and phosphodiesterase inhibitors on rat liver phosphorylase activity and adenosine 3',5'-cyclic monophosphate levels, *Biochem. Pharmacol.*, 23, 2139, 1974.

59. Kramer, G. L. and Wells, J. N., Xanthines and skeletal muscle: Lack of relationship between phosphodiesterase inhibition and increased twitch tension in rat diaphragms, *Mol. Pharmacol.*, 17, 73, 1980.
60. Kihlman, B. and Overgaard-Hansen, K., Inhibition of muscle phosphorylase by methylated oxypurines, *Exp. Cell Res.*, 8, 252, 1955.
61. Kasvinsky, P. J., Shechosky, S. and Fletterick, R. J., Synergistic regulation of phosphorylase a by glucose and caffeine, *J. Biol. Chem.*, 253, 9102, 1978.
62. Kobayashi, M., Soman, G. and Graves, D. J., A comparison of the activator sites of liver and muscle glycogen phosphorylase b, *J. Biol. Chem.*, 257, 14041, 1982.
63. Kasvinsky, P. J., Madsen, N. B., Sygusch, J. and Fletterick, R. J., The regulation of glycogen phosphorylase alpha by nucleotide derivatives. Kinetic and x-ray crystallographic studies, *J. Biol. Chem.*, 253, 3343, 1978.
64. Martin, W. H., Hoover, D. J., Armento, S. J., Stock, I. A., McPherson, R. K., Danley, D. E., Stevenson, R. W., Barrett, E. J. and Treadway, J. L., Discovery of a human liver glycogen phosphorylase inhibitor that lowers blood glucose *in vivo*, *Proc. Natl. Acad. Sci. U.S.A.*, 95, 1776, 1998.
65. Ercan-Fang, N. and Nuttall, F. Q., The effect of caffeine and caffeine analogs on rat liver phosphorylase a activity, *J. Pharmacol. Exp. Ther.*, 280, 1312, 1997.
66. Rush, J. W. and Spriet, L. L., Skeletal muscle glycogen phosphorylase a kinetics: effects of adenine nucleotides and caffeine, *J. Appl. Physiol.*, 91, 2071, 2001.
67. Fredholm, B. B., Are methylxanthine effects due to antagonism of endogenous adenosine? *Trends Pharmacol. Sci.*, 1, 129, 1980.
68. Fredholm, B. B., On the mechanism of action of theophylline and caffeine, *Acta Med. Scand.*, 217, 149, 1985.
69. Fredholm, B. B., Astra Award Lecture. Adenosine, adenosine receptors and the actions of caffeine, *Pharmacol. Toxicol.*, 76, 93, 1995.
70. Rall, T. W., The 1982 Theodore Weicker Memorial Award Oration: Evolution of the mechanism of action of methylxanthines: From calcium mobilizers to antagonists of adenosine receptors, *Pharmacologist*, 24, 277, 1982.
71. Sattin, A. and Rall, T. W., The effect of adenosine and adenine nucleotides on the cyclic adenosine 3', 5'-phosphate content of guinea pig cerebral cortex slices, *Mol. Pharmacol.*, 6, 13, 1970.
72. Daly, J. W., Adenosine receptors: targets for future drugs, *J. Med. Chem.*, 25, 197, 1982.
73. Fredholm, B. B., Ijzerman, A. P., Jacobson, K. A., Klotz, K. N. and Linden, J., International Union of Pharmacology. XXV. Nomenclature and classification of adenosine receptors, *Pharmacol. Rev.*, 53, 527, 2001.
74. Fredholm, B. B., Irenius, E., Kull, B. and Schulte, G., Comparison of the potency of adenosine as an agonist at human adenosine receptors expressed in Chinese hamster ovary cells, *Biochem. Pharmacol.*, 61, 443, 2001.
75. Fredholm, B. B., Arslan, G., Halldner, L., Kull, B., Schulte, G. and Wasserman, W., Structure and function of adenosine receptors and their genes, *Naunyn. Schmiedebergs Arch. Pharmacol.*, 362, 364, 2000.
76. Tong, E. Y., Bittar, E. E., Chen, S. S. and Danielson, B. G., The influence of caffeine on the sodium efflux in barnacle muscle fibers, *Experientia*, 28, 1031, 1972.
77. Bittar, E. E., Hift, H., Huddart, H. and Tong, E., The effects of caffeine on sodium transport, membrane potential, mechanical tension and ultrastructure in barnacle muscle fibers, *J. Physiol.*, 242, 1, 1974.

78. Hawke, T. J., Willmets, R. G. and Lindinger, M. I., K⁺ transport in resting rat hind-limb skeletal muscle in response to paraxanthine, a caffeine metabolite, *Can. J. Physiol. Pharmacol.*, 77, 835, 1999.
79. Lindinger, M. I., Willmets, R. G. and Hawke, T. J., Stimulation of Na⁺, K⁺-pump activity in skeletal muscle by methylxanthines: Evidence and proposed mechanisms, *Acta Physiol. Scand.*, 156, 347, 1996.
80. Honeyman, T. W., Strohsnitter, W., Scheid, C. R. and Schimmel, R. J., Phosphatidic acid and phosphatidylinositol labelling in adipose tissue. Relationship to the metabolic effects of insulin and insulin-like agents, *Biochem. J.*, 212, 489, 1983.
81. Buckley, J. T., Properties of human erythrocyte phosphatidylinositol kinase and inhibition by adenosine, ADP and related compounds, *Biochim. Biophys. Acta*, 498, 1, 1977.
82. Foukas, L. C., Daniele, N., Ktori, C. Anderson, K. E., Jensen, J. and Shepherd, P. R., Direct effects of caffeine and theophylline on p110δ and other phosphoinositide 3-kinases. Differential effects on lipid kinase and protein kinase activities, *J. Biol. Chem.*, 277, 37124, 2002.
83. Sarkaria, J. N., Busby, E. C., Tibbetts, R. S., Roos, P., Taya, Y., Karnitz, L. M. and Abraham, R. T., Inhibition of ATM and ATR kinase activities by the radiosensitizing agent, caffeine, *Cancer Res.*, 59, 4375, 1999.
84. Blasina, A., Price, B. D., Turenne, G. A. and McGowan, C. H., Caffeine inhibits the checkpoint kinase ATM, *Curr. Biol.*, 9, 1135, 1999.
85. Shepherd, P. R., Withers, D. J. and Siddle, K., Phosphoinositide 3-kinase: The key switch mechanism in insulin signalling, *Biochem. J.*, 333, 471, 1998.
86. Levi, L., The effect of coffee on the function of the sympatho-adrenomedullary system in man, *Acta Med. Scand.*, 181, 431, 1967.
87. Bellet, S., Roman, L., DeCastro, O., Kim, K. E. and Kershbaum, A., Effect of coffee ingestion on catecholamine release, *Metabolism*, 18, 288, 1969.
88. Robertson, D., Frolich, J. C., Carr, R. K., Watson, J. T., Hollifield, J. W., Shand, D. G. and Oates, J. A., Effects of caffeine on plasma renin activity, catecholamines and blood pressure, *N. Engl. J. Med.*, 298, 181, 1978.
89. Robertson, D., Wade, D., Workman, R., Woosley, R. L. and Oates, J. A., Tolerance to the humoral and hemodynamic effects of caffeine in man, *J. Clin. Invest.*, 67, 1111, 1981.
90. MacNaughton, K. W., Sathasivam, P., Vallerand, A. L. and Graham, T. E., Influence of caffeine on metabolic responses of men at rest in 28 and 5 degrees C, *J. Appl. Physiol.*, 68, 1889, 1990.
91. Takiyyuddin, M. A., Cervenka, J. H., Sullivan, P. A., Pandian, M. R., Parmer, R. J., Barbosa, J. A. and O'Connor, D. T., Is physiologic sympathoadrenal catecholamine release exocytotic in humans?, *Circulation*, 81, 185, 1990.
92. Takiyyuddin, M. A., Brown, M. R., Dinh, T. Q., Cervenka, J. H., Braun, S. D., Parmer, R. J., Kennedy, B. and O'Connor, D. T., Sympatho-adrenal secretion in humans: Factors governing catecholamine and storage vesicle peptide co-release, *J. Auton. Pharmacol.*, 14, 187, 1994.
93. Graham, T. E., Sathasivam, P., Rowland, M., Marko, N., Greer, F. and Battram, D., Caffeine ingestion elevates plasma insulin response in humans during an oral glucose tolerance test, *Can. J. Physiol. Pharmacol.*, 79, 559, 2001.
94. Jackman, M., Wendling, P., Friars, D. and Graham, T. E., Metabolic catecholamine and endurance responses to caffeine during intense exercise, *J. Appl. Physiol.*, 81, 1658, 1996.

95. Greer, F., McLean, C. and Graham, T. E., Caffeine, performance and metabolism during repeated Wingate exercise tests, *J. Appl. Physiol.*, 85, 1502, 1998.
96. Graham, T. E., Helge, J. W., MacLean, D. A., Kiens, B. and Richter, E. A., Caffeine ingestion does not alter carbohydrate or fat metabolism in human skeletal muscle during exercise, *J. Physiol.*, 529, 837, 2000.
97. Greer, F., Friars, D. and Graham, T. E., Comparison of caffeine and theophylline ingestion: Exercise metabolism and endurance, *J. Appl. Physiol.*, 89, 1837, 2000.
98. Spriet, L. L., MacLean, D. A., Dyck, D. J., Hultman, E., Cederblad, G. and Graham, T. E., Caffeine ingestion and muscle metabolism during prolonged exercise in humans, *Am. J. Physiol.*, 262, E891, 1992.
99. Graham, T. E., Hibbert, E. and Sathasivam, P., Metabolic and exercise endurance effects of coffee and caffeine ingestion, *J. Appl. Physiol.*, 85, 883, 1998.
100. Bangsbo, J., Jacobsen, K., Nordberg, N., Christensen, N. J. and Graham, T., Acute and habitual caffeine ingestion and metabolic responses to steady-state exercise, *J. Appl. Physiol.*, 72, 1297, 1992.
101. Graham, T. E., Sathasivam, P. and MacNaughton, K. W., Influence of cold, exercise and caffeine on catecholamines and metabolism in men, *J. Appl. Physiol.*, 70, 2052, 1991.
102. Collomp, K., Ahmaidi, S., Audran, M., Chanal, J. L. and Prefaut, C., Effects of caffeine ingestion on performance and anaerobic metabolism during the Wingate Test, *Int. J. Sports Med.*, 12, 439, 1991.
103. Sasaki, H., Takaoka, I. and Ishiko, T., Effects of sucrose or caffeine ingestion on running performance and biochemical responses to endurance running, *Int. J. Sports Med.*, 8, 203, 1987.
104. Mazzeo, R. S., Catecholamine responses to acute and chronic exercise, *Med. Sci. Sports Exerc.*, 23, 839, 1991.
105. Kjaer, M., Neuroendocrine regulation during exercise, in *Biochemistry of Exercise*, Hargreaves, M. and Thompson, M., Eds., Human Kinetics, Champaign, IL, 1999, 47.
106. LeBlanc, J., Jobin, M., Cote, J., Samson, P. and Labrie, A., Enhanced metabolic response to caffeine in exercise-trained human subjects, *J. Appl. Physiol.*, 59, 832, 1985.
107. Bellet, S., Kershbaum, A. and Aspe, J., The effect of caffeine on free fatty acids, *Arch. Intern. Med.*, 116, 750, 1965.
108. Bellet, S., Kershbaum, A. and Finck, E. M., Response of free fatty acids to coffee and caffeine, *Metabolism*, 17, 702, 1968.
109. Dodd, S. L., Brooks, E., Powers, S. K. and Tulley, R., The effects of caffeine on graded exercise performance in caffeine naive versus habituated subjects, *Eur. J. Appl. Physiol. Occup. Physiol.*, 62, 424, 1991.
110. Chad, K. and Quigley, B., The effects of substrate utilization, manipulated by caffeine, on post-exercise oxygen consumption in untrained female subjects, *Eur. J. Appl. Physiol. Occup. Physiol.*, 59, 48, 1989.
111. Poehlman, E. T., Despres, J. P., Bessette, H., Fontaine, E., Tremblay, A. and Bouchard, C., Influence of caffeine on the resting metabolic rate of exercise-trained and inactive subjects, *Med. Sci. Sports Exerc.*, 17, 689, 1985.
112. Flinn, S., Gregory, J., McNaughton, L. R., Tristram, S. and Davies, P., Caffeine ingestion prior to incremental cycling to exhaustion in recreational cyclists, *Int. J. Sports Med.*, 11, 188, 1990.

113. Trice, I. and Haymes, E. M., Effects of caffeine ingestion on exercise-induced changes during high-intensity, intermittent exercise, *Int. J. Sport Nutr.*, 5, 37, 1995.
114. Tarnopolsky, M. A., Atkinson, S. A., MacDougall, J. D., Sale, D. G. and Sutton, J. R., Physiological responses to caffeine during endurance running in habitual caffeine users, *Med. Sci. Sports Exerc.*, 21, 418, 1989.
115. Donelly, K. and McNaughton, L., The effects of two levels of caffeine ingestion on excess postexercise oxygen consumption in untrained women, *Eur. J. Appl. Physiol. Occup. Physiol.*, 65, 459, 1992.
116. Hetzler, R. K., Knowlton, R. G., Somani, S. M., Brown, D. D. and Perkins, R. M., 3rd, Effect of paraxanthine on FFA mobilization after intravenous caffeine administration in humans, *J. Appl. Physiol.*, 68, 44, 1990.
117. Cole, K. J., Costill, D. L., Starling, R. D., Goodpaster, B. H., Trappe, S. W. and Fink, W. J., Effect of caffeine ingestion on perception of effort and subsequent work production, *Int. J. Sport Nutr.*, 6, 14, 1996.
118. Casal, D. C. and Leon, A. S., Failure of caffeine to affect substrate utilization during prolonged running, *Med. Sci. Sports Exerc.*, 17, 174, 1985.
119. Giles, D. and MacLaren, D., Effects of caffeine and glucose ingestion on metabolic and respiratory functions during prolonged exercise, *J. Sports Sci.*, 2, 35, 1984.
120. Cadarette, B. S., Levine, L., Berube, C. L., Posner, B. M. and Evans, W. J., Effects of varied dosages of caffeine on endurance exercise to fatigue, in *Biochemistry of Exercise*, Knuttgen, H. G., Vogel, J. A. and Poortmans, J., Eds., Human Kinetics Publishers, Inc., Champaign, IL, 1983, 871.
121. Sasaki, H., Maeda, J., Usui, S. and Ishiko, T., Effect of sucrose and caffeine ingestion on performance of prolonged strenuous running, *Int. J. Sports Med.*, 8, 261, 1987.
122. Van Soeren, M. H. and Graham, T. E., Effect of caffeine on metabolism, exercise endurance and catecholamine responses after withdrawal, *J. Appl. Physiol.*, 85, 1493, 1998.
123. Knapik, J. J., Jones, B. H., Toner, M. M., Daniels, W. L. and Evans, W. J., Influence of caffeine on serum substrate changes during running in trained and untrained individuals, in *Biochemistry of Exercise*, Knuttgen, H. G., Vogel, J. A. and Poortmans, J., Eds., Human Kinetics Publishers, Inc., Champaign, IL, 1983, 514.
124. Ivy, J. L., Costill, D. L., Fink, W. J. and Lower, R. W., Influence of caffeine and carbohydrate feedings on endurance performance, *Med. Sci. Sports*, 11, 6, 1979.
125. Mohr, T., Van Soeren, M., Graham, T. E. and Kjaer, M., Caffeine ingestion and metabolic responses of tetraplegic humans during electrical cycling, *J. Appl. Physiol.*, 85, 979, 1998.
126. Costill, D. L., Dalsky, G. P. and Fink, W. J., Effects of caffeine ingestion on metabolism and exercise performance, *Med. Sci. Sports*, 10, 155, 1978.
127. Van Baak, M. A. and Saris, W. H., The effect of caffeine on endurance performance after nonselective β-adrenergic blockade, *Med. Sci. Sports Exerc.*, 32, 499, 2000.
128. Wells, C. L., Schrader, T. A., Stern, J. R. and Krahenbuhl, G. S., Physiological responses to a 20-mile run under three fluid replacement treatments, *Med. Sci. Sports Exerc.*, 17, 364, 1985.
129. Titlow, L. W., Ishee, J. H. and Riggs, C. E., Failure of caffeine to affect metabolism during 60 min submaximal exercise, *J. Sports Sci.*, 9, 15, 1991.

130. McNaughton, L. R., Two levels of caffeine ingestion on blood lactate and free fatty acid responses during incremental exercise, *Res. Q. Exerc. Sport*, 58, 255, 1987.
131. Raguso, C. A., Coggan, A. R., Sidossis, L. S., Gastaldelli, A. and Wolfe, R. R., Effect of theophylline on substrate metabolism during exercise, *Metabolism*, 45, 1153, 1996.
132. Engels, H. J., Wirth, J. C., Celik, S. and Dorsey, J. L., Influence of caffeine on metabolic and cardiovascular functions during sustained light intensity cycling and at rest, *Int. J. Sport Nutr.*, 9, 361, 1999.
133. Weir, J., Noakes, T. D., Myburgh, K. and Adams, B., A high carbohydrate diet negates the metabolic effects of caffeine during exercise, *Med. Sci. Sports Exerc.*, 19, 100, 1987.
134. Hermansen, L., Hultman, E. and Saltin, B., Muscle glycogen during prolonged severe exercise, *Acta Physiol. Scand.*, 71, 129, 1967.
135. Bergstrom, J., Hermansen, L., Hultman, E. and Saltin, B., Diet, muscle glycogen and physical performance, *Acta Physiol. Scand.*, 71, 140, 1967.
136. Essig, D., Costill, D. L. and VanHandel, P. J., Effects of caffeine ingestion on utilization of muscle glycogen and lipid during leg ergometer cycling, *Int. J. Sports Med.*, 1, 86, 1980.
137. Erickson, M. A., Schwarzkopf, R. J. and McKenzie, R. D., Effects of caffeine, fructose and glucose ingestion on muscle glycogen utilization during exercise, *Med. Sci. Sports Exerc.*, 19, 579, 1987.
138. Bergstrom, J. and Hultman, E., A study of the glycogen metabolism during exercise in man, *Scand. J. Clin. Lab. Invest.*, 19, 218, 1967.
139. Chesley, A., Howlett, R. A., Heigenhauser, G. J., Hultman, E. and Spriet, L. L., Regulation of muscle glycogenolytic flux during intense aerobic exercise after caffeine ingestion, *Am. J. Physiol.*, 275, R596, 1998.
140. Laurent, D., Schneider, K. E., Prusaczyk, W. K., Franklin, C., Vogel, S. M., Krssak, M., Petersen, K. F., Goforth, H. W. and Shulman, G. I., Effects of caffeine on muscle glycogen utilization and the neuroendocrine axis during exercise, *J. Clin. Endocrinol. Metab.*, 85, 2170, 2000.
141. Graham, T. E., Caffeine, coffee and ephedrine: Impact on exercise performance and metabolism, *Can. J. Appl. Physiol.*, 26 Suppl, S103, 2001.
142. Winder, W. W., Effect of intravenous caffeine on liver glycogenolysis during prolonged exercise, *Med. Sci. Sports Exerc.*, 18, 192, 1986.
143. Arogyasami, J., Yang, H. T. and Winder, W. W., Effect of intravenous caffeine on muscle glycogenolysis in fasted exercising rats, *Med. Sci. Sports Exerc.*, 21, 167, 1989.
144. Arogyasami, J., Yang, H. T. and Winder, W. W., Effect of caffeine on glycogenolysis during exercise in endurance trained rats, *Med. Sci. Sports Exerc.*, 21, 173, 1989.
145. Tarnopolsky, M. A., Caffeine and endurance performance, *Sports Med.*, 18, 109, 1994.
146. Lopes, J. M., Aubier, M., Jardim, J., Aranda, J. V. and Macklem, P. T., Effect of caffeine on skeletal muscle function before and after fatigue, *J. Appl. Physiol.*, 54, 1303, 1983.
147. Tarnopolsky, M. and Cupido, C., Caffeine potentiates low frequency skeletal muscle force in habitual and nonhabitual caffeine consumers, *J. Appl. Physiol.*, 89, 1719, 2000.

148. Walton, C., Kalmar, J. M. and Cafarelli, E., Effect of caffeine on self-sustained firing in human motor units, *J. Physiol.*, 545, 671, 2002.
149. Eke-Okoro, S. T., The H-reflex studied in the presence of alcohol, aspirin, caffeine, force and fatigue, *Electromyogr. Clin. Neurophysiol.*, 22, 579, 1982.
150. Jacobson, B. H., Weber, M. D., Claypool, L. and Hunt, L. E., Effect of caffeine on maximal strength and power in elite male athletes, *Br. J. Sports Med.*, 26, 276, 1992.
151. Lindinger, M. I., Graham, T. E. and Spriet, L. L., Caffeine attenuates the exercise-induced increase in plasma [K+] in humans, *J. Appl. Physiol.*, 74, 1149, 1993.
152. Rivers, W. H. R. and Webber, H. N., The action of caffeine on the capacity for muscular work, *J. Physiol.*, 36, 33, 1907.
153. Weiss, B. and Laties, V. G., Enhancement of human performance by caffeine and the amphetamines, *Pharmacol. Rev.*, 14, 1, 1962.
154. Foltz, E. E., Ivy, A. C. and Barborka, C. J., The influence of amphetamine (benzedrine) sulfate, d-desoxyephedrine hydrochloride (pervitin) and caffeine upon work output and recovery when rapidly exhausting work is done by trained subjects, *J. Lab. Clin. Med.*, 28, 603, 1943.
155. Foltz, E. E., Schiffrin, M. J. and Ivy, A. C., The influence of amphetamine (benzedrine) sulfate and caffeine on the performance of rapidly exhausting work by untrained subjects, *J. Lab. Clin. Med.*, 28, 601, 1943.
156. Sinclair, C. J. and Geiger, J. D., Caffeine use in sports. A pharmacological review, *J. Sports Med. Phys. Fitness*, 40, 71, 2000.
157. Armstrong, L. E., Caffeine, body fluid-electrolyte balance and exercise performance, *Int. J. Sport Nutr. Exerc. Metab.*, 12, 189, 2002.
158. Butts, N. K. and Crowell, D., Effect of caffeine ingestion on cardiorespiratory endurance in men and women, *Res. Q. Exerc. Sport*, 56, 301, 1985.
159. Falk, B., Burstein, R., Rosenblum, J., Shapiro, Y., Zylber-Katz, E. and Bashan, N., Effects of caffeine ingestion on body fluid balance and thermoregulation during exercise, *Can. J. Physiol. Pharmacol.*, 68, 889, 1990.
160. Fulco, C. S., Rock, P. B., Trad, L. A., Rose, M. S., Forte, V. A., Jr., Young, P. M. and Cymerman, A., Effect of caffeine on submaximal exercise performance at altitude, *Aviat. Space Environ. Med.*, 65, 539, 1994.
161. Kovacs, E. M., Stegen, J. and Brouns, F., Effect of caffeinated drinks on substrate metabolism, caffeine excretion and performance, *J. Appl. Physiol.*, 85, 709, 1998.
162. MacIntosh, B. R. and Wright, B. M., Caffeine ingestion and performance of a 1,500-metre swim, *Can. J. Appl. Physiol.*, 20, 168, 1995.
163. Cohen, B. S., Nelson, A. G., Prevost, M. C., Thompson, G. D., Marx, B. D. and Morris, G. S., Effects of caffeine ingestion on endurance racing in heat and humidity, *Eur. J. Appl. Physiol. Occup. Physiol.*, 73, 358, 1996.
164. Hunter, A. M., St Clair Gibson, A., Collins, M., Lambert, M. and Noakes, T. D., Caffeine ingestion does not alter performance during a 100-km cycling time-trial performance, *Int. J. Sport Nutr. Exerc. Metab.*, 12, 438, 2002.
165. Berglund, B. and Hemmingsson, P., Effects of caffeine ingestion on exercise performance at low and high altitudes in cross-country skiers, *Int. J. Sports Med.*, 3, 234, 1982.
166. Slavin, J. L. and Joensen, D. J., Caffeine and sports performance, *Phys. Sportsmed.*, 13, 191, 1985.
167. Wemple, R. D., Lamb, D. R. and McKeever, K. H., Caffeine vs caffeine-free sports drinks: Effects on urine production at rest and during prolonged exercise, *Int. J. Sports Med.*, 18, 40, 1997.

168. Falk, B., Burstein, R., Ashkenazi, I., Spilberg, O., Alter, J., Zylber-Katz, E., Rubinstein, A., Bashan, N. and Shapiro, Y., The effect of caffeine ingestion on physical performance after prolonged exercise, *Eur. J. Appl. Physiol. Occup. Physiol.*, 59, 168, 1989.
169. Anselme, F., Collomp, K., Mercier, B., Ahmaidi, S. and Prefaut, C., Caffeine increases maximal anaerobic power and blood lactate concentration, *Eur. J. Appl. Physiol. Occup. Physiol.*, 65, 188, 1992.
170. Anderson, M. E., Bruce, C. R., Fraser, S. F., Stepto, N. K., Klein, R., Hopkins, W. G. and Hawley, J. A., Improved 2000-meter rowing performance in competitive oarswomen after caffeine ingestion, *Int. J. Sport Nutr. Exerc. Metab.*, 10, 464, 2000.
171. Bruce, C. R. anderson, M. E., Fraser, S. F., Stepto, N. K., Klein, R., Hopkins, W. G. and Hawley, J. A., Enhancement of 2000-m rowing performance after caffeine ingestion, *Med. Sci. Sports Exerc.*, 32, 1958, 2000.
172. Collomp, K., Candau, R., Millet, G., Mucci, P., Borrani, F., Prefaut, C. and De Ceaurriz, J., Effects of salbutamol and caffeine ingestion on exercise metabolism and performance, *Int. J. Sports Med.*, 23, 549, 2002.
173. Wiles, J. D., Bird, S. R., Hopkins, J. and Riley, M., Effect of caffeinated coffee on running speed, respiratory factors, blood lactate and perceived exertion during 1500-m treadmill running, *Br. J. Sports Med.*, 26, 116, 1992.
174. Collomp, K., Ahmaidi, S., Chatard, J. C., Audran, M. and Prefaut, C., Benefits of caffeine ingestion on sprint performance in trained and untrained swimmers, *Eur. J. Appl. Physiol. Occup. Physiol.*, 64, 377, 1992.
175. Perkins, R. and Williams, M. H., Effect of caffeine upon maximal muscular endurance of females, *Med. Sci. Sports*, 7, 221, 1975.
176. Powers, S. K., Byrd, R. J., Tulley, R. and Callender, T., Effects of caffeine ingestion on metabolism and performance during graded exercise, *Eur. J. Appl. Physiol. Occup. Physiol.*, 50, 301, 1983.
177. Gaesser, G. A. and Rich, R. G., Influence of caffeine on blood lactate response during incremental exercise, *Int. J. Sports Med.*, 6, 207, 1985.
178. French, C., McNaughton, L., Davies, P. and Tristram, S., Caffeine ingestion during exercise to exhaustion in elite distance runners. Revision, *J. Sports Med. Phys. Fitness*, 31, 425, 1991.
179. Bucci, L. R., Selected herbals and human exercise performance, *Am. J. Clin. Nutr.*, 72, 624S, 2000.
180. Bell, D. G., Jacobs, I. and Zamecnik, J., Effects of caffeine, ephedrine and their combination on time to exhaustion during high-intensity exercise, *Eur. J. Appl. Physiol. Occup. Physiol.*, 77, 427, 1998.
181. Bell, D. G., Jacobs, I., McLellan, T. M. and Zamecnik, J., Reducing the dose of combined caffeine and ephedrine preserves the ergogenic effect, *Aviat. Space Environ. Med.*, 71, 415, 2000.
182. Bell, D. G. and Jacobs, I., Combined caffeine and ephedrine ingestion improves run times of Canadian Forces Warrior Test, *Aviat. Space Environ. Med.*, 70, 325, 1999.
183. Bell, D. G., McLellan, T. M. and Sabiston, C. M., Effect of ingesting caffeine and ephedrine on 10-km run performance, *Med. Sci. Sports Exerc.*, 34, 344, 2002.
184. Bell, D. G., Jacobs, I. and Ellerington, K., Effect of caffeine and ephedrine ingestion on anaerobic exercise performance, *Med. Sci. Sports Exerc.*, 33, 1399, 2001.

185. Jacobs, I., Pasternak, H. and Bell, D. G., Effects of ephedrine, caffeine and their combination on muscular endurance, *Med. Sci. Sports Exerc.*, 35, 987, 2003.
186. Morton, R. H., Effects of caffeine, ephedrine and their combinations on time to exhaustion during high-intensity exercise. Correspondence, *Eur. J. Appl. Physiol. Occup. Physiol.*, 79, 379, 1999.
187. Morton, R. H., Effects of caffeine, ephedrine and their combination on time to exhaustion during high-intensity exercise. Correspondence, *Eur. J. Appl. Physiol. Occup. Physiol.*, 80, 610, 1999.
188. Goldberg, L., Elliot, D. and Kuehl, K., Effect of caffeine and ephedrine ingestion on anaerobic exercise performance. Correspondence, *Med. Sci. Sports Exerc.*, 34, 181, 2002.
189. Haller, C. A. and Benowitz, N. L., Adverse cardiovascular and central nervous system events associated with dietary supplements containing ephedra alkaloids, *N. Engl. J. Med.*, 343, 1833, 2000.
190. CDC, Adverse events associated with ephedrine-containing products-Texas, December 1993-September 1995, *Morb. Mortal. Wkly. Rep.*, 45, 689, 1996.
191. Shekelle, P. G., Hardy, M. L., Morton, S. C., Maglione, M., Mojica, W. A., Suttorp, M. J., Rhodes, S. L., Jungvig, L. and Gagne, J., Efficacy and safety of ephedra and ephedrine for weight loss and athletic performance: a meta-analysis, *J. Am. Med. Assoc.*, 289, 1537, 2003.

18
Carotenoids

Maria Stacewicz-Sapuntzakis and Veda Diwadkar-Navsariwala

CONTENTS

I. Introduction 325
II. Functions of Carotenoids in the Human Body 327
    A. β-Carotene and Other Provitamin A Carotenoids 327
    B. Lutein and Zeaxanthin 332
    C. Lycopene 333
III. Dietary Sources of Carotenoids 334
    A. Amounts in Various Foodstuffs 334
    B. Cooking Techniques and Bioavailability 335
    C. Carotenoid Supplements 337
IV. Body Reserves, Intake and Athletic Performance 337
    A. Measurement of Carotenoid Levels in Tissues and
       Plasma 337
    B. Epidemiological Studies 339
    C. Human Intervention Studies 340
V. Recommended Intake 342
VII. Summary 344
References 346

I. Introduction

Carotenoids are long-chained hydrocarbon compounds of plant origin that provide brilliant red, orange and yellow colors to many flowers, fruits and vegetables.[1] Many animals can absorb carotenoids from their food supply and use them to color their own body (pink feathers of flamingo, yellow skin of vultures, egg yolk). The intense colors provided by carotenoids may serve to attract members of the opposite sex, or to warn and repel predators. Although humans are less conspicuous in nature, we do have the ability to absorb a wide range of carotenoids, but there are individuals and populations

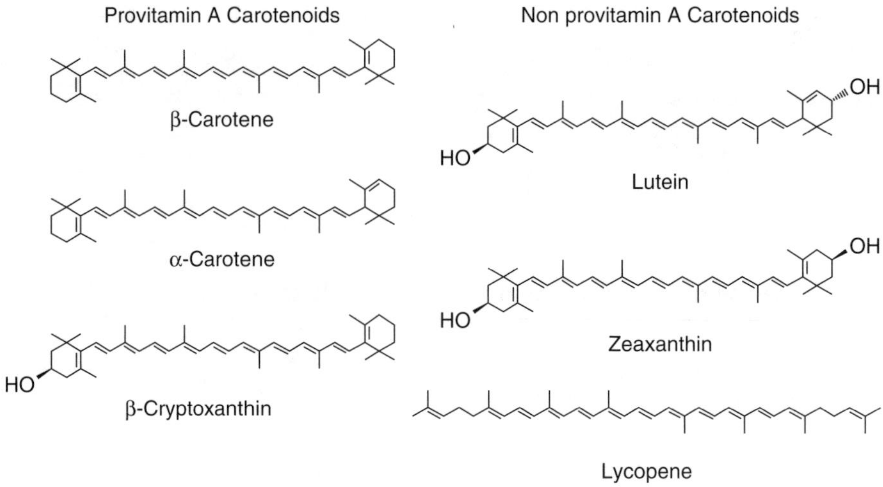

FIGURE 18.1

that consume negligible amounts of these compounds. Are carotenoids truly necessary for optimal health and performance of the human body? Can they serve as ergogenic aids in any capacity? This chapter explores the existing knowledge about carotenoids in an attempt to answer these questions and provide intake recommendations for athletes and physically active adults.

Carotenoids belong to terpenes, which are constructed of multiples of the 5-carbon isoprene units. Four such units in characteristic head-to-tail arrangement join with another C_{20} molecule (geranylgeraniol) by tail-to-tail central bond to produce the 40-carbon skeleton of carotenoids (tetraterpenes).[1] Fat-soluble vitamins A, E and K are also terpenes; another terpene, squalene (C_{30}), is a precursor of steroids. Plants, bacteria and fungi are capable of carotenoid synthesis, but animals produce only steroids and therefore require dietary sources of vitamin A, E and K. In plants, long-chain hydrocarbon carotenoid (lycopene) may undergo cyclization on the ends, forming α-carotene and β-carotene (Figure 18.1), which, in turn, produce oxygen-containing, more polar-derivative carotenoids with hydroxyl-, keto- or epoxy-substitute groups (xanthophylls).

The main role of carotenoids in photosynthetic plants is the protection of chlorophyll against light damage in green tissues, and also facilitation of photosynthesis under low light conditions.[2] Carotenoid molecules dissipate excessive energy as heat and capture sunlight in deep shade or under water. Those carotenoids are found in chloroplasts in very orderly arrangement with chlorophyll (photosynthetic antenna systems), and are responsible for conversion of light into chemical potential. Much higher concentrations of carotenoids give the flowers and fruits of some plants bright colors, attracting pollinators and fruit-consuming animals and thus aiding in the reproduction and dissemination of plant species. Human preference for colorful vegetables produced orange carrots and sweet potatoes by continuous breeding of strains with desired properties. In these plant tissues, carotenoids are packaged in chromoplasts, distinguished by their striking colors and filled with crystallized carotenoids.

Various plants have specific patterns of carotenoid synthesis, especially in their brightly colored tissues, and thus accumulate different carotenoids.[3] Tomatoes and watermelons are rich in lycopene, carrots and pumpkins have an abundance of α-carotene and β-carotene, β-cryptoxanthin is present in oranges, plums and peaches, lutein accompanies chlorophyll in all dark green vegetables such as broccoli and spinach (Table 18.1). Animals also differ in their ability to absorb carotenoids from food. Many species do not absorb intact carotenoids, but humans are capable of absorbing both hydrocarbons (lycopene, α- and β-carotene) and xanthophylls (lutein, β-cryptoxanthin), although more substituted carotenoids, with many hydroxyl-, keto- and epoxy-groups, cannot cross the intestinal mucosa and may be found in feces. Therefore, it is quite reliable to infer the consumption of various fruits and vegetables from analysis of individual serum samples. Serum carotenoid values reflect recent intake and can be greatly increased by consumption of particular foods rich in carotenoids.[4-6] Epidemiological studies confirm the correlation of fruit and vegetable intake with serum carotenoid levels, although it is not as strong as in intervention trials, partly because the researchers have to rely on food-intake interviews. People may not remember the quantity and frequency of their consumption or can exaggerate it for various reasons, and the food-composition tables are only an approximation of average carotenoid content in diet.[7]

II. Functions of Carotenoids in the Human Body

A. β-Carotene and Other Provitamin A Carotenoids

Among 600 existing carotenoids[1] about 50 are regularly consumed by humans in their food supply, but only a small number possess provitamin A activity. A cyclical structure of β-ionone ring, devoid of any substitutions, is a required feature of provitamin A carotenoids. β-Carotene is a symmetrical molecule with two β-ionone rings, one on each end, and thus yields two molecules of vitamin A aldehyde (retinal) by central cleavage. Many animals and humans possess genes for carotene 15,15′-monooxygenase enzyme, which cleaves C_{40} chain in the middle, but it is mainly expressed in the intestinal mucosa and liver. The enzyme will also produce retinal from other carotenoids, like α-carotene and β-cryptoxanthin, which have one intact β-ionone ring. Provitamin A carotenoids may also form vitamin A by asymmetric cleavage, a less efficient pathway that may yield only one molecule of retinal from β-carotene. The nutritional value of provitamin A carotenoids is based on their conversion rate to vitamin A in the human body.

TABLE 18.1

Carotenoid Content of Foods

Food Description	Carotenoid Content mg/100 g Edible Portion		
	β-carotene	Lutein + Zeaxanthin	Lycopene
Fruits			
Apricots (raw)	2.6	0	Trace
Apricots (canned, heavy syrup)	6.6	0	0.07
Cantaloupe, raw	1.6	0.04	0
Grapefruit (raw, pink and red)	0.6	0.01	1.5
Guava [61]	0.8	0	5.3
Mangos (canned, drained)	13.1	N/A	N/A
Papaya (fresh)	0.3	0	2.7[60]
Watermelon (raw)	0.3	0.02	4.9
Raw Vegetables			
Broccoli	0.8	2.4	0
Brussels sprouts	0.5	1.6	0
Baby Carrots	7.3	0.4	0
Carrots	8.8	N/A	N/A
Collards	3.3	N/A	N/A
Lettuce, cos or romaine	1.3	2.6	0
Peppers, sweet, red	2.4	N/A	N/A
Spinach	5.6	11.9	0
Squash (winter, butternut)	4.2	N/A	N/A
Sweet potato	9.2	0	0
Tomatoes (red, ripe, raw, year round average)	0.4	0.1	3.0
Vegetables (Cooked, Boiled, Drained)			
Broccoli	1.0	2.2	0
Brussels sprouts	0.5	1.3	0
Carrots	8.0	N/A	N/A
Collards	4.4	8.1	0
Corn (sweet)	N/A	1.8	N/A
Peas (canned, drained)	0.3	1.4	0
Peppers, sweet, red	2.2	N/A	N/A
Pumpkin (canned)	6.9	0	0
Spinach	5.2	7.0	0
Squash (winter, butternut, baked)	4.6	N/A	N/A
Sweet potato (baked in skin)	9.5	0	0
Tomatoes	0.3	0.2	4.4
Turnip greens	4.6	8.4	0

(continued)

TABLE 18.1 (CONTINUED)
Carotenoid Content of Foods

	Carotenoid Content mg/100 g Edible Portion		
Food Description	β-carotene	Lutein + Zeaxanthin	Lycopene
Other Foods			
Beef stew with vegetables (including potatoes and carrots)	1.8	0.06	0.3
Butter, light, stick with salt	1.2	N/A	N/A
Catsup	0.7	0	17.0
Chicken pot pie, with carrots, potatoes and peas, frozen	1.0	0.1	0
Egg, whole, raw, fresh (2 eggs = 100 g)[34]	0	1.0	0
Lasagna with meat and tomato sauce, frozen entrée, cooked	0.2	0.1	7.8
Meatloaf with mashed potatoes and tomato-based gravy, low-fat frozen entrée, cooked	0.1	0	0.9
Pasta in tomato sauce with cheese, canned	0.1	0	3.2
Pizza with pepperoni, cheese and sauce, thin crust, frozen	0.3	0.05	4.4
Sauce, pasta, spaghetti with marinara, ready-to-serve	0.4	0.2	16.0
Soup, minestrone, canned, condensed, commercial	0.9	0.2	1.5
Soup, tomato, canned, condensed, commercial	0.2	0.09	10.9
Tomato paste, canned with salt added	1.2	0.2	29.3
Vegetable juice cocktail, canned	0.8	0.08	9.7

Note: N/A = not available.
Source: Adapted from USDA-NCC carotenoid database for U.S. foods, 1998.
(http://www.nal.usda.gov/fnic/foodcomp/Data/car98/car98.html), except as noted.

It depends on many factors, among them the food matrix, the presence of fat in the diet, efficiency of absorption and health of the individual.[8]

The definitions of carotenoid conversion were recently changed to reflect the prevailing findings of low vitamin A activity of plant-derived food in combating vitamin A deficiency. A new term, retinol activity equivalent (RAE), was introduced to express the vitamin A activity of carotenoids:[9]

1 μg RAE = 1 μg dietary or supplemental vitamin A (retinol)
= 2 μg β-carotene in oil or supplements
= 12 μg dietary β-carotene (fruits and vegetables)
= 24 μg other provitamin A carotenoids in diet

Example: A diet contains 500 μg retinol, 2400 μg β-carotene, 480 μg α-carotene and 240 μg β-cryptoxanthin. Using the above-listed conversion factors, it adds up to: 500 + 2400/12 + 480/24 + 240/24 = 500 + 200 + 20 + 10 = 730 μg RAE.

Previously, vitamin A activity of dietary carotenoids was overestimated by a factor of 2, since retinol equivalent (RE) was equal to 6 μg β-carotene, or 12 μg of other provitamin A carotenoids.[10] The vitamin manufacturers and many nutritionists prefer to use International Units (IU) to express vitamin A activity (1 IU = 0.3 μg retinol, or 0.6 μg supplemental β-carotene, or 3.6 μg β-carotene from fruits and vegetables). It may become very confusing when reporting the total vitamin A intake from mixed foods, therefore it is preferable to specify the amounts of each carotenoid.

The estimated average requirement (EAR) is 625 μg RAE/day for men and 500 μg RAE/day for women. The recommended dietary allowance (RDA) for men is 900 μg RAE/day and for women 700 μg RAE/day, except in pregnancy (770 μg/day) and lactation (1300 μg/day). The tolerable upper intake level (UL) for adults is set at 3000 μg/day.[9]

Vitamin A is a necessary part of our photoreceptors in the retina, maintains differentiation of epithelial tissues, regulates normal reproduction, fetal development and growth of young children and adolescents.[11] Adequate vitamin A nutrition is therefore required for optimal health in conditions of strenuous exercise,[12] but large doses of preformed vitamin A should not be used as ergogenic aids. Vitamin A is extremely toxic and may cause acute poisoning when consumed in excess of UL for a prolonged time. The symptoms include bone pain, redness of skin, headache, nausea and disorientation. Fortunately, it is not possible to cause hypervitaminosis A by high intake of provitamin A carotenoids, because the conversion efficiency seems to be very limited and is regulated by the vitamin A status of the subject.[13] Most of the ingested carotenoids are not absorbed by the intestinal mucosa, but rather eliminated in the feces.[14] The absorbed provitamin A carotenoids are mostly transported to the liver and deposited in various tissues (adipose, skin, adrenal, ovaries, testes).[15] Habitual high intake of β-carotene may produce carotenemia,[16] a harmless condition characterized by yellowing of the skin, high circulating plasma levels and bright yellow ovaries in women.

Apart from its provitamin A function, β-carotene may act in other less established ways. Some of these actions may not require β-carotene or its provitamin A activity specifically, but may be also fulfilled by other carotenoids or even other lipophilic nutrients, like vitamin E. β-Carotene and other carotenoids are efficient antioxidants, especially at low partial pressure of oxygen, as is present in animal tissues, quenching singlet oxygen and preventing lipid peroxidation[17] *in vitro*. However, it is difficult to prove this antioxidant activity in healthy human subjects[18] unless they are depleted of dietary carotenoids for a prolonged period.[15]

Supplementation with β-carotene decreased indicators of lipid oxidation (breath pentane, serum lipid peroxide levels) of depleted young men[19] and adult women (serum malondialdehyde).[20] Nondiabetic individuals with heightened insulin resistance had lower plasma levels of α- and β-carotene and higher lipid hydroperoxides than insulin-sensitive subjects.[21] In conditions associated with increased oxidative stress, such as smoking, cystic fibrosis and diabetes, supplemental β-carotene seemed to be effective in some trials, but not in others.

It must be pointed out that plasma carotenoid levels significantly decrease in subjects under increased oxidative stress due to smoking, working environment or inflammatory diseases. Provitamin A carotenoids are especially affected by oxidative stress and were found to be depleted in smokers[22] and also in their nonsmoking wives.[23] They may rise quite rapidly after cessation of smoking.[24] In low-density lipoproteins (LDL), α- and β-carotene are the main targets for nitric oxide, which could make them important for prevention of atherosclerosis.[25] Serum β-carotene levels are inversely associated with C-reactive protein and white blood cell count, which are indicators of inflammation.[26] It does not necessarily mean that β-carotene has an anti-inflammatory action, but the elevated oxidative processes consume carotenoids at greater rate in inflammatory diseases (in smokers, pancreatitis, tuberculosis) and even in apparently healthy subjects. Reduced pulmonary function was associated with lower serum levels of carotenoids, especially β-cryptoxanthin, in a large New York state population study.[27]

There is some evidence that β-carotene may enhance the immune function of natural killer cells by increasing their lytic activity.[28] The natural killer cells constitute about 20% of white blood cells and are very important in fighting tumors and leukemia. Low dietary intake of carotenoids, especially β-carotene, α-carotene and lutein, is associated with a double risk of developing breast cancer.[29] Similar results were obtained from epidemiological studies of many other forms of cancer in various populations, especially for lung, oral cavity, pharynx and larynx cancers.[15] However, intervention trials using high-dose β-carotene supplements were not successful in finding a protective role, smokers and asbestos workers developed more lung cancers on supplements than those on placebo.[30]

There is little doubt that β-carotene protects skin against sunburn. Supplementation with β-carotene alone or in combination with other carotenoids (lutein and lycopene) increased the skin levels of these pigments and decreased the intensity of erythema induced by UV irradiation.[31] Even more important is the photoprotective effect of β-carotene in patients with erythropoietic protoporphyria, a congenital defect in which protoporphyrin accumulation causes excessive light sensitivity. Daily doses up to 150 mg β-carotene greatly increase tolerance to sunlight and decrease itching, burning and ulceration of skin in these patients by quenching excited species of porphyrins.[32]

B. Lutein and Zeaxanthin

Lutein and zeaxanthin are more polar than α- and β-carotene due to the presence of hydroxyl groups on each ring. They are derived from α- and β-carotene, respectively, and play an important role in protecting green plant tissues from sun damage. High concentrations are found in corn kernels and marigold petals, which are used as a source of colorant for poultry. Birds absorb hydroxycarotenoids from their food and accumulate them in their skin, fat and egg yolks, as well as their retinas. Humans also absorb hydroxycarotenoids more rapidly than α- and β-carotene and easily utilize lutein esters,[33] which are present in fruits and flowers. The most bioavailable source of lutein and zeaxanthin is egg yolk, due to the fine dispersion of carotenoids in fat.[34] The presence of these carotenoids in egg yolk is crucial for the development of bird embryos and the health of young birds. It is quite possible that carotenoids in general may also have an important role in human reproduction, as their levels change during menstrual cycle[35] and they accumulate in the ovary and most notably in corpus luteum.[36] High dietary intake of lutein is common in preindustrial societies such as those of South Pacific Island nations,[37] and may be protective against some cancers, especially lung cancer in smokers.[38]

The best-documented action of lutein and zeaxanthin is their possible role in protecting our vision from damage due to light and oxygen and in maintaining visual acuity by absorbing short-wave light.[39] Human retina has a yellow spot, called the macula lutea, about 0.5 cm in diameter, where the light passing through the lens is focused (fovea) and color vision and acuity are best developed. The macula is yellow because it contains macular pigment composed of zeaxanthin and lutein, with the former mostly in the center of the macula and the latter more abundant on the perimeter. Small amounts of these pigments can also be found in the lens of the eye.

Humans widely differ in the amount of macular pigment in the fovea, which can be increased by dietary intake of foods rich in lutein or zeaxanthin, such as spinach, corn[40] or eggs.[34] People with low levels of macular pigment may be at increased risk of developing macular degeneration with advancing age, a leading cause of blindness in the U.S. among persons older than 65 years. Serum carotenoid levels, except lycopene, were significantly lower in subjects with neovascular age-related macular degeneration[41] than in matched controls. The dietary intake of lutein and zeaxanthin, particularly from spinach and collard greens, was also found to be significantly lower in cases of macular degeneration. The additional oxidative stress produced by smoking greatly increased the risk of macular degeneration in persons with low intake of lutein/zeaxanthin, but the highest levels of intake ameliorated the effect of smoking.[42] As an additional bonus, lutein and zeaxanthin seem to promote clearer lenses, decreasing the risk of developing cataracts,[39] which are thought to develop as a consequence of accumulating oxidative damage to lens proteins.

C. Lycopene

Lycopene is a precursor of β-carotene in the biosynthetic pathway of plants, but, as an aliphatic hydrocarbon with 11 conjugated carbon–carbon double bonds, it has no provitamin A activity. It gives red color to tomatoes, watermelon, papaya, guava and pink grapefruit [43] and is very common in the American diet, but nearly absent in the East Asian food supply (China, Korea, Japan) and rather low in northern Europe (Finland). It is often the most abundant plasma carotenoid, especially in young American adults, who indulge in consumption of vast quantities of pizza, ketchup, salsa and spaghetti sauce, often neglecting other fruits and vegetables. Prolonged excessive intake of lycopene may cause lycopenemia, a yellow-orange pigmentation of skin, and even liver deposits of this carotenoid.[44,45]

Lycopene is the most efficient singlet oxygen quencher among carotenoids.[46] In skin, it is destroyed by ultraviolet irradiation preferentially over β-carotene.[47] It seems more effective than other carotenoids in reducing lipid peroxidation of circulating LDL, since the indicators of LDL oxidation were improved in men after 2 weeks on a tomato-juice diet, but not by carrot juice or a spinach-powder drink.[48] Similar results were found in young Japanese women when tomato juice was added to their diet.[49] When relationships of carotenoid levels in adipose tissue biopsies and acute myocardial infarction risk were studied in ten European countries, high levels of lycopene seemed to be protective.[50] In a Finnish study of a population with a very high rate of atherosclerosis, the men with the lowest plasma lycopene had much more risk of heart attack and stroke.[51] The carotid arteries of men with low plasma lycopene were narrower due to the increased thickness of the intima media,[52] which is associated with oxidative modification of LDL particles, causing formation of autoantibodies and inflammatory response.

Lycopene may also protect DNA from oxidative damage. White blood cells from healthy women were more resistant to *ex vivo* oxidative stress after short (2–3 weeks) supplementation of the subjects with tomato puree.[53,54] Oxidative damage to DNA may lead to point mutations if DNA repair enzymes fail to excise oxidized bases like 8-hydroxy-2'-deoxyguanosine (8-OHdG). Such mutations may initiate and promote various forms of cancer, and tomato consumption, with accompanying increased serum lycopene, shows promise of reducing cancer in the prostate, breast, lung, pancreas and digestive tract,[55,56] according to epidemiological surveys of diverse populations. Men who consume 4–5 servings of tomato products per week have lower risk of prostate cancer, especially in its aggressive form,[57] a finding that was confirmed by association of elevated plasma lycopene with decreased prostate cancer incidence.[58] When tomato sauce-based pasta dishes were consumed every day for 3 weeks by prostate cancer patients, their DNA damage in white blood cells decreased significantly and the prostate DNA also showed less 8-OHdG than comparable prostate tissue from other patients.[59] Their serum level of prostate-specific antigen (PSA), which indicates risk and severity of prostate cancer, as well as the success of treatment, was also significantly reduced by this short-term dietary intervention. Serum lycopene concentrations doubled during the study period and prostate tissue lycopene levels even tripled, with signs of increased apoptosis (programmed cell death) present in hyperplastic and neoplastic cells.[4] These promising results require further research and current clinical trials with lycopene supplements will hopefully clarify the role of lycopene in cancer prevention.

III. Dietary Sources of Carotenoids

A. Amounts in Various Foodstuffs

The most abundant carotenoids in the U.S. diet are β-carotene, lutein and lycopene (Table 18.2), with α-carotene, β-cryptoxanthin and zeaxanthin present in much lower quantities.[15] Lutein is usually assayed together with zeaxanthin, therefore these two carotenoids are reported as a sum of both in tables 18.1, 18.2 and 18.3.

Table 18.1 shows the distribution of β-carotene, lutein with zeaxanthin and lycopene in commonly consumed fruits, vegetables and mixed dishes. According to the United States Department of Agriculture (USDA) carotenoid database, β-carotene is the predominant carotenoid found in orange-yellow-colored vegetables and fruits such as carrots, pumpkin, sweet potato, apricots and mango; while green vegetables such as broccoli, spinach, collards and turnip greens contain higher amounts of lutein with zeaxanthin. Unlike other carotenoids, lycopene is contained in only a few food sources, including tomatoes, watermelon, pink grapefruits, guava and papaya. A large proportion of the lycopene consumed in the United States is derived mainly from cooked tomato products. Mixed dishes such as lasagna, pasta, pizza and tomato soup, which are popular among both children and adults, are excellent sources of lycopene.

Although the USDA database is commonly used to calculate carotenoid content of foods and individual intake, the users should be aware of its limitations. The values were obtained from various references, with both food samples and analytical methodology greatly differing in quality. While certain items were very thoroughly evaluated (carrots, tomatoes) by extensive sampling and careful analytical procedures, other values may be based

TABLE 18.2

Mean Intake of Carotenoids in the U.S. Population

Carotenoid	Mean intake (mg/day)	Range (mg)
α-carotene	0.04	0–7.4
β-carotene	1.7	0–13.2
β-cryptoxanthin	0.02	0–1.4
Lutein + zeaxanthin	1.5	0–7.5
Lycopene	2.2	0–123.3

Source: Adapted from the Third National Health and Nutrition Examination Survey (NHANES III), 1988–1994.[9]

TABLE 18.3

Serum Carotenoid Concentrations in U.S. Population (µmol/L)

Carotenoid	Children and Adolescents[a]		Adults[b]	
	Median	Range	Median	Range
α-carotene	0.05	0.01–0.88	0.06	0.00–3.00
β-carotene	0.25	0.02–1.77	0.27	0.00–12.56
β-cryptoxanthin	0.15	0.02–1.95	0.13	0.00–2.60
Lutein + zeaxanthin	0.28	0.07–1.74	0.32	0.00–8.40
Lycopene	0.42	0.02–1.71	0.41	0.00–2.31

[a] 6–16 years old, n = 4232. Adapted from Ford et al. (2002).[84]
[b] ≥ 20 years old, n = 14,914. Adapted from Ford (2000).[85]
[c] The median is used here as a measure of central tendency because carotenoid distribution is skewed (low values are more common)

on a single or imperfect determination. To illustrate the point, lutein and zeaxanthin content of eggs was corrected in Table 18.1 by using published values of 300 µg lutein and 200 µg zeaxanthin in one egg yolk.[34] The USDA original value is only 55 µg of lutein with zeaxanthin in 100 g of eggs (edible portion). According to the USDA database, the main carotenoid in papaya fruit is β-cryptoxanthin (761 µg/100 g) and there is no lycopene. However, Brazilian papayas were found to contain an average 2.65 mg of lycopene/100 g,[60] and our unpublished data confirm it. Guava fruit is not included in the USDA database, although it is a good source of carotenoids and its consumption is quite popular among some ethnic groups. It contains a considerable amount of lycopene (5.3 mg/100 g)[61] — more than fresh tomatoes.

The main dietary sources of β-cryptoxanthin are tangerines (0.5 mg/100 g) and orange juice (0.32 mg/100 g), but red peppers contain as much as 2.2 mg/100 g and persimmons 1.4 mg/100 g. α-Carotene is abundant only in carrots (4.6 mg/100 g), canned pumpkin (4.8 mg/100 g) and baked squash (1.1 mg/100 g). In general, a diet that includes a wide variety of fruits and vegetables can provide the best combination of all carotenoids.

Plants are not the only source of dietary carotenoids. Some animal products provide them in very bioavailable form, dispersed in digestible fat. Since cattle absorb β-carotene, milk, butter and beef liver contain considerable amounts of it, and some hard cheeses are colored with additional carotenoids (β-carotene or annatto) during processing. Chicken fat, skin, liver and eggs are rich in lutein and zeaxanthin, because the birds absorb them efficiently from feed, which is supplemented with marigold preparations. Salmon contains canthaxanthin and astaxanthin, absorbed by the fish from plankton in the wild and added to their feed in aquaculture to provide the desired flesh color. Canthaxanthin is well absorbed by humans, [62] while astaxanthin bioavailability is currently under investigation.

B. Cooking Techniques and Bioavailability

Considerable losses of carotenoids or alterations in their properties may occur in cooking and processing of fruits and vegetables. To obtain maximum health benefits, it is important to reduce the loss of carotenoids from foods in storage and processing. In plants, carotenoids are naturally protected in the tissue matrix; however, peeling, cutting, drying and pulping of fruits and vegetables increase exposure to oxygen and release enzymes that catalyze carotenoid degradation.[63] In home preparations, sautéeing of vegetables produces the greatest losses, followed by boiling, steaming and microwave cooking. In addition, prolonged cooking procedures such as deep-frying also result in substantial losses. Decreasing processing time, as well as cooking temperatures, can maximize carotenoid retention in foods. On the other hand, raw plant sources of carotenoids have limited bioavailability, while processing of fruits and vegetables breaks down the plant cell walls, thereby enhancing the release of carotenoids from the food matrix. Significant improvements in bioavailability of carotenoids have been observed following cooking. In one study, serum lycopene concentration rose considerably following the intake of heated tomato paste, but not of uncooked tomato juice.[64] In addition, due to their lipid-soluble nature, absorption of carotenoids is greatly enhanced by the presence of dietary fat. Therefore, including small amounts of dressing with vegetable and fruit salads can greatly improve carotenoid bioavailability.[65]

Prior to their absorption, carotenoids have to be released from the food matrix. The carotenes in plant foods such as carrots and tomatoes are present mainly in the crystalline form in membrane-bound chromoplasts.[2] Due to the stability of the crystals and the hydrophobicity of the carotenes, the major challenge for absorption is their dissolution in the aqueous contents of the intestinal lumen. Following release from the food matrix, carotenoids are incorporated into mixed micelles, which are tiny aggregates of hydrolyzed lipids and bile salts.[66] The hydrophobic carotenoids are assumed to move from the food matrix into the dietary lipid droplets and form fine lipid emulsions due to the shearing forces in the gastrointestinal tract. Next, they are transferred from the lipid emulsions to the mixed bile salt micelles. The solubilization of carotenoids is accomplished by their interaction with the detergent-like bile acids, which are part of the mixed micelles. Dietary fat aids in the formation of bile salt micelles by stimulating bile flow from the gall bladder. Following absorption into the enterocytes, intact carotenoids and their cleaved products (vitamin A) are incorporated into chylomicron particles and then carried via the lymphatic system into the blood stream. Since no specific carotenoid binding proteins have been identified, lipoproteins are assumed to be exclusively involved in the transport of carotenoids within the body.[66] In humans, chylomicrons are mainly composed of triglycerides. Once they enter the blood stream, chylomicrons are taken up by the liver, where the carotenoids are either stored, or repackaged and released with other lipoprotein particles. The physical characteristics of the different carotenoids seem to influence their distribution within the lipoprotein particles. The more apolar carotenoids, such as lycopene and β-carotene, are associated mainly with the LDL particles (80%), while the polar xanthophylls such as lutein and zeaxanthin distribute almost equally between LDL (44%)

and high-density lipoprotein (HDL) particles (38%).[67] It is hypothesized that the lipophilic carotenoids, lycopene and β-carotene are embedded in the triglyceride core of the lipoproteins, and triglyceride lipolysis is critical for their release. The polar xanthophylls are located on the surface of the lipoproteins and are therefore more prone to exchange with other lipoproteins in the circulation.[66]

C. Carotenoid Supplements

Industrial synthesis has succeeded in the production of three carotenoids, β-carotene, lycopene and zeaxanthin, that are used as supplements for humans. In addition, canthaxanthin and astaxanthin are produced for poultry supplementation and aquaculture. All these carotenoids have symmetrical structures.[68] Other carotenoids available in supplements (lutein, β-cryptoxanthin and α-carotene) must be extracted from natural sources. The supplements contain various amounts of carotenoids, alone or in combination with other vitamins and phytochemicals. The most bioavailable are capsules filled with carotenoid beadlets, which are water-dispersible, and capsules filled with oil or oleoresin-containing carotenoids.

Two exotic food sources approach supplements in their extremely high concentration of carotenoids. One of them, red palm oil, contains very high concentrations of α-carotene (67 mg/100 g), β-carotene (120.5 mg/100 g) and lycopene (20 mg/100 g).[69] People consuming food cooked with palm oil (Malaysia, Nigeria) may have extremely elevated circulating serum carotenoids, but their retinol levels remain in normal range. Another source, oily *gac* fruit (*Momordica cochinchinensis*) from Vietnam[70] contains 17.5 mg β-carotene and 80 mg lycopene per 100 g. Both red palm oil and *gac* fruit are studied as potential safe sources for improving vitamin A and carotenoid status of malnourished children in India and Vietnam,[13] but could also be used as supplements in the developed countries.

IV. Body Reserves, Intake and Athletic Performance

A. Measurement of Carotenoid Levels in Tissues and Plasma

There is still a paucity of data on the body reserves of carotenoids. However, the limited data indicate a distinct distribution of the different carotenoids in certain tissues. This tissue-specific accumulation suggests that carotenoids may differ in the extent of transport or utilization and may exert definite biological effects in specific tissue. Among the different tissues, carotenoids are found to be most concentrated in the liver, testes and adrenals,[71] and, within the same subject, liver concentrations of carotenoids are higher than in lung and kidneys.[72] Carotenoids have been found in pancreas, spleen, heart, thyroid and ovary.[73] In a recent clinical trial, substantial accumulation of lycopene was noted in the prostate tissue, following a 3-week intervention with 30 mg lycopene provided as tomato-based pasta dishes to prostate cancer patients.[59] This intervention trial indicates that chronic supplementation with carotenoids from food sources can produce elevated tissue levels.

In many epidemiological and intervention studies, carotenoid levels have been measured in buccal mucosal cells or fat tissue biopsies, as a reflection of tissue levels. Carotenoid concentrations can be accurately measured in serum and tissues by high performance liquid chromatography (HPLC) after extraction with organic solvents. The concentration of carotenoids is not particularly high in the adipose tissue, but it serves as a significant pool in the body because of its large mass. Concentrations of various carotenoids were determined in middle-aged and elderly men and women from eight European countries and Israel. Overall, adipose tissue carotenoid levels in men were 50–76% of those in women. Body mass index was inversely correlated with lycopene, and waist circumference was inversely related to both β-carotene and lycopene adipose tissue content.[74] The older subjects had more β-carotene and the younger more lycopene. Alcohol consumption was inversely associated with β-carotene levels in adipose tissue, but the correlation was positive for lycopene. In another study, β-carotene and lycopene were found to be the predominant carotenoids in abdominal adipose tissues of adults undergoing corrective surgery,[75] and the total carotenoid concentration showed a great (40-fold) variability between individuals.

Few studies have investigated the usefulness of adipose tissue carotenoid concentrations as estimates of usual dietary intake. In one study, although plasma concentrations of β-carotene were not correlated with dietary intake, adipose tissue levels were shown to increase 6-fold after 6 months of daily supplementation with 30 mg β-carotene and were well correlated with dietary intake after 4 months.[76] However, taking into account the interindividual variations in absorption of carotenoids, in this study both plasma and adipose tissue levels were considered to be more useful as markers of body reserves than of dietary intake. In another study, the relative distribution of carotenoids in the plasma, rather than in adipose tissue, was found to be similar to their abundance in the diet, but the authors concluded that both adipose tissue and plasma are useful indicators of intake.[77]

Increases in carotenoid levels of buccal mucosal cells have been observed following supplementation with various sources such as oral β-carotene supplements, vegetable juice and tomato juice.[78,79] This is a relatively noninvasive method in which cells can be obtained easily by scraping the inside of the cheek with a soft toothbrush. However, it is unknown whether these cells reflect the carotenoid levels in other tissues.

Macular pigment in the eyes is measured by noninvasive optical techniques of heterochromatic flicker photometry,[80] reflectometry,[81] or Raman spectroscopy.[82] All these techniques assess total carotenoids in the macula. Flicker photometry is a subjective psychophysical test requiring active involvement of the subject. The reflectometry and Raman spectroscopy are more objective methods of evaluating the density of carotenoids in the macula. The macular carotenoid pigment declined with age in a large population study, which used Raman spectroscopy and was significantly lower in macular degeneration cases, unless the patients were supplemented with lutein.[82] Supplementation with lutein and zeaxanthin (2.4 – 30 mg/d) increased macular pigment (measured by flicker-photometry) in a dose-dependent manner and correlated with serum concentration of these carotenoids.[80] There are considerable differences between the subjects in their baseline macular pigment density and response to supplementation, but generally people with more body fat require longer supplementation at higher doses. A relatively small and short (five weeks) supplementation with 9 mg lutein/day failed to change both macular pigment density (assessed by reflectometry) and adipose tissue lutein concentration in young and old subjects, although their serum and buccal mucosa levels increased significantly.[81] In this study, young and old subjects did not show initial differences in their blood or tissue carotenoid levels.

Raman spectroscopy can also detect the total amount of carotenoids in skin[83] by a completely noninvasive and safe method, amounting to 20 seconds laser exposure of a very small surface. The results are very promising for evaluating quickly a large number of subjects, or different skin regions, with the highest carotenoid concentration found in the palm of the hand.

Carotenoid status of individuals is usually assessed by their measurement in serum or plasma, accompanied by the estimates of dietary intake. The dietary intake is calculated from questionnaires, interviews or diaries, in which people report their eating habits, frequency of consumption and portion sizes of common food items. The results depend on accuracy of subjective reporting and on inclusiveness of the interview. The best correlations with serum carotenoids are found in feeding trials, when subjects consume the exact diet prepared by a research facility. Tables 18.2 and 18.3 converge on lycopene as the most prominent carotenoid both in U.S. diet and serum of children [84] and adults [85] alike, while β-carotene and lutein with zeaxanthin are tied for second place. The calculated mean intakes of α-carotene and β-cryptoxanthin are too low to account for their serum concentrations in U.S. population, which attests to poor quality of database for these carotenoids, or the less likely possibility of better absorption from scarce supply. The ranges in both tables display a great variability among the surveyed population, with prevailing low intake and corresponding low serum levels. Interesting differences were found among the intake and carotenoids serum concentrations in major ethnic groups, with Mexican Americans having the highest serum levels of α-carotene, β-cryptoxanthin and lutein with zeaxanthin, due to much higher frequency of vegetable consumption than European Americans or African Americans.

B. Epidemiological Studies

Dietary intake and serum carotenoid concentrations in athletes have not been well documented. The carotenoid status of physically active subjects and some athletes has been described in a comprehensive review evaluating the role of carotenoids in sports nutrition.[12] Few studies demonstrated that exercise and physical activity may reduce serum carotenoid concentrations. In a recent study, 30 seconds of sprint anaerobic exercise (Wingate test) were found to significantly decrease plasma β-carotene levels in physical education students, which lead the authors to suggest that athletes may be protected against oxidative stress induced by exercise if baseline levels of antioxidants are maintained by supplementation.[86] In a group of rural Japanese men and women frequency of hard physical activity was negatively associated with serum β-carotene, especially in men.[87]

In general, many studies suggest that physically active individuals tend to consume more fruits and vegetables. In the Netherlands, women who were more active were found to consume greater amounts of fruits and vegetables compared to sedentary women and consequently had significantly higher β-carotene levels.[88] Physically active older U.S. male veterans, participating in the Golden Age Games were three times more active than their sedentary counterparts, had significantly higher β-carotene intake from diet, but not supplements and had almost twofold higher levels of serum β-carotene.[89,90] In a Spanish cohort of healthy volunteers, physical activity was associated with an increase in the intake by 15.9% in vegetables, 6.7% in fruit, 19.7% in total carotenoids, 40.1% in α-carotene, 20.4% in β-carotene, 11.2% in lycopene and 26.1% in lutein.[91] Among sportsmen, eating habits were found to be much better in former Finnish world class athletes, who were more likely to eat fruits and vegetables and avoid vitamin supplements compared to the non-competitive control group.[92] From these studies it appears that any type of active lifestyle in individuals of all ages is often associated with healthy eating habits, especially in older subjects.

Many young women and men, including athletes, are extremely weight conscious and some seek radical weight loss methods in order to maintain their physical appearance or to enter a sport competition with strict weight limits. In a group of adolescent athletes, 81% of the subjects were unhappy with their present weight and 73% of the subjects wanted to lose weight. This self-image seemed to influence the food practices of the young athletes more than their nutrition knowledge.[93] In a recent study, adolescents using moderate methods of weight control were found to adopt better eating and exercise behaviors than extreme dieters or nondieters.[94] Extreme dieters were less likely to eat fruits and vegetables than were moderate dieters. The direct effect of such food practices on carotenoid status in these young individuals is unknown; however, extreme dietary measures such as avoiding fruits and vegetables or fat, which helps with carotenoid absorption, may have severe effects on body carotenoid reserves.

C. Human Intervention Studies

Many recent studies have suggested that reactive chemical species called free radicals may be elevated during exercise due to increases in oxygen consumption. This has led to a growing interest in the use of antioxidant supplements for individuals who exercise regularly — especially athletes. Although carotenoids are considered to be promising antioxidants, data on serum or tissue concentrations of carotenoids following physical exercise are limited and the effect of single or a combination of carotenoids on physical performance has not been much investigated. The question arises whether carotenoids should be included in the daily supplementation plan of athletes as ergogenic aids, either to improve performance or reduce exercise-induced damage.

Among all the carotenoids, β-carotene has been most extensively investigated for its ability to reduce exercise-induced lipid peroxidation, although usually in combination with other antioxidants, such as vitamin E and C. The existing data on antioxidant effects in athletes are ambiguous. In one study, daily supplementation of 592 mg of α-tocopherol equivalents, 1000 mg ascorbic acid and 30 mg of β-carotene in young healthy males lowered breath pentane and serum malondialdehyde levels, at rest and after exercise, but exercise-induced rise of these oxidation markers was not prevented. The exercise regimen consisted of 30 min treadmill running at 60% maximal oxygen consumption, followed by 5 min running at pace.[95] In another study, the urinary output of oxidatively damaged nucleic acid base, 8-OHdG, after 3 consecutive days of submaximal exercise in moderately trained subjects was not found to be different with or without supplementation with 553 mg of α-tocopherol, 1000 mg vitamin C and 10 mg β-carotene.[96] However, an antioxidant mixture of similar composition, with slightly higher levels of vitamin E (800 mg) was able to increase the antioxidant potential of the blood glutathione system, along with significant reductions in the indicators of muscle damage, including lactate dehydrogenase and creatine phosphokinase.[97] On the other hand, supplementation with 1000 mg vitamin E, 1250 mg vitamin C and 37.5 mg β-carotene was ineffective in protecting against exercise-induced lipid peroxidation or muscle damage.[98] In basketball players, 35 days of antioxidant treatment composed of 600 mg α-tocopherol, 1000 mg ascorbic acid and 32 mg β-carotene decreased plasma lipid peroxide levels by 28%.[99] Moreover, a significant decrease in lactate dehydrogenase at 24 hours after training was also observed, suggesting a beneficial effect of antioxidants against exercise-induced muscle damage.

In a recent study, daily supplementation with vitamin E (500 mg) and β-carotene (30 mg) for 90 days, along with vitamin C (1000 mg) during the last 15 days, enhanced the antioxidant enzyme activity of superoxide dismutase and catalase in neutrophils of well-trained sportsmen.[100] A tenfold increase in plasma β-carotene was observed in these men compared with the placebo group. In light of the negative effects of high-intensity training on the functionality of the immune system in sportsmen, this study suggests that β-carotene in an antioxidant mixture is able to maintain neutrophil function, possibly by protecting against auto-oxidation by free radical production.

Antioxidants have also been shown to protect lung function. Supplementation with 75 mg vitamin E, 15 mg β-carotene and 650 mg vitamin C provided daily for 3 months in amateur cyclists from the Netherlands was able to protect lung function against the acute effects of ozone.[101] It is postulated that antioxidants can modulate the reactivity of oxidants such as ozone before they cause injury to the lung tissue. In another study, supplementation with 30 mg of lycopene for 7 days protected 55% of patients against exercise-induced asthma.[102] Lycopene was supplemented between the exercise settings, which included a 7 min run on a motorized Quinton treadmill, followed by 8 min rest. Although the supplemental source of lycopene provided in this study also included small amounts of other antioxidant compounds (6% lycopene, 1.6% tocopherols, 1% phytoene and phytofluene, 0.25% β-carotene and other phytochemicals extracted from tomatoes), the patients showed a significant rise only in lycopene serum levels.

A group of U.S. Marines undergoing cold-weather field training received a complex mixture of antioxidants that included 12 mg β-carotene, 330 mg vitamin C, 650 IU of tocopherols ($\alpha, \beta, \gamma, \delta$), 167 µg selenium, 13.2 mg catechin from green tea, 0.5 mg lutein, 0.1 mg lycopene, 181 mg N-acetyl L-cysteine, 5 mg pomegranate extract and 100 mg vegetable concentrate, which supplied unspecified amounts of carotenoids.[103] Two such doses were taken daily, but failed to reduce an increased level of oxidative stress, compared with placebo. The 24-day training in the cold environment was quite strenuous; the men lost an average 6 kg body weight and their breath pentane, serum lipid peroxides and urine 8-OHdG increased significantly.

The described studies do not provide clear evidence of a beneficial role for antioxidants in exercise-induced lipid peroxidation, nor in their ability to improve physical performance, at least when the subjects were not carotenoid deficient. However, the lack of consistency could be related to differences in the experimental protocols of these studies, such as intensity or duration of the exercise, sampling time, baseline antioxidant status and training level of subjects. More systematic and well controlled investigations need to be undertaken with carotenoids as the single supplements before they can be prescribed as ergogenic aids.

V. Recommended Intake

Optimal nutrition enhances athletic performance and recovery from exercise, according to the position statement of the American Dietetic Association, Dietitians of Canada and American College of Sports Medicine (2000).[104] Fruits and vegetables are a necessary part of optimal nutrition and athletes should consume 5–9 servings daily, not counting white potatoes. Only 20% of Americans meet the minimum requirement of five servings, and many of them are older people (over 65) who consume traditional home-prepared meals. The great majority (89%) of teenage girls eat fewer fruits and vegetables.[105] Since so many epidemiological studies indicate better health and longevity for people consuming plenty of fruits and vegetables, athletes should be very concerned about including large amounts of produce in their diet. A wide variety of fruits and vegetables will provide the most benefit, and appetizing colors of red, orange, yellow and dark green make the choice more enjoyable. Cooking with a healthy amount of fat (at least 15% of total energy intake) increases both palatability and bioavailability of nutrients in plants, especially carotenoids. Carotenoids are among the best markers of fruit and vegetable intake, because they can be relatively easily identified in serum and tissues, but plants are a source of many other phytochemicals that may be beneficial for human health. Many of them are still not identified or not understood, because of the complexity of their interactions in the human body. The potential synergistic effects that can be produced by a combination of all these components in fruits and vegetables may surpass the benefit that can be obtained from a single compound in supplements. Therefore, the safest course for optimal health, a necessary prerogative for superb and sustained athletic performance, is a complex diet of whole foods.[106]

It must be emphasized that a panel of experts reevaluating dietary reference intakes of vitamins and minerals[9,15] recently decided that no EAR, RDA or UL for carotenoids could be established or even proposed at this time. The panel firmly supported the recommendation for consumption of five or more servings of fruits and vegetables per day, but concluded that β-carotene supplements are not advisable for the general population.

Although safety studies conducted with synthetic β-carotene and lycopene failed to discover any adverse effects in animals,[107] there are inherent limitations in extending these conclusions to human populations. Laboratory animals (rodents, dogs) do not absorb carotenoids like humans, and extremely high doses must be used to produce serum and tissue levels comparable to normal human values. Animals also do not engage in harmful practices like smoking and drinking excessive amounts of alcohol, but their tolerance for β-carotene was put to the test when researchers exposed them to tobacco smoke and alcohol in their available water supply. β-carotene beadlets in the diet increased alcohol-induced liver damage in rats,[108] and high doses of β-carotene supplementation produced pathological changes in the lungs of smoke-exposed ferrets.[109] Two large trials of high β-carotene supplement (20–30 mg/day for 4–6 years) reported an increase in lung cancer among cigarette smokers in Finland and among smokers and asbestos-exposed workers in U.S., especially those who also drank alcohol.[110, 111] Similar trials of antioxidants for the prevention of colorectal adenomas (polyps) found that β-carotene supplements slightly increased the risk of polyp recurrence in smokers or alcohol drinkers, but in supplemented subjects who indulged on both counts, the risk was doubled.[112] In participants who neither smoked nor drank alcohol, β-carotene supplements markedly reduced recurrence of colorectal adenoma. It is tempting to conclude that β-carotene supplements are probably harmless for healthy individuals whose lifestyle excludes smoking and drinking, as all competitive athletes are advised to embrace. However, carotenoid supplements may instill a false sense of security and induce neglect of a nutritious diet rich in fruits and vegetables.

Excessive intake of a single carotenoid may cause crystal-like deposits to accumulate in certain tissues. Canthaxanthin was used as a photoprotector of skin and a "tanning pill," till golden yellow crystals were observed in the retina of subjects taking more than 30 mg per day. The same deposits were observed in macaque monkeys treated with canthaxanthin supplements[113] for 30 months. Although the presence of canthaxanthin crystals did not seem to affect visual function, the "tanning pills" were banned. A recent class-action lawsuit accused supermarkets of not informing the customers about the presence of "artificial coloring," i.e., synthetic canthaxanthin, in farmed salmon,[114] although wild salmon contains both canthaxanthin and astaxanthin from marine algae and krill in the diet. In case of lycopene, liver deposits of pigment were found to develop and were blamed for inflicting considerable abdominal pain.[44,45] These symptoms were caused not by lycopene supplements, but by excessive intake of tomato products (2 liters of tomato juice daily, or 4–5 big tomatoes and pasta with tomato sauce each day for several years) and could have been due to other phytochemicals in tomatoes. It is very dangerous to idolize any particular nutrient or food and consume it in excessive quantities on a rigid daily schedule.

A reasonable balance of dietary lycopene, lutein and β-carotene, about 6 mg of each per day, seems to be optimal for maintaining health and avoiding chronic diseases.[15] Comparing this goal with mean intake in the U.S. population (Table 18.2), it is obvious that most people should greatly increase their consumption of colorful fruits and vegetables, since less than 5% of the interviewed subjects had enough β-carotene and lutein in their diet. We are much more enthusiastic about lycopene, because more than 25% of the population exceed 9 mg lycopene/day and 10% consume more than 20 mg/day,[15] due to a ubiquitous preference for tomato products in mixed dishes and condiments.

VII. Summary

The most unassailable function of some carotenoids is their role as precursors of vitamin A, but it is doubtful if all humans are able to produce enough vitamin A from dietary carotenoids to sustain optimal health and athletic performance. The low conversion rate of carotenoids from fruits and vegetables may necessitate supplementation with more bioavailable β-carotene in oil (palm oil, *gac* fruit) or synthetic preparations for avowed vegans. To ensure optimal intake of vitamin A, it should be ingested as retinyl esters from animal sources or fortified food products according to the latest RDA (700–900 µg RAE/day), if there is a suspicion of vitamin A deficiency. Most athletes have ample stores of vitamin A in the liver, and recommended high intake of provitamin A carotenoids (6 mg/day) could provide up to half of their requirement for vitamin A.

Less established is the action of carotenoids as antioxidants *in vivo*, although they seem to be helpful in many conditions associated with increased oxidative stress, which definitely lower the body stores of carotenoids. There are reports of carotenoids decreasing oxidative damage to DNA, lipids and proteins, but many interventions failed to see any positive changes. Most convincing is the increase in macular pigment caused by intake of lutein and zeaxanthin, which seems to protect the retina from light damage. Very promising are investigations of the role of carotenoids in maintaining cell communication and in gene regulation.

Can carotenoids serve as ergogenic aids apart from their provitamin A role? Many athletes are exposed to high levels of oxidative stress due to increase in oxygen consumption during intense physical training or competition. High intake of carotenoids may reduce the resulting damage, hasten the recovery, possibly ameliorate chronic inflammation associated with sport injuries and stimulate the immune response. In outdoor sports, carotenoids may protect the skin from sunburn and the eyes from light damage, ensure better acuity of vision and reduce glare. All these effects add up to improved condition of the athlete, general good health, stamina and the length of an individual's sport career, extending the ability to enjoy strenuous activities into the advanced age by preventing chronic diseases. Carotenoids serve as guides to a healthful and varied diet, because a choice of colorful fruits, vegetables and animal products will provide optimal nutrition, with other phytochemicals acting in synergistic beneficial ways. Reduction of this natural bounty to single carotenoid supplements is not advisable, since large doses may be harmful in the long range, especially in habitual smokers and drinkers.

To provide sound advice for athletes and their coaches, future research must solve numerous riddles and inconsistencies contained in enormous amounts of data about the health effects of carotenoids. Evaluation of athletic performance in conjunction with a carotenoid-deficient diet, single and combined carotenoid supplements, as well as controlled fruit- and vegetable-containing diets, is most essential to test putative benefits and to establish optimal intake. Adequate biomarkers of oxidative damage must be developed to accompany these intervention trials. The carotenoid status of athletes should be investigated, preferably by objective measures of serum concentrations to correlate parameters of intervention and in observational studies of athletes from various populations. The final word may belong to the fast-growing science of genetics, which could explain individual differences in utilization of carotenoids, whether due to mutations or altered gene expression.

References

1. Britton, G., Liaaen-Jensen, S. and Pfander, H., Isolation and analysis, in *Carotenoids*, Vol. 1 A, Birkhäuser-Verlag, Basel, 1995.
2. Demmig-Adams, B., Gilmore, A. M. and Adams, W. W., In vivo functions of carotenoids in higher plants, *FASEB J.*, 10, 403, 1996.
3. Cunningham, F. X. Jr., Regulation of carotenoid synthesis and accumulation in plants, *Pure Appl. Chem.*, 74, 1409, 2002.
4. Bowen, P., Chen, L., Stacewicz-Sapuntzakis, M., Duncan, C., Sharifi, R., Ghosh, L., Kim, H-S., Christov-Tzelkov, K. and van Breemen, R., Tomato sauce supplementation and prostate cancer: Lycopene accumulation and modulation of biomarkers of carcinogenesis, *Exp. Biol. Med.*, 227, 886, 2002.
5. Edwards, A. J., Vinyard, B. T., Wiley, E. R., Brown, E. D., Collins, J. K., Perkins-Veazie, P., Baker, R. A. and Clevidence, B. A., Consumption of watermelon juice increases plasma concentrations of lycopene and β-carotene in humans, *J. Nutr.*, 133, 1043, 2003.
6. Wingerath, T., Stahl, W. and Sies, H., β-Cryptoxanthin selectively increases in human chylomicrons upon ingestion of tangerine concentrate rich in β-cryptoxanthin esters, *Arch. Biochem. Biophys.*, 324, 384, 1995.
7. Heinonen, M., Valsta, L., Anttolainen, M., Ovaskainen, M-L., Hyvönen, L. and Mutanen, M., Comparisons between analyzed and calculated food composition data: carotenoids, retinoids, tocopherols, tocotrienols, fat, fatty acids and sterols, *J. Food Compos. Anal.*, 10, 3, 1997.
8. van Lieshout, M., West, C. E., Muhilal, Permaesih, D., Wang, Y., Xu, X., van Breemen, R. B., Creemers, A. F. L., Verhoeven, M.A. and Lugtenburg, J., Bioefficacy of β-carotene dissolved in oil studied in children in Indonesia, *Am. J. Clin. Nutr.*, 73, 949, 2001.
9. Food and Nutrition Board, Institute of Medicine, *Dietary Reference Intakes for Vitamin A, Vitamin K, Arsenic, Boron, Chromium, Copper, Iodine, Iron, Manganese, Molybdenum, Nickel, Silicon, Vanadium and Zinc*, National Academy Press, Washington D.C., 2001.
10. National Research Council, *Recommended Dietary Allowances*, National Academy Press, Washington D.C., 1989.
11. Underwood, B., Vitamin A in animal and human nutrition, in *The Retinoids*, Sporn, M.B., Roberts, A.B. and Goodman, D.S., Eds., New York Academic Press, 1984, 281.
12. Stacewicz-Sapuntzakis, M., Vitamin A and carotenoids, in *Sports Nutrition: Vitamins and Trace Elements*, Wolinsky, I. and Driskell, J.A., Eds., Boca Raton, FL, CRC Press, 1997,101.
13. Solomons, N. W., Carotenes as dietary precursors of vitamin A: Their past and their future, *Sight and Life Newsletter*, 2002/3, 87, 2002.
14. Bowen, P. E., Mobarhan, S. and Smith, J. C. Jr., Carotenoid absorption in humans, *Methods Enzymol.*, 214B, 3, 1993.
15. Food and Nutrition Board, Institute of Medicine, *Dietary Reference Intakes for Vitamin C, Vitamin E, Selenium and Carotenoids*, National Academy Press, Washington D.C, 2000, 325.
16. Leung, A. K. C., Carotenemia, *Adv. Pediatr.*, 34, 223, 1987.

17. Burton, G. W. and Ingold, K. U., β-Carotene: An unusual type of lipid antioxidant, *Science*, 224, 569, 1984.
18. Borel, P., Grolier, P., Boirie, Y., Simonet, L., Verdier, E., Rochette, Y., Alexandre-Gouabau, M-C., Beaufrere, B., Lairon, D. and Azais-Braesco, V., Oxidative stress status and antioxidant status are apparently not related to carotenoid status in healthy subjects, *J. Lab. Clin. Med.*, 132, 61, 1998.
19. Gottlieb, K., Zarling, E. J., Mobarhan, S., Bowen, P. and Sugerman, S., Beta-carotene decreases markers of lipid peroxidation in healthy volunteers, *Nutr. Cancer*, 19, 207, 1993.
20. Dixon, Z. R., Burri, B. J., Clifford, A., Frankel, E. N., Schneeman, B. O., Parks, E., Keim, N. L., Barbieri, T., Wu, M., Fong, A. K., Kretsch, M. J., Sowell, A. L. and Erdman, J. W. Jr., Effects of a carotene-deficient diet on measures of oxidative susceptibility and superoxide dismutase activity in adult women, *Free Radic. Biol. Med.*, 17, 544, 1994.
21. Facchini, F. S., Humphreys, M. H., DoNascimento, C. A., Abbasi, F. and Reaven, G. M., Relation between insulin resistance and plasma concentration of lipid hydroperoxides, carotenoids and tocopherols, *Am. J. Clin. Nutr.*, 72, 776, 2000.
22. Marangon, K., Herbeth, B., Lecomte, E., Paul-Dauphin, A., Grolier, P., Chancerelle, Y., Artur, Y. and Siest, G., Diet, antioxidant status and smoking habits in French men, *Am. J. Clin. Nutr.*, 67, 231, 1998.
23. Farchi, S., Forastiere, F., Pistelli, R., Baldacci, S., Simoni, M., Perucci, C. A. and Viegi, G., Exposure to environmental tobacco smoke is associated with lower plasma β-carotene levels among nonsmoking women married to a smoker, *Cancer Epidemiol. Biomarkers Prev.*, 10, 907, 2001.
24. Polidori, M. C., Stahl, W., Eichler, O., Niestroj, J. and Sies, H., Profiles of antioxidants in human plasma, *Free Radic. Biol. Med.*, 30, 456, 2001.
25. Kontush, A., Weber, W. and Beisiegel, U., Alpha- and β-carotenes in low-density lipoprotein are the preferred target for nitric oxide-induced oxidation, *Atherosclerosis*, 148, 87, 2000.
26. Erlinger, T. P., Guallar, E., Miller III, E. R., Stolzenberg-Solomon, R. and Appel, L. J., Relationship between systemic markers of inflammation and serum β-carotene levels, *Arch. Intern. Med.*, 161, 1903, 2001.
27. Schünemann, H. J., Grant, B. J. B., Freudenheim, J. L., Muti, P., Browne, R. W., Drake, J. A., Klocke, R. A. and Trevisan, M., The relation of serum levels of antioxidant vitamins C and E, retinol and carotenoids with pulmonary function in the general population, *Am. J. Respir. Crit. Care Med.*, 163, 1246, 2001.
28. Santos, M. S., Gaziano, J. M., Leka, L. S., Beharka, A. A., Hennekens, C. H. and Meydani, S. N., β-Carotene-induced enhancement of natural killer cells activity in elderly men: An investigation of the role of cytokines, *Am. J. Clin. Nutr.*, 68, 164, 1998.
29. Toniolo, P., Van Kappel, A. L., Akhmedkhanov, A., Ferrari, P., Kato, I., Shore, R. E. and Riboli, E., Serum carotenoids and breast cancer, *Am. J. Epidemiol.*, 153, 1148, 2001.
30. Omenn, G. S., Chemoprevention of lung cancer: The rise and fall of beta-carotene, *Ann. Rev. Public Health*, 19, 73, 1998.
31. Heinrich, U., Gärtner, C., Wiebusch, M., Eichler, O., Sies, H., Tronnier, H. and Stahl, W., Supplementation with β-carotene or similar amount of mixed carotenoids protects humans from UV-induced erythema, *J. Nutr.*, 133, 98, 2003.
32. Mathews-Roth, M. M., Carotenoids in erythropoietic protoporphyria and other photosensitivity diseases, *Annals N. Y. Acad. Sci.*, 691, 127, 1993.

33. Bowen, P. E., Herbst-Espinosa, S. M., Hussain, E. A. and Stacewicz-Sapuntzakis, M., Esterification does not impair lutein bioavailability in humans, *J. Nutr.*, 132, 3668, 2002.
34. Handelman, G. J., Nightingale, Z. D., Lichtenstein, A. H., Schaefer, E. J. and Blumberg, J. B., Lutein and zeaxanthin concentrations in plasma after dietary supplementation with egg yolk, *Am. J. Clin. Nutr.*, 70, 247, 1999.
35. Forman, M. R., Beecher, G. R., Muesing, R., Lanza, E., Olson, B., Campbell, W. S., McAdam, P., Raymond, E., Schulman, J. D. and Graubard, B. I., The fluctuation of plasma carotenoid concentrations by phase of the menstrual cycle: A controlled diet study, *Am. J. Clin. Nutr.*, 64, 559, 1996.
36. Heber, D., Plasma carotenoids and the menstrual cycle, *Am. J. Clin. Nutr.*, 64, 640, 1996.
37. Khachik, F., Beecher, G. R. and Smith, J. C., Lutein, lycopene and their oxidative metabolites in chemoprevention of cancer, *J. Cell Biochem.*, 236, 1995.
38. Le Marchand, L., Hankin, J. H., Kolonel, L. N., Beecher, G. R., Wilkens, L. R. and Zhao, L. P., Intake of specific carotenoids and lung cancer risk, *Cancer Epidemiol. Biomarkers Prev.*, 2, 183, 1993.
39. Hammond, B. R. Jr., Wooten, B. R. and Curran-Celentano, J., Carotenoids in the retina and lens: Possible acute and chronic effects on human visual performance, *Arch. Biochem. Biophys.*, 385, 41, 2001.
40. Hammond, B. R. Jr., Johnson, E. J., Russell, R. M., Krinsky, N. I., Yeum, K-J., Edwards, R. B. and Snodderly, D. M., Dietary modification of human macular pigment density, *Invest. Opthalmol. Vis. Sci.*, 38, 1795, 1997.
41. Eye Disease Case-Control Study Group, Antioxidant status and neovascular age-related macular degeneration, *Arch. Opthalmol.*, 111, 104, 1993.
42. Eye Disease Case-Control Study Group, Dietary carotenoids, vitamins A, C and E and advanced age-related macular degeneration, *J. Amer. Med. Assoc.*, 272, 1413, 1994.
43. Nguyen, M. L. and Schwartz, S. J., Lycopene: Chemical and biological properties, *Food Technol.*, 53, 38, 1999.
44. Reich, P., Shwachman, H. and Craig, J. M., Lycopenemia. A variant of carotenemia, *New Eng. J. Med.*, 262, 263, 1960.
45. La Placa, M., Pazzaglia, M. and Tosti, A., Lycopenaemia, *J. Eur. Acad. Dermatol. Venerol.*, 14, 311, 2000.
46. Di Mascio, P., Kaiser, S. and Sies, H., Lycopene as the most efficient biological carotenoid singlet oxygen quencher, *Arch. Biochem. Biophys.*, 274, 532, 1989.
47. Ribaya-Mercado, J. D., Garmyn, M., Gilchert, B. A. and Russell, R. M., Skin lycopene is destroyed preferentially over β-carotene during ultraviolet irradiation in humans, *J. Nutr.*, 125, 1854, 1995.
48. Bub, A., Watzl, B., Abrahamse, L., Delincée, H., Adam, S., Wever, J., Müller, H. and Rechkemmer, G., Moderate intervention with carotenoid-rich vegetable products reduces lipid peroxidation in men, *J. Nutr.*, 130, 2200, 2000.
49. Maruyama, C., Imamura, K., Oshima, S., Suzukawa, M., Egami, S., Tonomoto, M., Baba, N., Harada, M., Ayaori, M., Inakuma, T. and Ishikawa, T., Effects of tomato juice consumption on plasma and lipoprotein carotenoid concentrations and the susceptibility of low density lipoprotein to oxidative modification, *J. Nutr. Sci. Vitaminol.*, 47, 213, 2001.

50. Kohlmeier, L., Clark, J. D., Gomez-Garcia, E., Martin, B. C., Steck, S. E., Kardinaal, A. F., Ringstad, J., Thamm, M., Masaev, V., Riemersma, R., Martin-Moreno, J. M., Huttunen, J. K. and Kok, F. J., Lycopene and myocardial infarction risk in the EURAMIC study, *Am. J. Epidemiol.*, 146, 618, 1997.
51. Rissanen, T. H., Voutilainen, S., Nyyssönen, K., Lakka, T. A., Sivenius, J., Salonen, R., Kaplan, G. A. and Salonen, J. T., Low serum lycopene concentration is associated with an excess incidence of acute coronary events and stroke: The Kuopio Ischaemic Heart Disease Risk Factor Study, *Brit. J. Nutr.*, 85, 749, 2001.
52. Rissanen, T. H., Voutilainen, S., Nyyssönen, K., Salonen, R., Kaplan, G. A. and Salonen, J. T., Serum lycopene concentrations and carotid atherosclerosis: the Kuopio Ischaemic Heart Disease Risk Factor Study, *Am. J. Clin. Nutr.*, 77, 133, 2003.
53. Riso, P., Pinder, A., Santangelo, A. and Porrini, M., Does tomato consumption effectively increase the resistance of lymphocyte DNA to oxidative damage? *Am. J. Clin. Nutr.*, 69, 712, 1999.
54. Porrini, M. and Riso, P., Lymphocyte lycopene concentration and DNA protection from oxidative damage is increased in women after a short period of tomato consumption, *J. Nutr.*, 130, 189, 2000.
55. Hwang, E-S. and Bowen, P. E., Can consumption of tomatoes or lycopene reduce cancer risk? *Integrative Cancer Ther.*, 1, 121, 2002.
56. La Vecchia, C., Tomatoes, lycopene intake and digestive tract and female hormone-related neoplasms, *Exp. Biol. Med.*, 227, 860, 2002.
57. Giovannucci, E., Ascherio, A., Rimm, E. B., Stampfer, M. J., Colditz, G. A. and Willett, W. C., Intake of carotenoids and retinol in relation to risk of prostate cancer, *J. Natl Cancer Inst.*, 87, 1767, 1995.
58. Gann, P. H., Ma, J., Giovannucci, E., Willett, W., Sacks, F. M., Hennekens, C. H. and Stampfer, M. J., Lower prostate cancer risk in men with elevated plasma lycopene levels: Results of a prospective analysis, *Cancer Res.*, 59, 1225, 1999.
59. Chen, L., Stacewicz-Sapuntzakis, M., Duncan, C., Sharifi, R., Ghosh, L., van Breemen, R., Ashton, D. and Bowen, P. E., Oxidative DNA damage in prostate cancer patients consuming tomato sauce-based entrees as a whole-food intervention, *J. Natl Cancer Inst.*, 93, 1872, 2001.
60. Kimura, M., Rodriguez-Amaya, D. B. and Yokoyama, S. M., Cultivar differences and geographic effects on carotenoid composition and vitamin A value of papaya, *Lebens. Wissen. Technol.*, 24, 415, 1991.
61. Padula, M. and Rodriguez-Amaya, D. B., Characterization of carotenoids and assessment of the vitamin A value of Brazilian guavas (*Psidium guajava* L.), *Food Chem.*, 20, 11, 1986.
62. White, W. S., Stacewicz-Sapuntzakis, M., Erdman, J. W. Jr. and Bowen, P. E., Pharmacokinetics of β-carotene and canthaxanthin after ingestion of individual and combined doses by human subjects, *J. Am. Coll. Nutr.*, 13, 665, 1994.
63. Rodriguez-Amaya, D. B., Effect of processing and storage on food carotenoids, *Sight and Life Newsletter*, 2002/3, 25, 2002/3.
64. Stahl, W. and Sies, H., Uptake of lycopene and its geometrical isomers is greater from heat-processed than from unprocessed tomato juice in humans, *J. Nutr.*, 122, 2161, 1992.
65. Brown, M. J., Ferruzzi, M. G., Nguyen, M. L., Cooper, D. A., Eldridge, A. L. and Schwartz, S. J. and White, W. S., The bioavailability of carotenoids is higher in salads ingested with full-fat versus fat-reduced salad dressings as measured by using electrochemical detection. *Am. J. Clin. Nutr.*, 2004 (in press).

66. Furr, H. C. and Clark, R. M., Intestinal absorption and tissue distribution of carotenoids, *J. Nutr. Biochem.*, 8, 364, 1997.
67. Romanchik, J. E., Morel, D. W. and Harrison, E. H., Distribution of carotenoids and α-tocopherol among lipoproteins do not change when human plasma is incubated *in vitro*, *J. Nutr.*, 125, 2610, 1995.
68. Ernst, H., Recent advances in industrial carotenoid synthesis, *Pure Appl. Chem.*, 74, 1369, 2002.
69. Micozzi, M. S., Beecher, G. R., Taylor, P. R. and Khachik, F., Carotenoid analyses of selected raw and cooked foods associated with a lower risk for cancer, *J. Natl Cancer Inst.*, 82, 282, 1990.
70. Vuong, L. T., Dueker, S. R. and Murphy, S. P., Plasma β-carotene and retinol concentrations of children increase after a 30-d supplementation with the fruit *Momordica cochinchinensis* (gac), *Am. J. Clin. Nutr.*, 75, 872, 2002.
71. Stahl, W., Schwarz, W., Sundquist, A. R. and Sies, H., *Cis-trans* isomers of lycopene and beta-carotene in human serum and tissues, *Arch. Biochem. Biophys.*, 294, 173, 1992.
72. Kaplan, L. A., Lau, J. M. and Stein, E. A., Carotenoid composition, concentrations and relationships in various human organs, *Clin. Physiol. Biochem.*, 8, 1, 1990.
73. Schmitz, H. H., Poor, C. L., Wellman, R. B. and Erdman, J. W., Concentration of selected carotenoids and vitamin A in human liver, kidney and lung tissue, *J. Nutr.*, 121, 1613, 1991.
74. Virtanen, S. M., van't Veer, P., Kok, F., Kardinaal, A. F. and Aro, A., Predictors of adipose tissue carotenoid and retinol levels in nine countries. The EURAMIC study, *Am. J. Epidemiol.*, 144, 968, 1996.
75. Parker, R. S., Carotenoid and tocopherol composition of human adipose tissue, *Am. J. Clin. Nutr.*, 47, 33, 1988.
76. Kardinaal, A. F., van't Veer, P., Brants, H. A., van den Berg, H., van Schoonhoven, J. and Hermus, R. J., Relations between antioxidant vitamins in adipose tissue, plasma and diet, *Am. J. Epidemiol.*, 141, 440, 1995.
77. El-Sohemy, A., Baylin, A., Kabagambe, E., Ascherio, A., Spiegelman, D. and Campos, H., Individual carotenoid concentrations in adipose tissue and plasma as biomarkers of dietary intake, *Am. J. Clin. Nutr.*, 76, 172, 2002.
78. Peng, Y-S., Peng, Y-M., McGee, D. L. and Alberts, D. S., Carotenoids, tocopherols and retinoids in human buccal mucosal cells: Intra- and interindividual variability and storage stability, *Am. J. Clin. Nutr.*, 59, 636, 1994.
79. Paetau, I., Rao, D., Wiley, E. R., Brown, E. D. and Clevidence, B. A., Carotenoids in human buccal mucosa cells after 4 wk of supplementation with tomato juice or lycopene supplements, *Am. J. Clin. Nutr.*, 70, 490, 1999.
80. Bone, R. A., Landrum, J. T., Guerra, L. H. and Ruiz, C. A., Lutein and zeaxanthin dietary supplements raise macular pigment density and serum concentrations of these carotenoids in humans, *J. Nutr.*, 133, 992, 2003.
81. Cardinault, N., Gorrand, J., Tyssandier, V., Grolier, P., Rock, E. and Borel, P., Short-term supplementation with lutein affects biomarkers of lutein status similarly in young and elderly subjects, *Exp. Gerontol.*, 38, 573, 2003.
82. Bernstein, P. S., New insights into the role of the macular carotenoids in age-related macular degeneration. Resonance Raman studies, *Pure Appl. Chem.*, 74, 1419, 2002.
83. Hata, T. R., Scholz, T. A., Ermakov, I. V., McClane, R. W., Khachik, F., Gellermann, W. and Pershing, L. K., Non-invasive Raman spectroscopic detection of carotenoids in human skin, *J. Invest. Dermatol.*, 115, 441, 2000.

84. Ford, E. S., Gillespie, C., Ballew, C., Sowell, A. L. and Mannino, D. M., Serum carotenoid concentrations in U.S. children and adolescents, *Am. J. Clin. Nutr.*, 76, 818, 2002.
85. Ford, E. S., Variations in serum carotenoid concentrations among United States adults in ethnicity and sex, *Ethn. Dis.*, 10, 208, 2000.
86. Groussard, C., Machefer, G., Rannou, F., Faure, H., Zouhal, H., Sergent, O., Chevanne, M., Cillard, J. and Gratas-Delamarche, A., Physical fitness and plasma non-enzymatic antioxidant status at rest and after a Wingate test, *Can. J. Appl. Physiol.*, 28, 79, 2003.
87. Takatsuka, N., Kawakami, N., Ohwaki, A., Ito, Y., Matsushita, Y., Ido, M. and Shimizu, H., Frequent hard physical activity lowered serum beta-carotene level in a population study of rural city of Japan, *Tohoku J. Exp. Med.*, 176, 131, 1995.
88. Voorrips, L. E., van Staveren, W. A. and Hautvast, J. G. A. J., Are physically active elderly women in a better nutritional condition than their sedentary peers?, *Eur. J. Clin. Nutr.*, 45, 545, 1991.
89. Connor-Bote, E., Murphy, P. A., Orloff, S. B., Ottosen, W., Rothschild, R. L., Sullivan, J. M. and Iber, F. L., Dietary intake in active veterans participating in the Golden Age Games compared with healthy sedentary veterans, in *National Veterans Golden Age Games Research Monograph*, Langbein, W. E., Wyman D. J. and Osis A. Eds., Edward Hines, Jr., VA Hospital, Hines, IL, 1995, 11
90. Kazi, N., Murphy, P. A., Connor, E. S., Bowen, P. E., Stacewicz-Sapuntzakis, M. and Iber, F. L., Serum antioxidant and retinol levels in physically active versus physically inactive elderly veterans, in *National Veterans Golden Age Games Research Monograph*, Langbein, W. E., Wyman D. J. and Osis A. Eds., Edward Hines, Jr., VA Hospital, Hines, IL, 1995, 50
91. Tormo, M. J., Navarro, C., Chirlaque, M. D., Barber, X., Argilaga, S., Agudo, A., Amiano, P., Barricarte, A., Beguiristain, J. M., Dorronsoro, M., Gonzalez, C. A., Martinez, C., Quiros, J. R. and Rodriguez, M., Physical sports activity during leisure time and dietary intake of foods and nutrients in a large Spanish cohort, *Int. J. Sport Nutr. Exerc. Metab.*, 13, 47, 2003.
92. Fogelholm, M., Kaprio, J. and Sarna, S., Healthy lifestyles of former Finnish world class athletes, *Med. Sci. Sports Exerc.*, 26, 224, 1994
93. Perron, M. and Endres, J., Knowledge, attitudes and dietary practices of female athletes, *J. Am. Diet. Assoc.*, 85, 573, 1985.
94. Story, M., Neumark-Sztainer, D., Sherwood, N., Stang, J. and Murray, D., Dieting status and its relationship to eating and physical activity behaviors in a representative sample of U.S. adolescents, *J. Am. Diet. Assoc.*, 99, 410, 1998.
95. Kanter, M. M., Nolte, L. A. and Holloszy, J. O., Effects of an antioxidant vitamin mixture on lipid peroxidation at rest and postexercise, *J. Appl. Physiol.*, 74, 965, 1993.
96. Witt, E. H., Reznick, A. Z., Viguie, C. A., Starke-Reed, P. and Packer, L., Exercise, oxidative damage and effects of antioxidant manipulation, *J. Nutr.*, 122, 766, 1992.
97. Viguie, C. A., Packer, L. and Brooks, G. A., Antioxidant supplementation affects indices of muscle trauma and oxidant stress in human blood during exercise, *Med. Sci. Sports Exerc.*, 21, S16, 1989.

98. Kanter, M. M. and Eddy, D. E., Effect of antioxidant supplementation on serum markers of lipid peroxidation and skeletal muscle damage following eccentric exercise, *Med. Sci. Sports Exerc.*, 24, S17, 1992.
99. Schroder, H., Navarro, E., Mora, J., Galiano, D. and Tramullas, A., Effects of alpha-tocopherol, beta-carotene and ascorbic acid on oxidative, hormonal and enzymatic exercise stress marker in habitual training activity of professional basketball players, *Eur. J. Nutr.*, 40, 178, 2001.
100. Tauler, P., Aguilo, A., Fuentespina, E. and Tur, J. A., Diet supplementation with vitamin E, vitamin C and β-carotene cocktail enhances basal neutrophil antioxidant enzymes in athletes, *Eur. J. Physiol.*, 443, 791, 2002.
101. Grievink, L., Jansen, S. M. A., van't Veer, P. and Brunekreef, B., Acute effects of ozone on pulmonary function of cyclists receiving antioxidant supplements, *Occup. Environ. Med.*, 55, 13, 1998.
102. Nahum, N. H. and Ben-Amotz, A., Reduction of exercise-induced asthma oxidative stress by lycopene, a natural antioxidant, *Allergy*, 55, 1184, 2000.
103. Schmidt, M. C., Askew, E. W., Roberts, D. E., Prior, R. L., Ensign, W. Y. Jr. and Hesslink, R. E., Oxidative stress in humans training in a cold, moderate altitude environment and their response to a phytochemical antioxidant supplement, *Wilderness Environ. Med.*, 13, 94, 2002.
104. Manore, M. M., Barr, S. I. and Butterfield, G. E., Position of the American Dietetic Association, Dietitians of Canada and the American College of Sports Medicine: nutrition and athletic performance, *J. Am. Diet. Assoc.*, 100, 1543, 2000.
105. Food Institute Report, Few American families eat recommended five daily fruit and vegetable servings, 75, 4, 2003.
106. Messina, M., Lampe, J. W., Birt, D. F., Appel, L. J., Pivonka, E., Berry, B. and Jacobs, D. R. Jr., Reductionism and the narrowing nutrition perspective: time for reevaluation and emphasis on food synergy, *J. Am. Diet. Assoc.*, 101, 1416, 2001.
107. McClain, R. M. and Bausch, J., Summary of safety studies conducted with synthetic lycopene, *Regulat. Toxicol. Pharmacol.*, 37, 274, 2003.
108. Leo, M. A., Aleynik, S. I., Aleynik, M. K. and Lieber, C. S., β-Carotene beadlets potentiate hepatotoxicity of alcohol, *Am. J. Clin. Nutr.*, 66, 1461, 1997.
109. Liu, C., Wang, X-D., Bronson, R. T., Smith, D. E., Krinsky, N. I. and Russell, R. M., Effect of physiological versus pharmacological β-carotene supplementation on cell proliferation and histopathological changes in the lungs of cigarette smoke-exposed ferrets, *Carcinogenesis*, 21, 2245, 2000.
110. Albanes, D., Heinonen, O. P., Taylor, P. R., Virtamo, J., Edwards, B. K., Rautalahti, M., Hartman, A. M., Palmgren, J., Freedman, L. S., Haapokaski, J., Barrett, M. J., Pietinen, P., Malila, N., Tala, E., Liippo, K., Salomaa, E. R., Tangrea, J. A., Teppo, L., Askin, F. B., Taskinen, E., Erozan, Y., Greenwald, P. and Huttunen, J. K., α-Tocopherol and β-carotene supplements and lung cancer incidence in the Alpha-Tocopherol Beta-Carotene Prevention Study: effects of baseline characteristics and study compliance, *J. Natl. Cancer Inst.*, 88, 1560, 1996.
111. Omenn, G. S., Goodman, G. E., Thornquist, M. D., Balmes, J., Cullen, M. R., Glass, A., Keogh, J. P. Jr., Meyskens, F. L. Jr., Valanis, B., Williams, J. H. Jr., Barnhart, S., Cherniack, M. G., Brodkin, C. A. and Hammar, S., Risk factors for lung cancer and for intervention effects in CARET, the Beta-carotene and Retinol Efficacy Trial, *J. Natl. Cancer Inst.*, 88, 1550, 1996.

112. Baron, J. A., Cole, B. F., Mott, L., Haile, R., Grau, M., Church, T. R., Beck, G. J. and Greenberg, E. R., Neoplastic and antineoplastic effects of β-carotene on colorectal adenoma recurrence: results of a randomized trial, *J. Natl. Cancer Inst.*, 95, 717, 2003.
113. Goralczyk, R., Buser, S., Bausch, J., Bee, W., Zühlke, U. and Barker, F. M., Occurrence of birefringent retinal inclusions in cynomolgus monkeys after high doses of canthaxanthin, *Invest. Opthalmol. Vis. Sci.*, 38, 741, 1997.
114. Burros, M., Farmed salmon looking less rosy. *The New York Times*, May 28, 2003.

19

Coenzyme Q_{10}

Shi Zhou

CONTENTS

I. Introduction 355
II. Chemical Structure 357
III. Functions 357
   A. Energy Coupling in ATP Synthesis 357
   B. Antioxidant 358
   C. Other Functions 360
   D. Effects of Supplementation on Exercise Performance 360
IV. Body Reserves 364
V. Dietary and Supplemental Sources 365
VI. Interactions with Other Nutrients and Drugs 366
VII. Recommendations for Dietary Supplementation 368
VIII. Future Research Directions or Needs 370
IX. Summary 371
References 371

I. Introduction

Coenzyme Q_{10} (CoQ_{10}), also known as ubiquinone or ubidecarenone, is a vitamin-like, lipid-soluble compound existing in all cells. It was initially isolated from animal tissues in the mid-1950s and found to be capable of undergoing reversible oxidation and reduction.[1-3] This property is the key to the crucial roles of CoQ_{10} in the body, as a redox electron carrier in the mitochondria that is coupled to energy transfer and as an essential antioxidant.[4,5] It is also known that CoQ_{10} is involved in several other cellular functions— assisting in regeneration of other antioxidants, influencing stability, fluidity and permeability of membranes, stimulating cell growth and inhibiting cell death.[4,6] Approximately one half of CoQ_{10} is found in the mitochondria inner membrane and the remaining is distributed in the nucleus, microsomal and other components of the cell.[7]

Coenzyme Q_{10} can be synthesized in the cell, therefore it is not regarded as a *bona fide* vitamin (which must be obtained from food), although the term "vitamin Q" has been used in some publications with reference to the beneficial effects of dietary supplementation.[8] Increased metabolic demand and certain hormones may stimulate biosynthesis of CoQ_{10}.[8,9] However, because CoQ_{10} is localized in the central hydrophobic portion of the membrane, its concentration cannot increase freely within the membrane due to potential destabilization and physical limits of the bilayer.[9] Normally, CoQ_{10} is saturated in the membrane. However, decreased biosynthesis and increased degradation may cause deficiency.[4] Decreased CoQ_{10} content in tissues has been found in certain health disorders,[10-13] in the elderly[14] and in some athletes under high-intensity training.[12,15]

The effects of CoQ_{10} as a complementary therapeutic drug have been investigated since the mid-1960s. Supplementation with exogenous CoQ can effectively correct its deficiency in plasma and has shown therapeutic effects for various types of health disorders. For instance, CoQ_{10} has been used in treatment of cardiovascular disorders such as congestive heart failure, ischemic heart disease, diastolic dysfunction, cardiomyopathy, hypertension, ventricular arrhythmia and patients undergoing cardiovascular surgery. It has also be used in the treatment of cancer, neurodegenerative diseases such as Parkinson's and Huntington's, mitochondrial encephalomyopathies, endocrine and hormonal disorders such as diabetes, immune disorders including AIDS, metabolic and nutritional disorders, reproductive disorders, retinal degeneration, and muscular dystrophy. A recent search of the MEDLINE database (in May 2003) found close to 4000 publications on CoQ_{10}-related topics. A cumulative number of randomized, placebo-controlled trials have evaluated the roles of CoQ_{10} in health and disease. The outcomes of such research and clinical experience have been frequently summarized in scholarly reviews, for example, on CoQ_{10}'s biochemical function and synthesis,[4,8,9,16-21] its efficacy in treatment of cardiovascular disorders[7,22-26] and its roles in health and specific diseases.[3,5,27-30] A number of international symposia on CoQ_{10} have been held since the mid-1970s, including the recent conference of the International Coenzyme Q_{10} Association (London, November 2002).

In addition to the prescribed use for treatment of diseases, CoQ has been commercially marketed as a dietary supplement for improving general heart function, enhancing antioxidant capacity, delaying the aging process, enhancing the immune system and improving endurance and physical performance. Based on the current understanding of its function, particularly in energy coupling and as an antioxidant, and some evidence of deficiency during intensive training, the effects of CoQ supplementation in healthy and athletic populations have been examined in a relatively small number of studies, with expectations of improving physical performance,[31-37] reducing oxidative damage to tissues during intensive training[38-40] or facilitating

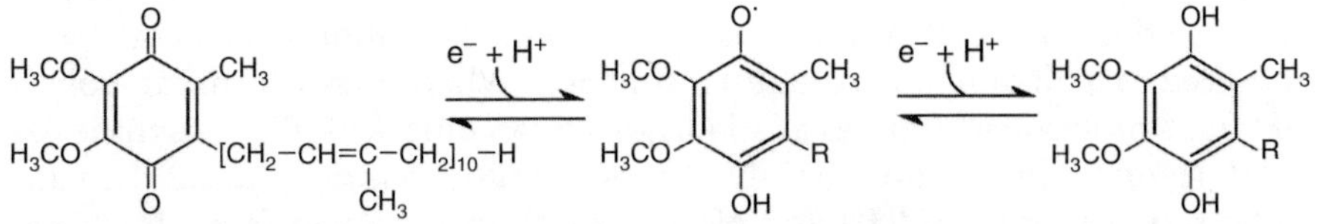

FIGURE 19.1
Coenzyme Q_{10} in oxidized form (ubiquinone, on the left), reduced form (ubiquinol, on the right) and intermediate form (semiquinone).

recovery process after exercise.[12,37] However, limited and conflicting evidence has been reported in the literature regarding these hypothesized benefits. In general, current evidence does not seem to support that supplementation of CoQ_{10} has a significant performance-enhancing effect in healthy, nondeficient individuals.

II. Chemical Structure

The primary structure of CoQ_{10} is a benzoquinone with a side chain that contains a number of isoprene units. Several homologues of CoQ are found in microorganisms, plants and animals. For example, the primary form found in rodents contains nine isoprene units (CoQ_9), and Q_6, Q_7 and Q_8 are found in yeast and bacteria. The form found in humans has 10 units of isoprene, which is why it is named coenzyme Q_{10}, or chemically defined as 2,3-dimethoxy-5-methyl-6-decaprenyl-1,4-benzoquinone.[3] In the body, CoQ_{10} may exist as oxidized (ubiquinone), semiquinone intermediate or reduced (ubiquinol) forms (Figure 19.1). Considering that it needs an easy reference to these forms and that CoQ_{10} has more functions than just being a coenzyme, in 1975 "ubiquinone" was recommended as the "official" name for this compound by the International Union of Pure and Applied Chemistry and International Union of Biochemistry and Molecular Biology (IUPAC-IUB) Commission on Biochemical Nomenclature. "Ubi-" also indicates the universal existence of this quinoid structure in all cells.[5,8]

III. Functions

A. Energy coupling in ATP synthesis

As a coenzyme, CoQ_{10} plays a crucial role in mitochondrial adenosine triphosphate (ATP) synthesis (oxidative phosphorylation). ATP is a universal energy "currency" required in many cellular processes, for instance, muscle contraction and active transport of substances across cell membrane. When ATP is broken down to adenosine diphosphate (ADP) and inorganic phosphate (Pi), energy is released from the phosphate bond and used by the cell. Because the storage of ATP in cell is very limited, continuous resynthesis is required to maintain normal cellular function. When energy from catabolism of "fuel" molecules becomes available, ADP can bind with Pi to resynthesize ATP. The ATP production mainly occurs in mitochondria. During the catabolic process, protons (H^+) and electrons (e^-) are removed from the "fuel" molecules via either the glycolysis pathway that is located in the cytosol, or the tricarboxylic acid (TCA) cycle pathway located in the central space of mitochondria (termed matrix). The protons and electrons are subsequently transported by coenzymes into the electron transport chain that resides on the mitochondrial inner membrane. The electron transport chain consists of a set of proton and electron carriers (coenzymes) including nicotinamide adenine dinucleotide (NAD), flavin adenine dinucleotide (FAD), CoQ_{10} and a number of cytochromes. Most of these coenzymes are associated with proteins (enzymes) on the mitochondrial inner membrane, forming complexes, whereas cytochrome c and CoQ are not combined with proteins, therefore regarded as mobile carriers.[41]

In these complexes the electrons carried by NADH (the reduced form of NAD) and $FADH_2$ (the reduced form of FAD) from Complexes I and II are passed onto CoQ_{10}, converting it from oxidized form to reduced form, which in turn passing the electrons to the cytochromes in Complex III while CoQ_{10} being oxidized again. The oxidized CoQ_{10} migrates in the membrane back to Complexes I and II to start the next "shuttle" movement. At the same time, protons are transported from the matrix side across the inner mitochondrial membrane. This movement of protons creates an electrochemical gradient as the space outside of the inner membrane having more protons and a positive electrical charge. When the protons moving back into the matrix, driven by the gradient, energy is released which is used for the synthesis of ATP. The protons and electrons combine with oxygen to form water at the end of the electron transport chain, to complete the process.[41] This aerobic energy transfer process will not work without an adequate level of CoQ_{10}.

With an understanding of CoQ_{10}'s role in oxidative phosphorylation, it is not surprising that a positive correlation is found between muscle CoQ_{10} content and percentage of slow-twitch oxidative muscle fiber, exercise capacity and/or marathon performance.[42]

B. Antioxidant

The reduced form of CoQ_{10} (ubiquinol) can act as a non-specific antioxidant to neutralize and scavenge free radicals.[4,6,9] Free radicals are atomic or molecular species containing an unpaired electron. They are generally highly reactive and tend to either lose or gain an electron. Free radicals can be produced in reactions initiated by drugs, alcohol and other foreign toxins, or during cellular respiratory processes.[43] Major types of free radicals and their derivatives include reactive oxygen species (ROS) and reactive nitrogen species. During the oxidative phosphorylation process as discussed previously, molecular oxygen meets protons and electrons to form water. Four electrons are required but oxygen can only receive one electron at a time. Addition of one, two, or three electrons to molecular oxygen leads to the transient production of free radicals of superoxide anion (O_2^-), hydrogen peroxide (H_2O_2) and hydroxyl radical (OH·), respectively.[43] The enzymatic reactions involved in the glycolysis, TCA and electron transport chain pathways can consume majority of the ROS. However, 2–5% of the total electron flux during the normal metabolism may leak out of the mitochondrion, via pathways involving semiquinone and NADH dehydrogenase, to generate free radicals.[7,43] There are several other potential sources of ROS involving the activities of xanthine oxidase and xanthine dehydrogenase, lipid peroxidase and hydroperoxidases and activated neutrophil, etc.[7]

At moderate concentrations, free radicals and their derivative species play an important role as regulatory mediators in signaling processes. However, at high concentrations, these species are hazardous for living organisms and damage all major cellular constituents.[44] Strenuous exercise may increase the production of free radicals, which have been implicated in the development of exercise-induced tissue damage.[43] To protect against such potentially damaging effects of free radicals, cells possess several antioxidants, including some enzymes. CoQ_{10} and vitamins E and C (their reduced forms) are among those regarded as antioxidant nutrients.[7,43,45] CoQ_{10} in its reduced form is the only known lipid-soluble antioxidant that is located in the cell membrane and for which the body has enzyme systems capable of regenerating the reduced form of CoQ_{10}.[24,46]

The cell membrane mainly consists of phospholipids and proteins. It is believed that ubiquinol plays an important role in protecting the membrane from lipid peroxidation (deterioration). It may act as either an independent antioxidant or a co-antioxidant with vitamin E and C.[5,20] It has been suggested that ubiquinol may prevent both the initiation and propagation of lipid peroxidation, possibly because of its lipidphilic property and location in the membrane that allows its access to the proton-motive Q cycle. Vitamin E acts exclusively in inhibition of the propagation of lipid peroxidation.[8] After quenching a free radical, CoQ_{10} can be recycled within plasma membranes and cytosol by quinone reductase.[5] It has also been shown that ubiquinol provides protection for proteins and DNA against oxidative damage.[8]

Ubiquinol assists in converting tocopheryl (oxidized form of vitamin E) radical back to tocopherol (reduced form of vitamin E).[4,47] Therefore, CoQ_{10} supplementation has a sparing effect on tocopherol.[48,49] The presence of CoQ_{10} is also required for regeneration of vitamin C (ascorbate) from ascorbate radical outside of the cell.[4] These antioxidants play important roles in dealing with oxidative stress.

Oxidative stress can be defined as an imbalance between oxidant production and the antioxidant capacity of the cell that is biased toward greater pro-oxidant activity.[5,50,51] Oxidative stress has been suggested to be a relevant factor in aging, as well as in different pathological conditions.[51] The ratio of ubiquinol to total ubiquinone content in plasma has been proposed as a marker of oxidative stress.[52] There has been evidence that this ratio decreases immediately after a VO_{2max} test and recovers within 30 minutes. It has been speculated that this change in antioxidant reflects its role in the elimination of exercise-derived free radicals.[53]

C. Other Functions

CoQ_{10} influences the stabilization, fluidity and permeability of the membrane.[9] The reduced form of CoQ_{10} is more hydrophilic, therefore the hydroxyl (OH) group can lie closer to the surface of the membrane. The change of its position with oxidation/reduction may modify structural or enzymatic properties of the membrane.[4]

It has also been proposed that CoQ_{10} may play a role in regulation of cell growth and death.[4] Walter et al.[21] reviewed the current studies on the role of CoQ and other quinone analogs as modulators of the permeability transition pore (PTP) on mitochondrial inner membrane. The PTP is a Ca^{2+}-sensitive channel that plays a key role (when it is open) in different models of cell death. CoQ_{10} at a certain concentration is one of the quinones that act as inhibitors of PTP. However, interestingly, at higher concentrations, with the coexistence of high levels of Ca^{2+}, CoQ_{10} behaves like an inducer rather than an inhibitor of PTP.

The regulatory role of CoQ in cell signaling and gene expression has drawn increased attention.[4] Hydrogen peroxide (H_2O_2) can be generated during autooxidation of semiquinone formed during electron transport activity in the membrane. Linnane et al.[54] reported that CoQ_{10} supplementation appeared to have a regulatory effect on global gene expression in skeletal muscles, proteins and muscle fiber types in relation to aging. The authors speculated that the regulation is achieved via superoxide formation, with H_2O_2 as a second messenger to the nucleus. Although the published data was from a small number of subjects and lacked pre-treatment data and statistical analysis, the hypothesis raised warrants further investigation.

D. Effects of Supplementation on Exercise Performance

It has been suggested that the CoQ_{10} redox "shuttle" might be a rate-limiting step in oxidative phosphorylation,[32] although this has not been commonly accepted.[18] There also has been evidence that athletes undergoing heavy physical training may develop plasma CoQ_{10} deficiency.[12] In some clinical trials, CoQ_{10} treatment was shown to enhance exercise tolerance in cardiac and other types of patients whose functional capacity or CoQ_{10} level were lower than normal.[55-58] It was then hypothesized that dietary supplementation of CoQ_{10} might enhance mitochondrial oxidative phosphorylation capacity that would lead to increased aerobic exercise performance in healthy individuals and athletes.[32,36] However, limited and conflicting evidence has been found in the literature.

On the positive side, there have been reports that daily supplementation of 60 to 100 mg of CoQ_{10} for 4 to 8 weeks enhanced exercise performance in healthy sedentary individuals as well as trained athletes. In one study, 8 weeks of CoQ_{10} supplementation at a dosage of 60 mg per day resulted in an increased exercise performance in six healthy sedentary individuals.[59] Another study used the same dosage for 5 weeks in a group of young cyclists and found a significant improvement in VO_{2max}.[60] The authors commented that the plasma concentration of CoQ_{10} prior to the supplementation period in these athletes was significantly lower than that of the same athletes during the winter time, as well as lower than that of untrained counterparts. It was suggested there might be an accelerated turnover or a higher muscular request for CoQ_{10} during intensive training. Alternatively, it would imply that the supplementation might be more beneficial for those athletes who are experiencing CoQ_{10} deficiency. Improved running performance of a group of trained runners was found after 40 days of supplementation at 100 mg per day.[61] The authors speculated that CoQ_{10} supplementation improved athletic performance via a better aerobic fuel utilization. In another study, 30 days of CoQ_{10} supplementation at a dose of 100 mg per day resulted in a significant increase in VO_{2max} and physical work capacity, in volleyball players as well as sedentary subjects.[62]

Using a double-blind crossover design, Ylikoski and colleagues investigated the effect of CoQ_{10} supplementation on exercise performance of 25 elite Finnish cross-country skiers. Daily supplementation of 90 mg CoQ_{10} for 6 weeks significantly improved both aerobic and anaerobic thresholds and VO_{2max}, and 94% of the athletes felt the preparation was beneficial in improving performance and recovery during the supplementation period, but only 33% of the athletes reported similar perception during the placebo period.[37]

Interestingly, Faff and colleagues reported that supplementation at 100 mg per day for 30 days increased performance in an all out 30 second cycle test, indicating an enhanced anaerobic power.[63]

On the contrary, a number of studies failed to demonstrate any performance-enhancing effect with CoQ_{10} supplementation. A couple of studies involved untrained subjects. In an open-label trial, 4 weeks of CoQ_{10} supplementation at a daily dose of 100 mg did not improve performance of 60-minute cycling at 50% VO_{2max} followed by incremental cycling to exhaustion in 12 healthy untrained young men.[64] Porter and colleagues used 150 mg CoQ_{10} per day for 2 months in 15 middle-aged men (44.7 years on average). With a single-blind design, those who had lower blood CoQ_{10} levels were assigned to the supplement group (n = 8) and others (n = 7) to the placebo group. No treatment effect was found for VO_{2max}, lactate threshold, forearm oxygen uptake, blood flow and handgrip exercise.[33]

Several studies investigated the effects of supplementation on athletes. Braun et al. investigated college-age male cyclists who were given 100 mg of CoQ_{10} per day (n = 6) or placebo (n = 6) for 8 weeks in a double-blind manner. No significant difference between the two groups was found in cycling performance, VO_{2max}, submaximal physiological parameters or lipid peroxidation. The authors speculated that the chosen graded cycle testing protocol might not be sensitive enough to detect the potential ergogenic benefits of CoQ_{10} which might only be effective at relatively light workloads.[32]

Laaksonen and associates investigated effects of 6 weeks of CoQ_{10} supplementation at 120 mg per day in young (22–38 years of age) and older (60–74 years) trained men, in a double-blind placebo-controlled crossover design. There was a significant difference in VO_{2max} between the young and elderly, however no effect of CoQ_{10} supplementation was detected in either VO_{2max} or prolonged exercise (60 minutes cycling at 60% VO_{2max} followed by graded exercise to exhaustion). The performance in the prolonged exercise was even better after the placebo treatment than that after the CoQ_{10} treatment.[65]

Weston and colleagues used a dosage of 1 mg CoQ_{10} per kilogram body mass (70 mg on average) per day for 4 weeks, in a double-blind placebo-controlled trial. Eighteen trained young male cyclists were paired and randomly allocated to CoQ_{10} or placebo group. The results showed no significant effect on VO_{2max}, ventilation threshold and other measured variables. The authors noted a trend that a small number of subjects who had lower than normal plasma CoQ_{10} level prior to supplementation showed greater response than those who had no prior deficiency, and speculated that oral CoQ_{10} supplementation might be beneficial only to athletes who experience a deficiency.[36]

Using a single-blind design, Bonetti and colleagues reported that 8 weeks of CoQ_{10} supplementation at the dose of 100 mg per day did not elevate VO_{2max} and anaerobic threshold levels for middle-aged cyclists (40 years old on average, n = 14 in each of placebo and CoQ_{10} group), although the maximum workload completed at exhaustion was increased in the CoQ_{10} group. The authors speculated that the increased muscular performance might be due to antioxidant effect or potential action of CoQ_{10} on CNS, but this would require confirmation in further investigations.[31]

One study utilized a much higher dose on 11 well-trained triathletes in a double-blind placebo-controlled crossover trial.[34] The supplement contained CoQ_{10} (100 mg), cytochrome c (500 mg), inosine (100 mg) and vitamin E (200 IU) in each capsule. An exhaustive performance test, consisting of 90 minutes of running on treadmill at the intensity of 70% VO_{2max} followed by cycling at 70% VO_{2max} till exhaustion, was conducted after each treatment period. Subjects were given six capsules (containing 600 mg CoQ_{10}) prior to and three capsules (300 mg CoQ_{10}) following each workout, with a minimum of 10 workouts per week for 4 weeks. The 4 weeks of supplementation did not significantly enhance the exercise time to exhaustion, as compared with the same period of placebo treatment.

Furthermore, there has been one report that CoQ_{10} supplementation produced less increment in performance than placebo treatment. The study involved 22 days of supplementation with a dose of 120 mg per day for CoQ_{10} group (n = 9) and a placebo group (n = 9). Five days of high-intensity anaerobic training were inserted during the trial. At the end of the trial, the performance of the placebo group was found better than that of the CoQ_{10} group.[39]

Apparently, with the available evidence, it is difficult to make a conclusion on the effects of exogenous CoQ_{10} on exercise performance. In an overview, most studies that engaged a double-blind placebo-control design do not seem to support that CoQ_{10} supplementation has performance-enhancing effect in healthy non-CoQ_{10}-deficient individuals.

Deficiency can be caused by decreased rate of synthesis, increased rate of breakdown, certain diseases and in rare cases by genetic factors.[4] Limited information is available on whether and to what extent athletes in training would experience CoQ_{10} deficiency. Karlsson et al. reported that endurance-trained male athletes had lower plasma CoQ_{10} (0.69 µg/ml) than sedentary subjects (0.82 µg/ml).[66] Swedish national downhill skiers who used normal diet during a 1-week training camp demonstrated significantly decreased plasma and muscle CoQ_{10}, while those who used a special diet rich in CoQ_{10} and carbohydrates maintained normal CoQ_{10} levels.[12] Also, an investigation of the Swedish national soccer team found that 16 of the 24 players showed lower levels of CoQ_{10} than the average of healthy population.[12]

Little is known about the regulation of CoQ_{10} biosynthesis and intracellular redistribution.[4,9] The biosynthesis of CoQ_{10} may be regulated according to the oxidative status and can be elevated in response to oxidative stress.[8] Faff suggested that, although supplementation of antioxidant vitamins including CoQ_{10} have no apparent effect on athletic performance or detectable changes of lipid peroxidation parameters in the blood, it is possible that the antioxidants prevent oxidative tissue damage, improve the physical health of athletes and other physically hard-working individuals.[67] However, little evidence is available in the current literature that supports that supplementation of CoQ_{10} would reduce cell damage during exercise training in humans.

Plasma creatine kinase (CK) and lactate dehydrogenase (LDH) have been used as indicators of cell damage (enzymes leaking out from cell). A study on rats found that the serum CK and LDH levels increased in the control group but not in the CoQ_{10}-supplementation group after 90 minutes of downhill treadmill running, indicating a protective effect of CoQ_{10} on exercise-induced muscular injury. Furthermore, these enzyme activities increased to the same levels in both groups 40 hours post exercise, suggesting CoQ_{10} supplementation did not mediate the inflammatory processes after exercise. The muscle CoQ_{10} content increased by the treatment, but CoQ_9 content was unchanged. This indicates the supplemented CoQ_{10} has, at least in part, incorporated into the muscle.[68]

A similar effect has been found in a study on humans.[69] Eight subjects in each of placebo, vitamin E (900 IU daily) or CoQ_{10} (270 mg daily) groups were given the supplement for 13 days before and 1 day after a downhill run. Before and after the treatment, maximal aerobic (incremental to exhaustion) and anaerobic (20-second sprint) cycling tests were performed. The results showed that the previous eccentric exercise (downhill run) significantly reduced subsequent performance in maximal aerobic exercise and elevated CK level in subjects treated with placebo, while those who took supplementation of CoQ_{10} or vitamin E were not affected, possibly due to the antioxidant effects that attenuated muscle damage. The anaerobic performance was unaffected by the supplementation.

In a double-blind placebo-controlled trial the effect of combined CoQ_{10} and vitamin E supplementation on exercise-induced lipid peroxidation and muscular damage was examined in 37 moderately trained marathon runners.[40] The results indicated that 3 weeks of supplementation (90 mg CoQ_{10} and 13.5 mg vitamin E daily) prior to a marathon run enhanced the antioxidative capacity of plasma and the proportion of ubiquinol to total CoQ_{10}, but produced no effect on lipid peroxidation or muscular damage induced by exhaustive exercise, as indicated by serum CK activity and lactate. Similarly, Burstein and colleagues provided young healthy volunteers with 150 mg CoQ_{10} daily for 14 weeks in a double-blind placebo-controlled trial during a physical training period. At the end of this period, a 45 km march was performed. The group using CoQ_{10} (n = 26) showed no difference in muscle damage, as indicated by elevated CPK, compared with the placebo group (n = 25).[38]

Again, currently there is no sufficient evidence available in the literature to confirm the protective role of CoQ_{10} supplementation against oxidative damage associated with heavy exercise and training. There has been one report that CoQ_{10} supplementation together with high-intensity exercise seemed to increase cell damage, because CK concentration was increased in plasma, possibly due to increased free radical production.[70]

IV. Body Reserves

Coenzyme Q_{10} can be endogenously synthesized in all cells of healthy individuals.[4] It is synthesized by the convergence of two metabolic pathways. The quinone moiety is derived from tyrosine or phenylalanine and the isoprenoid units are from acetyl-coenzyme A via the mevalonate pathway.[8] It appears that the synthesis of CoQ_{10} begins in the endoplasmic reticulum and is completed in the Golgi apparatus, from where CoQ_{10} is transported to other locations in the cell. A limited amount is discharged from the liver into blood. It seems that endogenous CoQ_{10} is not distributed among different tissues through circulation.[8] In tissues with unimpaired synthetic capacity, CoQ_{10} reaches saturation level in each membrane.[4]

The total body content of CoQ_{10} is estimated to be approximately 0.5–1.5 grams.[7] Within the cell, 40–50% of CoQ_{10} are found in the mitochondrial inner membrane, 25–30% in the nucleus, 15–20% in the microsomal and 5–10% in the cytosol.[7] The distribution of CoQ_{10} in tissues is related to the lipid content as well as the metabolic activity, with relatively higher concentrations found in the heart (~70 mg/kg wet weight), liver (~60 mg/kg) and skeletal muscle (~40 mg/kg) and low concentration in the blood

(~1 mg/liter).[12] The muscle and serum levels of CoQ_{10} may vary greatly in healthy subjects and the serum levels largely depend on the quantities of circulating lipoproteins that contain CoQ_{10}.[5,65]

The plasma CoQ_{10} concentration of healthy individuals is normally around 0.75–1.00 μg/ml.[3,71] The major factors that may alter, or are found to be correlated with plasma CoQ_{10} levels, include age, gender (slightly higher in male than in female), alcohol consumption, diet, physical exercise and certain diseases, as well as serum levels of cholesterol, triglycerides and γ-glutamyl-transferase.[4,71] Kalén and colleagues analyzed CoQ_{10} levels in human tissues (6–8 samples each) and found that the peak values occurred at approximately 20 years of age in most tissues (lung, heart, spleen, liver and kidney), then fall slowly thereafter, except that in the pancreas and adrenal gland the highest value was found at 1 year of age.[14] The decrease of CoQ_{10} with aging may account for the age-related increase in oxidative damage to proteins and DNA, as well as increased incidence of degenerative diseases such as cancers and cardiovascular disorders.[8] Coenzyme Q_{10} appears to have a high turnover rate in all tissues and its half-life time $(T_{1/2})$ varies between 50 (muscle) and 125 (kidney) hours depending on the tissue.[8,9] Tissue CoQ_{10} level may increase under the influence of oxidative stress, such as physical exercise, cold adaptation and thyroid hormone treatment.[8]

Karlsson reported a reciprocal relationship between plasma and muscle CoQ_{10} in healthy subjects and explained this phenomenon by physical conditioning levels.[12] Muscle lipids increase with training while the blood lipids decrease. The muscle CoQ_{10} was positively correlated with percent slow-twitch (ST) muscle fibers, however, the plasma CoQ_{10} showed a reversed relationship. Therefore, physically active individuals who are using well-balanced diet may have unsaturated plasma CoQ_{10} levels, indicating a decreased level of antioxidant that is due to high turnover and breakdown rate.[12] Laaksonen et al. also found that physical activity significantly affected muscle CoQ_{10} levels. However, no correlation between serum and muscle levels of CoQ_{10} was established, suggesting different regulatory mechanisms.[72] Investigation on rats showed an increased CoQ and cytochrome *c* reductase activity in red quadriceps and soleus muscles and adipose tissues, but not in cardiac and white quadriceps muscles, in response to endurance exercise training.[73]

There has been a report that there are differences in plasma CoQ_{10} concentration between ethnic groups that may relate to the risk to coronary heart disease.[74] However, the reason for such differences, e.g., whether they are related to genetic or dietary factors, has not been examined.

V. Dietary and Supplemental Sources

Kamei and colleagues from Japan analyzed the CoQ_{10} content in a variety of foods (mostly one sample from each type of food, some with 2–4 samples).[75] Some selected CoQ_{10}-rich foods include meat, e.g., pork, beef and chicken (containing ~20–40 micrograms per gram wet weight), certain types of fish, e.g., sardine, mackerel and cuttlefish (~20–60 µg per gram), and some vegetables, e.g., spinach and broccoli (~8–10 µg per gram). Soybean oil (~90 µg per gram), rapeseed oil (~70 µg per gram) and sesame oil (~30 µg per gram) also contain a high level of CoQ_{10}. Other reports showed that pork heart contained the highest level of CoQ_{10} (200 µg per gram) while beef heart contained 40 µg per gram.[3,76]

Limited information is available on the daily dietary intake of CoQ_{10} in different populations. The Swedish National Food Administration has computed the daily intake of 2 to 20 mg per day based on the above-mentioned food content.[12,75] The Japanese consume 5–10 mg per day of CoQ_{10} through food products.[77] Coenzyme Q_{10} intake in the Danish population is estimated to be 3–5 mg per day.[76]

Correction for deficiencies and to raise plasma CoQ_{10} to the therapeutic level (around 1.5–2 µg/ml)[12,78] requires a much higher level of intake than what is available from a normal diet. There are manufactured CoQ_{10} products that are synthesized chemically from certain plant materials or by microorganisms.[26,77] Many pharmaceutical companies provide CoQ_{10} products as either single content or mixed with other supplements. They are available in different forms, including capsules, tablets, softgels, or liquid and in various doses, e.g., 30 mg, 50 mg or 100 mg. CoQ_{10} products are readily available over the counter in some countries such as the United States and Australia, while a prescription is required in some other countries, such as Japan and Italy.[12]

Studies on the bioavailability of CoQ_{10} in different preparations, including emulsified, granule-based and oil-based, showed no difference in the plasma CoQ_{10} increment after the supplementation over 1 to 8 weeks.[12,79] However, it has also been reported that plasma CoQ_{10} in response to a single dose (100 mg) seems to be affected by the formulations, and that the preparation of emulsion in soft gelatin capsules demonstrated much higher bioavailability than that of powder in hard gelatin capsules within 36 hours.[80]

The pharmacokinetics studies indicate that the plasma level of CoQ_{10} reaches a peak approximately 6 hours after ingestion and a second peak may occur at 24 hours after dosing, that may indicate enterohepatic recycling. The plasma half-life is approximately 31–33 hours.[81–83] Approximately 60% of CoQ_{10} appears to be eliminated in the feces.[81] Studies on rats showed that about half of injected labeled CoQ was excreted without modification.[84]

VI. Interactions with Other Nutrients and Drugs

There have reportedly been a few interactions of CoQ_{10} with other nutrients and drugs, including statins, adriamycin, warfarin, anabolic androgen steroids and vitamin E.

Statins that are used to control cholesterol levels, such as lovastatin, simvastatin or pravastatin, have shown an effect of decreasing CoQ_{10} levels in serum.[26,85] Lovastatin reduces cholesterol level by inhibiting an enzyme in the mevalonate pathway (HMG-CoA reductase). Because the synthesis of CoQ_{10} also uses this pathway, the biosynthesis of CoQ_{10} is decreased. Therefore, although lovastatin may reduce the cholesterol level, a new risk of cardiac disease may occur because of decreased CoQ_{10} synthesis.[86] Supplementation of CoQ_{10} may correct this condition.[87] A newly developed cholesterol-managing drug, squalestatin 1, acts on a lower site of the mevalonate pathway that does not affect CoQ_{10} synthesis (even increases CoQ_{10} level) while inhibiting cholesterol synthesis.[8] Cholestyramine or fibrate derivatives also do not appear to affect CoQ_{10} concentrations.[26]

There have been several reported cases that CoQ_{10} interacts with warfarin, which is an anticoagulant drug.[88] Because CoQ_{10} is structurally related to menaquinone (vitamin K) it may have a procoagulant effect.[5,26,88] The patients whose condition was stabilized by warfarin showed reversed effect after addition of CoQ_{10} and the previous condition resumed when the use of CoQ_{10} was discontinued. It has been recommended that, until more is known about the effect, the concomitant use of CoQ_{10} with warfarin should be avoided due to the potential risk of thrombotic complications.[88] However, Kato et al. found that CoQ_{10} reduced blood viscosity in ischemia heart-disease patients.[89] Karlsson also reported decreased blood viscosity as a side effect of CoQ_{10} and vitamins E and F1 supplementation.[12] These conflicting effects on blood viscosity and coagulation require further clarification.

Coenzyme Q_{10} has been used to treat the oxidative damage-induced cardiotoxicity caused by antineoplastic drugs. However, a study on mice indicated that CoQ_{10} affected the metabolism of an anticancer agent, doxorubicin (adriamycin), by increasing the concentrations of a putatively toxic adriamycin metabolite.[90] Elevated levels of this doxorubicin metabolite (aglycone 1) are associated with decreased survival in murine models. Therefore, clinical application of CoQ_{10} concomitant with antitumor drugs requires special caution and CoQ_{10} should not be undertaken during chemotherapy with doxorubicin. However, use of CoQ_{10} after cessation of chemotherapy has shown beneficial effects.[5]

There has been a report that male power athletes who were abusing anabolic androgenic steroids demonstrated increased serum CoQ_{10} concentration and the ration of CoQ_{10} to low-density lipoprotein (LDL).[91] The authors suggest that the drug has an effect on the by-products of the mevalonate pathway.

Co-supplementation of CoQ_{10} with vitamin E and other nutrients has been frequently used. Recently, Kaikkonen and colleagues reported that a daily dose of 200 mg resulted in a significant increase in plasma CoQ_{10} level.[17] However, simultaneous supplementation of CoQ_{10} with vitamin E resulted in a much smaller increase. The authors speculated that there might be competitive absorption between CoQ_{10} and vitamin E, or alternatively, the plasma concentrations are regulated by a tocopherol-binding protein in the liver that might regulate the transport of orally ingested CoQ_{10} into lipoproteins.

Few adverse effects of CoQ_{10} supplementation have been reported in the literature. In a multicenter study in Italy,[22] daily doses of 50 to 150 mg CoQ_{10} (78% of the participants received 100 mg per day) was provided for 90 days among 2359 patients in New York Heart Association classes II and III who had previously received at least 6 months of conventional therapy. The treatment showed improvement of clinical signs and symptoms in the majority of the patients, while there was fewer than 1% (n = 22) of the participants reported adverse effects that were related to the treatment. The adverse effects were mainly discomfort in the gastrointestinal tract (nausea, n = 19). In this and other reports, other adverse effects may include diarrhea, appetite suppression, photophobia, rashes or irritability. High doses (100 mg or higher) in the evening may cause mild insomnia in some individuals. The side effects tend to diminish when the dosage is decreased.[30,92] There was a report that a small number (4 out of 10) of Huntington's disease patients experienced mild side effects such as headache, heartburn, fatigue and increased involuntary movements. These were probably related to high doses of CoQ_{10} supplementation at 600 to 1200 mg.[93]

There were reports on clinical relapse on withdrawal of CoQ_{10}; however, reinstatement of therapy resulted in improvement.[26]

VII. Recommendations for Dietary Supplementation

No recommended daily allowance has been proposed for CoQ_{10}. The rationale for supplementation may be quite different, for instance, as a complementary therapy for various diseases, improvement of general health or potential enhancement of physical performance.

In general, unimpaired endogenous biosynthesis together with balanced diet appears to be able to maintain a normal level of CoQ_{10} in tissues. Populations susceptible to CoQ_{10} deficiency are those with genetic failure of biosynthesis, cardiac diseases or those using drugs that inhibit synthesis of CoQ_{10}, the elderly and athletes under training. Oral supplementation of CoQ_{10} can effectively correct deficiency.[11,94,95] Daily dosages of 30–60 mg are recommended to prevent CoQ_{10} deficiency.[26] It has been recommended that blood levels of CoQ_{10} should reach at least 1.5–2.0 µg/ml to elicit therapeutic effects.[12,78] These levels can normally be achieved by using daily dosages of 60–150 mg for 1 week or longer.[31,36,37,78,96–98] It is suggested that dosages over 100 mg/day should be divided, e.g., two or three 50 mg dosages, each administered with a meal.[26]

In addition, for correction of deficiency, the elevated level of CoQ_{10} in blood may serve several important functions, such as an enhanced protection of LDL from lipid peroxidation, a prevention of free-radical damage caused by neutrophil in inflammatory diseases and a prevention of oxidative injury by endothelial cells resulting from ischemia reperfusion. These and other protective functions may account for the majority of reported beneficial effects of CoQ_{10} administration in experimental and clinical trials.[8] However, it seems that exogenous supplementation does not always increase skeletal muscle CoQ_{10} level above normal[4,9] in either healthy subjects[35] or patients with mitochondrial myopathies but without CoQ deficiency,[99] possibly because the membranes are normally saturated with CoQ_{10}. The plasma concentration of CoQ_{10} is much lower than that in nondeficient tissues, even after a high dose of supplementation. Dietary CoQ_{10} is found to have been taken up by liver, spleen, kidney and brain, but not by other organs in animals.[9] How the exogenous CoQ_{10} is selectively taken by different organs is not clear. With the beneficial effects of increased plasma concentration, it is speculated that CoQ_{10} may be able to moderate receptors at the plasma membrane and exert a certain effect without entering into the cell.[9]

Although higher doses have been used for treatment of diseases[93,100] as well as for athletes[34] without known adverse effect, the commonly used daily dose is between 30–150 mg. In Australia, the Complementary Medicines Evaluation Committee (1999) recommended a maximum daily dose of not more than 150 mg.[101]

The effects of CoQ_{10} supplementation on pregnancy and lactation has not been examined. Because of its hemodynamic, bioenergetic and immunogenic effects, caution should be exercised when CoQ_{10} is used during pregnancy.[5]

Whether CoQ_{10} could induce a performance-enhancing effect or not, it has been promoted as a beneficial nutritional supplement for general health. For example, it has been suggested that CoQ_{10} has profound beneficial effects in the following aspects:[16]

- Acts as a novel antioxidant
- Enhances stamina, endurance and energy levels
- Helps to reduce body weight
- Normalizes blood pressure
- Attenuates immune function
- Protects against cardiovascular dysfunction
- Reverses periodontal disease
- Increases effectiveness of various chemotherapeutic agents and antimalarial drugs

Many of these suggested benefits still need further research to be scientifically defined. Current clinical experience and evidence from research trials do not allow a precise prescription based on the dose–effect relationship for these suggested effects of CoQ_{10}.

For legal and ethical reasons, CoQ_{10} is not included in the list of Prohibited Classes of Substances and Prohibited Methods in the Olympic Movement Anti-Doping Code, which was updated on January 1, 2003.[102]

VIII. Future Research Directions or Needs

During the past three decades or so, a large number of laboratory investigations and clinical trials have been carried out to investigate the roles of CoQ_{10} in normal and abnormal body functions. However, it is still difficult to draw definitive conclusions regarding the effect of CoQ_{10} supplementation on each individual health condition without additional large, well-designed clinical trials. Issues concerning optimum target dosages, potential interactions, monitoring parameters and the role of CoQ_{10} as a monotherapeutic agent need to be further investigated, as suggested by Tran and colleagues in a recent review on use of CoQ_{10} in cardiovascular diseases.[26] The metabolism of CoQ_{10} itself is also not yet well understood.[9] How its rate of synthesis and breakdown is regulated in response to altered metabolic demand, such as exercise, is not clear.

The current literature does not seem to support the hypothesized effects of CoQ_{10} supplementation in performance enhancement and protection of cells from exercise-induced oxidative damage, either used alone or with other supplements. A few related issues need to be further investigated.

To enhance the capacity of oxidative phosphorylation it is expected that exogenous CoQ_{10} would be taken up by the mitochondria in muscles. However, evidence in this regard is controversial.[11,35] Case studies on myopathic patients showed that muscular deficiency of CoQ_{10} can be corrected by supplementation.[103] Karlsson et al. also reported an increased muscle CoQ_{10} in healthy subjects as a result of supplementation.[11] However, Svensson et al. reported that oral supplementation did not increase muscle and mitochondrial concentration of CoQ_{10}.[35] A similar result was also found in our laboratory; 4 weeks of supplementation did not significantly elevate skeletal muscle CoQ_{10}.[104] Dietary CoQ_{10} appears in the circulation and may be taken up by mononuclear but not polynuclear cells.[84] Labeled CoQ injected into rats is found to be taken up from the circulation by liver, spleen, adrenals, ovaries, thymus and heart but not by muscle, brain and kidney.[84] The conflicting findings and mechanisms for the selective uptake of CoQ by different tissues require further research.

It has been suggested that many of the methods currently used to detect exercise-induced free radical production, lipid peroxidation and muscle-tissue damage are not sensitive or specific. These warrant development and use of more specific *in vivo* markers of free-radical-mediated lipid peroxidation in studies involving exercising subjects.[43] The same is true for the indicators of aerobic performance, such as VO_{2max} and ventilation or lactate threshold, because many factors can affect these variables.[32,105] In cardiac patients, CoQ_{10} improves exercise tolerance by enhancing myocardium function. In patients with skeletal muscle disorders, improved muscle energy metabolism by CoQ_{10} can be shown in the ratio of Pi/PCr, PCr/ATP and muscle pH, as detected by nuclear magnetic resonance (^{31}P-NMR).[106,107] However, in healthy nondeficient individuals, whether CoQ_{10} supplementation affects cardiac or skeletal muscle metabolic capacity is not clear.

Most of the published studies on CoQ_{10} intervention for exercise performance typically involved small numbers of subjects and used trial periods of 1 to 3 months. Further studies with various supplementation periods, dosages and larger numbers of participants are needed to provide more conclusive evidence on the ergogenic effects of CoQ_{10}.

The functions of CoQ_{10} in cellular signaling, regulation of gene expression and control of cell proliferation and apoptosis have been speculated on but not yet well explained.[4] These are of particular interest in relation to exercise-induced cell damage, death or growth, and should be examined.

There has been little information on the acute change of plasma and muscle CoQ_{10} level in response to a single bout of exercise.[53] Further evidence on the changes of CoQ_{10} and other antioxidants in response to exercise of various forms and intensities is needed for a better understanding of the roles in fatigue and recovery processes in exercise and training.

IX. Summary

Coenzyme Q_{10} has been effectively used as a complementary therapeutic agent for certain health disorders and as a nutritional supplementation for prevention of deficiency and enhancement of antioxidant capacity. However, equivocal evidence has been reported on the ergogenic effect of CoQ_{10} supplementation in healthy and athletic populations. Unless a deficiency status is proven, supplementation of CoQ_{10} might not demonstrate significant effects in exercise performance or prevention of oxidative damage to cells, although it is generally safe to use in prevention of potential deficiency caused by increased degradation in prolonged strenuous physical activities.

References

1. Crane, F. L., Hatefi, Y., Lester, R. I. and Widmer, C., Isolation of a quinone from beef heart mitochondria, *Biochim. Biophys. Acta* 25, 2201, 1957.
2. Morton, R. A., Wilson, G. M., Lowe, J. S. and Leat, W. M. F., Ubiquinone, in *Chemical Industry*, December, pp. 1649–50, 1957.
3. Overvad, K., Diamant, B., Holm, L., Hølmer, G., Mortensen, S. A. and Stender, S., Coenzyme Q_{10} in health and disease, *Eur. J. Clin. Nutr.* 53 (10), 764–70, 1999.
4. Crane, F. L., Biochemical functions of coenzyme Q_{10}, *J. Am. Coll. Nutr.* 20 (6), 591–598, 2001.
5. Jones, K., Hughes, K., Mischley, L. and McKenna, D. J., Coenzyme Q-10: Efficacy, safety and use, *Altern. Ther. Health Med.* 8 (3), 42–55, 2002.
6. Niki, E., Mechanisms and dynamics of antioxidant action of ubiquinol, *Mol. Aspects Med.* 18 (Suppl), S63–70, 1997.

7. Greenberg, S. and Frishman, W. H., Co-enzyme Q_{10}: A new drug for cardiovascular disease, *J. Clin. Pharmacol.* 30 (7), 596-608, 1990.
8. Ernster, L. and Dallner, G., Biochemical, physiological and medical aspects of ubiquinone function, *Biochim. Biophys. Acta* 1, 195-204, 1995.
9. Turunen, M., Swiezewska, E., Chojnacki, T., Sindelar, P. and Dallner, G., Regulatory aspects of coenzyme Q metabolism, *Free Radic. Res.* 36 (4), 437–443, 2002.
10. Folkers, K., Littarru, G. P., Ho, L., Runge, T. M., Havanonda, S. and Cooley, D., Evidence for a deficiency of coenzyme Q_{10} in human heart disease, *Int. Z. Vitaminforsch.* 40 (3), 380–90, 1970.
11. Karlsson, J., Diamant, B., Folkers, K. and Lund, B., Muscle fiber types, ubiquinone content and exercise capacity in hypertension and effort angina, *Ann. Med.* 23 (3), 339–44, 1991.
12. Karlsson, J., *Antioxidants and Exercise*, Human Kinetics, Champaign, IL, 1997.
13. Mortensen, S. A., Vadhanavikit, S., Muratsu, K. and Folkers, K., Coenzyme Q_{10}: Clinical benefits with biochemical correlates suggesting a scientific breakthrough in the management of chronic heart failure, *Int. J. Tissue React.* 12 (3), 155–162, 1990.
14. Kalén, A., Appelkvist, E. L. and Dallner, G., Age-related changes in the lipid compositions of rat and human tissues, *Lipids* 24 (7), 579–584, 1989.
15. Karlsson, J., Diamant, B., Folkers, K., Åström, H., Gunnes, S., Liska, J. and Semb, B., Ischaemic heart disease, skeletal muscle fibers and exercise capacity, *Eur. Heart J.* 13 (6), 758–762, 1992.
16. Bagchi, D., Coenzyme Q_{10}: A novel cardiac antioxidant, *J. Orthomol. Med.* 12 (1), 4–10, 1997.
17. Kaikkonen, J., Tuomainen, T. P., Nyyssönen, K. and Salonen, J. T., Coenzyme Q_{10}: Absorption, antioxidative properties, determinants and plasma levels, *Free Radic. Res.* 36 (4), 389–397, 2002.
18. Lenaz, G., Fato, R., Di Bernardo, S., Jarreta, D., Costa, A., Genova, M. L. and Castelli, G. P., Localization and mobility of coenzyme Q in lipid bilayers and membranes, *Biofactors* 9 (2–4), 87–93, 1999.
19. Sinatra, S. T., Alternative medicine for the conventional cardiologist, *Heart Dis.* 2 (1), 16–30, 2000.
20. Villalba, J. M. and Navas, P., Plasma membrane redox system in the control of stress-induced apoptosis, *Antioxd. Redox Signal.* 2 (2), 213–230, 2000.
21. Walter, L., Miyoshi, H., Leverve, X., Bernard, P. and Fontaine, E., Regulation of the mitochondrial permeability transition pore by ubiquinone analogs. A progress report, *Free Radic. Res.* 36 (4), 405–412, 2002.
22. Baggio, E., Gandini, R., Plancher, A. C., Passeri, M. and Carmosino, G., Italian multicenter study on the safety and efficacy of coenzyme Q_{10} as adjunctive therapy in heart failure. CoQ_{10} Drug Surveillance Investigators, *Mol. Aspects Med.* 15 (Suppl), S287–S294, 1994.
23. Langsjoen, H., Langsjoen, P., Willis, R. and Folkers, K., Usefulness of coenzyme Q_{10} in clinical cardiology: a long-term study, *Mol. Aspects Med.* 15 (Suppl), S165–S175, 1994.
24. Langsjoen, P. H. and Langsjoen, A. M., Overview of the use of CoQ_{10} in cardiovascular disease, *Biofactors* 9 (2–4), 273–284, 1999.
25. Soja, A. M. and Mortensen, S. A., Treatment of congestive heart failure with coenzyme Q_{10} illuminated by meta-analyses of clinical trials, *Mol. Aspects Med.* 18 (Suppl), S159–S168, 1997.

26. Tran, M. T., Mitchell, T. M., Kennedy, D. T. and Giles, J. T., Role of coenzyme Q_{10} in chronic heart failure, angina and hypertension, *Pharmacotherapy* 21 (7), 797–806, 2001.
27. Beal, M. F., Coenzyme Q_{10} as a possible treatment for neurodegenerative diseases, *Free Radic. Res.* 36 (4), 455–460, 2002.
28. Ebadi, M., Govitrapong, P., Sharma, S., Muralikrishnan, D., Shavali, S., Pellett, L., Schafer, R., Albano, C. and Eken, J., Ubiquinone (coenzyme Q_{10}) and mitochondria in oxidative stress of Parkinson's disease, *Biol. Signals Recept.* 10 (3–4), 224–53, 2001.
29. Hodges, S., Hertz, N., Lockwood, K. and Lister, R., CoQ_{10}: Could it have a role in cancer management? *Biofactors* 9 (2–4), 365–370, 1999.
30. Pepping, J., Coenzyme Q_{10}, *Am. J. Health Syst. Pharm.* 56 (6), 519–21, 1999.
31. Bonetti, A., Solito, F., Carmosino, G., Bargossi, A. M. and Fiorella, P. L., Effect of ubidecarenone oral treatment on aerobic power in middle-aged trained subjects, *J. Sports Med. Phys. Fitness* 40 (1), 51–57, 2000.
32. Braun, B., Clarkson, P. M., Freedson, P. S. and Kohl, R. L., Effects of coenzyme Q_{10} supplementation on exercise performance, VO_{2max} and lipid peroxidation in trained cyclists, *Int. J. Sport Nutr.* 1 (4), 353–365, 1991.
33. Porter, D. A., Costill, D. L., Zachwieja, J. J., Krzeminski, K., Fink, W. J., Wagner, E. and Folkers, K., The effect of oral coenzyme Q_{10} on the exercise tolerance of middle-aged, untrained men, *Int. J. Sports Med.* 16 (7), 421–427, 1995.
34. Snider, I. P., Bazzarre, T. L., Murdoch, S. D. and Goldfarb, A., Effects of coenzyme athletic performance system as an ergogenic aid on endurance performance to exhaustion, *Int. J. Sport Nutr.* 2 (3), 272–286, 1992.
35. Svensson, M., Malm, C., Tonkonogi, M., Ekblom, B., Sjödin, B. and Sahlin, K., Effect of Q_{10} supplementation on tissue Q_{10} levels and adenine nucleotide catabolism during high-intensity exercise, *Int. J. Sport Nutr.* 9 (2), 166–180, 1999.
36. Weston, S. B., Zhou, S., Weatherby, R. P. and Robson, S. J., Does exogenous coenzyme Q_{10} affect aerobic capacity in endurance athletes? *Int. J. Sport Nutr.* 7, 197–206, 1997.
37. Ylikoski, T., Piirainen, J., Hanninen, O. and Penttinen, J., The effect of coenzyme Q_{10} on the exercise performance of cross-country skiers, *Mol. Aspects Med.* 18 (Suppl), S283–S290, 1997.
38. Burstein, R., Frankel, M. and Kalmovitz, B., Ubiquinone as a potential agent to minimize muscle membrane damage induced by exercise, *Med. Sci. Sports Exerc.* 27 (Suppl), 1138, 1995.
39. Malm, C., Svensson, M., Ekblom, B. and Sjödin, B., Effects of ubiquinone-10 supplementation and high intensity training on physical performance in humans, *Acta Physiol. Scand.* 161 (3), 379–384, 1997.
40. Kaikkonen, J., Kosonen, L., Nyyssönen, K., Porkkala-Sarataho, E., Salonen, R., Korpela, H. and Salonen, J. T., Effect of combined coenzyme Q_{10} and d-alpha-tocopheryl acetate supplementation on exercise-induced lipid peroxidation and muscular damage: A placebo-controlled double-blind study in marathon runners, *Free Radic. Res.* 29 (1), 85–92, 1998.
41. Garrett, R. H. and Grisham, C. M., *Biochemistry*, Saunders College Publishing, Harcourt Brace College Publishers, Sydney, 1995.
42. Karlsson, J., Lin, L., Sylvén, C. and Jansson, E., Muscle ubiquinone in healthy physically active males, *Mol. Cell. Biochem.* 156 (2), 169–172, 1996.
43. Kanter, M. M., Free radicals, exercise and antioxidant supplementation, *Int. J. Sport Nutr.* 4, 205–220, 1994.

44. Dröge, W., Free radicals in the physiological control of cell function, *Physiol. Rev.* 82, 47–95, 2002.
45. Gillham, B., Papachristodoulou, D. K. and Thomas, J. H., *Wills' Biochemical Basis of Medicine*, 3rd ed., Butterworth Heinemann, Oxford, 2000.
46. Ernster, L. and Forsmark-Andree, P., Ubiquinol: An endogenous antioxidant in aerobic organisms, *Clin. Investig.* 71 (Suppl), S60–S65, 1993.
47. Wang, X. and Quinn, P. J., Vitamin E and its function in membranes, *Prog. Lipid Res.* 38 (4), 309–336, 1999.
48. Lass, A., Forster, M. J. and Sohal, R. S., Effects of coenzyme Q_{10} and alpha-tocopherol administration on their tissue levels in the mouse: elevation of mitochondrial alpha-tocopherol by coenzyme Q_{10}, *Free Radic. Biol. Med.* 26 (11–12), 1375–1382, 1999.
49. Thomas, S. R., Neuzil, J. and Stocker, R., Cosupplementation with coenzyme Q prevents the prooxidant effect of (alpha-tocopherol) and increases the resistance of LDL to transition metal-dependent oxidation initiation, *Arterioscler. Thromb. Vasc. Biol.* 16 (5), 687–696, 1996.
50. Thannickal, V. J. and Fanburg, B. L., Reactive oxygen species in cell signaling, *Am. J. Physiol. Lung Cell Mol. Physiol.* 279, L1005–1028, 2000.
51. Yamamoto, Y. and Yamashita, S., Plasma ratio of ubiquinol and ubiquinone as a marker of oxidative stress, *Mol. Aspects Med.* 18 (Suppl), S79–S84, 1997.
52. Yamashita, S. and Yamamoto, Y., Simultaneous detection of ubiquinol and ubiquinone in human plasma as a marker of oxidative stress, *Anal. Biochem.* 250 (1), 66–73, 1997.
53. Okamoto, T., Mizuta, K., Mizobuchi, S., Usui, A., Takahashi, T., Fujimoto, S. and Kishi, T., Decreased serum ubiquinol-10 levels in healthy subjects during exercise at maximal oxygen uptake, *Biofactors* 11 (1–2), 31–33, 2000.
54. Linnane, A. W., Kopsidas, G., Zhang, C., Yarovaya, N., Kovalenko, S., Papakostopoulos, P., Eastwood, H., Graves, S. and Richardson, M., Cellular redox activity of coenzyme Q_{10}: effect of CoQ_{10} supplementation on human skeletal muscle, *Free Radic. Res.* 36 (4), 445–53, 2002.
55. Kamikawa, T., Kobayashi, A., Yamashita, T., Hayashi, H. and Yamazaki, N., Effects of coenzyme Q_{10} on exercise tolerance in chronic stable angina pectoris, *Am. J. Cardiol.* 56 (4), 247–251, 1985.
56. Satta, A., Grandi, M., Landoni, C. V., Migliori, G. B., Spanevello, A., Vocaturo, G. and Neri, M., Effects of ubidecarenone in an exercise training program for patients with chronic obstructive pulmonary diseases, *Clin. Ther.* 13 (6), 754–757, 1991.
57. Vanfraechem, J., Picalausa, C. and Folkers, K., Coenzyme Q_{10} and Physical Performance in Myocardial Failure, in *Biomedical and Clinical Aspects of Coenzyme Q*, Folkers, K. and Yamamura, Y., Eds., Elsevier Science Publishers, Amsterdam, 1984, pp. 281–289.
58. Yamabe, H. and Fukuzaki, H., The beneficial effect of coenzyme Q_{10} on the impaired aerobic function in middle aged women without organic disease, in *Biomedical and Clinical Aspects of Coenzyme Q*, Folkers, K., Littarru, G. P. and Yamagami, T., Eds., Elsevier Science Publishers, Amsterdam, 1991, pp. 535–540.
59. Vanfraechem, J. and Folkers, K., Coenzyme Q_{10} and Physical Performance, in *Biomedical and Clinical Aspects of Coenzyme Q*, Folkers, K. and Yamamura, Y., Eds., Elsevier North Holland Biomedical Press, Amsterdam, 1981, pp. 235–241.

60. Guerra, G., Bargossi, A., Fiorella, P. and Piazzi, S., Effect of the administration of ubidecarenone over the maximum consumption of oxygen and on the physical performance in a group of young cyclists, *Med. Sport (Roma)* 40, 359–364, 1987.
61. Fiorella, P. L., Bargossi, A. M., Grossi, G., Motta, R., Senaldi, R., Battino, M., Sassi, S., Sprovieri, G. and Lubich, T., Metabolic effects of coenzyme Q_{10} treatment in high level athletes, in *Biomedical and Clinical Aspects of Coenzyme Q*, Folkers, K., Littarru, G. P. and Yamagami, T., Eds., Elsevier Science Publishers, Amsterdam, 1991, pp. 513–520.
62. Zeppilli, P., Merlino, B., De Luca, A., Palmieri, V., Sanrini, C., Vannicelli, R., La Rosa Gangi, M., Caccese, R., Cameli, S., Servidei, S., Ricci, E., Silvestri, G., Lippa, S., Oradei, A. and Littarru, G. P., Influence of coenzyme-Q_{10} on physical work capacity in athletes, sedentary people and patients with mitochondrial disease, in *Biomedical and Clinical Aspects of Coenzyme Q*, Folkers, K., Littarru, G. P. and Yamagami, T., Eds., Elsevier Science Publishers, Amsterdam, 1991, pp. 541–545.
63. Faff, J., The influence of ubiqinone of the intense work capacity and on serum activities of creatine kinase and aspartate amino-transferase, *Biol. Sport* 14 (1), 37–44, 1997.
64. Zuliani, U., Bonetti, A., Campana, M., Cerioli, G., Solito, F. and Novarini, A., The influence of ubiquinone (CoQ_{10}) on the metabolic response to work, *J. Sports Med. Phys. Fitness* 29 (1), 57–62, 1989.
65. Laaksonen, R., Fogelholm, M., Himberg, J. J., Laakso, J. and Salorinne, Y., Ubiquinone supplementation and exercise capacity in trained young and older men, *Eur. J. Appl. Physiol.* 72 (1–2), 95–100, 1995.
66. Karlsson, J., Diamant, B., Edlund, P. O., Lund, B., Folkers, K. and Theorell, H., Plasma ubiquinone, alpha-tocopherol and cholesterol in man, *Int. J. Vit. Nutr. Res.* 62 (2), 160–164, 1992.
67. Faff, J., Effects of the antioxidant supplementation in athletes on the exercise-induced oxidative stress, *Biol. Sport* 18 (1), 3–20, 2001.
68. Shimomura, Y., Suzuki, M., Sugiyama, S., Hanaki, Y. and Ozawa, T., Protective effect of coenzyme Q_{10} on exercise induced muscular injury, *Biochem. Biophys. Res. Commun.* 176 (1), 349–355, 1991.
69. Clarke, R., Hodges, S. J. and Sellens, M., Effects of coenzyme Q_{10} and vitamin E on maximal aerobic and anaerobic exercise following eccentric-induced muscle damage, in *Third Conference of the International Coenzyme Q_{10} Association*, London, UK, http://wwwcsi.unian.it/coenzymeQ/thefirst.html, 2002, pp. 116.
70. Malm, C., Svensson, M., Sjöberg, B., Ekblom, B. and Sjödin, B., Supplementation with ubiquinone-10 causes cellular damage during intense exercise, *Acta Physiol. Scand.* 157, 511–512, 1996.
71. Kaikkonen, J., Nyyssönen, K., Tuomainen, T. P., Ristonmaa, U. and Salonen, J. T., Determinants of plasma coenzyme Q_{10} in humans, *FEBS Lett.* 443 (2), 163–166, 1999.
72. Laaksonen, R., Riihimaki, A., Laitila, J., Martensson, K., Tikkanen, M. J. and Himberg, J.-J., Serum and muscle tissue ubiquinone levels in healthy subjects, *J. Lab. Clin. Med.* 125 (4), 517–521, 1995.
73. Gohil, K., Rothfuss, L., Lang, J. and Packer, L., Effect of exercise training on tissue vitamin E and ubiquinone content, *J. Appl. Physiol.* 63 (4), 1638–1641, 1987.

74. Hughes, K., Lee, B. L., Feng, X., Lee, J. and Ong, C. N., Coenzyme Q_{10} and differences in coronary heart disease risk in Asian Indians and Chinese, *Free Radic. Biol. Med.* 32 (2), 132–138, 2002.
75. Kamei, M., Fujita, T., Kanbe, T., Sasaki, K., Oshiba, K., Otani, S., Matsui-Yuasa, I. and Morisawa, S., The distribution and content of ubiquinone in foods, *Int. J. Vit. Nutr. Res.* 56 (1), 57–63, 1986.
76. Weber, C., Bysted, A. and Hølmer, G., The coenzyme Q_{10} content of the average Danish diet, *Int. J. Vitam. Nutr. Res.* 67 (2), 123–129, 1997.
77. Coach Industries, Inc., Nisshin Pharma Ubidecarenone: The original coenzyme Q-10 defining the standard for excellence, http://www.coachindustries.com/coQ10.html, May 2003.
78. Langsjoen, P. H., Vadhanavikit, S. and Folkers, K., Response of patients in classes III and IV of cardiomyopathy to therapy in a blind and crossover trial with coenzyme Q_{10}, *Proc. Natl. Acad. Sci. U. S. A.* 82 (12), 4240–4244, 1985.
79. Lyon, W., Van, d. B. O., Pepe, S., Wowk, M., Marasco, S. and Rosenfeldt, F. L., Similar therapeutic serum levels attained with emulsified and oil-based preparations of coenzyme Q_{10}, *Asia Pac. J. Clin. Nutr.* 10 (3), 212–215, 2001.
80. Wahlqvist, M. L., Wattanapenpaiboon, N., Savige, G. S. and Kannar, D., Bioavailability of two different formulations of coenzyme Q_{10} in healthy subjects, *Asia Pac. J. Clin. Nutr.* 7 (1), 37–40, 1998.
81. Lücker, P. W., Wetzelsberger, N., Hennings, G. and Rehn, D., Pharmacokinetics of coenzyme ubidecarenone in healthy volunteers, in *Biomedical and Clinical Aspects of Coenzyme Q*, Folkers, K. and Yamamura, Y., Eds., Elsevier, Amsterdam, 1984, pp. 143–151.
82. Tomono, Y., Hasegawa, J., Seki, T., Motegi, K. and Morishita, N., Pharmacokinetic study of deuterium-labelled coenzyme Q_{10} in man, *Int. J. Clin. Pharmacol. Ther. Toxicol.* 24 (10), 536–541, 1986.
83. Weis, M., Mortensen, S. A., Rassing, M. R., Moller-Sonnergaard, J., Poulsen, G. and Rasmussen, S. N., Bioavailability of four oral coenzyme Q_{10} formulations in healthy volunteers, *Mol. Aspects Med.* 15 (Suppl), S273–S280, 1994.
84. Dallner, G., Regulation of coenzyme Q metabolism, in *Third Conference of the International Coenzyme Q_{10} Association* London, UK, http://wwwcsi.unian.it/coenzymeQ/thefirst.html, 2002, pp. 9.
85. Bliznakov, E. G., Lipid-lowering drugs (statins), cholesterol and coenzyme Q_{10}. The Baycol case: A modern Pandora's box, *Biomed. Pharmacother.* 56 (1), 56–59, 2002.
86. Folkers, K., Langsjoen, P., Willis, R., Richardson, P., Xia, L. J., Ye, C. Q. and Tamagawa, H., Lovastatin decreases coenzyme Q levels in humans, *Proc. Natl. Acad. Sci. U. S. A.* 87 (22), 8931–8934, 1990.
87. Palomäki, A., Malminiemi, K., Solakivi, T. and Malminiemi, O., Ubiquinone supplementation during lovastatin treatment: effect on LDL oxidation *ex vivo*, *J. Lipid Res.* 39 (7), 1430–1437, 1998.
88. Heck, A. M., DeWitt, B. A. and Lukes, A. L., Potential interactions between alternative therapies and warfarin, *Am. J. Health Syst. Pharm.* 57 (13), 1221–1227, quiz 1228–30, 2000.
89. Kato, T., Yoneda, S., Kako, T., Koketsu, M., Hayano, I. and Fujinami, T., Reduction in blood viscosity by treatment with coenzyme Q_{10} in patients with ischemic heart disease, *Int. J. Clin. Pharmacol. Ther. Toxicol.* 28 (3), 123–126, 1990.

90. Shinozawa, S., Gomita, Y. and Araki, Y., Tissue concentration of doxorubicin (adriamycin) in mouse pretreated with alpha-tocopherol or coenzyme Q_{10}, *Acta Med. Okayama* 45 (3), 195–199, 1991.
91. Karila, T., Laaksonen, R., Jokelainen, K., Himberg, J. J. and Seppala, T., The effects of anabolic androgenic steroids on serum ubiquinone and dolichol levels among steroid abusers, *Metab. Clin. Exp.* 45 (7), 844–847, 1996.
92. Chagan, L., Ioselovich, A., Asherova, L. and Cheng, J. W., Use of alternative pharmacotherapy in management of cardiovascular diseases, *Am. J. Manag. Care* 8 (3), 270–285, 2002.
93. Feigin, A., Kieburtz, K., Como, P., Hickey, C., Claude, K., Abwender, D., Zimmerman, C., Steinberg, K. and Shoulson, I., Assessment of coenzyme Q_{10} tolerability in Huntington's disease, *Mov. Disord.* 11 (3), 321–323, 1996.
94. Folkers, K., Vadhanavikit, S. and Mortensen, S. A., Biochemical rationale and myocardial tissue data on the effective therapy of cardiomyopathy with coenzyme Q_{10}, *Proc. Natl. Acad. Sci. U. S. A.* 82 (3), 901–904, 1985.
95. Nobuyoshi, M., Saito, T., Takahira, H., Yamano, Y. and Kanazawa, T., Levels of Coenzyme Q_{10} in biopsies of left ventricular muscle and influence of administration of Coenzyme Q_{10}, in *Biomedical and Clinical Aspects of Coenzyme Q*, Folkers, K. and Yamamura, Y., Eds., Elsevier, Amsterdam, 1984, pp. 222–229.
96. Burke, B. E., Neuenschwander, R. and Olson, R. D., Randomized, double-blind, placebo-controlled trial of coenzyme Q_{10} in isolated systolic hypertension, *South. Med. J.* 94 (11), 1112–1117, 2001.
97. Chello, M., Mastroroberto, P., Romano, R., Bevacqua, E., Pantaleo, D., Ascione, R., Marchese, A. R. and Spampinato, N., Protection by coenzyme Q_{10} from myocardial reperfusion injury during coronary artery bypass grafting, *Ann. Thorac. Surg.* 58 (5), 1427–1432, 1994.
98. Weber, C., Sejersgard, J. T., Mortensen, S. A., Paulsen, G. and Hølmer, G., Antioxidative effect of dietary coenzyme Q_{10} in human blood plasma, *Int. J. Vit. Nutr. Res.* 64 (4), 311–315, 1994.
99. Zierz, S., von Wersebe, O., Bleistein, J. and Jerusalem, F., Exogenous coenzyme Q (CoQ) fails to increase CoQ in skeletal muscle of two patients with mitochondrial myopathies, *J. Neurol. Sci.* 95 (3), 283–290, 1990.
100. Shults, C. W., Beal, M. F., Fontaine, D., Nakano, K. and Haas, R. H., Absorption, tolerability and effects on mitochondrial activity of oral coenzyme Q_{10} in parkinsonian patients, *Neurology* 50 (3), 793–795, 1998.
101. Australian Therapeutic Goods Administration, Chemicals & Non-Prescription Medicines, *TGA News*, 29, May 1999, http://www.tga.health.gov.au/docs/html/tganews/news29/chem.htm
102. World Anti-Doping Agency, Olympic Movement Anti-Doping Code. Appendix A: Prohibited classes of substances and prohibited methods, http://www.wada-ama.org, 2003.
103. Di Giovanni, S., Mirabella, M., Spinazzola, A., Crociani, P., Silvestri, G., Broccolini, A., Tonali, P., Di Mauro, S. and Servidei, S., Coenzyme Q_{10} reverses pathological phenotype and reduces apoptosis in familial CoQ_{10} deficiency, *Neurology* 57 (3), 515–518, 2001.
104. Zhou, S., Zhang, Y., Hu, H., Davie, A., Marshall-Gradisnik, S. and Brushett, D., Muscle and plasma coenzyme Q10 concentrations and aerobic power in response to four weeks supplementation. *J. Sci. Med. Sport* 6 (4 Suppl), 78, 2003.

105. Zhou, S. and Weston, S., Reliability of using D-max method to define physiological responses to incremental exercise testing, *Physiol. Meas.* 18 (2), 145–154, 1997.
106. Mizuno, M., Quistorff, B., Theorell, H., Theorell, M. and Chance, B., Effects of oral supplementation of coenzyme Q_{10} on ^{31}P-NMR detected skeletal muscle energy metabolism in middle-aged post-polio subjects and normal volunteers, *Mol. Aspects Med.* 18 (Suppl), S291–S298, 1997.
107. Barbiroli, B., Iotti, S. and Lodi, R., Improved brain and muscle mitochondrial respiration with CoQ. An *in vivo* study by ^{31}P-MR spectroscopy in patients with mitochondrial cytopathies, *Biofactors* 9 (2–4), 253–260, 1999.

20
Ginseng

Luke R. Bucci, Amy A. Turpin, Christina Beer and Jeff Feliciano*

CONTENTS

I. Introduction 379
II. Human Studies — Physical Performance 380
III. Human Studies — Mental Effects 384
IV. Human Study Variables Delineating Conditions of Efficacy 384
V. Other Adaptogenic Herbs 390
    A. Cordyceps 390
    B. Eleutherococcus 391
    C. Rhodiola species 395
VI. Unresolved Issues and Guide for Future Research 399
VII. Summary 400
Acknowledgments 401
References 401

I. Introduction

The research on Asian ginseng (*Panax ginseng*) and human performance is decidedly mixed. Cogent arguments for efficacy and lack of efficacy can be made, depending upon which studies or outcomes are emphasized. This chapter will explore the differences between studies showing efficacy or lack thereof for *Panax ginseng* preparations, and propose that the glass is half full. Also, several herbs are being marketed with "ginseng alternatives" as ergogenic aids. We feel that this chapter represents the most objective and

---

* The authors are employed by a dietary supplement company that does not sell products containing the herbs discussed in this chapter, except for minor amounts of Korean ginseng powder in one herbal combination (efficacy and claims for this product are not based on ginseng content). Thus, the authors do not have a true vested interest in the herbs discussed in this chapter.

comprehensive review to date on ergogenic effects of ginseng-like herbs (*Cordyceps sinensis, Eleutherococcus senticosus,* and *Rhodiola* species).

Several recent reviews on ergogenic effects of ginseng have expressed equivocal conclusions.[1-9] This chapter will focus on explaining key variables from human studies that seem to delineate what conditions, if any, are required for ergogenic effects of Asian ginseng products.

Ergogenic effects of Asian ginseng (*Panax ginseng*) can be divided into physical performance parameters during exercise (exercise outcomes, physiological changes, metabolic measures and hormone levels) and mental performance parameters (psychomotor measures, reaction time, mood, cognition, memory and accuracy of repetitive tasks). Why examine mental effects of ginseng in relation to ergogenicity? Physical performance has a large psychological component, as evidenced by the need for placebo controls in human studies. Thus, an ergogenic effect of ginseng may be possible by affecting mental aspects without affecting any physical parameters, in essence, generating an effect akin to a placebo effect.

II. Human Studies — Physical Performance

Table 20.1 updates the list of controlled human studies on ergogenicity of Asian ginseng for physical performance.[1-39] Uncontrolled studies, those measuring hormone levels only and those with subjects having serious diseases were not listed.[40-53] The prodigious Chinese literature on medical uses of ginseng and its combinations was excluded since it does not relate directly to ergogenicity. Taken as a whole, studies with equivalent designs and dosages appear to have divergent outcomes. Some reports found lack of effect from Asian ginseng on any measurements of physical performance, whereas some reports found one or more significant effects along with nonsignificant effects. Only a few studies found significant changes in all parameters tested. Thus, a reviewer of this field can deduce that Asian ginseng is effective, or not, as an ergogenic aid, depending upon which studies or outcomes are emphasized. Nevertheless, it is appropriate to deduce that simply taking any Asian ginseng product as directed will not guarantee any ergogenic effect on physical performance beyond a placebo effect.

American ginseng (*Panax quinquefolius* L.) is known to contain different proportions of ginsenosides and is described as having different properties, effects and uses from Asian ginseng.[54-56] One investigation examined the effect of American ginseng on physical performance of eight volunteer subjects.[57,58] No effects on cycle time to exhaustion and other physiological responses to exercise were seen for doses of 8 and 16 mg/kg. This study was crippled with a low subject number (four per group) and short duration (seven days), questioning reliability of results. Another study found no changes in body composition, maximal oxygen uptake or resting energy expenditure after 35 days of either placebo or 1.125 grams of North American

Ginseng

TABLE 20.1
Results of Controlled Human Studies with *Panax Ginseng* on Physical Performance

Study (Reference)	Subject n[a]	Study Design	Subjects (Age Range)	Daily Dose	Preparation Type	Study Duration	Effects (Statistically Significant Unless Otherwise Stated)
Dörling[10]	60	dbpc	Volunteers 22–80 y	200 mg	G115[b]	90 d	Improved postexercise recovery (stair climbing), oxygen consumption.
Forgo[12]	120	dbpc	8 groups males/females 30–39 y 40–60 y	200 mg	G115	12 wk	Improved vital capacity, forced expiration volume, maximum expiratory flow, maximal breathing capacity, work output. NS: serum LH, FSH, testosterone, estradiol, blood chemistries.
Knapik[13]	11	dbpc	West Point cadet marathoners	2000 mg	1.5% glycosides	4 wk	NS: R values, glucose, lactate, free fatty acids, glycerol, insulin, cortisol, and growth hormone.
Teves[14]	12	dbpc	Marathon runners 22 ± 1 y	2000 mg	1.5% glycosides	4 wk	NS: run time to exhaustion, aerobic capacity, heart rate, V_E, and RPE.
Forgo[15]	30	rdbpc	Elite athletes 19–31 y	200 mg + 400 mg E or 200 mg 7%	G115 + Vit. E G115[b] (7%)	9 wk	Improved oxygen uptake (21%), maximal breathing capacity, vital capacity, and forced expiration volume; reduced lactate, HR. NS: serum LH, testosterone, and cortisol.
Forgo[16]	28	dbpc	Pro soccer amateur athletes 20–30 y	200 mg	G115	60 d	Improved oxygen uptake, forced expiration volume, vital capacity, HR, visual reaction times. NS: time to exhaustion, glucose, and lactate.
Macareg[17]	12	rdbpcco	Marathon runners	?	?	?	
Tesch[18]	38	rdbpc	Healthy males 50–54 y	80 mg	Gericomplex[c]	8 wk	Improved HR (60, 90, 120, 150, 180W), RPE (60, 90, 120 W workloads) before vs. after (no change placebo); work capacity, maximum workload, decreased lactate @ 180W vs. placebo. NS: time to exertion.
McNaughton[19]	30	rdbpcco	Athletes 15 females, 15 males	1000 mg	Ginseng root powder	6 wk	Improved aerobic capacity, pectoral strength (27%), quadriceps strength (18%), post-exercise recovery. NS: grip strength.
Gribaudo[20]	12	rdbpc	Young males	1000 mg	+ 1 g fenugreek iron	15 d	Increased total work output (arm flexion). NS: lactate.
Gribaudo[21]	14	rdbpc	Trained male cyclists 22 y ave	1000 mg	+ 1 g fenugreek iron	30 d	Increased anaerobic threshold, blood ferritin, maximum work, RER, V_E max, VO_2max. NS: lactate.

(continued)

TABLE 20.1 (CONTINUED)
Results of Controlled Human Studies with *Panax Ginseng* on Physical Performance

Study (Reference)	Subject n[a]	Study Design	Subjects (Age Range)	Daily Dose	Preparation Type	Study Duration	Effects (Statistically Significant Unless Otherwise Stated)
Pieralisi[22]	49	rdbpcco	Healthy male sport teachers 21–47 y	200 mg	Geriatric Pharmaton[c]	6 wk 1 wk washout	Improved total work load, time to exhaustion, aerobic capacity, ventilation, oxygen consumption, carbon dioxide production, lactate, and heart rate. NS: RER.
van Schepdael[23]	43	rdbpcco	24–36 y female triathletes	400 mg	G115	10 wk periods	Prevented loss of physical fitness (VO_2max) after 10 wk in second period. (possible carry-over effect). NS: VO_2max, HR, lactate.
Sandberg[24]	50	rdbpc	Healthy sedentary Swedish men under stress	180 mg	Gerimax	8 wk	Decreased peak lactate during cycle ergometry. NS: VO_2max.
Engels[25,26]	19	rdbpc	Healthy females 26 ± 1 y	200 mg	G115	8 wk	NS: maximum work, O_2 consumption, RER, V_E, HR, lactate, daily activity levels
Cherdrungsi[27]	41	rdbpc	Active male students (4 groups) 19–26 y	300 mg	Standardized extract, Pharmagin, New Century Pharma, Seoul, Korea	8 wk	Decreased body fat both exercisers and non-exercisers; increased VO_2max, improved heart rate, leg strength anaerobic power in non-exercisers; NS differences between ginseng + exercise and placebo + exercise except for resting heart rate decreased by ginseng
Lifton[28]	11	rdbpcco	Trained cyclists 24 ± 3 y	3 g	Unspecified	13 d, 7 d wash	NS: max HR, VO_2max, RER, treadmill time (total workload) NOTE: inadequate statistical power
Engels[29,30]	31	rdbpc	YMCA males 26 ± 3 y	200 or 400 mg	G115	8 wk	NS: submaximal or maximal oxygen consumption, RER, RPE, peak lactate, HR during exercise.
Wesnes[31]	64	rdbpc	Healthy neurasthenia volunteers 40–65 y	160, 320, 640 mg	Ginkoba[d]	13 wk	Improved HR at max cycle ergometry (160 mg dose).

Study	n	Design	Subjects	Dose	Product	Duration	Results
Allen[32]	28	rdbpc	8 females, 20 males 23 ± 3 y	200 mg	G115S[b] (7% ginsenosides)	3 wk	NS: oxygen uptake, exercise time, workload, plasma lactate, hematocrit, HR, RPE at 150 W, 200 W or peak.
Kolokouri[34] Engels[33,35]	19	rdbpc	Healthy females 19–42 y	400 mg	G115	8 wk	NS: peak anaerobic power output, mean anaerobic power output, rate of fatigue, immediate postexercise recovery HR from all-out, 30 s leg cycle ergometry (Wingate protocol).
Ziemba[36]	15	rdbpc	male soccer players 19 ±1 y	350 mg	KRKA, Novo Mesto, Poland	6 wk	NS: VO_2max, lactate threshold.
Johnson[37]	10	rdbpcco	5 male, 5 female recreationally active 26 ±5 y	1800 mg	Panax ginseng, Eleutherococcus, enoki, reishi, cordyceps, citrus reticulata	2 wk periods; 1 wk washout	NS: lactate post-weightlifting (0-15 min), % lactate clearance
Engels[38,39]	27	rdbpc	Active healthy adults	400 mg	G115	8 wk	NS: salivary IgA, exercise recovery heart rate, physical performance

[a] Abbreviations: ? — data not listed or unavailable; BP — blood pressure; co — crossover; d — days; db — double-blind; DHEA — dehydroepiandrosterone; DHEAS — dehydroepiandrosterone sulfate; EtOH — ethanol, FSH — follicle stimulating hormone; hGH — human growth hormone; HR — heart rate; IgA — Immunoglobulin A; IGF-1 — Insulin-like growth factor-1; LH — luteinizing hormone; mo — months; n — subject number; NC — not controlled; NK — Natural Killer cells; NS — Not Significant; pc — placebo controlled; r — randomized; RBC — red blood cell count; R, RER — respiratory exchange ratio; RPE — ratings of perceived exertion; s = seconds; V_E — expiratory ventilation; V_E max — maximum expiratory ventilation; VO_2 — oxygen uptake; VO_2max — maximal oxygen uptake; W — watts; wk — weeks; y — years; (-) — negative effects.

[b] G115 refers to Ginsana G115®, a commercially available, proprietary *Panax ginseng* extract product containing 100 mg ginseng root extract standardized to 4% ginsenosides per softgel, equivalent to 500 mg powdered root. Ginsana G115S contained 100 mg of ginseng root extract standardized to 7% ginsenoside extract per softgel. Ginsana G115® is a registered trademark of GPL Ginsana Products, Lugano, Switzerland, and Pharmaton Ltd., Switzerland (owned by Boehringer Ingelheim Pharmaceuticals).

[c] Gericomplex® and Germiax® — G115 40 mg, Vitamin A 2667 IU, vitamin D3 200 IU, vitamin B1 1.4 mg, vitamin B2 1.6 mg, vitamin B3 18 mg, vitamin B6 2 mg, vitamin B12 1 mcg, biotin 150 mcg, vitamin C 60 mg, vitamin E 10 mg, phosphorus 100 mg, iron 10 mg, calcium 100 mg, copper 2 mg, magnesium 10 mg, manganese 2.5 mg, zinc 1 mg per tablet. Geriatric Pharmaton® was similar to Gericomplex® but with 26 mg dimethylaminoethanol per capsule. Gericomplex® and Geriatric Pharmaton® are registered trademarks of Boehringer Ingelheim Pharmaceuticals.

[d] Ginkoba® — GK501 (100 mg G115 + 60 mg standardized *Ginkgo biloba* extract (24% ginkgolides) per softgel. Ginkoba® is a registered trademark of Boehringer Ingelheim Pharmaceuticals.

ginseng supplementation.[59] Insulin levels were elevated by ginseng. Goel et al. tested 40 athletes given 400 mg daily for 28 days of Cold-FX®, a North American ginseng extract, for doping by an International Olympic Committee (IOC) testing lab.[60] North American ginseng in this particular product did not induce positive IOC doping control urinalysis tests.

III. Human Studies — Mental Effects

Table 20.2 updates the list of controlled human studies on mental effects of Asian ginseng that may affect ergogenicity.[1-9,12,16,24,31,36,56,61-88] Like Table 20.1, uncontrolled studies were not included.[89-91] Unlike some other reviews, studies investigating products containing ginseng along with other ingredients have been included.[31,66-68,71-73,76,77,80-82,84,86,88] The usual additions are Daily Value levels of vitamins and minerals (Gericomplex® or Geriatric Pharmaton®), or *Ginkgo biloba* standardized extracts (Gincosan® or Ginkoba®). Most studies actually used the additional ingredients in control groups or compared active ingredients singly and in combination, lending more acceptability to results pertaining to ginseng itself. Also, since these products are on the market and chosen by consumers as ginseng products, they have practical relevance — ignoring their contribution favors an incomplete picture of ginseng effects.

The majority of studies found at least one significant change in mental functions from Asian ginseng, but usually, only some of the measured parameters improved. Conditions for efficacy appeared to be older age (over 40 years), stressful conditions (both physical and psychological), and longer duration of administration. In general, reaction times to auditory or visual cues were improved, fatigue reduced, feelings of well-being improved, and errors on cognitive tasks reduced. During exercise, ratings of perceived exertion were usually not changed by ginseng. Table 20.2 suggests that the popularity and long-term marketplace acceptance of Asian ginseng supplements may rely more on mental effects (feelings of well-being) than any apparent physiological changes. In other words, people taking Asian ginseng supplements feel better (beyond a placebo effect), prompting continued use.

IV. Human Study Variables Delineating Conditions of Efficacy

Data to date indicate that Asian ginseng may have conditional efficacy as an ergogenic aid for aerobic and anaerobic performance. Simple Chi-square analysis[92] of key experimental variables frequently subpar in human physical performance studies suggests some conditions of efficacy. Table 20.3 illustrates the association of subject number with any statistically significant, non-mental, ergogenic effect, pooling 22 controlled studies from 27 reports

TABLE 20.2
Results of Human Studies on Mental Performance Related to Ergogenicity with *Panax Ginseng*

Study (reference)	Subject n[a]	Study Design	Subjects (Age Range)	Daily Dose	Preparation Type	Study Duration	Effects (Statistically Significant Unless Otherwise Stated)
Popov;[56] Bae[64]	32	dbpc	Male radio operators (21–23 y)	2 ml	40% ethanol tincture	Acute	Decreased errors in radio transmission of coded messages (17 vs 31%). NS: number of characters transmitted
Revers[62]	?	dbpc	elderly	80 mg	Geriatric Pharmaton	100 d	Improved vitality, alertness, rigidity, concentration, visual/motor coordination, positive outlook, visual and auditory reaction times.
Schmidt[63]	540	dbpc	?	?	Gericomplex and G115[c]	?	Improved subjective and objective parameters; normalized blood glucose, and BP.
Dörling[10]	60	dbpc	Volunteers 22–80 y	200 mg	G115	90 d	Improved visual and auditory reaction times, two-hand coordination, alertness, and subjective assessments.
Forgo[12]	120	db	30–60 y	200 mg	G115	12 wk	Improved reaction times, subjective assessments of mood, work output, sleep, concentration, vitality.
Hallstrom[65]	12	dbpcco	night shift nurses (21–27 y)	1200 mg	Korean white ginseng powder	3 d	Improved tapping rate test. Trends: mood, somatic symptoms, blood glucose. (-) sleep quality.
Garay Lillo[66]	?	dbpc	Middle-aged with asthenia	80 mg	Geriatric Pharmaton	6 mo	Improved appetite, feelings of depression, irritability.
Forgo[16]	28	dbpc	20–30 y Pro soccer, amateur athletes	200 mg	G115	60 d	Improved visual reaction times.
Le Faou[67]	?	dbpc	?	?	Gericomplex	?	Improved fatigue and recovery.
D'Angelo[68]	32	dbpc	20–24 y males	200 mg	G115	12 wk	Improved mental arithmetic calculations. Trend: attention, choice reaction time, auditory reaction time, tapping test, recognition, and visual reaction time.
Garay Lillo[69]	?	dbpc	?	80 mg	Geriatric Pharmaton	6 mo	Improved psychomotor capabilities.
Rosenfeld[70]	?	dbpc	Asthenia, neurological disorders	?	?	?	Improved neuropsychological measures.

(continued)

TABLE 20.2 (CONTINUED)
Results of Human Studies on Mental Performance Related to Ergogenicity with *Panax Ginseng*

Study (reference)	Subject n[a]	Study Design	Subjects (Age Range)	Daily Dose	Preparation Type	Study Duration	Effects (Statistically Significant Unless Otherwise Stated)
Garay Lillo[71]	103	open	Asthenia 40–60 y (51 y ave)	80 mg	Pharmaton Complex	1 y	Improved anxiety, depression, vulnerability, blood pressure, body weight.
Sandberg[24]	50	rdbpc	Healthy men	180 mg	Gerimax	8 wk	Improved ability to complete a detail-oriented editing task (letter-crossover cognition test) = less errors (−0.16 vs. +0.22, $P<0.025$), speed at which number of lines read in 5 min improved.
Wiklund[72,73]	390	rdbpc	middle-age 42 ±9 y	80 mg	Gericomplex G115	12 wk	Improved alertness, relaxation, appetite, overall score, and general well-being.
Smith[74]	19	rdbpc	Females 26 ± 1 y	200 mg	G115	8 wk	NS: POMS, PANAS (psychological tests), and RPE.
Sotaniemi[75]	36	rdbpc	Newly-diagnosed diabetics	100, 200 mg	G115	8 wk	Improved mood, psychophysical performance, physical activity; reduced fasting blood sugar, glycated hemoglobin and body weight.
Ussher[76]	95	rdbpc	Middle managers 39 ±10 y	80 mg	Gericomplex	8 wk	Improved overall stress (Multi-Modal Stress Questionnaire), vigor-activity (POMS). Trend: improved overall mood (POMS), physical stress and cognitive stress (MMSQ) NS: 10 other POMS, MMSQ measures, Satisfaction With Life Scale, General Affect Questionnaire, Self Esteem Questionnaire ratings. Changes favored ginseng group (marked positive skew), but variability too high for significance. Significant interaction between diet and positive effect.
Le Gal[77]	232	mcrdbpc	men/women long-time functional fatigue 25–60 y	80 mg	Gericomplex or G115	6 wk	Decreased Fatigue Score (2.8 vs. 3.7 from 9.3 each, $P = 0.019$); 5.7% ginseng but 15.2% placebo group reported fatigue ($P = 0.023$); efficacy rated better by subjects and physicians.
Caso Marasco[78]	625	mcrdbpc	Functional fatigue 18–65 y	200 mg	G115 (both groups got multiple vitamin/mineral)	12 wk	Improved quality of life (+11.9 vs. +6.4 — all 11 parameters sig by G115), prevention of increased body weight (+0.10 vs. +0.84), and diastolic BP: (−0.14 vs. +1.09) Took 3 mo for results to be significant. NS: Systolic BP, resting HR.

Author	N	Design	Subjects	Dose	Product	Duration	Results
Sorensen[79]	112	rdbpc	Healthy volunteers 40–70 y	400 mg	Gerimax standardized extract	8–9 wk	Improved auditory reaction times, 2/9 cognition tests. Trend: better abstract thinking. NS: memory, concentration, well-being.
Wesnes[31]	64	rdbpc	Healthy neurasthenia volunteers 40–65 y	160, 320, 640 mg	Ginkoba[d]	13 wk	Improved CDR computerized assessment system, global score from Symptom Checklist 90 Revised (640 mg dose).
Wiklund[80]	384	mcrdbpcpg	PMP women symptomatic	80 mg	Gericomplex	16 wk	Improved depression, well-being, health subscales. Trend for improved Psychological Well-Being Index ($P<0.10$). NS: Women's Health Questionnaire, Visual Analogue scale scores, vasomotor symptoms (hot flushes), physiological signs
Ziemba[36]	15	rdbpc	male soccer players 19±1 y	350 mg	KRKA, Novo Mesto, Poland	6 wk	Decreased reaction time before and during cycle exercise to exhaustion shortened (improved psychomotor performance).
Wesnes[81]	256	mcrdbpcpgra	Healthy middle-aged volunteers 56 y ave	320 mg o.d. or as b.i.d.	Ginkoba	12 wk + 2 wk followup	Improved Index of Memory Quality (from CDR computerized cognitive assessment system) 7.5%; working memory, long-term memory.
Ussher[82]	313	rdbpc	Healthy volunteers 49 ± 9 y	80 mg	Gericomplex	8 wk	Improved confusion bewilderment (POMS), behavioral strain (MMSQ). Restrained eaters subgroup had many improved results. Trend: most other variables NS: SF-36 pain scale
Kennedy[83]	20	rdbpcco	Healthy young	200, 400, 600 mg	G115	Acute, 1 wk washout	Improved quality of memory and secondary memory (400 mg dose). NS: 200, 600 mg doses. (-) for speed of attention (200, 600 mg), ratings of alertness at 6 h (200, 400 mg)
Kennedy[84]	20	rdbpcco	Healthy young	320, 640, 960 mg	Ginkoba	Acute, 1 wk washout 6 d tests	Improved performance of quality of memory for 960 mg; improved secondary memory; dose-dependent improvement in speed of attention.
Cardinal[85]	83	prdbpc	healthy 43 females 26 y ave	200, 400 mg	G115	8 wk	$P<0.016$ for significance (instead of usual $P<0.05$). NS: positive effect, negative effect, total mood disturbance.

(continued)

TABLE 20.2 (CONTINUED)
Results of Human Studies on Mental Performance Related to Ergogenicity with *Panax Ginseng*

Study (reference)	Subject n[1]	Study Design	Subjects (Age Range)	Daily Dose	Preparation Type	Study Duration	Effects (Statistically Significant Unless Otherwise Stated)
Scholey[86]	20	rdbpcco	Healthy young adults 21 y ave	3 doses each 320, 640, 960 mg	G115 and/or Ginkoba	Acute, 1 wk wash	Improved arithmetic tasks with differing cognitive loads, accuracy (significant time, dose, task-specific effects) (-) Slowed responses during Serial Sevens test. Ginkoba improved accuracy in Serial Sevens and Serial Threes test.
Kennedy[88]	20	rdbpcco	Healthy young adults 21 y ave	400 mg; 960 mg combo	G115; Ginkoba	Acute	Improved secondary performance on CDR battery, speed of performing mental tasks, accuracy of attentional tasks. NS: self-rated mood. Ginkoba improved secondary performance on CDR battery, subtraction tasks, self-rated mood.
Ellis and Reddy[87]	30	rdbpc	?	200 mg	?	8 wk	Improved social functioning, mental health, mental component summary from validated SF-36v2 Health Survey at 4, but not 8 wk. 58% of ginseng subjects thought they were getting active, only 17% of placebo ($P<0.05$).

[a] Abbreviations: ? — data not listed or unavailable; b.i.d. — twice daily; BP — blood pressure; CDR — Cognitive Drug Research; co — crossover; d — days; db — double-blind; DHEA — dehydroepiandrosterone; DHEAS — dehydroepiandrosterone sulfate; HR — heart rate; mc — multicenter; mo — months; n — subject number; NC — not controlled; NK — Natural Killer cells; NS — Not Significant; o.d. — once daily; PANAS — positive and negative affect schedule; pc — placebo controlled; pg — parallel groups; POMS — profile of mood survey; PMP — post-menopausal; r — randomized; RPE — ratings of perceived exertion; wk — weeks; y — years; (-) — negative effects;

[b] Gericomplex® — G115 40 mg, Vitamin A 2667 IU, vitamin B1 1.4 mg, vitamin B2 1.6 mg, vitamin B3 18 mg, vitamin B6 2 mg, vitamin B12 1 mcg, biotin 150 mcg, vitamin C 60 mg, vitamin D3 200 IU, vitamin E 10 mg, phosphorus 100 mg, iron 10 mg, calcium 100 mg, copper 2 mg, magnesium 10 mg, manganese 2.5 mg, zinc 1 mg per tablet. Geriatric Pharmaton® and Pharmaton complex was similar to Gericomplex® but with 26 mg dimethylaminoethanol per capsule. Gericomplex® and Geriatric Pharmaton® are registered trademarks of Boehringer Ingelheim Pharmaceuticals.

[c] G115® refers to a commercially available, proprietary *Panax ginseng* extract product containing 200 mg ginseng root extract standardized to 4% ginsenosides. G115® is a registered trademark of GPL Ginsana Products, Lugano, Switzerland, and Pharmaton Ltd., Switzerland (owned by Boehringer Ingelheim Pharmaceuticals).

[d] Ginkoba® — GK501 (100 mg G115 + 60 mg standardized *Ginkgo biloba* extract per softgel

Ginseng

TABLE 20.3

Asian Ginseng (*Panax ginseng*) — Physical Performance — Subject Number Effect Chi-Square 2 × 4 Contingency Table

Subject Number/Group	0–6	7–12	13–20	≥21	
Any significant findings	0	2	5	4	11
No significant findings	2	7	2	0	11
Total	2	9	7	4	22

Note: H_o = Subject number per group has no relation to finding significant effects. $\chi^2 = 10.064$, therefore reject H_o, $P<0.025$. Studies with lower subject numbers (and thus, lower statistical power) find significant effects less often.

TABLE 20.4

Asian Ginseng (*Panax ginseng*) — Mental and Physical Performance — Length of Administration Effect Chi-Square 2 × 4 Contingency Table

Length of Administration (Weeks)	< 4	4–8	8–12	> 12	
Any significant findings	2	5	11	13	31
No significant findings	3	3	5	0	11
Total	5	8	16	13	42

Note: H_o = Length of administration has no relation to finding significant effects. $\chi^2 = 8.308$, therefore reject H_o, $P<0.05$. Studies with longer duration (> 12 weeks) find significant effects more often.

listed in Table 20.1.[10-15,17-19,22-39] The null hypothesis tested and rejected was: Subject number per group has no relation to finding significant results. It is evident that low subject number is associated with lack of ergogenic effects. Since subject number relates directly to statistical power, especially for measurements as notoriously variable as human performance, it can be deduced that studies with low statistical power had a greater chance of committing a Type II statistical error (more difficult time finding any significant changes if they were present).[92] If these studies are then excluded, then almost two-thirds of controlled studies (11/17 or 65%) found some ergogenic effects for Asian ginseng.

Table 20.4 illustrates Chi-Square analyses of the effect of length of administration. Both mental and physical performance studies (42 total) were included in order to get an overview of the entire body of literature pertinent to ginseng.[10-16,18-39,62,65,66,68,69,71-82,85,87] Acute administration studies were excluded. The null hypothesis tested was: Study length has no relation to finding any significant results. Breaking the lengths of administration into commonly used monthly (four week) intervals found a significant relation between duration and finding any significant effects ($P<0.05$). In fact, all studies lasting 12 weeks or longer (13) found at least one significant outcome, whereas all studies finding no significant changes (11) lasted eight weeks or less. Thus, duration of study length is important for finding effects on mental and physical performance with Asian ginseng. This finding fits the traditional use of Asian ginseng as a tonic to be taken chronically over long periods of time.

Findings from this crude analysis mean that study quality has limited previous conclusions of authors and reviewers investigating Asian ginseng ergogenicity. Conclusions of previous reviews were based on studies not finding significant changes having equivalent weight. It is now evident that many studies on Asian ginseng may have not found any evidence of efficacy (Type II statistical error) for two simple reasons: 1) low statistical power (from low subject number); and 2) insufficient duration. If studies with low subject numbers and shorter durations are given less weight, then a majority of studies (31/42, 74%) support some ergogenic effects from standardized Asian ginseng.

V. Other Adaptogenic Herbs

In the sports nutrition marketplace, herbal products targeted for endurance performance have been switching from *Panax ginseng* to "newer" ginseng-like herbs such as *Cordyceps sinensis, Eleutherococcus (Acanthopanax) senticosus* and *Rhodiola* species. Thus, it is important to consider the most popular "ginseng alternatives." Human clinical studies on exercise performance with these herbs are scarce, but significant amounts of research from China, Russia, Japan and other countries have established putative active components, mechanisms of actions, effects on animals and some medical uses in humans.[2,3,8,40,93-109]

A. Cordyceps

Cordyceps sinensis (Brek.) Saccardo catapulted into the American marketplace in 1994 following reports of unprecedented, world record-breaking running performances by Chinese athletes.[100] Their coach attributed some of their success to a mixture of traditional Chinese herbs, including caterpillar fungus, *Cordyceps sinensis*.

Cordyceps is an unusual "herb" — it is actually the fruiting body of a fungus that is parasitic for certain species of caterpillars (*Hepialus armoricanus*).[96] Obviously, harvesting *Cordyceps* in the wild limited its use in traditional Chinese medicine to the elite, and *Cordyceps* was used mostly as a tonic for withering and wasting of old age, restoring Kidney Yang (Life Fire). *Cordyceps* has been cultivated outside of caterpillars to satisfy pharmaceutical demand in China, and commercial sources for dietary supplements are almost always extracted from these sources. Cultured *Cordyceps* mycelial extracts have been shown to possess the same purported active ingredients as the wild strain. In China, *Cordyceps* is an herbal pharmaceutical under investigation for treatment of immune deficiencies, cardiovascular diseases, diabetes, cancer and inflammatory diseases.[96,97]

Five controlled human studies on *Cordyceps* and exercise (all reports are abstracts) have appeared.[110-114] All studies appeared to use the same product (CordyMax Cs-4, Pharmanex, Simi Valley, CA) at 3 g/d for 4-12 wks, mirroring Chinese herbal pharmaceutical doses. VO_2max and anaerobic threshold were increased from baseline by *Cordyceps*, but not by placebo, in elderly Chinese adults.[110] Analysis of differences between groups was not reported. Highly fit athletes were given 4.5 g *Cordyceps*/d or placebo for six weeks.[111] VO_2, ventilatory threshold and VCO_2 were not different between groups after a maximal treadmill running test. During a submaximal running test, *Cordyceps* led to increases in oxygen pulse and decreases in respiratory exchange ratio and blood lactate. Similar testing in ten athletes found an 11% increase in maximal aerobic capacity from 4 wks of *Cordyceps* administration, similar to the study by Xiao.[112] However, the change was not statistically significant because of the wide variability and low subject number. Another study examined effects of *Cordyceps* in 110 normal sedentary subjects over a 12 wk period.[113] Significant improvements in time to walk one mile (33 sec difference from placebo group), work output (8.4% difference from placebo in Jeukendrup bike test), VO_2peak, exercise time to VO_2peak and a decrease in respiratory exchange ratio were found for the *Cordyceps* group, but not the placebo group. Finally, after six wks of administration to competitive cyclists, neither *Cordyceps* nor placebo affected VO_2max, ventilatory threshold, or a work-based time trial.[114]

Similar to *Panax ginseng* studies, larger subject numbers and longer times of administration were associated with more significant outcomes. At least each study appeared to use the same material, which allows for closer comparison than for other herbals. It appears that *Cordyceps* mycelial extract supplementation at 3 g/d may be associated with some improvements in exercise performance in untrained persons, but cannot be routinely expected for trained persons. These data indicate that *Cordyceps* may assist adaptation to exercise in untrained persons.

B. Eleutherococcus

Eleutherococcus senticosus is more commonly known as Siberian ginseng, and also by its traditional Chinese medicine name of Ciwujia (or variations in spelling thereof). *Acanthopanax senticosus* is an older term that is regaining popularity. However, *Panax ginseng* and *Panax quinquefolium* growers, suppliers and researchers have conducted a campaign to remove the word ginseng from non-*Panax* species in commercial use, so the favored term is now Eleutherococcus, not Siberian ginseng.[115]

In 1996, Eleutherococcus became popular with endurance athletes due to promotion of a product called Endurox®, which claimed a carbohydrate sparing action and increased utilization of fat for energy during exercise,

TABLE 20.5
Results of Human Studies with *Eleutherococcus senticosus* (Siberian Ginseng, Ciwujia) on Physical and Mental Performance

Study (Reference)	Subject n[a]	Study Design	Subjects (Age Range)	Daily Dose	Preparation Type	Study Duration	Effects (Statistically Significant Unless Otherwise Stated)
Blokhin[129] Walker[53]	71	Open	Weightlifters, gymnasts	?	? root and leaf extract	?	"...collected observations indicated that the adaptogen improved the general mental and physical state of the test subjects, increased their work capacity, and decreased their fatigue."[53]
Brekhman[40] Schezin[131] Walker[53]	60,000	Open	Test drivers	?	?	10 y	Decreased lost work time 25%, flu cases by 40%, improved overall general health.
Asano[130]	6	sbco	Baseball players 21–22 y	4 ml/d	Liquid extract = 300 mg dry extract/d	8 d	Improved VO_2max 12%, oxygen pulse 8%, total work 23%, time to exhaustion 16%. Possible order effect.
McNaughton[19]	30	rdbpcco	Healthy athletes	1000 mg/d	Root powder (Bioglan Laboratories)	6 wk	Pectoral strength increased 13%, quadriceps strength increased 15%. NS: VO_2max, HR recovery post-exercise, grip strength.
Bajevskij[142]	10	dbpcco	healthy males 20–30 y	7 caps/d	50 mg *Rhodiola rosea*/100 mg Eleutherococcus/150 mg Schizandra extracts/cap	7 d	Increased PWC170 (928 vs. 774); improved all cardiac rhythm measures (15); decreased HR; activated sympathetic nervous system (mobilizes energy). Order effect!
Goncharov[143]	11	dbpcco	healthy males 18–35 y	7 caps/d	50 mg *Rhodiola rosea*/100 mg Eleutherococcus/150 mg Schizandra extracts/cap	7 d order effect?	Decreased eye strain, mistakes on visual recognition of useful signals among distractors on CRT screen (at 12 and 24 h), improved some mental activity measures, EEG during early morning hours (omega potential), mood. NS reaction time, accuracy of task performance for 1st 12 h period.
Dowling[132]	20	Rdbpc	Elite distance 37 ± 8 y	3.4 ml/d	NutraPharm EtOH extract	6 wk	NS: time to exhaustion, HR, RPE, RER, VO_2, V_E, V_E/VO_2 lactate. Statistical power 0.16 to 0.52, meaning results inconclusive.

Ginseng

Ref	N	Design	Subjects	Dose	Form	Duration	Results
Wu[116,117] Campbell[119] Kaman[118]	8	Control group	Healthy males 25–35 y	800 mg/d	Endurox (1.6% eleutherosides)	2 wk	Decreased RQ, lactate by 31%, increased fat utilization 43%, improved HR recovery, increased anaerobic threshold 12%, oxygen utilization efficiency. No statistical analysis reported, possible order effects.
Azizov[133]	?	?	High class athletes 18–28 y	?	Elton	20 d	Decreased blood coagulation potential; normalized fibrinolysis; increased working capacity, recovery.
Azizov[134]	44	?	Athletes	?	Elton	20 d	Decreased urine chemiluminescence (antioxidant activity). urine MDA levels; increased physical working capacity.
Blumenthal[135]	NA	NA	NA	2–3 g/d	Root or equivalent extracts	NA	Tonic for invigoration and fortification in times of fatigue, debility or declining capacity for work and concentration; convalescence (no supportive data).
Dustman[121]; Plowman[122]	10	Rdbpcco	Healthy volunteers 24 y	800 mg/d	Endurox	10 d	NS: RER, HR, VO_2 lactate (Stairmaster).
Smeltzer[123]	1	ABAB	Healthy male 37 y	800 mg/d	Endurox	8 wk total	NS: RER, HR, VO_2 lactate (treadmill test, 2 wk periods).
Stout[124]	9	Rdbpcco	College male athletes 19 ± 2 y	800 mg/d	Endurox	10 d	NS: critical power (W), work (kJ) during anaerobic cycling.
Wu[120]	13	Control group	Healthy volunteers 50–57 y	800 mg/d	Endurox	2 wk	Decreased RQ (27% increase in lipid utilization as fuel during exercise), HR 9%; increased O_2 uptake per heartbeat 16%.
Cheuvront[126,127]	10	rdbpcco	male college students 24 ± 4 y	800 mg/d	Endurox	7 d	NS: HR, RPE, RER, VO_2, V_E, V_E/VO_2, energy, fat oxidation, plasma glycerol, lactate (cycle ergometer protocol).
Eschbach[127,128]	9	rdbco	Male cyclist volunteers 28 ± 2 y	1200 mg/d	Endurox	7 d	NS: O_2 consumption, RER, HR, RPE, lactate, glucose and 10 km cycling time (17.83 vs. 18.1 min).
Chase[136]	11	rdbpcco	Healthy volunteers 20–24 y	900 mg/d	?	14 d	Decreased lactate at rest. NS: time to exhaustion (treadmill), HR, RPE, RER, VO_2, V_E, V_E/VO_2, lactate.
Szolomicki[51]	35	NC	Healthy volunteers	75 drops	Taiga Wurzel liquid	1 mo	Improved oxygen plateau.

(continued)

TABLE 20.5 (CONTINUED)
Results of Human Studies with *Eleutherococcus senticosus* (Siberian Ginseng, Ciwujia) on Physical and Mental Performance

Study (Reference)	Subject n[a]	Study Design	Subjects (Age Range)	Daily Dose	Preparation Type	Study Duration	Effects (Statistically Significant Unless Otherwise Stated)
Gaffney[43]	18	rdbpc	Endurance club athletes 18-40 y	8 ml	Mediherb 35% EtOH extract, equivalent to 4 g/d dried root	6 wk	Decreased testosterone/cortisol ratio 29%. Trend: increased cortisol. NS: testosterone, T cell counts, CD4/CD8, NK cell count, lymphocyte count, B cell count.
Johnson[37]	10	rdbpcco	5 male, 5 female recreationally active 26 ± 5 y	1800 mg	Panax ginseng, Eleutherococcus, enoki, reishi, cordyceps, citrus reticulata	2 wk periods; 1 wk washout	NS: lactate post-weightlifting (0-15 min), % lactate clearance.

[a] Abbreviations: ? — not listed or unavailable; co — crossover; d — days; db — double-blind; EtOH — ethanol; HR — heart rate; kJ — kilojoules; MDA — malondialdehyde; n — subject number; NA — Not Applicable; NK — Natural Killer cells; NS — Not Significant; pc — placebo controlled; r — randomized; RER — respiratory exchange ratio; RPE — ratings of perceived exertion; RQ — respiratory quotient; V_E — expiratory ventilation; VO_2 — oxygen uptake; VO_{2max} — maximal oxygen uptake; W — watts; wk — weeks; y — years.

based on Chinese studies that have still not been published in an English-language, peer-reviewed journal (see www.endurox.com).[116-120] No studies from the manufacturer reported improvements in actual performance measurements — only metabolic parameters. Other investigators examined effects of Endurox® on performance and metabolic parameters during exercise, and did not find any improvements over placebo effects (see Table 20.5).[121-128]

The disparity of different preparations, dosages, amounts of putative active agents, study designs, subject numbers, training status, subject ages, measurements, lengths of administration, and low statistical power all conspire to make an uncertain consensus (Table 20.5). Most studies are characterized by very short durations and low subject numbers. However, the majority of controlled studies do not support improvements in physical performance from Eleutherococcus preparations.[2,3,8,19,37,40,43,116-136]

C. Rhodiola species

Rhodiola species are hardy shrubs that grow in circumpolar or high altitude regions.[93-95] The root is golden in color and aromatic, spawning popular names of Goldenroot, roseroot and Arctic ginseng. *Rhodiola rosea* root/rhizome extracts have been used as traditional medicine in Scandinavian and Slavic countries for centuries for increasing physical endurance, work productivity, longevity, altitude sickness, fatigue, infections, gastrointestinal disorders, impotence and fertility.[93-95] Other *Rhodiola* species have been used in Tibetan and Chinese traditional medicine practices (*R. crenulata*, *R. kirilowii*). Interestingly, only *Rhodiola rosea*, and not other *Rhodiola* species, possesses a unique class of cinammyl alcohols called rosavin, rosin and rosarin.[93-95]

After 1960, systematic scientific investigations by Russian and Scandinavian researchers produced a prodigious literature on *Rhodiola rosea*, most of which is not in English. The Soviet Ministry of Health settled on standardization of *Rhodiola rosea* root extracts to 3% rosavin and 0.8-1.0% salidroside contents, and in 1975, approved Rhodiola Extract Liquid (in 40% ethanol) for use as a stimulant in fatigue (asthenia), infectious illnesses, psychiatric conditions, and in healthy persons to relieve fatigue, improve memory, improve attention span and work productivity.[93-95] Sweden and Denmark also have approved *Rhodiola* extracts (SHR-5) as antifatigue stimulants and adaptogens.[93,137]

Table 20.6 lists available data on human studies using *Rhodiola* extracts for physical and mental performance.[93-95,137-150] It appears that *Rhodiola* extracts studied can be divided into three categories: 1) the original Russian ethanolic tincture: 2) the current Danish and Swedish herbal product SHR-5; and 3) and non-*rosea* species. Taken together, the results indicate that some parameters of physical and mental performance were improved by *Rhodiola* extract administration: increased work,[138,141,142,146,147] increased run time to exhaustion[141], heart rate recovery post-exercise[138,142] and lessening of fatigue

TABLE 20.6
Results of Human Studies with Rhodiola (*Rhodiola rosea* Extracts; Other *Rhodiola* species) on Physical and Mental Performance

Study (Reference)	Subject n[a]	Study Design	Subjects (Age Range)	Daily Dose	Preparation Type	Study Duration	Effects (Statistically Significant Unless Otherwise Stated)
Brown[93] Saratikov[138]	42	pc	Master level biathlete skiers 20–25 y	15 drops (150 mg 3% rosavins)	Ethanolic *Rhodiola rosea* liquid extract (3% rosavins)	Acute 30–60 min pre-race	Improved shooting accuracy, coordination, recovery time, strength, cardiovascular measures; decreased arm tremor; post race HR (30 min 105% baseline vs. 129% baseline ($P<0.02$)).
Germano[94] Saratikov[138]	52	pc	normal males 18–24 y	15 drops	Ethanolic *Rhodiola rosea* liquid extract (3% rosavins)	Acute	Improved maximal work 9% vs placebo, recovery HR (67 bpm vs. 87 bpm placebo).
Germano[94] Saratikov[138]	27	?	healthy volunteers 19–46 y	5–10 drops b.i.d.	Ethanolic *Rhodiola rosea* liquid extract (3% rosavins)	2–3 wk	Prevented asthenic decompensation in 55% of subjects; improved mental work performance under stressful conditions.
Saratikov[139]	?	?	healthy volunteers 19–30 y	5–10 drops b.i.d.	Ethanolic *Rhodiola rosea* liquid extract (3% rosavins)	2–3 wk	Normalized BP, decreased breathing rate at rest.
Saratikov[139]	37	?	Factory workers and n = 6 pilots	5–10 drops b.i.d.	Ethanolic *Rhodiola rosea* liquid extract (3% rosavins)	2 wk	Improved hearing (increased air, bone conduction of speech tones).
Germano[94] Saratikov[139]	?	Clinical observations	Healthy persons, athletes, military	5–10 drops b.i.d.	Ethanolic *Rhodiola rosea* liquid extract (3% rosavins)	?	Approved as stimulative remedy for healthy people during fatigue, work asthenia; sports medicine uses against flight exhaustion during intense muscular work; accelerate recovery from intense training.
Saratikov[139] Zhang[140]	?	?	Highland villagers	?	Rhodiola kirilowii	?	Prevented cardiopulmonary function abnormalities moving from 2500m to 4475m altitude.
Qian[141]	26	Control group (n = 15)	professional athletes 18–25 y	1.5 g/d	Rhodiola crenulata crude herb	75 d	Increased V_E, VO_2max, work capacity (treadmill), sports scores compared to controls.
Bajevskij[142]	10	dbpcco Order effect!	healthy males 20–30 y	7 caps/d	50 mg *Rhodiola rosea*/100 mg Eleuterococcus/150 mg Schizandra extracts/cap	7 d	Increased PWC170 (928 vs. 774); improved all cardiac rhythm measures (15); decreased HR; activated sympathetic nervous system (mobilizes energy).

Study	N	Design	Subjects	Dose	Preparation	Duration	Results
Goncharov[143]	11	dbpcco	healthy males 18–35 y	7 caps/d	50 mg *Rhodiola rosea*/100 mg Eleutherococcus/150 mg Schizandra extracts/cap	7 d order effect?	Decreased eye strain, mistakes on visual recognition of useful signals among distractors on CRT screen (at 12 and 24 h), improved some mental activity measures, EEG during early morning hours (omega potential), mood. NS reaction time, accuracy of task performance for 1st 12 h period.
Ramazanov[144] Anonymous[145]	77	pc	Normal volunteers	200 mg	*Rhodiola rosea* standardized extract or rosavin or salidroside	Acute	Increased serum FFAs post-exercise (30 min walk) from 100 to 116% vs. 100 to 104% for placebo group. Animal studies showed Rosavin, not salidroside, was active, but *R. Rosea* extract (with both) most active for serum FFAs post-exercise.
Spasov[146,147]	40	rdbpcco	Foreign male high school students (from India in Russia)	50 mg b.i.d.	185 mg/cap *Rhodiola rosea* SHR-5 extract (Rhodaxon) — 3.6% rosavin, 1.6% salidroside	20 d	Improved test scores (3.47 vs. 3.20 out of 4.00), physical fitness (cycle ergometry work), mental fatigue, neuromotoric (kinesthesiometric sensitivity) tests, self-assessed general well-being, psychic fatigue, situational anxiety. NS: correction of text tests, neuromuscular tapping test (dose suboptimal).
Darbinyan[148]	56	rdbpcco	Night duty physicians 24–35 y	1 cap	185 mg/cap *Rhodiola rosea* SHR-5 extract	2 wk periods	Improved 5 test Fatigue Index in 1st 2 wk period, reduced general fatigue under stressful conditions (night shift duty). NS: 2nd 2 wk period.
Xiu[149]	40	pc	Normal subjects	2 g/d	*Rhodiola crenulata* standardized extract (2% salidroside)	1 mo	Increased blood oxygen 12.6%, serum testosterone 76% vs. no change for placebo; improved subjective reports of memory, physical strength, sexual performance, sleep; serum testosterone increased 76% vs. no change from placebo. NS: serum estradiol, SOD.

(continued)

TABLE 20.6 (CONTINUED)
Results of Human Studies with Rhodiola (*Rhodiola rosea* Extracts; Other *Rhodiola* species) on Physical and Mental Performance

Study (Reference)	Subject n[a]	Study Design	Subjects (Age Range)	Daily Dose	Preparation Type	Study Duration	Effects (Statistically Significant Unless Otherwise Stated)
Shevtsov[137]	121	rdbpc w ctrl group	Military cadets night duty 19–21 y	2 or 3 caps	185 mg/cap *Rhodiola rosea* SHR-5 extract	Acute	Improved antifatigue index for mental work while fatigued; decreased errors on ring attention and numbers tests. Trend: feelings of well-being, short-term memory test ($P = 0.08$), larger pulse pressure (systolic-diastolic BP). NS: HR (-2 vs. 0 bpm); differences between Rhodiola doses.
Wing[150]	15	rdbpc	Healthy volunteers made hypoxic (4600 m) 20–33 y	?	*Rhodiola rosea*	7 d	Trend: decreased LPO ($P = 0.10$) instead of increased control levels after 60 min hypoxic exposure. NS: MDA; SaO_2, PCO_2 decreased similarly.

[a] Abbreviations: ? = data not listed or unavailable; b.i.d. — twice daily; BP — Blood Pressure; cap — capsule; co — crossover; d — days; db — double-blind; EEG — electroencephalogram; FFAs — free fatty acids; HR — Heart Rate; LPO — lipid peroxides; MDA — malondialdehyde; n — number; NA — Not Applicable; NS — Not Significant; pc — placebo-controlled; PCO_2 — carbon dioxide pressure; PWC — peak work capcacity; r — randomized; RER — Respiratory Exchange Ratio; RPE — Ratings of Perceived Exertion; SaO_2 — oxygen saturation; SOD — superoxide dismutase; V_E — expiratory ventilation; VO_2 — Oxygen consumption; VO_2max — Maximal Oxygen Consumption; W — Watts; y — years.

(asthenia).[137,139,146-148] Mental performance measurements (EEG, memory, reduction of errors performing tasks, shooting accuracy, hearing, mood, high school test scores) improved in most, but not all, studies.[137-139,143,146-148] One American study on *Rhodiola* extract found a trend for reduced serum lipoperoxides, but not MDA, and no effect on oxygen or carbon dioxide saturation after exposure to hypoxia (simulated 4600m).[150] Most of the studies showing benefits examined persons under considerable stress loads.

Rhodiola extracts appear to have some positive effects on physical and mental performance during periods of increased stress, according to controlled studies currently available. However, there is a need for further research on commercially available *Rhodiola* extracts (such as SHR-5 from Sweden or extracts standardized to both rosavin and salidroside) performed with adequate statistical power along with metabolic measurements and performance outcomes for confirmation of the unclear credibility of earlier reports. At this point in time, *Rhodiola* extracts that match the standardized extracts used in human clinical trials show potential for improving some aspects of mental and physical performance in athletes and sedentary persons.

VI. Unresolved Issues and Guide for Future Research

It is apparent that the quality of research on ergogenicity of *Panax ginseng* and other herbals leaves much to be desired. Sweeping generalizations of lack of efficacy have relied on studies that are known to have serious flaws (low statistical power and short duration) and likelihood of Type II statistical errors.

Guidelines for future research on ginseng and human performance were pointed out by Bahrke and Morgan[2] and are reiterated here. First, identification of the material used, with independent assays of ginsenoside content, are critical for comparing studies. Fortunately, most studies used commercial products (Ginsana G115® and its family of products) that have been shown to contain consistent amounts of ginsenosides. There are several testing laboratories that will test for total and individual ginsenosides, so availability of testing is not an issue. Also, independent testing can rule out contaminating substances (such as caffeine) which have been shown to occur in a few ginseng products.

Second, calculation and reporting in publications of the statistical power is vital to assess the strength of the study in question, but rarely reported. For example, Dowling and coauthors did report low statistical power (0.16-0.52) in their study on six subjects per group given *Eleutherococcus*.[132] They also found no significant changes. Thus, the reader can conclude that there is a chance for a Type II error, and this knowledge can temper any overreliance on this study's results. On the other hand, Wiklund and coauthors calculated that a minimum number of subjects needed for a statistical power of 0.80 (with alpha of 0.01 instead of the normal 0.05) was 175 per group, a number which was exceeded at study's end.[72] This study did find significant effects of ginseng.

Third, duration of administration is now shown to be important for studying effects of Asian ginseng, and perhaps, other herbs. Thus, studies lasting eight weeks or less can be criticized for not being of sufficient duration to find potential effects, if none are found. Finally, the continuous need for rigor in experimental design and reporting of methodology (Good Clinical Practices) has not abated. Future studies in this area now need to recognize that duration and statistical power analysis are crucial for a credible study.

Although not considered in this review on performance, safety issues should be considered in human studies, and objective measurements of safety should be built in to future studies on herbals. In addition, some athletes are concerned about drug testing and sending samples used in research studies to testing laboratories for doping analysis would yield very practical advice for consumers and administrators alike. One study has already shown that an American ginseng product did not cause positive doping tests.[60]

VII. Summary

This chapter has endeavored to explore some of the reasons why there are such overall discrepancies in research results on ginseng and ginseng-like herbs. Most studies with low statistical power and short duration found no significant effects, skewing interpretations of lack of efficacy. There are simply too many well-designed studies that exhibited significant improvements for ginseng groups to be able to say ginseng products are ineffective. On the other hand, significant influences were not universal in studies finding ergogenic effects. In other words, some, but not all, measured parameters of performance were significantly improved. It is now obvious that study quality and overzealous generalizations have created two camps: believers and disbelievers. Each camp can quote apparently competent and reliable scientific research to support efficacy or lack of efficacy for Asian ginseng extracts, which has created confusion amongst consumers and investigators alike.

It is now clear that Asian ginseng extracts standardized to 4% or more total ginsenosides may lead to some improvements beyond a placebo effect for mental and physical performance under certain conditions. Longer times of administration (over 12 weeks) during periods of substandard performance (for a variety of reasons, including advancing years and stressful conditions) seem to augur for modest but not universal improvements in feelings of well-being, mood, reaction times, neuromuscular control, mental functions, work capacity and endurance performance. Short-term use of Asian ginseng (four weeks or less) appears to have no effects on mental or physical performance beyond a placebo effect. Adding other nutrients or herbs to Asian ginseng appears to produce significant effects more often or in shorter time periods. American ginseng does not appear to be ergogenic, but only a couple of small studies have been performed. Other ginseng species (*Panax notoginseng*, *P. japonicus*) remain unstudied.

Ginseng-like herbs are slowly replacing Asian ginseng products in the marketplace, but research is less well developed for the predominant alternatives: *Cordyceps*, *Eleutherococcus* and *Rhodiola*. To date, data are leaning towards some efficacy for improvement of physical and mental performance for standardized *Cordyceps* and *Rhodiola* herbal products. Improvements in physical and mental performance by *Eleutherococcus* preparations are not supported by the available research, although quality of human studies have been extremely low. Combinations are even less studied, although many products combine these herbs.

Acknowledgments

The authors wish to thank Brooke Bouwhuis, Camilla Kragius, Janet Sorenson, Stephanie Blum and Trudy Day for expert technical assistance.

References

1. Bahrke, M.S., and Morgan, W.P., Evaluation of the ergogenic properties of ginseng, *Sports Med.*, 18, 229, 1994.
2. Bahrke, M.S., and Morgan, W.P., Evaluation of the ergogenic properties of ginseng: An update, *Sports Med.*, 29, 113, 2000.
3. Bucci, L.R., Selected herbals and human exercise performance, *Am. J. Clin. Nutr.*, 72, 624S, 2000.
4. Carr, C.J., Natural plant products that enhance performance and endurance, in *Enhancers of Performance and Endurance*, Carr C.J., Jokl E., Eds., Lawrence Erlbaum Associates, Hillsdale, NJ, 1986, 138.
5. Coleman, C.I., Hebert, J.H., Reddy, P., The effects of *Panax ginseng* on quality of life, *J. Clin. Pharm. Ther.*, 28, 5, 2003.
6. Hobbs, C., *The Ginsengs. A User's Guide*, Botanica Press, Santa Cruz, CA, 1996.
7. Lieberman, H.R., The effects of ginseng, ephedrine, and caffeine on cognitive performance, mood and energy, *Nutr. Rev.*, 59, 91, 2001.
8. Vogler, B.K., Pittler, M.H., and Ernst, E., The efficacy of ginseng. A systematic review of randomized clinical trials, *Eur. J. Clin. Pharmacol.*, 55, 567, 1999.
9. World Health Organization, Radix Ginseng, in *WHO monographs on selected medicinal plants*, Vol. 1, World Health Organization, Geneva, Switzerland, 1999, 168.
10. Dörling, E., Kirchdorfer, A.M., and Ruckert, K.H., Haben Ginsenoside Einfluβ auf das Leistungsvermögen? Ergebnisse einer Doppelblindstudie. [Do ginsenosides influence the performance? Results of a double-blind study], *Notabene Med.*, 10, 241, 1980.

11. Sandberg, F., Vitalitet och senilitet — effekten av ginsengglykosider på prestationsförmågan, [Vitality and senility — the effects of the ginsenosides on performance], *Svensk. Farmaceut. Tidskrift,* 84, 499, 1980.
12. Forgo, I., Kayasseh, L., and Straub, J.J., Einfluß eines standardisierten Ginseng-Extraktes auf das Allgemeinbefinden, die Reaktionsfähigkeit, Lungenfunktion und die gonadalen Hormone, [Influence of a standardized ginseng extract on general well-being, reaction capacity, pulmonary function and gonadal hormones], *Med. Welt,* 32, 751, 1981.
13. Knapik, J.J., Wright, J.E., Welch, M.J., Patton, J.F., Suek, L.L., Mello, R.P., Rock, P.B., and Teves, M.A., The influence of *Panax ginseng* on indices of substrate utilization during repeated, exhaustive exercise in man, *Fed. Proc.,* 42, 336, 1983. (abstract #257)
14. Teves, M.A., Wright, J.E., Welch, M.J., Patton, J.F., Mello, R.P., Rock, P.B., Knapik, J.J., Vogel, J.A., and der Mardosian, A., Effects of ginseng on repeated bouts of exhaustive exercise, *Med. Sci. Sports. Exer.,* 15, 162, 1983. (abstract #17)
15. Forgo, I., Wirkung von Pharmaka auf Körperliche Leistung und Hormonsystem von Sportlern. 2. Mitteilung. [Effect of drugs on physical performance and hormone system of sportsmen. 2.], *Münch. Med. Woch.,* 125, 822, 1983.
16. Forgo, I., and Schimert, G., Zur Frage der Wirkungsdauer des standardisierten Ginseng-Extraktes G115 bei gesunden Leistungssportlern. [The duration of effect of the standardized ginseng extract G115 in healthy competitive athletes], *Notabene Med.,* 15, 636, 1985.
17. Macareg, P.V.J., and Ramos, E., The effect of ginseng on exercise duration, *The XXIII FIMS World Congress of Sports Medicine Abstracts,* Federation Internationale de Medicine Sportive, Champaign, IL, 1986, 151. (abstract)
18. Tesch, P.A., Johansson, H., and Kaise, P., Effekten av ginseng, vitaminer och mineraler på fysisk arbetsförmåga hos medelålders män, [Effect of ginseng, vitamins and minerals on work capacity in middle aged men], *Läkartidningen,* 84, 4326, 1987.
19. McNaughton, L., Egan, G., Caelli, G., A comparison of Chinese and Russian ginseng as ergogenic aids to improve various facets of physical fitness, *Int. Clin. Nutr. Rev.,* 9, 32, 1989.
20. Gribaudo, C.G., Ganzit, G.P., and Verzini, E.F., Effetti sulla forza e sulla fatica muscolare di un prodotto ergogenico di origine naturale, *Med. Sport,* 43, 241, 1990.
21. Gribaudo, C.G., Ganzit, G.P., Biancotti, P.P., Astegiano, P., Effetti della somministrazione di un prodotto naturale ergogenico sulle doti aerobiche di ciclista agonisti, *Med. Sport,* 44, 335, 1991.
22. Pieralisi, G., Ripari, P., and Vecchiet, L., Effects of a standardized ginseng extract combined with dimethylaminoethanol bitartrate, vitamins, minerals, and trace elements on physical performance during exercise, *Clin. Ther.,* 13, 373, 1991.
23. Van Schepdael, P., Les effects du ginseng G115 sur la capacité physique de sportifs d'endurance, *Acta Ther.,* 19, 337, 1993.
24. Sandberg, F., and Dencker, L., Experimental and clinical tests on ginseng, *Z. Phytother.,* 15, 38, 1994.
25. Engels, H.J., Said, J., Wirth, J.C., and Zhu, W., Effect of chronic ginseng intake on metabolic responses during and in the recovery from graded maximal exercise, *Med. Sci. Sports Exerc.,* 27, S147, 1995. (abstract)

26. Engels, H.J., Said, J.M., and Wirth, J.C., Failure of chronic ginseng supplementation to affect work performance and energy metabolism in healthy adult females, *Nutr. Res.*, 16, 1295, 1996.
27. Cherdrungsi, P., and Rungroeng, K., Effects of standardized ginseng extract and exercise training on aerobic and anaerobic exercise capacities in humans, *Korean J. Ginseng Sci.*, 19, 93, 1995.
28. Lifton, B., Otto, R.M., and Wygand, J., The effect of ginseng on acute maximal aerobic exercise, *Med. Sci. Sports. Exer.*, 29, S249, 1997. (abstract #1414)
29. Engels, H.J., and Wirth, J.C., No ergogenic effects of ginseng (*Panax ginseng* C.A. Meyer) during graded maximal aerobic exercise, *J. Am. Diet Assoc.*, 97, 1110, 1997.
30. Engels, J.H., and Wirth, J.C., No ergogenic effects of ginseng (Panax ginseng C.A. Meyer) during graded maximal aerobic exercise, *Int. J. Sport Nutr.*, 8, 202, 1998. (abstract)
31. Wesnes, K.A., Faleni, R.A., Hefting, N.R., Hoogsteen, G., Houben, J.J., Jenkins, E., Jonkman, J.H., Leonard, J., Petrini, O., and van Lier, J.J., The cognitive, subjective, and physical effects of a ginkgo biloba/panax ginseng combination in healthy volunteers with neurasthenic complaints, *Psychopharmacol. Bull.*, 33, 677, 1997.
32. Allen, J.D., McLung, J., Nelson, A.G., and Welsch M., Ginseng supplementation does not enhance healthy young adult's peak aerobic exercise performance, *J. Am. Coll. Nutr.*, 17, 462, 1998.
33. Engels, H.J., Feng, S., and Chen, Y., Effect of ginseng (G115) on maximal aerobic exercise performance in aerobically fit young adults, *The XXVI FIMS World Congress of Sports Medicine Abstracts*, Federation Internationale de Medicine Sportive, Champaign, IL, 1998, 33. (abstract)
34. Kolokouri, I., Engels, H.J., Cieslak, T., and Wirth, J.C.., Effect of chronic ginseng supplementation on short duration, supramaximal exercise test performance, *Med. Sci. Sports Exer.*, 31, S117, 1999. (abstract #445)
35. Engels, H.J., Kolokouri. I., Cieslak, T.J., and Wirth, J.C., Effects of ginseng supplementation on supramaximal exercise performance and short-term recovery, *J. Strength Cond. Res.*, 15, 290, 2001.
36. Ziemba, A.W., Chmura, J., Kaciuba-Uscilko, H., Nazar, K., Wisnik, P., and Gawronski, W., Ginseng treatment improves psychomotor performance at rest and during graded exercise in young athletes, *Int. J. Sport Nutr.*, 9, 371, 1999.
37. Johnson, S.N., Plowman, S.A., DeLancey, M.R., Larson, B.M., Rudie, L., Miller, B., and Rzeutko, K., Effects of a ginseng and mushroom based herbal supplement on lactate response to resistance exercise, *Med. Sci. Sports Exerc.*, 34, S232, 2002. (abstract #1300)
38. Engels, H.J., Fahlman, M., Wirth, J.C., and Cieslak, T.J., Effects of ginseng on secretory immunoglobulin A, performance, and recovery from repetitive exhaustive exercise, *Med. Sci. Sports Exerc.*, 34, S233, 2002. (abstract #1306)
39. Engels, H.J., Fahlman, M.M., and Wirth, J.C., Effects of ginseng on secretory IgA, performance, and recovery from interval exercise, *Med. Sci. Sports Exerc.*, 35, 690, 2003.
40. Brekhman, I.I., and Dardymov, I.V., New substances of plant origin which increase nonspecific resistance, *Annu. Rev. Pharmacol.*, 9, 419, 1969.

41. Forgo, I., and Kirchdorfer, A.M., Ginseng steigert die körperliche Leistung. Kreislaufphysiologische Untersuchungen an Spitzensportlern beweisen: Der Stoffwechsel wird aktiviert. [Ginseng increases physical capacity of the body. Metabolic physiology examinations on world class athletes prove: metabolism is activated], *Ärztl. Prax.*, 33, 1784, 1981.
42. Forgo, I., and Kirchdorfer, A.M., Die Wirkung unterschiedlicher Ginsenosid-Konzentrationen auf die körperliche Leistungsfähigkeit. [The effect of different ginsenoside concentrations on physical work capacity], *Notabene Med.*, 12, 721, 1982.
43. Gaffney, B.T., Hugel, H.M., and Rich, P.A., The effects of Eleutherococcus senticosus and Panax ginseng on steroidal hormone indices of stress and lymphocyte subset numbers in endurance athletes, *Life Sci.*, 70, 431, 2001.
44. Gross, D., Shnekman, Z., Bleiberg, B., Dayan, M., Gittelson, M., and Efrat, R., Ginseng improves pulmonary functions and exercise capacity in patients with COPD, *Monaldi Arch. Chest Dis.*, 57, 242, 2002.
45. Johnson, A., et al. Whole ginseng effects on human response to demands for performance, *Proceedings of the 3rd International Ginseng Symposium*, Korea Ginseng Research Institute, 1980.
46. Kang, H.Y., Endogenous anabolic hormonal and growth factor responses to resistance exercise in carbohydrate and/or ginseng consumption, *Med. Sci. Sports Exerc.*, 31, S125, 1999. (abstract #496)
47. Kang H.Y., Kim S.H., Lee W.J., and Byrne, H.K., Effects of ginseng ingestion on growth hormone, testosterone, cortisol and insulin-like growth factor 1 responses to acute resistance exercise, *J. Strength Cond. Res.*, 16, 179, 2002.
48. Kiesewetter, H., Jung, F., Mrowietz, C., and Wenzel, E., Hemorrheological and circulatory effects of Gincosan, *Int. J. Clin. Pharmacol. Ther. Toxicol.*; 30, 97, 1992.
49. Murano, S., and Lo Russo, R., Experiencia con ARM 229. *Prensa Med. Argent.*, 71, 178, 1984.
50. Ng C.C., and Ng, Y.K., The effect of ginseng: a traditional Chinese medicine on physical capacity. *The XXIII FIMS World Congress of Sports Medicine Abstracts*, Federation Internationale de Medicine Sportive, Champaign, IL, 1986, 196. (abstract)
51. Tuttle, B.D., Dietary supplement for increasing energy, strength and immune function, US Patent 6,465,018, October 15, 2002.
52. von Ardenne, M., Measurements of the increase in the difference between the arterial and venous Hb-oxygen saturation obtained with daily administration of 200 mg standardized ginseng extract G115 for four weeks: Long-term increase of the oxygen transport into the organs and tissues of the organism through biologically active substances, *Panminerva Med.*, 29, 143, 1987.
53. Walker, M., Adaptogens: Nature's answer to stress, *Townsend Letter for Doctors*, 132, 751, 1994.
54. Awang, D.V.C., The anti-stress potential of North American ginseng (*Panax quinquefolius* L.), *J. Herbs Spices Medicinal Plants*, 6, 87, 1998.
55. Kitts, D., and Hu, C., Efficacy and safety of ginseng, *Public Health Nutr.* 3(4A), 473, 2000.
56. Popov, I.M., and Goldwag, W.J., A review of the properties and clinical effects of ginseng, *Am. J. Chin. Med.*, 1, 263, 1973.
57. Morris, A.C., Jacobs, I., Klugerman, A., and McLellan, T.M., No ergogenic effect of ginseng extract ingestion, Med. Sci. Sports Exer., 26, 56, 1994. (abstract #35)

58. Morris, A.C., Jacobs, I., McLellan, T.M., Klugerman, A., Wang, L.C., and Zamecnik, J., No ergogenic effect of ginseng ingestion, Int. J. Sport Nutr.,6, 263, 1996.
59. Humphreys, J.D., Robbins, S.J., Harber, V.J., Field, C.J., and McCargar, L.J., North American ginseng alters insulin response in exercising men, Can. J. Appl. Physiol., 26, S256, 2001.
60. Goel, D.P., Geiger, J.D., Kriellaars, D., Ayotte, C., and Pierce, G.N., Doping control urinalysis of a ginseng extract, Cold-FX®, in Canadian athletes, Can. J. Appl. Physiol., 26, S254, 2001. (abstract)
61. Sandberg, F., [Clinical effects of ginseng preparations], Z. Präk. Klin. Geriat., 4, 264, 1974.
62. Revers, W., Simon, W.C.M., Popp, F., Revers, R.C., Psychologische Wirkung eines Geriatricums auf alte Menschen, [Psychological effects of a geriatric preparation in the aged], Z. Präk. Klin. Geriat., 9, 418, 1976.
63. Schmidt, U.J., Kalbe, J., Schulz, F.H., and Brüscke, G., Pharmacotherapy and so-called basic therapy in old age. XI International Congress of Gerontology, Tokyo, August 20-25, 1978.
64. Hong, S.S., The clinical effects of Panax ginseng, in Korean Ginseng, 2nd ed., Bae, H.W., Ed., Korean Ginseng Research Institute, Korea, 1978, 162.
65. Hallstrom, C., Fulder, S., and Carruthers, M., Effects of ginseng on the performance of nurses on night duty, Comparat. Med. East West, 6, 277, 1982.
66. Garay Lillo, J., and de la Torre Adabia, A., Estudio clínico doble ciego de Pharmaton Complex en los procesos de involución. [Double-blind clinical study of Geriatric Pharmaton in the process of involution.], Geriátrika, 1, 29, 1984.
67. Le Faou, M., The effect of Geriatric Pharmaton versus placebo on physical fatigue and recovery, Sport et Medecine, 41, 34, 1985.
68. D'Angelo, L., Grimaldi, R., Caravaggi, M., Marcoli, M., Perucca, E., Lecchini, S., Frigo, G.M., and Crema, A., A double-blind, placebo-controlled clinical study on the effect of a standardized ginseng extract on psychomotor performance in healthy volunteers, J. Ethnopharmacol., 16, 15, 1986.
69. Garay Lillo, J., Estudio de farmacovigilancia con Pharmaton Complex durante seis menses. [Study over 6 months on the efficacy and tolerability of Pharmaton complex.], Geriátrika, 3, 441, 1987.
70. Rosenfeld, M.S., Nachtajler, S.P., Schwartz, G.T., and Sikorsky, N M , Evaluation of the efficacy of a standardised ginseng extract in patients with psychological asthenia and neurological disorders, La Semana Medica, 173, 148, 1989.
71. Garay Lillo, J., Garcia, J.C.C., Mauricio, G.C., Rubio, J.F., Gonzalez y Gonzalez, J.A., Manzanares, C.M., and de la Torre Abadia, A., Estudio multi-centrico a largo plazo con Pharmacon complex en pacientes adultos. [Long-term, multicenter study with Pharmaton Complex in adult patients.] Geriátrika, 8, 290, 1992.
72. Wiklund, I., Karlberg, J., and Lund, B., A double-blind comparison of the effect on quality of life of a combination of vital substances including standardized ginseng G115 and placebo, Curr. Ther. Res., 55, 32, 1994.
73. Wiklund, I., Karlberg, J., and Lund, B., Forbattrad livskvalitet av ginsengpreparat? Positiva effekter pa friska forvarvsarbetande, [Improved quality of life with ginseng preparations? Positive effects in healthy working people], Lakartidningen, 92, 3196, 1995.

74. Smith, K., Engels, H.J., Martin, J., and Wirth, J.C., Efficacy of a standardized ginseng extract to alter psychological function characteristics at rest and during exercise, *Med. Sci. Sports Exerc.*, 27, S147, 1995. (abstract)
75. Sotaniemi, E.A., Haapakoski, E., and Rautio, A., Ginseng therapy in non-insulin-dependent diabetic patients, *Diabetes Care*, 18, 1373, 1995.
76. Ussher, J.M., Dewberry, C., Malson, H., and Noakes, J., The relationship between health related quality of life and dietary supplementation in British middle managers: a double blind placebo controlled study, *Psychol. Health*, 10, 97, 1995.
77. Le Gal, M., Cathebras, P., and Strüby, K., Pharmaton capsules in the treatment of functional fatigue: a double-blind study versus placebo evaluated by a new methodology, *Phytother. Res.*, 10, 49, 1996.
78. Caso Marasco, A., Vargas Ruiz, R., Salas Villagomez, A., and Begoña Infante, C., Double-blind study of a multivitamin complex supplemented with ginseng extract, *Drugs Exptl. Clin. Res.*, 22, 323, 1996.
79. Sorensen, H., and Sonne, J., A double-masked study of the effects of ginseng on cognitive functions, *Curr. Ther. Res.*, 57, 959, 1996.
80. Wiklund, I.K., Mattsson, L.A., Lindgren, R., and Limoni, C., Effects of a standardized ginseng extract on quality of life and physiological parameters in symptomatic postmenopausal women: a double blind, placebo-controlled trial. Swedish Alternative Medicine Group, *Int. J. Clin. Pharmacol. Res.*, 19, 89, 1999.
81. Wesnes KA, Ward T, McGinty A, Petrini O. The memory enhancing effects of a *Ginkgo biloba/Panax ginseng* combination in healthy middle-aged volunteers. *Psychopharmacology (Berl)* 2000 Nov; 152(4):353-361.
82. Ussher, J.M., and Swann, C., A double blind placebo controlled trial examining the relationship between Health-Related Quality of Life and dietary supplements, *Br. J. Health Psychol.*, 5, 173, 2000.
83. Kennedy, D.O., Scholeya, A.B., and Wesnes, K.A., Dose dependent changes in cognitive performance and mood following acute administration of ginseng to healthy young volunteers, *Nutr. Neurosci.*, 4, 295, 2001.
84. Kennedy, D.O., Scholey, A.B., and Wesnes, K.A., Differential, dose-dependent changes in cognitive performance following acute administration of a Ginkgo biloba/Panax ginseng combination to healthy young volunteers, *Nutr. Neurosci.*, 4, 399, 2001.
85. Cardinal, B.J., and Engels, H.J., Ginseng does not enhance psychological well-being in healthy, young adults: results of a double-blind, placebo-controlled, randomized clinical trial, *J. Am. Diet. Assoc.*, 101, 655, 2001.
86. Scholey, A.B., and Kennedy, D.O., Acute, dose-dependent cognitive effects of Ginkgo biloba, Panax ginseng and their combination in healthy young volunteers: differential interactions with cognitive demand, *Hum. Psychopharmacol.*, 17, 35, 2002.
87. Ellis, J.M., and Reddy, P., Effects of *Panax ginseng* on quality of life, *Ann. Pharmacother.*, 36, 375, 2002.
88. Kennedy, D.O., Scholey, A.B., and Wesnes, K.A., Modulation of cognition and mood following administration of single doses of Ginkgo biloba, ginseng, and a ginkgo/ginseng combination to healthy young adults, *Physiol. Behav.*, 75, 739, 2002.

89. Simon, W., Kirchdorfer, A.M., and Dahse, G., Effizienzkontrolle eines ginsenghaltigen Geriatrikums mit der Kraepelinschen Arbeitsprobe, [Efficiency control of a gero-therapeutic containing ginseng by means of Kraepelin's working test], *Med. Mschr.*, 31, 39, 1977.
90. Wang, W.K., Chen H.L., Hsu, T.L., and Wang, Y.Y., Alteration of pulse in human subjects by three Chinese herbs, *Am. J. Chin. Med.*, 22, 197, 1994.
91. Tode, T., Kikuchi, Y., Hirata, J., Kita, T., Nakata, H., and Nagata, I., Effect of Korean red ginseng on psychological functions in patients with severe climacteric syndromes, *Int. J. Gynaecol. Obstet.*, 67, 169, 1999.
92. Zar, J.H., Contingency tables, *Biostatistical Analysis*, Prentice-Hall, Inc., Englewood Cliffs, NJ, 1974, 59.
93. Brown, R.P., Gerbarg, P.L., and Ramazanov, Z., Rhodiola rosea: a phytomedicinal overview. *J. Am. Botanical Council* (Herbalgram), 56, 40, 2002.
94. Germano, C., and Ramazanov, Z., Rhodiola rosea and human performance, in *Arctic Root (Rhodiola rosea). The Powerful New Ginseng Alternative*, Kensington Publishing Corp., New York, NY, 1999, 112.
95. Kelly, G.S., *Rhodiola rosea*: A possible plant adaptogen, *Altern. Med. Rev.*, 6, 293, 2001.
96. Zhu, J.S., Halpern, G.M., and Jones, K., The scientific discovery of an ancient Chinese herbal medicine: *Cordyceps sinensis* Part I, *J. Altern. Complement. Med.*, 4, 289, 1998.
97. Zhu, J.S., Halpern, G.M., and Jones, K., The scientific rediscovery of a precious ancient Chinese herbal regimen: *Cordyceps sinensis* Part II, *J. Altern. Complement. Med.*, 4, 429, 1998.
98. Talbott, S.M., Cordyceps sinensis, in *A Guide to Understanding Dietary Supplements*, Haworth Press, New York, 2003, 143.
99. Talbott, S.M., Rhodiola, in *A Guide to Understanding Dietary Supplements*, Haworth Press, New York, 2003, 203.
100. Steinkraus, D.C., and Whitfield, J.B., Chinese caterpillar fungus and world record runners, *Am. Entomologist*, Winter, 235, 1994.
101. Li, S.P., Li, P., Dong, T.T., and Tsim, K.W., Anti-oxidation activity of different types of natural Cordyceps sinensis and cultured Cordyceps mycelia, *Phytomedicine*, 8, 207, 2001.
102. Yamaguchi, Y., Kagota, S., Nakamura, K., Shinozuka, K., and Kunitomo, M., Antioxidant activity of the extracts from fruiting bodies of cultured Cordyceps sinensis, *Phytother. Res.*, 14, 647, 2000.
103. Huang, B.M., Hsu, C.C., Tsai, S.J., Sheu, C.C., and Leu, S.F., Effects of Cordyceps sinensis on testosterone production in normal mouse Leydig cells, *Life Sci.*, 69, 2593, 2001.
104. Wang, S.M., Lee, L.J., Lin, W.W., and Chang, C.M., Effects of a water-soluble extract of Cordyceps sinensis on steroidogenesis and capsular morphology of lipid droplets in cultured rat adrenocortical cells, *J. Cell. Biochem.*, 69, 483, 1998.
105. Manabe, N., Sugimoto, M., Azuma, Y., Taketono, N., Yamashita, A., Tsuboi, H., Tsunoo, A., Kinjo, N., Nian-Lai, H., and Miyamoto, H., Effects of mycelial extract of cultured *Cordyceps sinensis* on *in vivo* hepatic energy metabolism in the mouse, *Jpn. J. Pharmacol.*, 70, 85, 1996.
106. Manabe, N., Azuma, Y., Sugimoto, M., Uchio, K., Miyamoto, M., Taketomo, N., Tsuchita, H., and Miyamoto, H., Effects of the mycelial extract of cultured *Cordyceps sinensis* on *in vivo* hepatic energy metabolism and blood flow in dietary hypoferric anaemic mice, *Br. J. Nutr.*, 83, 197, 2000.

107. Dai, G., Bao, T., Xu, C., Cooper, R., and Zhu, J.S., CordyMax Cs-4 improves steady-state bioenergy status in mouse liver, *J. Altern. Complement. Med.*, 7, 231, 2001.
108. Balon, T.W., Jasman, A.P., and Zhu, J.S., A fermentation product of Cordyceps sinensis increases whole-body insulin sensitivity in rats, *J. Altern. Complement. Med.*, 8, 315, 2002.
109. Koh, J.H., Kim, K.M., Kim, J.M., Song, J.C., and Suh, H.J., Antifatigue and antistress effect of the hot-water fraction from mycelia of *Cordyceps sinensis*, *Biol. Pharm. Bull.*, 26, 961, 2003.
110. Xiao, Y., Huang, X.Z., Chen, G., Wang, M.B., Zhu, J.S., and Cooper, C.B., Increased aerobic capacity in healthy elderly humans given a fermentation product of Cordyceps Cs-4, *Med. Sci. Sports Exerc.*, 31, S174, 1999. (abstract 774)
111. Nicodemus, K.J., Hagan, R.D., Zue, J.S., and Baker, C., Supplementation with Cordyceps Cs-4 fermentation product promotes fat metabolism during prolonged exercise, *Med. Sci. Sports Exerc.*, 33, S164, 2001. (abstract 928)
112. Buchanan, A.R., Roberts, J.D., Smales, T.M., and Jones, N.T., Short term administration of CordyMax™ Cs-4 as a means to enhance maximal aerobic capacity, *Can. J. Appl. Physiol.*, 26, S248, 2001.
113. Talbott, S.M., Zhu, J.S., and Rippe, J.M., CordyMax™ Cs-4 enhances endurance in sedentary individuals, *Am. J. Clin. Nutr.*, 75, 401S, 2002. (abstract P200)
114. Parcell, A.C., Smith, J.M., Schulthies, S.S., Myrer, J.W., and Fellingham, G., Cordyceps sinensis supplementation does not improve endurance performance in competitive cyclists, *Med. Sci. Sports Exerc.*, 34, S231, 2002. (abstract 1295)
115. Davydov, M., and Krikorian, A.D., *Eleutherococcus senticosus* (Rupr. & Maxim) Maxim. (Araliacaea) as an adaptogen: a closer look, *J. Ethnopharmacol.*, 72, 345, 2000.
116. Wu, Y.N., Wang, X.Q., Zhao, Y.F., Wang, J.Z., Chen, H.J., Liu, H.Z., Li, R.W., Campbell, T.C., and Chen, J.S., Effect of Ciwujia (*Radix Acanthopanax senticosus*) preparation on human stamina, *J. Hyg. Res.*, 25, 57, 1996.
117. Wu, Y.N., Xuqing, W., Yunfeng, Z., and Kaman, R., Nutrition Science News. The effect of Endurox on exercise performance, Recent articles, www.endurox.com, last accessed May 1998.
118. Kaman, R., Endurox™: a novel agent that increases workout performance, Laboratory Report, Pacific Health Laboratories, 1996.
119. Campbell, T.C., Wu, Y.N., Lu, C.Q., Li, M., and Kaman, R.L., Effects of Radix Acanthopanax senticosus (Ciwujia) on exercise, *J. Strength Cond. Res.*, 11, 278, 1997. (abstract)
120. Wu, Y., Wang, X., Li, M., and Campbell, T.C., [Effect of Ciwujia (Radix Acanthopanax senticosus) preparation on exercise performance under constant endurance load for elderly], *Wei Sheng Yan Jiu*, 27, 421, 1998.
121. Dustman, K., Plowman, S.A., McCarthy, K., Ehlers, G., Bramer, A., Coreless, C., De Vantier, N., Freimuth, M., and Walicek, H., The effects of Endurox on the physiological responses to stair-stepping exercise, *Med. Sci. Sports Exer.*, 30, S323, 1998. (abstract #1839)
122. Plowman, S.A., Dustman, K., Walicek, H., Corless, C., Ehlers, G., The effects of Endurox on the physiological responses to stair-stepping exercise, *Res. Q Exerc. Sport*, 70, 385, 1999.
123. Smeltzer, K.D., and Gretebeck, R.J., Effect of Radix Acanthopanax senticosus on submaximal running performance, *Med. Sci. Sports Exerc.*, 30, S278, 1998. (abstract #1578)

124. Stout, J.R., Eckerson, J.M., and Yee, J.C., The effect of Endurox on parameters of the critical power test, *Res. Q Exer. Sport*, 69, A28, 1998. (abstract)
125. Cheuvront, S.N., Moffatt, R.J., Biggerstaff, K.D., Bearden, S., and McDonough, P., Effect of Endurox on various metabolic responses to exercise, *Med. Sci. Sports Exer.*, 30, S323, 1998. (abstract #1838)
126. Cheuvront, S.N., Moffatt, R.J., Biggerstaff, K.D., Bearden, S., and McDonough, P., Effect of Endurox on metabolic responses to submaximal exercise, *Int. J. Sport Nutr.*, 9, 434, 1999.
127. Eschbach, L.C., Webster, M.J., Boyd, J.C., Elmer, A.J., McArthur, P.D., Zoeller, R.F., Krebs, G.V., and Angelopoulos, T.J., The effect of Eleutherococcus senticosus (Siberian Ginseng) on substrate utilization and performance during prolonged cycling, *Med. Sci. Sports Exer.*, 31, S117, 1999. (abstract #444)
128. Eschbach, L.F., Webster, M.J., Boyd, J.C., McArthur, P.D., and Evetovich, T.K., The effect of Siberian ginseng (*Eleutherococcus senticosus*) on substrate utilization and performance, *Int. J. Sport. Nutr. Exerc. Metab.*, 10, 444, 2000.
129. Blokhin, B.N., The influence of eleutherococcus root and leaf extract on human work capacity under static and dynamic workloads, in *Eleutherococcus and Other Adaptogens from Far East Plants*, Siberian Department of the Academy of Sciences of the USSR, Vladivostok, USSR, 1966, 191.
130. Asano, K., Takahashi, T., Miyashita, M., Matsuzaka, A., Muramatsu, S., Kuboyama, M., Kugo, H., and Imai, J., Effect of *Eleutherococcus senticosus* extract on human physical working capacity, *Planta. Med.*, 48, 175, 1986.
131. Schezin, A.K., Zinkovich, V.I., and Galanova, L.K., Eleutherococcus in prevention of influenza, hypertonia, and ischemia in drivers of the Volzhsky Automobile Factory, in *New Data on Eleutherococcus and Other Adaptogens: Proceedings of the First International Symposium on Eleutherococcus, Hamburg, 1980*, Far East Scientific Center of the Academy of Sciences of the USSR, Vladivostok, USSR, 1981, 93.
132. Dowling, E.A., Redondo, D.R., Branch, J.D., Jones, S., McNabb, G., and Williams, M.H., Effect of *Eleutherococcus senticosus* on submaximal and maximal exercise performance, *Med Sci Sports Exer.*, 28, 482, 1996.
133. Azizov, A.P., [Effects of Eleutherococcus, Elton, Leuzea and Leveton on the blood coagulation system during training in athletes], *Eksp. Klin. Farmakol.*, 60, 58, 1997.
134. Azizov, A.P., Seifulla, R.D., Ankudinova, I.A., Kondrat'eva, I.I., and Roisova, I.G., [The effect of the antioxidants elton and leveton on the physical work capacity of athletes], *Eksp. Klin. Farmakol.*, 61, 60, 1998.
135. Blumenthal, M., Eleuthero (Siberian Ginseng) root. *The Complete German Commission E Monographs. Therapeutic Guide to Herbal Medicines*, American Botanical Council, Austin, TX, 1998, 124.
136. Chase, P.J., Darby, L.A., Liang, M.T.C., and Morgan, A.L., Efficacy of *Eleutherococcus senticosus* (Siberian Ginseng) as an ergogenic supplement, *Med. Sci. Sports Exerc.*, 32, S61. 2000. (abstract)
137. Shevtsov, V.A., Zholus, B.I., Shervarly, V.I., Vol'skij, V.B., Korovin, Y.P., Khristich, M.P., Roslyakova, N.A., and Wikman, G., A randomized trial of two different doses of a SHR-5 Rhodiola rosea extract versus placebo and control of capacity for mental work, *Phytomedicine*, 10, 95, 2003.
138. Saratikov, A.S., and Krasnov, E.A., [Stimulative properties of Rhodiola rosea], in *Rhodiola Rosea is a Valuable Medicinal Plant (Golden Root)*, Tomsk University Press, Tomsk, USSR, 1987, 69.

139. Saratikov A.S., and Krasnov, E.A., [Stimulation of energy production by Rhodiola rosea: biochemical mechanisms], in *Rhodiola rosea — medicinal plant and properties*, Tomsk University Press, Tomsk, USSR, 1987, 252.
140. Zhang, Z.H., Feng, S.H., Hu, G.D., Cao, Z.K., and Wang, L.Y., [Effect of *Rhodiola kirilowii* (Regel.) Maxim on preventing high altitude reactions. A comparison of cardiopulmonary function in villagers at various altitudes], *Chung Kuo Chung Yao Tsa Chih*, 14, 687, 1989.
141. Qian, J., Zhang, H., Yang, G., and Wang, B., Protective effects of Rhodiola crenulata on rats under antiorthostatic position and professional athletes, *Space Med. Med. Eng.*, 6, 6, 1993.
142. Bajevskij, R.M., Goncharov, I.B., Nikulina, G.A., Barsukova, Zh.V., Khozyainova, E.V., and Bogatova, R.I., The response of cardio-vascular system to dosed physical load under the effect of herbal adaptogen (final report). Russian Federation Ministry of Health, Institute of Medical and Biological Problems, Contract 93-11-615, Phase I, Moscow, CIS, 1994.
143. Goncharov, I.N., Sal'nitskiy, V.P., and Bogatova, R.I., Experimental trials of herbal adaptogen effect on the quality of operator activity, mental and professional working capacity (final report). Russian Federation Ministry of Health, Institute of Medical and Biological Problems, Contract 93-11-615, Phase I, Moscow, CIS, 1994.
144. Ramazanov, Z., and del Mar, Bernal Suarez M., Rhodiola rosea activates lipase and enhances level of fatty acids in serum, U.S. Patent application, unpublished, 1997.
145. Anonymous, Slim-Happy Plus ™ brochure, Pharmline, Florida, NY, 1999.
146. Spasov, A.A., Mandrikov, V.B., and Mironova, I.A., [The effect of the preparation rodakson on the psychophysiological and physical adaptation of students to an academic load], *Eksp. Klin. Farmacol.*, 63, 76, 2000.
147. Spasov, A.A., Wikman, G.K., Mandrikov, V.B., Mironova, I.A, and Meumoin, V.V., A double-blind, placebo-controlled pilot study of the stimulating and adaptogenic effect of Rhodiola rosea SHR-5 extract on the fatigue of students caused by stress during an examination period with a repeated low-dose regimen, *Phytomedicine*, 7, 85, 2000.
148. Darbinyan, V., Kteyan, A., Panossian, A., Gabrielian, E., Wikman, G., and Wagner, H., Rhodiola rosea in stress induced fatigue — a double blind cross-over study of a standardized extract SHR-5 with a repeated low-dose regimen on the mental performance of healthy physicians during night duty, *Phytomedicine*, 7, 365, 2000.
149. Xiu, R., Rhodiola and uses thereof, US Patent 6,399,116, 2002
150. Wing, S.L., Askew, E.W., Luetkemeier, M.J., Ryujin, D.T., Kamimori, G.H., Grissom, C.K., Lack of effect of Rhodiola or oxygenated water supplementation on hypoxemia and oxidative stress, *Wilderness Environ. Med.*, 14, 9, 2003.
151. Szolomicki, J., Samochowiec, L., Wojcicki, J., Drozdzik, M., and Szolomocki, S., The influence of active components of Eleutherococcus senticosus on cellular defence and physical fitness in man, *Phytother. Res.*, 14, 30, 2000.

21
Lipoic Acid

Henry C. Lukaski*,**

CONTENTS

I. Introduction 412
II. α-Lipoic Acid 412
    A. Chemical Structure and Biochemical Functions 412
    B. Antioxidant Functions 413
        1. Free Radical Scavenging 413
        2. Metal Chelation 413
        3. Interaction with Other Antioxidants 414
        4. Gene Expression 416
III. Dietary Sources 416
    A. Analytical Methods 416
    B. Animal and Plant Sources 417
    C. Absorption and Bioavailability 418
IV. Uses of Supplemental α-Lipoic Acid 419
    A. Glucose Disposal 419
        1. Animal Studies 419
        2. Human Studies 420
    B. Exercise Metabolism 421
    C. Prevention of Diabetes Complications 422
V. Tolerance and Safety 423
VI. Recommendations for Future Research 423
VII. Summary 424
References 424

---

* Mention of a trademark or proprietary product does not constitute a guarantee of the product by the United States Department of Agriculture and does not imply its approval to the exclusion of other products that may also be suitable..

** U.S. Department of Agriculture, Agricultural Research, Northern Plains Area is an equal opportunity/affirmative action employer and all agency services are available without discrimination.

I. Introduction

All cells in the body are exposed chronically to oxidants from both endogenous and exogenous sources but are equipped with antioxidant defense systems.[1] Nutrients, both water- and lipid-soluble, compose an important feature of the antioxidant defense system with which humans have evolved. Antioxidant nutrients include β-carotene and other carotenoids, α-tocopherol or vitamin E, ascorbic acid or vitamin C and selenium. Other dietary constituents have either direct antioxidant activity, such as flavinoids, or indirect antioxidant activity such as zinc, copper and manganese as components of antioxidant enzymes such as the superoxide dismutases.[2]

Thiols are another group of antioxidant compounds with biological activity. These compounds, which are characterized by disulfide or S-S bonds, participate in many cellular functions including biosynthetic pathways, detoxification by conjugation and cell division. Studies of oxidative stress have highlighted the key roles of biothiols. Glutathione, the most abundant nonprotein thiol in the cell, is essential for detoxification of peroxides as cofactors of various selenium-dependent peroxidases. Lipoic acid, a constituent of some foods, is another biothiol and serves to bolster glutathione as an antioxidant.

This chapter summarizes the biological functions of lipoic acid. It describes the multifunctional roles of lipoic acid in various aspects of metabolism, support of oxidative defense and control of cell function. The effects of supplemental lipoic acid on facilitation of glucose utilization and substrate metabolism in skeletal muscle during physical exercise and attenuation of complications of diabetes are discussed.

II. α-Lipoic Acid

α-Lipoic acid is known by various chemical names including thioctic acid, 1,2-dithiolane-3-pentanoic acid and 1,2-dithiolane-3-valeric acid. It was discovered in 1937 as a growth factor for potatoes by Snell et al.[3] In the 1950s, Reed and co-workers[4,5] isolated and characterized the compound. Although initially described as a vitamin, α-lipoic acid was later discovered to be synthesized by plants and animals.[6,7]

A. Chemical Structure and Biochemical Functions

The chemical structures of α-lipoic acid and its reduced form, dihydrolipoic acid, are shown in Figure 21.1. In eukaryotes, lipoic acid is bound covalently to the ε-amino group of lysine residues of distinct mitochondrial proteins.[8] Many of these proteins are subunits of the pyruvate dehydrogenase, the

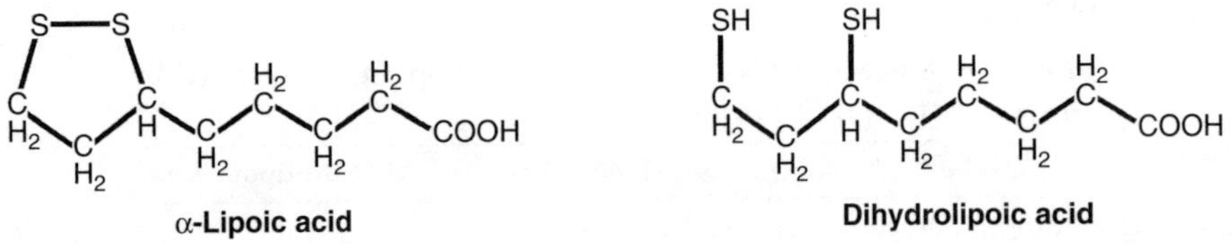

FIGURE 21.1
Molecular structures of α-lipoic acid (1,2-dithiolane-3-pentanoic acid) and dihydrolipoic acid.

branched chain keto acid dehydrogenase and the a-ketoglutarate dehydrogenase complexes.[9] These complexes catalyze the oxidative decarboxylation of pyruvate into acetyl-CoA and α-ketoglutarate into succinyl-CoA, respectively. Thus, lipoate possesses a central position in metabolism; it regulates carbon flow into the Krebs cycle resulting in the production of ATP.[9] Although the complete biosynthetic pathway for α-lipoic acid is not fully elucidated, octanoate apparently is the immediate precursor for the 8-carbon fatty acid chain and cysteine is the source of sulfur.[10] The mitochondrion is the site of *de novo* α-lipoic acid synthesis and function.[8]

B. Antioxidant Functions

Specific criteria are used to determine the antioxidant potential of a compound.[2] Some characteristics include chemical and biochemical functions such as free radical quenching, metal chelation, amphiphilic character, interactions with other antioxidants and modulation of gene expression. Other important criteria to consider when evaluating preventive or therapeutic interventions include absorption and bioavailability, concentrations in tissues and cells and safety. The α-lipoic acid/dihydrolipoic acid redox pair (Figure 21.1), fulfills a number of these criteria and, thus, has been termed an "ideal antioxidant."[11]

1. Free Radical Scavenging

There is general agreement about the antioxidant properties of α-lipoic acid and dihydrolipoic acid (Table 21.1). The predominant form that interacts with reactive oxygen species (ROS) is dihydrolipoic acid but the oxidized form, α-lipoic acid, also can inactivate free radicals.

2. Metal Chelation

Lipoic acid may provide antioxidant activity by chelation of pro-oxidant minerals. Transition mineral elements, such as copper, iron, mercury and cadmium, can induce free radical damage by catalyzing the decomposition of hydroperoxides and generating highly toxic hydroxyl radicals. Early results found that lipoate formed stable complexes with Cu^{+2}, Mn^{+2} and Zn^{+2} in aqueous solution.[12] *In vitro* studies using a hydroxyl radical scavenging

TABLE 21.1
Summary of Reactive Oxygen and Nitrogen Species Scavenged by
α-Lipoic Acid and Dihydrolipoic Acid

Oxidant	α-Lipoic Acid	Dihydrolipoic Acid
Hydrogen peroxide	+	+
Singlet oxygen	+	–
Hydroxyl radical	+	+
Nitric oxide radical	+	+
Superoxide radical	–	+
Hypochlorous acid	+	+
Peroxynitrite	+	+
Peroxyl radical	–	+

Note: + = scavenges; – = does not scavenge.
Source: Adapted from Packer, L., Witt, E.H. and Tritschler, H.J., Alpha-lipoic acid as a biological antioxidant, *Free Radic. Biol. Med.*, 19:227–250, 1995.

assay in which deoxyribose is used as a detector molecule show that addition of α-lipoic acid to hepatic microsomes spares degradation of deoxyribose.[13] The α-lipoic acid apparently chelates Fe^{+2} and diminishes the amount of hydroxyl radical detectable by deoxyribose. Oxidative stress is induced when vitamin C and iron are incubated together. The vitamin C chelates the iron and reduces it to Fe^{+2}; subsequently Fe^{+2} transfers one electron to oxygen and promotes ROS generation. If vitamin C is present in a 50-fold excess of iron, α-lipoic acid is unable to compete with vitamin C for chelation, and lipid peroxidation occurs.[13] Similarly, complexation of Cu^{+2} by α-lipoic acid explains the protection in copper-induced lipid peroxidation.[14]

Complexation of metals by dihydrolipoic acid also may result in antioxidant activity. Studies in rat hepatocytes showed that addition of physiological concentrations of α-lipoic acid significantly reduced amounts of biochemical markers of lipid peroxidation associated with cadmium exposure.[15] *In vivo* evidence for antioxidant activity through metal chelation was reported in rodents. Supplemental α-lipoic acid prevented cadmium-induced lipid peroxidation in brain, heart and testes of rats without affecting tissue accumulation of cadmium.[16] Mice treated with cadmium (40 μg/kg intraperitoneal) and simultaneously with intraperitoneal (*i.p.*) α-lipoic acid, corresponding to a dose of α-lipoic acid/cadmium molar ratio of 5:1, or saline had significantly decreased lipid peroxidation and increased activity of glutathione peroxidase and catalase.[17] Importantly, α-lipoic acid reduced mortality from cadmium exposure from 70 to 15%. Thus, α-lipoic acid reduces toxicity of heavy metals by up-regulating the activities of antioxidant enzymes.

3. Interaction with Other Antioxidants

α-Lipoic acid is closely associated with the metabolism of other antioxidants. Rosenberg and Culik[6] first observed that α-lipoic acid prevented symptoms of vitamin C and vitamin E deficiency in rodents. Podda et al.[18] later reported

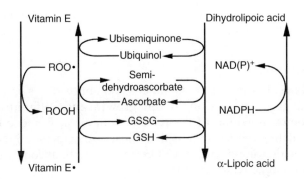

FIGURE 21.2
Summary of the antioxidant network highlighting the interactions among vitamin E, ubiquinol, vitamin C, glutathione (GSH) and α-lipoic acid oxidation-reduction cycles. Reactive species are identified with • to designate the unpaired electron or radical form. Vitamin E and ubiquinol are found in the hydrophobic component whereas ascorbic acid, GSH and α-lipoic acid are active in the aqueous components of cells.

that supplemental α-lipoic acid also prevented symptoms of vitamin E deficiency in hairless tocopherol-deficient mice. Furthermore, α-lipoic acid and dihydrolipoic acid were found in various mouse tissues, demonstrating that α-lipoic acid is reduced metabolically *in vivo*. Dihydrolipoic acid is a strong reductant that regenerates oxidized antioxidants such as ascorbate, glutathione, coenzyme Q and vitamin E.[19,20] These findings strongly support an interaction among various antioxidants known as the antioxidant network (Figure 21.2). When vitamin E scavenges a peroxyl radical, a tocopherol radical is produced. This radical may be reduced by several antioxidants such as ascorbate, ubiquinol and reduced glutathione. Dihydrolipoic acid can reduce all of these antioxidants and be regenerated by several enzymes, including lipoamide reductase, glutathione reductase and thioredoxin reductase. Thus, α-lipoic acid and dihydrolipoic acid have key roles in the antioxidant network. Also, α-lipoic acid has water- and lipid-soluble characteristics (e.g., amphiphilic) that facilitate the reduction of oxidized antioxidants at the boundary between water and lipid.

Supplementation with α-lipoic acid boosts glutathione concentrations *in vivo* and *in vitro*.[21,22] Glutathione is an important water-soluble endogenous antioxidant; it is linked to many physiological processes including detoxification of xenobiotics, modulation of signal transduction, regulation of immune response and enzyme activities. *In vitro* studies with human cells have provided some insight into mechanisms by which α-lipoic acid increases glutathione. Cysteine availability is the rate-limiting step in glutathione synthesis. Lipoic acid is rapidly taken up into cells, quickly reduced to dihydrolipoic acid and secreted into the medium, where it reduces cystine to cysteine. The cell accumulates cysteine 10 times faster than cystine, which enhances glutathione synthesis.[23] Thus, α–lipoic acid and dihydrolipoic acid act as antioxidants directly through radical quenching and metal chelation, but indirectly through recycling of other antioxidants and enhancement of intracellular glutathione.

4. Gene Expression

Growing interest in the effects of oxidants and antioxidants on signal transduction and gene expression in cell growth and differentiation have prompted the examination of the roles of α-lipoic acid and dihydrolipoic acid on modulating the redox-sensitive transcription factor κB (NF-κB). Several lines of evidence indicate that ROS may be the final common signal for many stimuli that activate NF-kB.[24]

Stressors that activate NF-κB also increase intracellular ROS. Some examples include tumor-necrosis factor (TNF-α), lipopolysacchrade, ultraviolet light and gamma radiation.[25,26] Administration of hydrogen peroxide directly stimulates NF-κB activation.[27] Many antioxidants, including vitamin E derivatives,[28] catechol derivatives,[29] and lipoate,[30] block NF-κB activation. In contrast, agents that deplete cellular thiol antioxidants have induced NF-κB activation.[31]

Redox mechanisms regulate NF-κB[32,33] and sulfhydryl groups are involved in this regulation.[34] There are two distinct steps in the activation of NF-κB, both of which may be affected by a thiol antioxidant such as α-lipoic acid. Early steps are activation of NF-κB and dissociation from an inhibitory subunit, I κB. These steps are partially under redox control, with oxidation stimulating activation and dissociation. The binding of activated NF-κB to DNA involves cysteine residues whose redox status is important because reduced cysteine apparently enhances binding. Thus, the effects of thiol-containing antioxidants is complex and not fully elucidated.

While it appears that α-lipoic acid or dihydrolipoic acid may influence gene expression at one or more levels, the exact mechanisms and significance have yet to be clearly determined.

III. Dietary Sources

Interest in exogenous sources of α-lipoic acid has grown because of its potential as a health supplement and as an ergogenic aid.[35] However, it is not known if protein-bound lipoic acid itself may act as an antioxidant or serve as a source of free lipoic acid. Thus, recent efforts have focused on determination of endogenous α-lipoic acid in the form of lipoyllysine in plant and animal tissues, which are common ingredients in the diets of humans.

A. Analytical Methods

Three analytical methods, high-performance liquid chromatography (HPLC), gas chromatography (GC) and GC mass spectroscopy (GC-MS) have been used to determine the contents of endogenous lipoic acid in biological specimens.[36-38] The GC and GC-MS methods are sensitive and specific for measurement of endogenous lipoic acid but they require preparation of derivatives, which may reduce recovery. Protein-bound lipoic acid must be liberated by acidic or basic hydrolysis of proteins at high temperatures, which leads to decomposition and eventually to decreased recovery of free lipoic acid.[37,38] Alternate approaches, such as enzyme immunoassay[38] or coupling HPLC with ultraviolet detection at 330 nm,[36] have not yielded improved analytical accuracy or sensitivity. HPLC with electrochemical detection (dual-gold electrode coated with triple-distilled mercury, operated at a 0.5 v potential) provides contemporary estimates of endogenous lipoic acid, as lipoyllysine, in plant- and animal-tissue samples.[39,40]

B. Animal and Plant Sources

The lipoyllysine content of animal tissues is variable (Table 21.2). Values ranged from 0.1 to 2.6:g/g dry weight, with the lowest and highest values in the lung and kidney, respectively. Lipoyllysine also was present in high concentrations in muscular tissues such as the heart, skeletal muscle and intestines.

The pattern of lipoyllysine content in plant tissues was species-specific (Table 21.3). The highest values, three- to fivefold higher than in broccoli and tomato, respectively, were found in spinach. Other "green" tissues, such as pea and brussel sprout, have appreciable but much lower contents of lipoyllysine, whereas rice bran has minimal contents.

Explanation of the higher reported lipoyllysine contents of plants compared with animal tissues focused on the fact that plants have two isoforms of keto acid dehydrogenase complex, in chloroplasts as well as mitochondria.[39] In spinach, mitochondria contain almost twofold more lipoyllysine (:g/g dry weight) than the chloroplasts; a similar relationship was found if the results are presented in terms of protein content. However, animal mitochondria (e.g., liver) exhibit a twofold increase in lipoyllysine concentration compared with plant tissues (1.62 vs. 0.78:g/mg protein). These differences reflect variations in metabolic activity of animals and plants.

C. Absorption and Bioavailability

Knowledge of the contents of α-lipoic acid in common food sources is useful in assessing intake of this antioxidant. However, awareness of the absorption and tissue distribution of α-lipoic acid, its *in vivo* reduction to dihydrolipoic acid and metabolism to shorter homologs is needed before recommendations can be made for intakes of α-lipoic acid to promote health.

α-Lipoic acid is absorbed from the diet, transported to tissues, taken up by cells where a large proportion is rapidly converted to dihydrolipoic acid.[23] When ^{14}C-labeled lipoic acid (0.5 mg/100 g body weight) was administered *i.p.* to rats, urinary excretion was maximal at 3–6 hr with 60% of administered radioactivity recovered within 24 hr. Respiratory $^{14}CO_2$ was maximal at 3 hr;

TABLE 21.2

Mean Lipoyllysine Content of Various Animal Tissues

	Concentrations	
Source	Dry Weight (μg/g)	Protein (ng/mg)
Kidney[a]	2.64	50.57
Heart[a]	1.51	41.42
Liver[a]	0.86	15.49
Spleen[a]	0.36	5.69
Brain[a]	0.27	4.85
Pancreas[a]	0.12	1.97
Skeletal muscle[b]	0.97	ND
Small intestine[b]	0.83	ND
Large intestine[b]	1.18	ND
Stomach[b]	0.63	ND

Note: ND = not determined
[a] Bovine acetone powders.
[b] Lyophilized rat tissue.
Source: Adapted from Lodge, J.K., Youn, H.-D., Handelman, G.J., Konishi, T., Matsugo, S., Mathur, V.V. and Packer, L., Natural sources of lipoic acid: Determination of lipoyllysine released from protease-digested tissues by high performance liquid chromatography incorporating electrochemical detection, *J. Appl. Nutr.*, 49:3–11, 1997.

TABLE 21.3

Average Lipoyllysine Content of Selected Plant Products

	Concentrations	
Source	Dry Weight (μg/g)	Protein (ng/mg)
Spinach	3.15	92.51
Broccoli	0.94	41.01
Tomato	0.56	48.61
Green pea	0.39	17.13
Brussel sprouts	0.39	18.39
Rice bran	0.16	4.44

Source: Adapted from Lodge, J.K., Youn, H.-D., Handelman, G.J., Konishi, T., Matsugo, S., Mathur, V.V. and Packer, L., Natural sources of lipoic acid: Determination of lipoyllysine released from protease-digested tissues by high performance liquid chromatography incorporating electrochemical detection, *J. Appl. Nutr.*, 49:3–11, 1997.

30% of the injected radioactivity was recovered within 24 hr. Excretion of radioactivity after lipoate dosing by gavage was similar to that measured after *i.p.* administration. Accumulation of radioactivity was greatest in liver, intestinal contents and skeletal muscle. Incubation of ^{14}C-lipoate with hepatic homogenates or mitochondrial preparations resulted in 80% of the administered radioactivity either excreted or found in the tissues.[41,42] α-Lipoic acid administered to a variety of cell and tissue systems appears in the media as dihydrolipoic acid.[43] α-Lipoic acid added to the media of human fibroblasts or Jurkat cells was quickly taken up by the cells and reached a peak concentration of 1.5 mM within 10 minutes, and the cells released dihydrolipoic acid into the media.[44]

Bioavailability of racemic lipoic acid has been studied in humans with single-dose administrations. Supplemental α-lipoic acid is absorbed quickly with a peak blood concentration achieved at 30 to 60 min in healthy adults; it also exhibits dose proportionality between 50 and 600 mg.[45] α-Lipoic acid was rapidly absorbed, with peak plasma concentrations observed between 30 and 60 min after ingestion. However, after oral intake of the racemic mixture, maximal plasma concentrations of the R-enantiomer were 40–50% higher than those of the S-enantiomer (50 mg:136 compared with 68 ng/mL for the R- and S-forms; 600 mg: 1812 versus 978 ng/mL for the R- and S-forms). The decline from peak plasma concentrations was very rapid, about 30 min. Lipoic acid was absorbed more efficiently from an aqueous solution than from galenic preparations. Concomitant food intake reduced the bioavailability of α-lipoic acid.[46] Therefore, α-lipoic acid was recommended to be ingested up to 30 min before a meal.

The absolute bioavailabilities (oral versus intravenous) of 200 mg of α-lipoic acid in aqueous solution have been determined to be 38% and 28% for the for the R- and S-forms, respectively.[47] Galenic forms had lower absolute bioavailabilities, with 25% for R-lipoic acid and 20% for S-lipoic acid. Another study reported no differences between the enantiomers with a similar bioavailability of 29% after ingestion of 200 mg of lipoic acid.[48] The apparent low bioavailability of α-lipoic acid may be the result of a high first-pass effect. Data indicate that α-lipoic acid is metabolized extensively in the liver. Studies in rats fed ^{14}C-labeled rac-α-lipoic acid show extensive metabolism into short-chain homologs formed through β-oxidation of lipoic acid.[42] Similar findings of metabolism of α-lipoic acid to short-chain metabolites were reported in healthy men and patients with diabetic neuropathy.[49]

IV. Uses of Supplemental α-Lipoic Acid

Knowledge of the potential beneficial effects of α-lipoic acid has prompted numerous studies to determine the utility and efficacy of this antioxidant to promote optimal glucose utilization and protection against exercise-induced ROS damage.

A. Glucose Disposal

1. Animal Studies

In addition to its antioxidant properties, α-lipoic acid has potent insulin-mimetic characteristics. It directly activates tyrosine and serine/thyronine kinases of the insulin signaling pathway in both skeletal muscle and adipocytes, which stimulates glucose uptake.[50,51]

Increasing evidence suggests that insulin resistance may be associated with increased oxidative stress and that supplementation with antioxidants may be beneficial.[52] *In vitro* studies consistently show that α-lipoic acid enhances glucose transport in isolated rat diaphragm[50] and skeletal muscle.[53] Subsequently, obese Zucker rats, an animal model of insulin resistance, were used to determine the effects of acute and chronic *i.p.* treatments with R-, S-lipoic acid on glucose transport in skeletal muscle.[54] α-Lipoic acid markedly increased net glucose uptake that was associated with increased glycogen synthesis. Another *in vitro* study confirmed that α-lipoic acid significantly increased glucose uptake into muscle from lean (insulin-sensitive) and obese (insulin-resistant) Zucker rats.[55] Acute treatment with *i.p.* R-lipoic acid increased insulin-mediated glucose transport by 64%, whereas the S form showed no significant effect. Chronic R-lipoic acid administration reduced plasma insulin and free fatty acids. Further, R-lipoic acid improved insulin-stimulated glycogen synthesis and glucose oxidation. Interestingly, the level of GLUT-4 transport protein was unaffected by chronic treatment with R-lipoic acid but was reduced by S-lipoic acid. Thus, glucose disposal is promoted by the R- but not the S-enantiomer of α-lipoic acid.

The effects of supplemental α-lipoic acid in another model of impaired glucose utilization also has been studied. When administered *in vivo*, α-lipoic acid reduced blood glucose in hyperglycemic streptozotocin-diabetic rats.[56] These findings in rodent models of type 1 and type 2 diabetes suggest a beneficial effect of α-lipoic acid supplementation on measures of glucose disposal.

The independent effects of exercise training and α-lipoic acid supplementation on glucose transport in skeletal muscle of obese Zucker rats has recently been studied.[57,58] During an oral glucose tolerance test, exercise training alone or in combination with R-lipoic acid (30 mg/kg/d for 6 wks) resulted in a significant decrease in glucose (26–32%) and insulin (29–30%) responses compared with sedentary rats. R-lipoic acid alone reduced (19%) the glucose-insulin index, an indicator of increased insulin sensitivity, which was reduced further (48–52%) in the combined exercise and lipoic acid group. Exercise or lipoic acid supplementation individually increased insulin-mediated glucose transport (44–57%) in soleus muscle. Supplemental α-lipoic acid improved glucose transport by decreasing protein carbonyls levels, whereas exercise training increased GLUT-4 protein contents. Thus, lipoic acid interacts additively with endurance exercise training to improve insulin action in insulin-resistant skeletal muscle.

2. Human Studies

Studies on the use of R-, S-lipoic acid on insulin-stimulated glucose disposal have been carried out in patients with type 2 diabetes. Acute intravenous (*i.v.*) administration of 1000 mg of α-lipoic acid significantly improved insulin-stimulated glucose disposal as assessed by the glucose clamp technique.[59] Improved insulin-stimulated glucose uptake of similar magnitude,

55% of treated compared with 26% of controls, also was found in 20 noninsulin-dependent diabetics after 10 d of 500 mg injections of lipoic acid.[60] Oral administration of α-lipoic acid supplements (1200 mg/d) to lean and obese diabetics decreased both serum lactate and pyruvate and improved insulin sensitivity after 4 wks of treatment.[61] In a multicenter trial, a 4-wk oral treatment with α-lipoic acid increased insulin-stimulated glucose disposal (62% of treated compared with 25% of placebo group) in patients with type 2 diabetes.[62] Interestingly, insulin-stimulated glucose utilization was optimal at a dose of 600 mg/d; increased dosages (1200 and 1800 mg/d) had no additional benefit on insulin sensitivity in these adult diabetics. These promising findings of a beneficial effect of α-lipoic acid supplementation on glucose disposal await confirmation in a large-scale randomized controlled trial.

B. Exercise Metabolism

Physical activity promotes health and induces oxidative stress. In 1978, Dillard et al.[63] reported that exercise of moderate intensity increased the content of pentane, a lipid peroxidation byproduct, in exhaled air. Davies et al.[64] showed for the first time that exhaustive treadmill exercise increased the free radical concentration of skeletal muscle and liver in rats two- to threefold. These initial studies led to more detailed investigations that concluded that muscle contractions with associated increased oxygen utilization increase the generation of ROS and possible tissue damage.[65]

Although it has been proposed that supplemental α-lipoic acid may improve mitochondrial function by facilitating the activity of lipoyllysine-containing enzymes,[66] there is no clear evidence to support this hypothesis. The first study to examine the effects of oral α-lipoic acid supplementation, as well as the effect of a single bout of strenuous exercise and endurance exercise training on the lipoyllysine content of skeletal muscle and liver tissues in rats produced interesting results.[67] Increased intake of lipoate did not increase the incorporation of the lipoyl moiety to tissue protein. However, endurance training significantly increased lipoyllysine content in the liver at rest; a bout of exhaustive exercise also increased hepatic lipoyllysine content. In vastus lateralis muscle, training did not increase lipoyllysine content. A single bout of exhaustive exercise, however, markedly increased the amount of lipoyllysine in this muscle. Comparison of tissue lipoyllysine data with results for free or loosely bound lipoate showed no association between these apparently related variables. Thus, the tightly protein-bound lipoyllysine pool in tissues is independent of the loosely bound or free lipoate status in the tissue.

In the first study of the efficacy of α-lipoic acid supplementation in exercise-induced oxidative stress, Khanna et al.[68] determined the effect of gavage-administered lipoate (150 mg/kg body weight for 8 wks) on lipid peroxidation and glutathione-dependent antioxidant defenses in some tissues of male Wistar rats. Lipoate supplementation significantly increased total glutathione in liver and blood, which showed that supplemental lipoate is a pro-glutathione agent in certain tissues *in vivo*. Lipoate supplementation, however, did not affect the total glutathione content of the kidney, heart and skeletal muscles. A lipoate-supplementation-dependent increase in the hepatic glutathione pool was related to increased resistance to lipid peroxidation. This beneficial effect against oxidative lipid damage was also observed in the heart and red gastrocnemius muscle. Lower lipid peroxide concentrations in certain tissues of lipoate-supplemented rats suggest a strengthening of the antioxidant defense network in these tissues.

Although these findings in rodent models strongly suggest a beneficial effect of α-lipoic acid supplementation on protection against oxidative damage, there is a conspicuous lack of data examining similar hypotheses in humans.

C. Prevention of Diabetes Complications

Accumulating evidence highlights the potential benefits of α-lipoic acid in amelioration of complications of diabetes. Oxidative stress has been found widely in diabetes.[69–71] Diabetic patients have increased concentrations of lipid-peroxidation products measured as thiobarbituric-acid-reactive substances, lipid peroxides, F_2-isoprostanes, oxidatively damaged DNA bases and reduced glutathione. Oxidative stress may determine the onset and progression of late diabetes complications.[70] Increased oxidative stress in diabetes apparently is the result of hyperglycemia, which stimulates the polyol pathway, formation of advanced glycated endproducts and subsequent formation of reactive oxygen radicals.

Cataract formation is common in advanced diabetes. The polyol pathway is considered the primary cause of cataract formation in diabetes. α-Lipoic acid can be protective in different ways. The reduction of R-lipoic acid by lipoamide reductase depends on availability of NADH. Thus, intramitochondrial reduction of R-lipoic acid can attenuate NADH surplus in diabetes. In an *in vitro* model of glucose-induced lens opacity, stereospecific protection by lipoic acid was observed.[72] Although R-lipoic acid completely protected the lens, addition of racemic lipoic acid decreased damage by only half, whereas S-lipoic acid potentiated deterioration of the lens. This result is consistent with the specific reduction of R-lipoic acid in mitochondria and its effect in enhancing glutathione synthesis. In newborn rats treated with known inhibitors of glutathione synthesis, α-lipoic acid supplements prevented cataract formation in more than 60% of the animals.[73]

Hyperglycemia and endoneural-hypoxia-causing oxidative stress have been implicated in the development of diabetic neuropathy. Hyperglycemia-induced oxidative stress was shown to induce programmed cell death of nerves and thus might contribute to the pathologies in diabetic neuropathy.[74] Lipoic acid has been reported to improve motor-nerve conduction velocity in experimental diabetic neuropathy[75] and to protect peripheral nerves from ischemia in rats.[76]

Treatment of painful neuropathic symptoms and improving quality of life is important in management of late-stage diabetic complications. Clinical trials have evaluated the efficacy of lipoic acid supplements in the treatment of polyneuropathy. In the α-Lipoic Acid Diabetic Neuropathy Study, 3-wk *i.v.* lipoic acid administrations of 600 and 1200 mg significantly ameliorated clinical symptoms of pain, numbness, paresthesia and burning.[77] The Third α-Lipoic Acid Diabetic Neuropathy Study found that injections of 600 mg of lipoic acid for 3 wks significantly improved the neuropathy impairment score and there was a trend for improved neuropathy impairment scores after 7 months in response to oral supplements of 1800 mg/d of α-lipoic acid, but no effect on neuropathic symptoms.[78] The authors attribute the unsuccessful outcome to intercenter variability in scoring symptoms for the impairment score.

V. Tolerance and Safety

Although neither animal nor human studies to date have shown serious side effects with administration of α-lipoic acid, there is a paucity of information available on this subject. Packer et al.[79] stated that the LD50 is approximately 400–500 mg/kg body weight following *i.v.* administration in rats and 400–500 mg/kg after oral dosing to dogs. There is no evidence of carcinogenic or teratogenic effects, but pregnant women are advised not to ingest supplemental α-lipoic acid until more data are available. A daily dose of 1800 mg α-lipoic acid was tolerated by diabetic adults for 7 months without adverse effects.[78] In humans, side effects include allergic skin reactions and possible hypoglycemia in diabetic patients as a consequence of improved glucose utilization with very high doses of α-lipoic acid.[79]

VI. Recommendations for Future Research

The alluring findings described in this chapter suggest many opportunities to more fully elucidate uses of α-lipoic acid in facilitating optimal physiological function and health.

There is a critical need to expand the food composition data base to include α-lipoic acid. Only a very limited number of foods, principally animal and plant tissues, have been analyzed for α-lipoic acid content. Expansion of analyses to include a greater variety of vegetables, grains, fruits, fish, meat and poultry is needed for use in determination of α-lipoic acid intakes of various segments of the population. This information will enable dietitians to determine the amounts of α-lipoic acid obtained from consumption of foods as well as the possibility of constructing specialty diets for physically active individuals and diabetic patients.

Another challenge is to determine whether improvements in glucose metabolism already demonstrated in animal models can be achieved, wholly or in part, in humans. Similar to other food components that are present in very low concentrations in foods and are speculated to be antidiabetogenic, such as chromium, α-lipoic acid may be a nutriceutical when provided in adequate amounts in the diet.

VII. Summary

α-Lipoic acid and its reduced form, dihydrolipoic acid, are unique and vital antioxidants. They quench a variety of ROS, inhibit reactive oxygen generators and spare other antioxidants. Experimental as well as clinical trials highlight the potential usefulness of α-lipoic acid as a promoter of glucose utilization independent of GLUT-4 and a therapeutic agent to ameliorate adverse effects of diabetes on cataract formation and neuropathy. Although enriching the α-lipoic acid content of diets boosts antioxidant capacities in liver and skeletal muscles of rodents, similar studies have not been performed in humans. Thus, there is no scientific evidence that supplemental α-lipoic acid enhances physical or sport performance in humans, despite promising findings in rodent models. Until convincing data from food fortification or supplementation trials with α-lipoic acid are available, it is merely speculative that this natural antioxidant promotes physical function and performance in humans. The antioxidant properties of α-lipoic acid and its interaction with other key antioxidants such as vitamin E, ascorbate and glutathione will provide numerous opportunities for future research activity in human health promotion and amelioration of disease symptoms.

References

1. Halliwell, B. and Gutteridge, J.M.C., *Free Radicals in Biology and Medicine*, Oxford, Clarendon, 2nd ed., 1989.
2. Halliwell, B., Antioxidants in human health and disease, *Ann. Rev. Nutr.*, 16:33–50, 1996.
3. Snell, E.E., Strong, F.M. and Peterson, W.H., Growth factors for bacteria. VI: Fractionation and properties of an accessory for lactic acid bacteria, *Biochem. J.*, 31:1789–1794, 1937.
4. Reed, L.J., DeBusk, B.G., Gunsalus, I.C. and Hornberger, C.S., Cristalline alpha-lipoic acid: a catalytic agent associated with pyruvate dehydrogenase, *Science*, 114:93–94, 1951.
5. Reed, L.J., The chemistry and function of lipoic acids, *Adv. Enzymol.*, 18:319–347, 1957.
6. Rosenberg, H.R., Culik R,. Effect of alpha-lipoic acid on vitamin C and vitamin E deficiencies, *Arch. Biochem. Biophys.*, 80:86–93, 1959.
7. Carreau, J.P., Biosynthesis of lipoic acid via unsaturated fatty acids, *Meth. Enzymol.*, 62:152–158, 1979.

8. Reed, L.J., Multienzyme systems, *Acc. Chem. Res.*, 7:40–46, 1974.
9. Patel, M.S. and Roche, T.E., Molecular biology and biochemistry of pyruvate dehydrogenase complexes, *FASEB J.*, 4:3224–3233, 1990.
10. Dupre, S., Spoto, G., Matarese, R.M., Orlando, M. and Cavallini, D., Biosynthesis of lipoic acid in the rat: incorporation of ^{35}S- and ^{14}C-labeled precursors, *Arch. Biochem. Biophys.*, 202:361–365, 1980.
11. Moini, H., Packer, L. and Saris, N.-E.L., Antioxidant and pro-oxidant activities of α-lipoic acid and dihydrolipoic acid, *Toxicol. Appl. Pharmacol.*, 182:84–90, 2002.
12. Sigel, H., Prijs, B., McCormick, D.B. and Shih, J.C.H., Stability of binary and ternary complexes of α-lipoate and lipoate derivatives with Cu^{+2}, Mn^{+2} and Zn^{+2} in solution, *Arch. Biochem. Biophys.*, 187:208–214, 1978.
13. Scott, B.C., Arouma, O.I., Evans, P.J., O'Neill, C., van der Vliet, A., Cross, C.E., Tritschler, H and Halliwell, B., Lipoic acid and dihydrolipoic acid as antioxidants: A critical evaluation, *Free Radical Res.*, 20:119–133, 1994.
14. Ou, P.M., Tritschler, H.J. and Wolff, S.P., Thioctic (lipoic) acid: A therapeutic metal-chelating antioxidant, *Biochem. Pharmac.*, 50:123–126, 1995.
15. Müller, L. and Menzel, H., Studies on the efficacy of lipoate and dihydrolipoate in the alteration of cadmium^{+2} toxicity in isolated hepatocytes, *Biochim. Biophys. Acta*, 1052:386–391, 1990.
16. Sumathi, R., Devi, V.K. and Varalakshmi, P., DL-Lipoic acid protection against cadmium-induced tissue lipid peroxidation, *Med. Sci. Res.*, 22:23–25, 1994.
17. Bludovska, M., Kotyzova, D., Koutensky, J. and Eybl, V., The influence of α-lipoic acid on the toxicity of cadmium, *Gen. Physiol. Biophys.*, 18:28–32, 1999.
18. Podda, M., Tritschler, H.J., Ulrich, H. and Packer, L., Alpha-lipoic acid supplementation prevents symptoms of vitamin E deficiency, *Biochem. Biophys. Res. Commun.*, 204:98–104, 1994.
19. Kagan, V.E., Shvedova, A., Serbinova, E., Khan, S., Swanson, C., Powell, R. and Packer, L., Dihydrolipoic acid: A universal antioxidant both in the membrane and in the aqueous phase, *Biochem. Pharmacol.*, 44:1637–1649, 1992.
20. Bast, A. and Haennen, G.R.M.M., Interplay between lipoic acid and glutathione in the protection against microsomal lipid peroxidation, *Biochim. Biophys. Acta*, 963:558–561, 1988.
21. Han, D., Handelman, G., Marcocci, L., Sen, C.K., Kobuchi, H., Tritschler, H.J., Flohe, L. and Packer, L., Lipoic acid increases *de novo* synthesis of cellular glutathione by improving cystine utilization, *Biofactors* 6:321–328, 1997.
22. Sen, C.K., Roy, S. and Packer, L., Regulation of cellular thiols in human lymphocytes by α-lipoic acid: A flow cytometric analysis, *Free Radic. Biol. Med.*, 22:1241–1257, 1997.
23. Sen, C.K., Nutritional biochemistry of cellular glutathione, *J. Nutr. Biochem.*, 8:660–672, 1997.
24. Brown, K., Park, S., Kanno,, T., Franzoso, G. and Siebenist, U., Mutual regulation of the transcription activator NF-κB and its inhibitor, I kappa B-a, *Proc. Natl. Acad. Sci.*, 90:2532–2536, 1993.
25. Geng, Y., Zhang, B. and Lotz, M., Protein tyrosine kinase activation is required for lipopolysaccharide induction of cytokines in human blood monocytes, *J. Immunol.*, 151:6692–6700, 1993.
26. Schreck, R., Albermann, K. and Baeuerle, M., NF-kb: an oxidative stress-responsive transcription factor of eukaryotic cells, *Rad. Res. Commun.*, 17:221–237, 1993.

27. Schreck, R., Rieber, P. and Baeuerle, M., Reactive oxygen intermediates are apparently widely used messengers in the activation of NF-κβ transcription factors, *EMBO J.*, 10:2247–2258, 1991.
28. Suzuki, Y.J. and Packer, L., Inhibition of NF-κβ activation by vitamin E derivatives, *Biochem. Biophys. Res. Commun.*, 193:277–283, 1993.
29. Suzuki, Y.J. and Packer, L., Inhibition of NF-κβ activation factor by catechol derivatives, *Biochem. Mol. Bio. Intl.*, 15:299–306, 1994.
30. Packer, L. and Suzuki, Y.J., Vitamin E and alpha-lipoate: Role in antioxidant recycling and activation of NF-κβ factor, *Mol. Aspects Med.*, 14: 229–239, 1993.
31. Staal, F.J., Roederer, M., Herzenberg, L.A. and Herzenberg, L.A., Intracellular thiols regulate activation of nuclear factor κβ and transcription of human immunodeficiency virus, *Proc. Natl. Acad. Sci.*, 87:9943–9947, 1990.
32. Baeuerle, P.A., The inducible transcription activator NF-κβ: regulation by distinct protein subunits, *Biochem. Biophys. Acta*, 1072:63–80, 1991.
33. Toledano, M.B. and Leonard, W.J., Modulation of transcription factor NF-κβ binding activity by oxidation reduction *in vitro*, *Proc. Natl. Acad. Sci.*, 88:4328–4332, 1991.
34. Matthews, J.R., Kakasugi, N., Virelizier, J.-L., Yodoi, J. and Hay, R.T., Thioredoxin regulates the DNA binding activity of NF-kb by reduction of as disulfide bond involving cys 62, *Nucleic Acid Res.*, 20:3821–3830, 1992.
35. Packer, L., Witt, E.H. and Tritschler, H.J., Alpha-lipoic acid as a biological antioxidant, *Free Radic. Biol. Med.*, 19:227–250, 1995.
36. Hayakawa, K. and Oizumi, J., Determination of lipoyllysine derived from enzymes by liquid chromatography, *J. Chromatogr.*, 490:33–41, 1989.
37. Kataoka, H., Hirabayashi, N and Makita, M., Analysis of lipoic acid in biological samples by gas chromatography with flame photometric detection, *J. Chromatogr.*, 615:197–202, 1993.
38. White, R.H., A gas chromatographic method for the analysis of lipoic acid in biological samples, *Anal. Biochem.*, 110:89–92, 1981.
39. Lodge, J.K., Youn, H.-D., Handelman, G.J., Konishi T., Matsugo, S., Mathur, V.V. and Packer, L., Natural sources of lipoic acid: Determination of lipoyllysine released from protease-digested tissues by high performance liquid chromatography incorporating electrochemical detection, *J. Appl. Nutr.*, 49:3–11, 1997.
40. Sen, C.K., Roy, S., Khanna, S. and Packer, L., Determination of oxidized and reduced lipoic acid using high-performance liquid chromatography and coulometric detection, *Meth. Enzymol.*, 299:239–246, 1999.
41. Harrison, E.H. and McCormick, D.B., The metabolism of dl-[1,6-^{14}C] lipoic acid in the rat, *Arch. Biochem. Biophys.*, 160:514–522, 1974.
42. Spence, J.T. and McCormick, D.B., Lipoic metabolism in the rat, *Arch. Biochem. Biophys.*, 174:13–19, 1976.
43. Peinado, J., Sies, H. and Akerboom, T.P.M., Hepatic lipoate uptake, *Arch. Biochem. Biophys.*, 273:389–395, 1989.
44. Handelman, G.J., Han, D., Tritschler, H. and Packer, L., α-Lipoic acid reduction by mammalian cells to the dithiol form and release into the culture medium, *Biochem. Pharmacol.*, 47:1725–1730, 1994.
45. Breithaupt-Grögler, K., Niebach, G., Schneider, E., Erb, K., Hermann, R., Schug, B.S. and Belz, G.G. Dose -proportionality of oral thiocytic acid — coincidence of assessments via pooled plasma and individual data, *Eur. J. Clin. Pharmacol. Sci.*, 8:57–65, 1999.

46. Gleiter, C.H., Schug, B.S., Hermann, R. Influence of food intake on the bioavailability of thiocytic acid enantiomers (letter), *Eur. J. Clin. Pharmacol.*, 50: 513, 1996.
47. Hermann, R., Niebach, G. and Borbe, H.O., Enantioselective pharmacokinetics and bioavailability of different racemic alpha-lipoic acid formulations in healthy volunteers, *Eur. J. Clin. Pharmacol. Sci.*, 4:167–176, 1996.
48. Teichert, J., Kern, J., Tritschler, H.J., Ulrich, H. and Preiss, R. Investigations on the pharmacokinetics of alpha-lipoic acid in healthy volunteers, *Int. J. Clin. Pharmacol. Ther.*, 36:625–628, 1998.
49. Teichert, J. and Preiss, R., High-performance liquid chromatography assay for alpha-lipoic acid and five of its metabolites in human plasma and urine, *J. Chromatogr. B Analyt. Biomed. Life Sci.*, 769:269–281, 2002.
50. Haugaard, N. and Haugaard, E.S., Stimulation of glucose utilization by thiocytic acid in rat diaphragm incubated *in vitro*, *Biochim. Biophys. Acta*, 222:583–586, 1970.
51. Yaworsky, K., Somwar, R., Ramlal, T., Tritschler, H.J. and Klip, A., Engagement of the insulin-sensitive pathway in the stimulation of glucose transport by alpha-lipoic acid in 3T3-L1 adipocytes, *Diabetologia*, 43:294–303, 2000.
52. Ceriello, A., Oxidative stress and glycemic regulation, *Metabolism*, 49:27–29, 2000.
53. Henriksen, E.J., Jacob, S., Streeper, R.S., Fogt, D.L., Hokama, J.Y. and Tritschler, H.J., Stimulation by a-lipoic acid of glucose transport activity in skeletal muscle of lean and obese Zucker rats, *Life Sci.*, 61: 805–812, 1997.
54. Jacob, S., Streeper, R.S., Fogt, D.L., Hokama, J.Y., Tritschler, H.J., Dietze, G.J. and Henriksen, E.J., The antioxidant a-lipoic acid enhances insulin stimulated glucose metabolism in insulin-resistant rat skeletal muscle, *Diabetes*, 45:1024–1029, 1996.
55. Streeper, R.S., Henriksen,E.J., Jacob, S, Hokama, J.Y., Fogt, D.L. and Tritschler, H.J., Differential effects of lipoic acid stereoisomers on glucose metabolism in insulin-resistant skeletal muscle, *Am. J. Physiol.*, 273:E185–E191, 1997.
56. Khamaisi, M., Potashnik, R., Tirosh, A., Demshchak, E., Rudich, A., Tritschler, H., Wessel, K. and Bashan, N., Lipoic acid reduced glycemia and increases muscle GLUT-4 content in streptozotocin-diabetic rats, *Metabolism*, 46:763–768, 1997.
57. Saengsirisuwan, V., Kinnick, T.R., Schmitt, M.B. and Henriksen, E.J., Interactions of exercise training and lipoic acid on skeletal muscle glucose transport in obese Zucker rats, *J. Appl. Physiol.*, 91: 145–153, 2001.
58. Saengsirisuwan, V., Perez, F.R., Kinnick, T.R. and Henriksen, E.J., Effects of exercise training and antioxidant R-ALA on glucose transport in insulin-sensitive rat skeletal muscle, *J. Appl. Physiol.*, 92: 50–58, 2002.
59. Jacob, S., Henriksen, E.J., Schiemann, A.L., Simon, I., Clancy, D.E., Tritschler, H.J., Jung, W.I., Augustin, H.J. and Dietze, G.J., Enhancement of glucose disposal in patients with type 2 diabetes by alpha-lipoic acid, *Arzneimittelforschung*, 45:872–874, 1995.
60. Jacob, S., Henriksen, E.J., Tritschler, H.J., Augustin, H.J. and Dietze, G.J., Improvement of insulin-stimulated glucose-disposal in type 2 diabetes after repeated parenteral administration of thiocytic acid, *Exp. Clin. Endocrinol. Diabetes*, 104:284–288, 1996.

61. Konrad, T., Vicini, P., Kusterer. K., Hoflich, A., Assadkhani, A., Bohles, H.J., Sewell, A., Tritschler, H.J., Cobelli, C., Usadel, K.H., α-Lipoic acid treatment decreases serum lactate and pyruvate concentrations and improves glucose effectiveness in lean and obese patients with type 2 diabetes, *Diabetes Care*, 22:280–287, 1999.
62. Jacob, S., Ruus, P., Henriksen, E.J., Tritschler, H.J., Maerker, E., Renn, W., Augustin, H.J., Dietze, G.J. and Rett, K., Oral administration of rac-α-lipoic acid modulates insulin sensitivity in patients with type 2 diabetes mellitus: A placebo-controlled pilot trial, *Free Radic. Biol. Med.*, 27:309–314, 1999.
63. Dillard, C.J., Litov, R.E., Savin, W.M., Dumelin, E.E. and Tappel, A.L., Effects of exercise, vitamin E and ozone on pulmonary function and lipid peroxidation, *J. Appl. Physiol.*, 45:927–932, 1978.
64. Davies, K.J., Quintanilha, A.T., Brooks, G.A. and Packer, L., Free radicals and tissue damage produced by exercise, *Biochem. Biophys. Res. Commun.*, 107:1198–1205, 1982.
65. Sen, C.K., Packer, L. and Hanninen, O., Eds., *Exercise and Oxygen Toxicity*, Elsevier Science Publishers BV, Amsterdam, 1994.
66. Bustamante, J., Lodge, J.K., Marcocci, L., Tritschler, H.J., Packer, L. and Rihn, B.H., Alpha-lipoic acid in liver metabolism and disease, *Free Radic. Biol. Med.*, 24:1223–1039, 1998.
67. Khanna, S., Atalay, M., Lodge, J.K. et al., Skeletal muscle and liver lipoyllysine content in response to exercise, training and dietary alpha-lipoic acid supplementation, *Biochem. Mol. Biol. Int.*, 46:297–306, 1998.
68. Khanna, S., Atalay, M., Laaksonen, D.E., Gul, M., Roy, S. and Sen, C.K., α-Lipoic acid supplementation: tissue glutathione homeostasis at rest and after exercise, *J. Appl. Physiol.*, 86:1191–1196, 1999.
69. Oberley, L.W., Free radicals and diabetes, *Free Radic. Biol. Med.*, 5:113–120, 1988.
70. Nourooz-Zadeh, J., Rahimi, A., Tajaddini-Sarmadi, J., Tritschler, H.J., Rosen, P., Halliwell, B. and Betteridge, D.J., Relationships between plasma measures of oxidative stress and metabolic control in NIDDM, *Diabetologia*, 40:647–654, 1997.
71. Borcea, V., Nourooz-Zadeh, J., Wolff, S.P., Klevesath, M., Hofmann, M., Urich, H., Wahl, P., Ziegler, R., Tritschler, H.J., Halliwell, B. and Nawroth, P.P., Alpha-lipoic acid decreases oxidative stress eeven in diabetic patients with poor glycemic control and albuminuria, *Free Radic. Biol. Med.*, 26:1495–1500, 1999.
72. Kilic,, F., Handelman, G.J., Traber, K, Tsang, K., Packer, L. and Trevithick, J.R., Modeling cortical cataractogenesis XX. *In vitro* effect of alpha-lipoic acid on glutathione concentrations in lens in model diabetic cataractogenesis, *Biochem. Mol. Biol. Int.*, 46:585–595, 1998.
73. Maitra, I., Serbinova, E., Tritschler, H.J. and Packer, L., Alpha-lipoic acid prevents buthionine sulfoximine-induced cataract formation in newborn rats, *Free Radic. Biol. Med.* 18:823–829, 1995.
74. Green, D.A., Stevens, M.J., Obrosova, I. and Feldman, E.I., Glucose-induced oxidative stress and programmed cell death in diabetic neuropathy, *Eur. J. Pharmacol.* 375:217–223, 1999.
75. Cameron, N.E., Cotter, M.A., Horrobin, D.H. and Tritschler, H.J., Effects of alpha-lipoic acid on neurovascular function in diabetic rats: Interaction with essential fatty acids, *Diabetologica* 41:390–399, 1998.

76. Mitsui, Y., Schmelzer, J.D., Zollman, P.J., Mitsui, M., Tritschler, H.J. and Low, P.A., Alpha-lipoic acid provides neuro-protection from ischemia-reperfusion injury of peripheral nerves, *J. Neurol. Sci.* 163:11–16, 1999.
77. Ziegler, D., Hanefeld, M., Ruhnau, K.J., Meissner, H.P., Lobisch, M., Schutte, K., Nehrdich, D., Dannehl, K. and Gries, F.A., Treatment of symptomatic diabetic peripheral neuropathy with antioxidant alpha-lipoic acid. A three-week multi-centre randomized controlled trial (ALADIN Study), *Diabetologia*, 38:1425–1433, 1995.
78. Ziegler, D., Hanefeld, M., Ruhnau, K.J., Hasche, H., Lobisch, M., Schutte, K., Kerum, G. and Malessa, R., Treatment of symptomatic diabetic polyneuropathy with the antioxidant alpha-lipoic acid: A seven-month multi-centre randomized controlled trial (ALADIN III Study). ALADIN III Study Group Alpha-Lipoic Acid in Diabetic Neuropathy, *Diabetes Care*, 22:1296–1302, 1999
79. Packer, L., Witt, E.H., Tritschler, H.J., Wessel, K. and Ulrich, H., Antioxidant properties and clinical implications of α-lipoic acid, in *Biothiols in Health and Disease*, Packer, L. and Cadenas, E., Eds., Marcel Dekker, New York, 1995, chap. 22.

22
Myo-Inositol and Pangamic Acid

Robert A. Wiswell and Hans C. Dreyer

CONTENTS

I. Introduction 431
II. Myo-Inositol 432
    A. Chemical Structure(s) and Purification or Synthesis. 433
    B. Metabolism 433
        1. Production 434
        2. Functions 434
    C. Dietary Sources 435
    D. Current Research on Myo-Inositol 435
        1. Myo-Inositol and Psychiatric Disorders 435
        2. Myo-Inositol and Cancer 437
        3. Myo-Inositol and Peripheral Neuropathy 437
    E. As an Ergogenic Aid — "Fortifying" Health Booster 437
    F. Conclusion 438
III. Pangamic Acid 438
    A. DMG: Chemical Structure(s) and Purification or
       Synthesis 439
    B. Sources 440
    C. Dosage 440
    D. An Ergogenic Aid — Metabolic Enhancer 441
IV. Summary 441
References 442

I. Introduction

Myo-inositol, once thought to be a vitamin, is actually an isomer of glucose. Initially discovered from muscle extract, in 1850 Scherer called it "inosit," a Greek root for muscle.[1] The term "cyclitol" is used to refer collectively to all nine inositols; however, when inositol is used alone without the prefix, it is inferred to be the "myo" isomer. It has been suggested that inositol may help to reduce the symptoms associated with mental conditions such as depression, panic disorder, obsessive-compulsive disorder (OCD), attention deficit disorder and Alzheimer's disease. Additionally, inositol has purported to be useful in combating the nerve damage associated with the progression of diabetes, as well as tumor formation associated with certain cancers of the lungs.

In the late 1930s, a substance that was water-soluble and seemed to have metabolic enhancement properties was isolated from apricot pits. This substance was later named pangamic acid ($C_{10}H_{19}O_8N$); and, on occasion, due to its water-solubility properties, was inappropriately called vitamin B_{15}. This substance was produced for human consumption under the name of B_{15} in 1943. In the early 1950s, several studies were reported in the Russian literature about the value of B_{15}. In 1965, V.N. Bukin edited a book entitled *Vitamin B_{15}* that was published by USSR Academy of Sciences.[2] Several articles (in Russian) that appeared touting the value of B_{15} were summarized by Gray and Titlow.[3] Under the name of pangamic acid, DMG was patented in 1949 and introduced as a natural food supplement in 1978. Over the years, a range of proposed benefits have been developed to include the value of pangamic acid as a metabolic enhancer, as cancer protective, as a means of enhancing cognitive behavior, etc.

Claims that have been investigated scientifically and published in peer-reviewed journals will be the subject of this chapter. Claims that cannot be substantiated through scientific work will be reported as such. It is up to the consumer to investigate fully to his or her satisfaction the claims made by manufacturers or providers of health-related supplements. Moreover, at all times, persons should seek the advice of their primary care physicians before taking any supplements, especially if they are currently being treated for a condition for which the supplement they wish to take is purported to "cure" or "combat."

II. Myo-Inositol

Myo-inositol (Alternative names vitamin B_8, inositol hexaphosphate, IP6) is by far the most prevalent in nature of the nine possible stereo isomers of hexahydroxycylcohexane, the chemical name for inositol.

Myo-inositol or inositol, unofficially and incorrectly referred to as "vitamin B_8," is present in all animal tissues, with the highest levels in the heart and brain.[4] Because inositol is a component of our bodies' cell membrane, it cannot qualify as a vitamin. Typically, vitamins are agents of metabolism that speed up or facilitate biochemical processes involving organic molecules. Myo-inositol was referred to as a "B" vitamin because of its solubility in water.

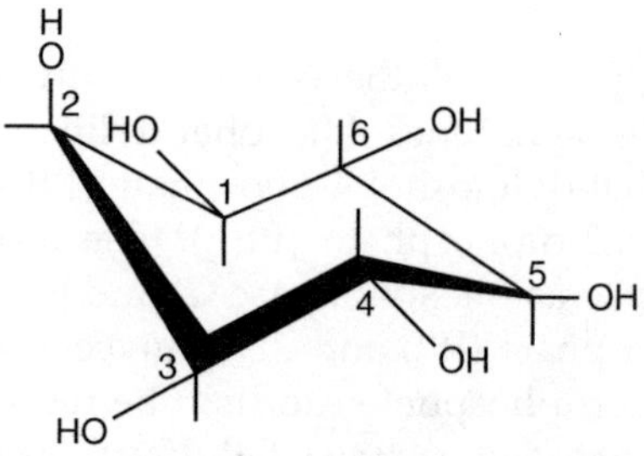

FIGURE 22.1

myo-inositol

Numerous providers of vitamins and supplements sell myo-inositol or inositol either alone or in combination with other "fortifying" health boosters. Inositol retailers tout their product to combat body ailments such as thinning hair, obesity and high cholesterol, and to improve vital organ health. This chapter will address the claims as well as review the literature to determine whether the claims made by these providers are substantiated by scientific research.

A. Chemical Structure(s) and Purification or Synthesis.

Myo-inositol depicted (Figure 22.1) has five free hydroxyl groups, of which as many as three have been found to be phosphorylated in mammalian cells in various combinations. Mathematically, there can be more than 60 inositol phosphates.[4] Myo-inositol is synthesized in our cells from glucose-6-phosphate by cyclic aldol condensation irreversibly by the enzyme D-inositol-3-phosphate synthase.[1]

Phosphorylated inositols are collectively called phosphoinositides or PIs.[4] Myo-inositol is water soluble; however, the addition of a hydrophobic compound such as diacylglycerol, will make it water-insoluble. The significance is that the water-soluble form may be free floating within the cytosol, whereas the lipid form will be found within the plasma membrane bi-layer.

B. Metabolism

The inositol(s) serve as important precursor molecules for cell function, including cell signaling involving the inositol phosphate pathway, membrane trafficking, actin cytoskeletal functioning, cell death signaling and cell adhesion.[1] Inositol phospholipids play a significant role in cell-to-cell communication in the central nervous system.[1] Myo-inositol can also function as an osmolyte[5] used to control the fluid volume of cells in various tissues such as the brain, endothelium and kidney.

1. Production

Phosphatidylinositol (PI) is synthesized from myo-inositol and diacylglycerol (DAG) in cell membranes. Phosphatidylinositol is phosphorylated (converted to) phosphatidylinositol 4-phosphate (PIP) and subsequently to phosphatidylinositol 4,5-diphosphate (PIP$_2$). The enzymes phospholipase Cβ_1 (typically referred to as phospholipase C) and β_2 catalyze the formation of inositol 1,4,5-triphosphate (IP$_3$) and diacylglycerol (breakdown products of PIP$_2$).[6] Inositol 1,4,5-triphosphate can then be recycled into myo-inositol (inositol) and DAG is recycled back to CDP-diacylglycerol. These two (myo-inositol and CDP-diacylglycerol) can then combine and form phosphatidylinositol again to complete the recycling process.[6]

Phosphatidylinositol can serve as an arachadonic acid pool in lipid membranes and as such can serve as substrate for prostaglandin synthesis. Prostaglandins are short-lived hormone-like substances found in almost all tissues. They act locally and are usually associated with the symptoms of pain, inflammation, fever, nausea and vomiting.

2. Functions

Phosphatidylinositol serves a prominent role in signal transduction through hormone, neurotransmitter and growth-factor activation of its receptor. Angiotensin II is an example of a hormone that exerts its effects via the IP$_3$-DAG pathway. Examples of neurotransmitters that utilize the IP$_3$-DAG pathway are acetylcholine, norepinephrine, serotonin (5-hydroxytryptamine, 5-HT) and gamma-aminobutyrate (GABA). The degradation of PI results in the formation of two second messengers: DAG and inositol triphosphate (IP$_3$).[7] Diacylglycerol activates protein kinase C, a Ca^{++}-dependent enzyme that promotes the cleaving of proteins, whereas IP$_3$ enhances intracellular calcium release from the endoplasmic reticulum. In fact, IP$_3$ is the major second messenger to cause increases in intracellular Ca^{++}. Intracellular inositol (IP$_3$) is short-lived, being rapidly dephosphorylated to inositol 1,4-bisphosphate (IP$_2$) and inositol 1-phosphate (IP), which are inactive forms of second messengers and are part of the recycling pathway described above.[6]

The combined effects of inositol 1,4,5-triphosphate and diacylglycerol act synergistically to increase the amount of phosphorylated proteins within the cell. Diacylglycerol activates protein kinase C, an enzyme that combines phosphate with an organic compound (proteins), and IP$_3$ elevates the intracellular Ca^{++}, thus activating calmodulin-dependent protein kinase. Calmodulin is the most widely distributed calcium binding proteins within our bodies' cells.

Membrane-bound PI can form a carbohydrate bridge to specific proteins such as alkaline phosphatase (a digestive enzyme found in the small intestine), acetylchoine esterase (an enzyme that degrades the neurotransmitter acetylchoine) and lipoprotein lipase (an enzyme that hydrolyzes triacylglycerol). Activation of membrane-bound PI is accomplished by phospholipase C, which, as mentioned above, activates protein kinase C.

Inositol, in addition to acting as a second messenger, can also act as an osmolyte and, together with sorbitol, make up the polyalcohols, one of three groups of osmolytes in mammalian cells,[5] the other two groups being methylamines and amino acids. Myo-inositol concentration can be regulated by Na$^+$-coupled transporter[5] or sodium myo-inositol transporter.[1] This transporter is upregulated[8] with increased cellular ionic concentration.[9] Swollen cells rapidly lose inositol from the cytoplasm through a volume-sensitive organic osmolyte anion channel.[1,10,11]

C. Dietary Sources

Inositol is a monosaccharide (a simple sugar) derivative and is considered a sugar alcohol. Other sugar alcohols include glycerol, manitol and sorbitol. Dietary sources of inositol are naturally abundant, found in many foods including beans, citrus fruit, nuts, rice, organ meats, pork, wheat germ and many plants.[4] Thus, inositol is ubiquitous in carbohydrate diets. Moreover, our cells have the ability to convert D-glucose-6-phosphate to D-inositol-3-phophate, which is easily dephosphorylated to myo-inositol. A typical diet includes about 1g/day[12] and is supplemented by endogenous synthesis of inositol by the kidney, averaging about 4g/day.[13] Other organs of the body, such as the brain and testes, synthesize inositol.[14]

D. Current Research on Myo-Inositol

1. Myo-Inositol and Psychiatric Disorders

The central nervous system (CNS) has the ability to synthesize myo-inositol as well as absorb it from the blood derived from dietary sources. Myo-inositol is believed to mediate its psychiatric effects by increasing the availability of inositol as a precursor to intracellular phosphatidylinositol second-messenger cycles. Moreover, several studies using inositol to treat disorders such as depression, panic disorder and OCD with inositol were found to have effects similar to antidepressant drugs, in particular the selective serotonin reuptake inhibitors. Myo-inositol appears to work by potentiateing the 5-HT availability as well as decreasing the desensitization of the 5-HT receptors over time in a log linear fashion.[15] Myo-inositol works differently from other antidepressants that affect the availability or lack of neurotransmitter selective serotonin reuptake inhibitors (SSRI).

a. Depression

Several studies have looked to see whether myo-inositol can attenuate the effects of clinical depression. Levine et al. (1995) demonstrated that, under double-blind conditions, 12 g/day of inositol (N = 13) or placebo (N = 15), inositol was effective in improving scores on the Hamilton Depression Rating Scale, whereas placebo was not. Additionally, blood samples collected demonstrated no adverse effects of the inositol on liver or kidney function.[16] However, in a follow-up study, the original subjects from the previous study who responded to inositol demonstrated a relapse of symptoms once the inositol was removed from their diets 10 to 12 months later.[17] Other studies have suggested no beneficial results from inositol supplementation in persons suffering from clinical depression who also responded unfavorably to SSRI treatments.[18] This suggests that depression sufferers who are unable to respond to conventional SSRI drugs may also fail to respond to inositol supplementation. When inositol plus SSRI was compared with placebo plus SSRI, no significance was observed in depression scores between groups.[19] This study suggests that inositol may not increase the potency or further decrease depression when added as a cocktail to SSRI drugs.

b. Panic Disorder

Several studies have demonstrated significant reductions in the number and severity of panic attacks in double-blinded controlled cross-over trials of myo-inositol and placebo[20] and myo-inositol and an established antidepressant, fluvoxamine, used for panic attacks.[21] These controlled studies indicate that myo-inositol is as effective in treating panic disorder as fluvoxamine. Myo-inositol was superior to fluvoxamine in reducing the incidence of panic attacks, $p = 0.049$.[21] Thus, available research suggests a positive response to myo-inositol within this patient population.

c. Obsessive-Compulsive Disorder

A double-blind placebo-controlled study indicated that persons suffering from obsessive-compulsive disorder (OCD) had significantly lower scores on the Yale-Brown Obsessive Compulsive Scale (YBOCS), a tool used by psychiatrists and researchers of human behavior to assess affective disorders such as OCD. These researchers conducted a crossover trial with 18 g/day of inositol or placebo for 6 weeks each, which means that both groups of OCD patients received the drug and the placebo, but at different blocks of time for 6 weeks each. Significant reductions in YBOCS scores were observed in subjects during those blocks of time in which they were being supplemented with inositol, demonstrating therapeutic benefits.[22]

d. Attention Deficit Disorder with Hyperactivity

Controlled studies have shown that myo-inositol treatment may exacerbate symptoms related to attention deficit disorder with hyperactivity (ADDH). Whereas antidepressant medication has proven effective in ameliorating the symptoms of ADDH, myo-inositol was ineffective and may have magnified the symptoms. This is in contrast to what previous studies have demonstrated with affective disorders that responded favorably to antidepressant medication and myo-inositol. The initial trial involved 11 children who ingested inositol or placebo (double-blinded cross-over trial) for 8 weeks at a dose of 200mg/kg of body weight.[23]

2. Myo-Inositol and Cancer

Myo-inositol has been tested in lab animals to determine whether the process of tumor formation due to cigarette smoke can be attenuated or reversed. One study demonstrated that myo-inositol in combination with dexamethasone reduced the incidence of lung tumor formation in rats subjected to 5 months of mainstream cigarette smoke at a concentration of 132 mg total suspended particulates per cubic meter.[24] Further studies by these authors concluded that myo-inositol and dexamethasone administration for 4 months following 5 consecutive months of cigarette smoke exposure significantly reduced the incidence of lung tumor formation.[25] These two studies indicate that myo-inositol and dexamethasone cocktail administered during cigarette smoke exposure or after cigarette smoke exposure significantly reduced the incidence of lung tumor formation in male mice.

3. Myo-Inositol and Peripheral Neuropathy

Twenty-eight young diabetics with short disease duration participated in a double-blind study by taking 6 g of myo-inositol or placebo daily for 2 months.[26] The aim was to demonstrate a possible beneficial effect of this compound on subclinical diabetic neuropathy. Measurement of vibratory perception threshold, motor and sensory conduction velocity and amplitude of nerve potential did not disclose any effect of the myo-inositol given. In accordance with these observations, no indication of a lack of myo-inositol in human diabetic blood or tissue could be found. The concentration of myo-inositol in the plasma and erythrocyte of four human diabetics was normal or high, even though the loss of urinary myo-inositol was greater than in the case of the four normal subjects. Further, an analysis of the content of free and lipid-bound myo-inositol in muscle biopsies taken from the four diabetics did not give any indication of deficiency. The content of myo-inositol in their muscle tissue remained uninfluenced by oral supplementation of myo-inositol.[26]

E. As an Ergogenic Aid — "Fortifying" Health Booster

Theresearch cited above indicates that inositol may ameliorate the symptoms associated with psychiatric disorders such as depression, panic disorder and OCD while having no effect on ADDH symptoms. Additionally, research points to myo-inositol as having positive effects on lung tumor formation in rats either during or post-exposure to cigarette smoke. However, it is from psychiatric research on myo-inositol that we can glean some clues as to myo-inositol as a possible "ergogenic aid." As part of the understanding of medications for depression, scientists have examined what effect a particular drug has on activity levels in animal models. Two studies[27,28] demonstrated that activity levels in rats increased when administered inositol either by injection or orally. Measurements of activity (number of rearings) in the inositol group (administered by injection) was significantly greater than in the control group.[28] This study was followed up some time later with an orally introduced inositol group that demonstrated a 30% increase in walking and a 60% increase in the number of rearings.[27] Additionally, researchers using reserpine to induce hypoactivity have demonstrated that rats injected with inositol simultaneously decreased their immobility time and reducing the "hypoactivity" effects of the reserpine drug.[7] From these findings it becomes conceivable that inositol administration may increase activity in humans and may very well be the source of the "ergogenic" effects of myo-inositol purported in some literature.

F. Conclusion

Although some animal model-derived data suggest that inositol may increase activity and as such may serve as a dietary supplement, more research is required before a definitive statement can be made regarding the use of myo-inositol as an ergogenic aid. This leads to the conclusion that myo-inositol may better serve the human population as a psychiatric tool, at least for the time being. As for myo-inositol's effects on lung-tumor formation in humans, more research is required before any conclusions can be made. However, the animal data are promising and certainly represent good evidence to support the need for human-based research on the effects of myo-inositol as an ergogenic aid. Moreover, several reviews[1,4,29,30] are available to help the reader learn more about the potential use of myo-inositol.

III. Pangamic Acid

Pangamic acid (referred to as DMG in this chapter) is also known as vitamin B_{15}, N,N-dimethylglycine, calcium pangamate. N,N-Dimethylglycine (DMG) is a modified amino acid produced within cells from choline and betaine and, as a result, is classified as an intermediary metabolite. Its role as an ergogenic aid is related to the purported role of DMG as a potent methyl donor. Furthermore, it has been suggested that DMG plays a crucial role in cellular respiration and oxygen transport and, as such, may help reduce the appearance of lactic acid and reduce fatigue. Over the past few years, several claims for DMG have been made that include:

- A potent scavenger of free radicals[31]
- Used in the management of AIDS and Cancer[32]
- Is an immuno-modulator[33]
- Lowers plasma levels of homocysteine (Cardiovascular benefits)[34–36]
- Increases running performance and reduces fatigue[37,38]

- Potentiates insulinotropic action[39]
- Used in the management of patients with autistic disorder[40,41]

A. DMG: Chemical Structure(s) and Purification or Synthesis

DMG is a critical intermediate in the process by which choline undergoes a series of steps that end with glycine and yield three methyl groups for other transmethylation processes. In the end, it has been proposed that these additional methyl groups are used to produce sulfur-adenosylmenthionine (SAMe), an important methyl donor in several important biochemical processes. The process can be depicted as follows:

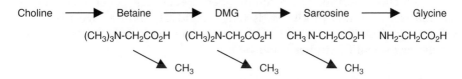

The result of this processing of choline yields three very distinct products. First and most obvious is the provision of a methyl source. Second and equally important, is the production of three very important metabolites (betaine, DMG and sarcosine). Each of these intermediaries is important. Betaine supplementation, for example, has been reported to decrease post-methionine hyperhomocysteinemia in chronic renal failure.[29] Furthermore, Look et al.[32] observed that the elevated levels of DMG and methyl glycine (MG) associated with AIDS patients decreased as a result of antiretroviral therapy. Third, the process yields a quantity of the end product glycine. Glycine has been used as a nutrient, a buffering agent and a retardant of rancidity in fats. It should be noted that of the 50 mmol of methyl groups ingested per day by humans, approximately 60% are derived from choline.[42] For specific methodologies and purification techniques, the reader is directed to references 43 to 45.

Because of the multiplicity of outcomes, it is very difficult to isolate the value of DMG per se. If it is a valuable nutrient, is it only because of its role as a methyl group supplier? Is it related to the elevated sarcosine produced? Are there specific reactions that are directly regulated by DMG? These questions must be answered in a scientific manner before substantive statements can be made about its value.

There are several more recent reports about the value of DMG in a clinical setting. McGregor et al.[34] reported that DMG accumulates during uremia and is associated with elevated plasma homocysteine concentrations. This finding is interesting in that elevated levels of homocysteine have been associated with risk of CHD.[46] Liet et al.[47] reported that DMG does not change resting oxygen consumption in children with Saguenay-Lac-Saint-Jean cytochrome-c oxidase deficiency (SLSJ-COX), a distinct form of congenital lactic acidosis.

This result is in contrast to the earlier findings reported by Meduski in the late 1980s using an animal model to suggest that DMG ingestion reduces activity- induced elevations in lactic acid.[48]

B. Sources

DMG can be found in brewer's yeast, Tortula yeast, brown rice, whole grains (rice bran, wheat germ and wheat bran), sunflower, pumpkin and sesame seeds and liver.

C. Dosage

In that there is no specific benefit derived from general use of DMG, it is very difficult to formulate a recommended dosage. The general recommendation from product producers is that individuals should ingest between 25–50 mg of DMG daily. Table 22.1 reports on the different strategies used in the few scientific studies published.

TABLE 22.1

Summary of Dosages Reported in the Literature

Reference	Purpose	Dosage	Duration	Substance
McGregor et al.[34] Schwahn et al.[35]	Influence of betain supplementation on hyperhomocysteinemia in chronic renal failure	4 g	Daily for 3 months	Glycine Betaine (GB)
Bolman and Richmond[41]	Crossover trial of low dose dimethylglycine in autistic disorder	125 mg	4 weeks, dosage 1–3 tablets depending on weight	DMG
Liet et al.[47]	Short-term use on oxygen consumption in children with cytochrome oxidase deficiency	125 mg	3 days	DMG
Gray and Titlow[3]	Pangamic acid utilization on maximal treadmill performance	300 mg	7 day/week for 3 weeks	Calcium Gluconate and N,N-Dimethylglycine
Pipes[37]	Influence of pangamic acid on treadmill running time	5 mg	Every day for 1 week	Pangamic Acid
Girandola et al.[49]	Effects of B_{15} ingestion on metabolic kinetics during treadmill running	2.4 g	1 a day for 2 weeks	Gluconate and N,N-dimethlglycine

D. An Ergogenic Aid — Metabolic Enhancer

In 1980, our laboratory conducted an experiment utilizing pangamic acid to improve oxygen uptake kinetics during an 8-minute submaximal cycle ergometer task (75% of VO_{2max}).[49] Only eight subjects were used in a single-blind study design. Our findings were in conflict with earlier reports,[50] suggesting that submaximal oxygen consumption would be lowered and post-exercise lactic acid would be reduced. There were absolutely no differences in any of the metabolic parameters we measured, leading to the conclusion that the ingestion of pangamic acid does not result in any metabolic or circulatory advantages for human subjects during short-term submaximal exercise.[49]

Our findings were also in direct opposition to the preliminary results reported by Pipes.[37] This study reported a 23% increase in treadmill running time to exhaustion and a 27% increase VO_{2max} as a result of 7 days of 5 mg ingestion of pangamic acid in 12 male track athletes. While this study is often cited to support the use of pangamic acid, it has not yet been published in a full-length peer-reviewed article.[37]

The only other studied reported in the literature about pangamic acid and treadmill performance was by Gray et al.[3] They studied the effect of ingesting 6–50 mg tablets of calcium gluconate and N, N-DMG taken daily for 3 weeks on short-term maximal treadmill performance in 16 male track athletes. The study design was very similar to that employed by Pipes.[37] They measured metabolic and cardiovascular parameters during the exercise as well as post-test glucose and lactic acid. From the results they concluded, "Ingestion of pangamic acid does not produce significant changes in short-term maximal treadmill performance."[3]

IV. Summary

Both myo-inositol and DMG are ubiquitous in nature and are involved indirectly in a myriad of biochemical reactions. To summarize the literature on athletic performance and ingestion of these two substances, one could say there is no human research data to validate their use. However, there seems to be a growing literature that may support the clinical and therapeutic uses of these substances. The variety of new research studies focusing on specific disease processes (including autism, depression, cancer and coronary heart disease) are utilizing controlled studies to investigate the efficacy of use and may allow for a more definitive answer to the question of the value of DMG and myo-inositol. To believe that these two substances have ergogenic value, one must conduct larger-sample clinical trials to evaluate their use, mechanism of use and potential side effects. It is obvious that DMG, for example, is most likely not a metabolic enhancer leading to lower oxygen consumption and decreased anaerobic metabolites during exercise. This does not preclude the possibility that its presence influences DNA production and long-term physiologic adaptations. The study of Sandu et al.[51] is the beginning of in-depth micro-level research that will shed light on the value of DMG on muscle function and exercise capacity. It is already evident that DMG and myo-inositol have clinical relevance and value; it is not clear that they have ergogenic value.

References

1. Fisher, S. K., Novak, J. E. and Agranoff, B. W. Inositol and higher inositol phosphates in neural tissues: Homeostasis, metabolism and functional significance. *J Neurochem, 82*: 736–754, 2002.
2. Bukin, V. N. *Vitamin B-15 (Pangamic Acid)*. Moscow: USSR Academy of Sciences, 1965.
3. Gray, M. E. and Titlow, L. W. The effect of pangamic acid on maximal treadmill performance. *Med Sci Sports Exerc, 14*: 424–427, 1982.
4. Irvine, R. F. and Schell, M. J. Back in the water: The return of the inositol phosphates. *Nat Rev Mol Cell Biol, 2*: 327–338, 2001.
5. Lang, F., Busch, G. L., Ritter, M., Volkl, H., Waldegger, S., Gulbins, E. and Haussinger, D. Functional significance of cell volume regulatory mechanisms. *Physiol Rev, 78*: 247–306, 1998.
6. Berridge, M. J. Inositol trisphosphate and diacylglycerol as second messengers. *Biochem J, 220*: 345–360, 1984.
7. Einat, H., Karbovski, H., Korik, J., Tsalah, D. and Belmaker, R. H. Inositol reduces depressive-like behaviors in two different animal models of depression. *Psychopharmacology* (Berl), 144: 158–162, 1999.
8. Nakanishi, T., Turner, R. J. and Burg, M. B. Osmoregulatory changes in myo-inositol transport by renal cells. *Proc Natl Acad Sci USA, 86*: 6002–6006, 1989.
9. Burg, M. B. Molecular basis for osmoregulation of organic osmolytes in renal medullary cells. *J Exp Zool, 268*: 171–175, 1994.
10. Kinne, R. K., Czekay, R. P., Grunewald, J. M., Mooren, F. C. and Kinne-Saffran, E. Hypotonicity-evoked release of organic osmolytes from distal renal cells: systems, signals and sidedness. *Ren Physiol Biochem, 16*: 66–78, 1993.
11. Ruhfus, B. and Kinne, R. K. Hypotonicity-activated efflux of taurine and myo-inositol in rat inner medullary collecting duct cells: Evidence for a major common pathway. *Kidney Blood Press Res, 19*: 317–324, 1996.
12. Holub, B. J. Metabolism and function of myo-inositol and inositol phospholipids. *Annu Rev Nutr, 6*: 563–597, 1986.
13. Clements, R. S., Jr. and Diethelm, A. G. The metabolism of myo-inositol by the human kidney. *J Lab Clin Med, 93*: 210–219, 1979.
14. Hauser, G., Arnold, M. and Finelli, V. N. The effect of the injection of 2-O,C-methylene myo-inositol into rats on inositol metabolism in brain and kidney slices. *Biochim Biophys Acta, 116*: 125–132, 1966.
15. Rahman, S. and Neuman, R. S. Myo-inositol reduces serotonin (5-HT2) receptor induced homologous and heterologous desensitization. *Brain Res, 631*: 349–351, 1993.
16. Levine, J., Barak, Y., Gonzalves, M., Szor, H., Elizur, A., Kofman, O. and Belmaker, R. H. Double-blind, controlled trial of inositol treatment of depression. *Am J Psychiatry, 152*: 792–794, 1995.
17. Levine, J., Barak, Y., Kofman, O. and Belmaker, R. H. Follow-up and relapse analysis of an inositol study of depression. *Israel J Psychiatry Relat Sci, 32*: 14–21, 1995.
18. Nemets, B., Mishory, A., Levine, J. and Belmaker, R. H. Inositol addition does not improve depression in SSRI treatment failures. *J Neural Transm, 106*: 795–798, 1999.

19. Levine, J., Mishori, A., Susnosky, M., Martin, M. and Belmaker, R. H. Combination of inositol and serotonin reuptake inhibitors in the treatment of depression. *Biol Psychiatry*, 45: 270–273, 1999.
20. Benjamin, J., Levine, J., Fux, M., Aviv, A., Levy, D. and Belmaker, R. H. Double-blind, placebo-controlled, crossover trial of inositol treatment for panic disorder. *Am J Psychiatry*, 152: 1084–1086, 1995.
21. Palatnik, A., Frolov, K., Fux, M. and Benjamin, J. Double-blind, controlled, crossover trial of inositol versus fluvoxamine for the treatment of panic disorder. *J Clin Psychopharmacol*, 21: 335–339, 2001.
22. Fux, M., Levine, J., Aviv, A. and Belmaker, R. H. Inositol treatment of obsessive-compulsive disorder. *Am J Psychiatry*, 153: 1219–1221, 1996.
23. Levine, J. Controlled trials of inositol in psychiatry. *Eur Neuropsychopharmacol*, 7: 147-155, 1997.
24. Witschi, H., Espiritu, I. and Uyeminami, D. Chemoprevention of tobacco smoke-induced lung tumors in A/J strain mice with dietary myo-inositol and dexamethasone. *Carcinogenesis*, 20: 1375–1378, 1999.
25. Witschi, H., Uyeminami, D., Moran, D. and Espiritu, I. Chemoprevention of tobacco-smoke lung carcinogenesis in mice after cessation of smoke exposure. *Carcinogenesis*, 21: 977–982, 2000.
26. Gregersen, G., Bertelsen, B., Harbo, H., Larsen, E. andersen, J. R., Helles, A., Schmiegelow, M. and Christensen, J. E. Oral supplementation of myo-inositol: effects on peripheral nerve function in human diabetics and on the concentration in plasma, erythrocytes, urine and muscle tissue in human diabetics and normals. *Acta Neurol Scand*, 67: 164–172, 1983.
27. Kofman, O., Agam, G., Shapiro, J. and Spencer, A. Chronic dietary inositol enhances locomotor activity and brain inositol levels in rats. *Psychopharmacology* (Berl), 139: 239–242, 1998.
28. Kofman, O., Bersudsky, Y., Vinnitsky, I., Alpert, C. and Belmaker, R. H. The effect of peripheral inositol injection on rat motor activity models of depression. *Isr J Med Sci*, 29: 580–586, 1993.
29. Vanhaesebroeck, B., Leevers, S. J., Ahmadi, K., Timms, J., Katso, R., Driscoll, P. C., Woscholski, R., Parker, P. J. and Waterfield, M. D. Synthesis and function of 3-phosphorylated inositol lipids. *Annu Rev Biochem*, 70: 535–602, 2001.
30. Fisher, S. K., Heacock, A. M. and Agranoff, B. W. Inositol lipids and signal transduction in the nervous system: an update. *J Neurochem*, 58: 18–38, 1992.
31. Hariganesh, K. and Prathiba, J. Effect of dimethylglycine on gastric ulcers in rats. *J Pharm Pharmacol*, 52: 1519–1522, 2000.
32. Look, M. P., Riezler, R., Berthold, H. K., Stabler, S. P., Schliefer, K., Allen, R. H., Sauerbruch, T. and Rockstroh, J. K. Decrease of elevated N,N-dimethylglycine and N-methylglycine in human immunodeficiency virus infection during short-term highly active antiretroviral therapy. *Metabolism*, 50: 1275–1281, 2001.
33. Graber, C. D., Goust, J. M., Glassman, A. D., Kendall, R. and Loadholt, C. B. Immunomodulating properties of dimethylglycine in humans. *J Infect Dis*, 143: 101–105, 1981.
34. McGregor, D. O., Dellow, W. J., Lever, M., George, P. M., Robson, R. A. and Chambers, S. T. Dimethylglycine accumulates in uremia and predicts elevated plasma homocysteine concentrations. *Kidney Int*, 59: 2267–2272, 2001.
35. McGregor, D. O., Dellow, W. J., Robson, R. A., Lever, M., George, P. M. and Chambers, S. T. Betaine supplementation decreases post-methionine hyperhomocysteinemia in chronic renal failure. *Kidney Int*, 61: 1040–1046, 2002.

36. Schwahn, B. C., Hafner, D., Hohlfeld, T., Balkenhol, N., Laryea, M. D. and Wendel, U. Pharmacokinetics of oral betaine in healthy subjects and patients with homocystinuria. *Br J Clin Pharmacol*, 55: 6–13, 2003.
37. Pipes, T. The effects of pangamic acid on performance in trained athletes. *Med Sci Sports Exerc*, 12: 98, 1980.
38. Tonda, M. E. and Hart, L. L. N,N dimethylglycine and L-carnitine as performance enhancers in athletes. *Ann Pharmacother*, 26: 935–937, 1992.
39. Cancelas, J., Villanueva-Penacarrillo, M. L., Valverde, I. and Malaisse, W. J. Potentiation by glutamic acid dimethyl ester of GLP-1 insulinotropic action in fed anaesthetized rats. *Int J Mol Med*, 8: 531–532, 2001.
40. Kidd, P. M. Autism, an extreme challenge to integrative medicine. Part: 1: The knowledge base. *Altern Med Rev*, 7: 292–316, 2002.
41. Bolman, W. M. and Richmond, J. A. A double-blind, placebo-controlled, crossover pilot trial of low dose dimethylglycine in patients with autistic disorder. *J Autism Dev Disord*, 29: 191–194, 1999.
42. Niculescu, M. D. and Zeisel, S. H. Diet, methyl donors and DNA methylation: interactions between dietary folate, methionine and choline. *J Nutr*, 132: 2333S–2335S, 2002.
43. Holm, P. I., Ueland, P. M., Kvalheim, G. and Lien, E. A. Determination of choline, betaine and dimethylglycine in plasma by a high-throughput method based on normal-phase chromatography-tandem mass spectrometry. *Clin Chem*, 49: 286–294, 2003.
44. Laryea, M. D., Steinhagen, F., Pawliczek, S. and Wendel, U. Simple method for the routine determination of betaine and N,N-dimethylglycine in blood and urine. *Clin Chem*, 44: 1937–1941, 1998.
45. Lundberg, P., Dudman, N. P., Kuchel, P. W. and Wilcken, D. E. 1H NMR determination of urinary betaine in patients with premature vascular disease and mild homocysteinemia. *Clin Chem*, 41: 275–283, 1995.
46. Kang, S. S., Wong, P. W. and Malinow, M. R. Hyperhomocyst(e)inemia as a risk factor for occlusive vascular disease. *Annu Rev Nutr*, 12: 279–298, 1992.
47. Liet, J. M., Pelletier, V., Robinson, B. H., Laryea, M. D., Wendel, U., Morneau, S., Morin, C., Mitchell, G. and Lacroix, J. The effect of short-term dimethylglycine treatment on oxygen consumption in cytochrome oxidase deficiency: a double-blind randomized crossover clinical trial. *J Pediatr*, 142: 62–66, 2003.
48. Meduski, J., Hyman, S., Kilz, R., Kim, K., Thein, P. and Yoshimoto, T. Vitamin B-15. In: Pacific Slope Biochemical Conference, University of California, San Diego, July 7–9, 1980. (personal communication)
49. Girandola, R. N., Wiswell, R. A. and Bulbulian, R. Effects of pangamic acid (B-15) ingestion on metabolic response to exercise. *Biochem Med*, 24: 218–222, 1980.
50. Stacpoole, P. W., Henderson, G. N., Yan, Z. and James, M. O. Clinical pharmacology and toxicology of dichloroacetate. *Environ Health Perspect*, 106 Suppl 4: 989–994, 1998.
51. Sandu, C., Nick, P., Hess, D., Schiltz, E., Garrow, T. A. and Brandsch, R. Association of betaine-homocysteine S-methyltransferase with microtubules. *Biol Chem*, 381: 619–622, 2000.

23

Pyruvate and Dihydroxyacetone

David J. Dyck

CONTENTS

I. Introduction.. 445
   A. Description and Reported Effects 445
   B. Metabolism.. 446
II. Weight Loss Effects... 447
   A. Initial Animal Studies 447
   B. Human Studies 448
   C. Summary 449
III. Athletic Performance.. 449
   A. Rationale and Possible Mechanisms 449
   B. Human Studies 450
   C. Summary 452
IV. Overall Summary and Recommendations 452
References 453

I. Introduction

A. Description and Reported Effects

Pyruvate and dihydroxyacetone are 3-carbon carbohydrate metabolites (trioses); specifically, dihydroxyacetone phosphate is formed from the cleavage of fructose-1, 6-bisphosphate, and pyruvate is produced by the dephosphorylation of phosphoenolpyruvate at the terminal stage of glycolysis. The fate of pyruvate may either be oxidation to yield carbon dioxide and acetyl CoA and cause a small increase in Krebs' cycle intermediates, or be reduced to lactate, depending on the cellular conditions. In combination, these trioses have been reported to provide numerous beneficial effects, including the reduction of body mass and fat, lowering of blood lipids and improved endurance during aerobic exercise. Among the first observed metabolic effects of these triose compounds was a protective effect against the development of fatty liver in rats receiving chronic ethanol feedings.[1,2] Subsequent studies demonstrated that dietary supplementation of pyruvate and dihydroxyacetone resulted in decreased body fat of rats and swine[3,4] and reduced body mass in genetically obese Zucker rats.[5]

Supplementation with pyruvate and dihydroxyacetone has also been shown to decrease the regain in body mass and fat in obese humans after an initial weight reduction.[6,7] The mechanism of this effect is unclear, although there is limited evidence in both humans[8] and rodents[3] that elevations in plasma thyroxin subsequent to supplementation may be responsible for an enhanced metabolic rate. Pyruvate supplementation has also been reported to improve insulin sensitivity in rodents[9] and myocardial contractility in guinea pigs,[10–12] and reduce heart rate and systolic blood pressure[7] in humans. Pyruvate supplementation has also been advocated as an ergogenic aid, although the scientific evidence to support this is scant. Two studies have demonstrated an improvement in aerobic performance following the consumption of trioses for 7 days,[8,13] possibly due to improved glucose utilization. In nearly all of these studies, however, the dosages of pyruvate and dihydroxyacetone ingested were very high. For example, in human studies, dosages of up to 90–100 g per day of trioses have been administered; however, pyruvate supplements are usually sold in 1-gram capsules, making such a dosage impractical. In particular with humans, an undesirable aspect of supplementation with high dosages of pyruvate is the accompanying mineral content, typically as either sodium or calcium. Thus, in virtually all studies, pyruvate is combined with dihydroxyacetone, usually in a 1:3 ratio. The relevancy of extremely high triose dosages, as well as the inability to often distinguish between the effects of pyruvate and dihydroxyacetone, must be considered when extrapolating from these studies to the real world. However, it is generally assumed — both in the scientific literature and in the lay press — that pyruvate is the active ingredient.

B. Metabolism

Pyruvate, a ketoacid, is susceptible to acid hydrolysis, with the resultant liberation of carbon dioxide gas. Thus, intuitively, one would expect that at least some of the ingested pyruvate would be degraded within the acidic environment of the stomach. However, this aspect of pyruvate metabolism is generally overlooked. Furthermore, there has been virtually no information to date regarding the efficacy of pyruvate or dihydroxyacetone supplementation. We recently measured whole blood and plasma pyruvate concentrations for a 4-hr period following the acute ingestion a range of oral dosages of pyruvate (7, 15 and 25 g) and found no changes in blood pyruvate levels.[14] This strongly suggests that pyruvate is either very poorly absorbed, or alternatively, is rapidly cleared or metabolized by peripheral tissues such as the liver, kidney or skeletal muscle. For example, 25 g of pyruvate represents 12.5 gram of glucose, which is less than 10% of the storage capacity of the liver for glycogen. Thus, it is entirely possible that the liver could rapidly clear any pyruvate absorbed into the blood. In our study,[14] pyruvate was barely detectable in the urine following ingestion and previous studies were unable to detect pyruvate in the feces.[6,8] Although skeletal muscle is also a potential site for pyruvate clearance, intravenous pyruvate infusion does not acutely increase muscle pyruvate content.[15] Not surprisingly, given the likelihood of decarboxylation of pyruvate in the stomach, the occurrence of gastrointestinal distress including borborygmous (rumbling of the stomach), flatulence and nausea has been reported in humans consuming high dosages of trioses,[8,13] as well as at more modest dosages.[14] Thus, until future studies are able to more clearly define the efficacy of pyruvate supplementation, it should be considered highly debatable as to whether appreciable amounts of ingested pyruvate can even be absorbed, let alone reach potential target tissues such as skeletal muscle.

The mechanisms underlying the potential ergogenic and weight loss effects of pyruvate (see below), while poorly elucidated at best, focus at least to some extent on the supposed ability of pyruvate to reach its major target tissues, including skeletal muscle, liver and possibly adipose tissue. However, given the inability to elevate blood pyruvate levels with acute supplementation, as well as a lack of effect on skeletal muscle pyruvate content, such proposed mechanisms cannot be substantiated.

II. Weight Loss Effects

A. Initial Animal Studies

The first demonstrated metabolic effect of pyruvate and dihydroxyacetone was a reduction in the development of hepatic steatosis in rats chronically fed ethanol.[1,2] This effect was presumably due to the ability of pyruvate to oxidize the NADH generated from ethanol metabolism, thus preventing triacylglycerol synthesis that requires NADH as a cofactor. Futhermore, Stanko et al.[2] also noted a reduction in abdominal fat, indicating that the metabolic effects of pyruvate or dihydroxyacetone were not restricted to hepatic lipogenesis, but rather, that these trioses might have a more generalized effect on lipid synthesis. In spite of the apparent ability of pyruvate to inhibit hepatic lipogenesis, there has been little indication that supplementation significantly reduces blood lipids in the rat. For example, there are reports of elevated,[1] unaltered[3] and reduced[5] plasma triacylglycerol concentrations following pyruvate supplementation. Cholesterol levels have generally been unaffected.

Subsequently, it was demonstrated that including pyruvate and dihydroxyacetone in the diets of rats[3,5] and pigs[4] reduced weight gain and carcass fat, increased the metabolic rate and shifted fuel utilization toward lipid oxidation. It was speculated that the increases in metabolic rate and lipid oxidation were secondary to an increase in plasma thyroxine and a decrease in insulin, respectively.[3]

B. Human Studies

Several studies using human subjects have demonstrated improvements in the loss of body mass and fat following the administration of high dosages of pyruvate and dihydroxyacetone[7,16–18] or reduced weight gain with the reintroduction of a hypercaloric diet.[6] The total dosage of trioses used in these studies has ranged from 22 to 90 g and the actual amount of pyruvate has varied from 15 to 53 g. In two of these studies,[7,16] morbidly obese women were confined to bed rest in a metabolic ward and consumed 500 to 1000 kcal/d. In these studies, women receiving trioses for 21 d lost an additional 0.9 to 1.6 and 0.8 to 1.3 kg of body mass and fat, respectively, than those receiving placebo treatment (about 6 and 4 kg loss, respectively). Although relatively small, these differences were statistically significant. These findings have been criticized on the premise that the imposed conditions (morbid obesity, bed confinement, extreme caloric deficit) are so unusual that they are of no relevance to the average person attempting to reduce body mass and fat.[19] Furthermore, the losses of body mass and fat reported in these studies are relatively small, and the additional weight loss induced by dietary trioses is actually less than the losses induced by caloric restriction alone in the placebo group. In fairness, though, it must also be pointed out that the body mass loss induced by pyruvate or dihydroxyacetone supplementation in these studies is similar to that reported in studies using other popular treatments generally considered effective, such as ephedrine stacked with methylxanthines and aspirin.[20–22] However, all of these initial studies have originated from the same laboratory. Few subsequent studies have examined the effects of smaller, more practical dosages of isolated pyruvate on body mass or fat loss[23,24] and the results have been equivocal.

The effects of pyruvate supplementation on resting blood lipids in humans have also been examined, albeit to a limited extent. While some beneficial effects have been observed, in general there is little effect of pyruvate supplementation on blood lipids. Small (~5%) reductions in LDL- and total cholesterol[17] have been observed in hyperlipidemic subjects receiving pyruvate or dihydroxyacetone while consuming a high-fat diet, but not when fed a low-fat diet.[18]

One of the major difficulties in interpreting and evaluating the aforementioned pyruvate studies is the lack of a postulated feasible mechanism. Similar to rodent studies that have provided evidence, albeit limited, for changes in thyroxin levels, one study has reported an increase in plasma thryoxin following 7 days of pyruvate and dihydroxyacetone supplementation in humans.[8] Thus, overall, there is very little evidence either in animals or humans that hormonal levels are altered following pyruvate supplementation, and absolutely no evidence of changes induced by practical dosages (e.g., up to ~3-6 g/d) in humans. At least two additional mechanisms have been postulated to account for improved loss of body mass and fat following pyruvate supplementation. These include (1) accelerated futile cycling between pyruvate and phosphoenolpyruvate in the muscle[25] and (2) reversal of electron transport in the liver, with pyruvate serving as a precursor for substrates (oxaloacetate and dihydroxyacetone phosphate) of the malate/aspartate and glycerol-3-phosphate shuttles.[26] The result in each case would be an increase in ATP turnover and therefore an increase in the metabolic rate. Both mechanisms are highly speculative and unproven. More importantly, for either of these mechanisms to be considered valid, it is essential that pyruvate actually reach its target tissue (skeletal muscle, liver) and increase its concentration in that tissue. However, as already indicated, there is absolutely no evidence that supplemental pyruvate becomes elevated in the systemic circulation, let alone peripheral tissues such as muscle and liver. Finally, the determination of body composition in all human studies reported to date have used bioelectric impedance, which may not be appropriate for determining relatively small changes in body fat in the obese population.[27]

C. Summary

Although the use of relatively high dosages of pyruvate and dihydroxyacetone has been shown to significantly enhance the loss of body mass and fat in rodents and humans, the beneficial effects of low dosages of isolated pyruvate have not been established. This is of considerable practical significance, since pyruvate is marketed in 500 mg or 1 g capsules and generally does not contain dihydroxyacetone. Changes in body composition have only been determined by bioelectric impedance. Validation with other techniques, including dual energy x-ray absorptiometry (DEXA), should be considered. Furthermore, there is no clear, proven mechanism by which pyruvate would increase metabolic rate and enhance loss of body mass or fat.

III. Athletic Performance

A. Rationale and Possible Mechanisms

Although pyruvate has been widely reported in the lay press as being an ergogenic aid, the rationale underlying any such effect is poorly established, and the actual evidence for performance enhancement is extremely limited. The first study to examine the potential performance-enhancing effects of pyruvate was conducted in rats.[28] Pyruvate administration, by means of an intravenous injection, actually decreased the time required to run to exhaustion (i.e., ergolytic) and coincided with a greater rate of muscle and liver glycogen stores. These results prompted the investigators to speculate that cabohydrate utilization in general, and specifically muscle glucose uptake, was increased following pyruvate administration. However, the effect of pyruvate supplementation on muscle glucose transporter (GLUT 4) content or translocation, which would alter muscle glucose uptake, has never been investigated.

B. Human Studies

Three studies that have examined the effects of chronic supplementation of pyruvate or dihydroxyacetone on aerobic endurance have been published.[8,13,14] The first two studies were conducted in the same laboratory using untrained subjects; both found a positive effect of combined pyruvate and dihydroxyacetone supplementation on aerobic performance. In their initial study, Stanko et al.[8] reported a 23% increase (i.e., improvement) in time to exhaustion during arm ergometer exercise in untrained individuals following 7 days of supplementation with 100 g of trioses (25 g pyruvate plus 75 g dihydroxyacetone) per day. Exercise was performed at a power output corresponding to 65% VO_2 peak during arm ergometry. The mechanism of improved endurance was not clearly elucidated in this study, but arm glucose extraction was increased about twofold during exercise following triose supplementation. Substrate utilization was assessed by indirect calorimetry, but respiratory quotient and exchange ratio values greater than 1.0 negated their use in determining the relative contributions of fat and carbohydrate to overall energy provision. A potentially confounding factor in this study was a 44% increase in resting glycogen concentrations following pyruvate and dihydroxyacetone supplementation that may have contributed to the improved performance. This is supported by the fact that muscle glycogen concentrations at exhaustion were similar in both trials, thus indicating greater glycogen utilization in the pyruvate/dihydroxyacetone trial (about 95 vs. 60 mmol/kg dry wt.).

In their second study, Stanko et al.[13] assessed endurance during cycling at 70% VO_{2peak} in untrained individuals following placebo and 100 g pyruvate/dihydroxyacetone supplementation for 7 days. Dietary carbohydrate content was increased to 70% of total caloric content in both trials, resulting in similar resting muscle glycogen concentrations prior to exercise, and eliminating this as a confounding factor. Similar to their previous findings,[8] endurance was improved by 20% with pyruvate/dihydroxyacetone and was accompanied by an increase in glucose extraction of the exercising muscles. However, contrary to their previous study, total muscle glycogen utilization was similar in both trials, indicating that the rate of glycogen utilization was actually decreased during the pyruvate/dihydroxyacetone trial. It was hypothesized that the performance-enhancing effect of pyruvate/dihydroxyacetone was due to enhanced muscle glucose uptake and oxidation. This was supported by a greater estimate of total glucose oxidation throughout the entire exercise trial, although disturbingly, the respiratory exchange ratio values measured at consecutive 10 min intervals were not significantly different between treatments.

Although both of the studies by Stanko et al.[8,13] demonstrate an ergogenic effect of pyruvate, several aspects that affect interpretation of their data must be considered. First, it is highly desirable to use well-trained subjects when conducting performance trials, since they are accustomed to exhaustive exercise and will reliably force themselves to fatigue in each trial, whereas untrained individuals may not.[29,30] Both of the Stanko studies indicate the usage of untrained individuals, although no specific information is given in the first Stanko study[8] regarding subjects' aerobic capacity. In the second study,[13] the average subject's VO_{2max} was only 47.3 ± 0.4 ml/kg/min, which is indicative of their relative lack of aerobic fitness. Thus, the reliability and value of the outcome of the performance trials in these studies must be questioned. Second, while the trials by Stanko do appear to be well designed, it must also be questioned as to whether the subjects were truly blinded to their treatment, which is absolutely essential to properly evaluate the efficacy of putative performance enhancers. Since high triose dosages cause significant gastrointestinal distress, as reported in the Stanko studies, it is likely that most of the subjects were able to correctly identify their treatment. Third, the total triose dosage used in these studies (100 g) is extremely high and completely impractical. Pyruvate is commonly marketed as 1 gram capsules, which are approximately 65 to 70% pure (although we have found levels of purity as low as 50% when analyzed in our laboratory). Thus, at 70% purity, it would take 142 capsules per day to consume 100 g of triose compounds, or 36 capsules per day to consume 25 g of pyruvate alone. Finally, it is impossible to identify whether it is actually the pyruvate or the dihydroxyacetone that is responsible for any ergogenic effect observed. Thus, given the fact that marketed pyruvate supplements often do not contain dihydroxyacetone, it is imperative to clearly establish pyruvate as the active ergogenic ingredient.

The third study to examine the effects of pyruvate supplementation on aerobic performance was conducted in our laboratory.[14] This is the only study to date to examine the isolated effects of pyruvate (i.e., no dihydroxyacetone) on aerobic endurance. Well-trained males (VO_{2max}, 62.3 ± 3.3 ml/kg/min) received 7 g per d of calcium pyruvate or placebo (dextrose) for 1 week (randomized crossover) and cycled to exhaustion at 75–80% of their VO_{2max}. There was no difference in performance time (about 90 min) between the trials. Importantly, most of the subjects could not correctly identify the supplementation they were receiving and none complained of gastrointestinal distress, verifying that the subjects were truly blinded to their treatment. Thus, under such conditions (proper blinding, modest dosages of isolated pyruvate, trained individuals), there was no discernable performance-enhancing effect of pyruvate.

C. Summary

Only two previous studies from the same laboratory,[8,13] have demonstrated an ergogenic effect of pyruvate. The effects were determined in untrained subjects at very high combined dosages of pyruvate and dihydroxyacetone. Due to the side effects of high dosages of pyruvate and dihydroxyacetone, it is debatable as to whether these subjects were actually blinded to their treatment. Supplementation of a more realistic dosage of pyruvate alone appears to be without any discernable side effects, but unfortunately, also does not improve aerobic endurance in well-trained subjects. Thus, given the very small number of published studies that have examined the ergogenic effects of pyruvate, and the lack of evidence of any effect in reliable, well-trained subjects consuming practical dosages, the use of pyruvate as an ergogenic aid has no scientific basis.

IV. Overall Summary and Recommendations

Initial studies performed in rodents and swine indicated the potential for pyruvate supplementation to enhance body mass and fat loss, as well as to improve insulin sensitivity and blood lipids. This was achieved with high dosages of pyruvate, generally in combination with dihydroxyacetone. In humans, the administration of high triose dosages also appears to result in enhanced mass and fat loss. However, all of these studies with humans originate from the same laboratory and utilize conditions that are not relevant to the general population. Very few studies have examined the effect of modest dosages of pyruvate alone on body mass and composition, and the findings have been equivocal. Two studies, also from the laboratory of Stanko,[8,13] have demonstrated ergogenic properties of high dosages of pyruvate and dihydroxyacetone in untrained individuals. However, in the only subsequent study, chronic low dosages of isolated pyruvate failed to enhance aerobic performance in well-trained individuals. Finally, there is no evidence indicating that pyruvate is effectively absorbed or increases its concentration in target tissues such as skeletal muscle or liver. Given this information and the lack of a proven plausible mechanism underlying any putative weight loss or ergogenic effect, it is impossible to recommend the use of pyruvate as a supplement.

References

1. Goheen, S. C., Pearson, E. E., Larkin, E. C. and Rao, G. A. The prevention of alcoholic fatty liver using dietary supplements: Dihydroxyacetone, pyruvate and riboflavin compared to arachidonic acid in pair-fed rats. *Lipids*, 16, 43, 1981.
2. Stanko, R. T., Mendelow, H., Shinozuka, H. and Adibi, S. A. Prevention of alcohol-induced fatty liver by natural metabolites and riboflavin. *J. Lab. Clin. Med.*, 91, 228, 1978.
3. Stanko, R. T. and Adibi, S. A. Inhibition of lipid accumulation and enhancement of energy expenditure by the addition of pyruvate and dihydroxyacetone to a rat diet. *Metabolism*, 35, 182, 1986.
4. Stanko, R. T., Ferguson, T. L., Newman, C. W. and Newman, R. K. Reduction of carcass fat in swine with dietary addition of dihydroxyacetone and pyruvate. *J. Anim. Sci.*, 67, 1272, 1989.
5. Cortez, M. Y., Torgan, C. E., Brozinick Jr., J. T., Miller, R. H. and Ivy, J. L. Effect of pyruvate and dihydroxyacetone consumption on the growth and metabolic state of obese Zucker rats. *Am. J. Clin. Nutr.*, 53, 847, 1991.
6. Stanko, R. T. and Arch, J. E. Inhibition of regain in body weight and fat with addition of 3-carbon compounds to the diet with hyperenergetic refeeding after weight reduction. *Int. J. Obes. Relat. Metab. Disord.*, 20, 925, 1996.
7. Stanko, R. T., Tietze, D. L. and Arch, J. E. Body composition, energy utilization and nitrogen metabolism with a severly restricted diet supplemented with dihydroxyacetone and pyruvate. *Am. J. Clin. Nutr.*, 55, 771, 1992.
8. tanko, R. T., Robertson, R. J., Spina, R. J., Reilly Jr., J. J., Greenawalt, K. D. and Goss, F. L. Enhancement of arm exercise endurance capacity with dihydroxyacetone and pyruvate. *J. Appl. Physiol.*, 68, 119, 1990.
9. Ivy, J. L., Cortez, M. Y., Chandler, R. M., Byrne, H. K. and Miller, R. H. Effects of pyruvate on the metabolism and insulin resistance of obese Zucker rats. *Am. J. Clin. Nutr.*, 59, 331, 1994.
10. Bunger, R., Mallet, R. T. and Hartman, D. A. Pyruvate-enhanced phosphorylation potential and inotropism in normoxic and postischemic isolated working heart. Near-complete prevention of reperfusion contractile failure. *Eur. J. Biochem.*, 180, 221, 1989.
11. Mallet, R. T. and Bunger, R. Metabolic protection of post-ischemic phosphorylation potential and ventricular performance. *Adv. Exp. Med. Biol.*, 346, 233, 1993.
12. Mentzer, R. M., Van Wylen, D. G. and Sodhi, J. Effect of pyruvate on regional ventricular function in normal and stunned myocardium. *Ann. Surg.*, 209, 629, 1989.
13. Stanko, R. T., Robertson, R. J., Galbreath, R. W., Reilly Jr., J. J., Greenawalt, K. D. and Goss, F. L. Enhanced leg exercise endurance with a high-carbohydrate diet and dihydroxyacetone and pyruvate. *J. Appl. Physiol.*, 69, 1651, 1990.
14. Morrison, M. A., Spriet, L. L. and Dyck, D. J. Pyruvate ingestion for 7 days does not improve aerobic performance in well-trained individuals. *J. Appl. Physiol.*, 89, 549, 2000.
15. Constantin-Teodosiu, D., Simpson, E. J. and Greenhaf, P. L. The importance of pyruvate availability to PDC activation and anaplerosis in human skeletal muscle. *Am. J. Physiol. (Endocrinol. Metab.)*, 276, E472, 1999.

16. Stanko, R. T., Tietze, D. L. and Arch, J. E. Body composition, energy utilization and nitrogen metabolism with a 4.25-MJ/d low-energy diet supplemented with pyruvate. *Am. J. Clin. Nutr.*, 56, 630, 1992.
17. Stanko, R. T., Reynolds, H. R., Lonchar, K. D. and Arch, J. E. Plasma lipid concentrations in hyperlipidemic patients consuming a high-fat diet supplemented with pyruvate for 6 wk. *Am. J. Clin. Nutr.*, 56, 950, 1992.
18. Stanko, R. T., Reynolds, H. R., Hoyson, R., Janosky, J. E. and Wolf, R. Pyruvate supplementation of a low-cholesterol, low-fat diet: Effects on plasma lipid concentrations and body composition in hyperlipidemic patients. *Am. J. Clin. Nutr.*, 59, 423, 1994.
19. Sukala, W. R. Pyruvate: Beyond the marketing hype. *Int. J. Sports Med.*, 8, 241, 1998.
20. Astrup, A., Breum, L., Toubro, S., Hein, P. and Quaade, F. The effect and safety of an ephedrine/caffine compound compared to ephedrine, caffeine and placebo in obese subjects on an energy restricted diet. A double blind trial. *Int. J. Obes.*, 16, 269, 1992.
21. Daly, P. A., Krieger, D. R., Dulloo, A. G., Young, J. B. and Landsberg, L. Ephedrine, caffeine and aspirin: safety and efficacy for treatment of human obesity. *Int. J. Obes.*, 17, S73, 1993.
22. Pasquali, R., Cesari, M. P., Melchionda, N., Stefanini, C. and Raitano, A. Does ephedrine promote weight loss in low-energy-adapted obese women? *Int. J. Obes.*, 11, 163, 1987.
23. Kalman, D., Roufs, J. and Maharam, L. G. Effects of exogenous pyruvate on body composition and energy levels. *Med. Sci. Sports Exer.*, 30, S156, 1998. (Abstract)
24. Stone, M. H., Sanborn, K., Smith, L. L., O'Bryant, H. S., Hoke, T., Utter, A. C., Johnson, R. L., Bros, R., Hruby, J., Pierce, K. C., Stone, M. E. and Garner, B. Efffects of in-season (5 weeks) creatine and pyruvate supplementation on anaerobic performance and body composition in American football players. *Int. J. Sport Nutr.*, 9, 146, 1999.
25. Ivy, J. L. Effect of pyruvate and dihydroxyacetone on metabolism and aerobic endurance capacity. *Med. Sci. Sports Exerc.*, 30, 837, 1998.
26. McCarty, M. F. The origins of western obesity: A role for animal protein? *Med. Hypoth.*, 54, 488, 2000.
27. Kushner, R. F., Gudivaka, R. and Schoeller, D. A. Clinical characteristics influencing bioelectrical impedance analysis measuremens. *Am. J. Clin. Nutr.*, 64, 423S, 1996.
28. Bagby, G. J., Green, H. J., Katsuta, S. and Gollnick, P. D. Glycogen depletion in exercising rats infused with glucose, lactate, or pyruvate. *J. Appl. Physiol.*, 45, 425, 1978.
29. Hickey, M. S., Costill, D. L., McConell, G. K., Widrick, J. J. and Tanaka, H. Day to day variation in time trial cycling performance. *Int. J. Sports. Med.*, 13, 467, 1992.
30. McLellan, T. M., Cheung, S. S. and Jacobs, I. Variability of time to exhaustion during submaximal exercise. *Can. J. Appl. Physiol.*, 20, 39, 1995.

24
Tannins

Jay Kandiah

CONTENTS
I. Overview of Plant Tannins 455
II. Chemistry and Constituents of Tea 456
III. Tea and Oxidative Stress 457
IV. Tea and Blood Pressure 458
V. Tea and Cardiovascular Disease 459
VI. Tea and Cancer 460
VII. Conclusion 461
References 462

I. Overview of Plant Tannins

Phenolic compounds include many distinct classes of low-molecular-weight flavanoids, phenolic acids and esters, as well as intermediate- to high-molecular-weight polyphenols commonly known as tannins. In plants, the two major groups of tannins include hydrolyzable and condensed tannins (proanthocyanidins). Hydrolyzable tannins are polymers of gallic or ellagic acid that are found mainly in berries and nuts. These tannins yield polyhydric alcohol and phenylcarboxylic acid in the presence of acid, base and hot water or by enzymatic action. Condensed tannins are polymers of catechins or epicatechins that are found mainly in beverages (tea, coffee, cocoa, red wine), certain fruits (bananas, apples, persimmons), vegetables (spinach, rhubarb) and legumes (peas). As their name indicates, these tannins are formed by auto-oxidative or enzymatic polymerization of flavan-3-ol (catechin, epicatechin, etc.) and flavan-3,4-diol units or a mixture of the two.[1]

Historically, tannins are noted for their ability to bind and precipitate carbohydrates and proteins, as well as to chelate metals. Tannins have a high affinity to bind with proline-rich proteins (e.g., gelatin) or salivary proline-rich proteins. When bound to mouth proteins, tannins produce a distinctively astringent taste sensation.[2] Unlike hydrolyzable tannins, condensed tannins are unabsorbed from the gastrointestinal tract, but are complexed to salivary proline-rich protein during passage through the alimentary canal. On the other hand, hydrozable tannins are absorbed after ingestion and are bound to proteins (e.g., serum albumin).

In recent years, tea polyphenols have gained much recognition in the scientific arena, especially in nutrition and health. Many studies have shown that tea consumption can provide protection against cancer, inflammation, coronary heart disease, hepatic necrosis and dental caries.[3,4] This chapter will speculate the role of tannins, specifically tea, as a possible ergogenic aid.

II. Chemistry and Constituents of Tea

After water, tea is the second-most-consumed beverage in the world, with per capita consumption of approximately 120 mL/d.[5] This popular beverage, consumed in the form of hot or cold infusions, is derived from the dried leaves of the plant *Camellia Sinensis*, a native of Southeast Asia. Derived from the evergreen shrub of the Theaceae family, the tea plant is cultivated in many countries around the world. The majority of the tea consumed in the world is black tea (78%), which is mostly consumed in Western countries (e.g., Americans and Europeans) and in some Asian countries (e.g. India, Sri Lanka). This is followed by green tea (20%) which is commonly consumed in some Asian (e.g., Japan, Korea, China), Middle Eastern and North African countries. Produced and consumed in smaller amounts is oolong tea (2%) which is mainly prevalent in Taiwan and southern China.[5]

Besides seasonal and climatic conditions, the type of species, age of the plant, plucking position of the leaves and horticultural practices influence the composition of the tea leaves.[6] Differences in green, black and oolong tea depend on the manufacturing processes. Green tea is manufactured by preventing fermentation. Freshly harvested leaves are either pan-fried or steamed to inactivate the enzymes, resulting in dry fresh tea leaves. The major polyphenolic compounds in green tea are epicatechins (e.g., (-) epigallocatechins-3-gallate (EGCG); (-) epigallocatechin (ECG); (-) epicatechin-3-gallate (ECG) and (-) epicatechin (EC)). These are commonly known as catechins and are responsible for the characteristic flavor and color of green tea. Catechins constitute approximately 30% of the dry weight of green tea.[7,8] The catechins that are most abundant in green tea are EGCG, which have received by far the most attention.

In manufacturing black tea, the fresh leaves are withered so that more than 50% of the moisture content of the leaf is reduced, which results in a high concentration of polyphenols in the leaves. The leaves are then crushed and rolled to enable fermentation of the polyphenols. This fermentation process causes green tea catechins to be fully oxidized and polymerized to form theaflavins (theaflavin, theaflavin-3-gallate, theaflavin 3′-gallate and theaflavin-3,3′- digallate) and thearubigens. Thus, the major polyphenols in black tea are theaflavins and thearubigens. Theaflavins are responsible for the taste and reddish color of black tea, while thearubigens are the most abundant phenolic fraction in black tea. Unlike green tea, catechins constitute only 9% of the dry weight of black tea.[7,8] The manufacturing of oolong tea is initially similar to that of black tea. After the leaves have been rolled, they are fired to terminate the oxidation. This results in a partially fermented tea that consists of both green tea catechins, black tea theaflavins and possibly thearubigens.[7] Fermentation of black and green tea leads to important differences in the polyphenlic composition of tea leaves.[8]

III. Tea and Oxidative Stress

Tea polyphenols have been widely recognized for their antioxidant properties, namely their ability to scavenge reactive oxygen species.[9] In addition to their free radical quenching and metal-chelating abilities, catechins prevent low-density lipoprotein (LDL) oxidation by recycling other antioxidants such as vitamin E.[10,11] In an *in vivo* study in animals, Reidl et al.[2] compared the oxidative stress in young rats fed diets of rat chow with water or green tea (the polyphenolic compound, EGCG). Duration of the study was for 6.5 weeks, during which time oxidative stress was imposed by acute exercise. Measurement of oxidative stress in various tissues of rats showed that those on the water diet exhibited oxidative stress in the kidney and liver. Interestingly, rats on green tea did not exhibit oxidative stress in either organs, suggesting that long-term ingestion of green tea may provide antioxidant protection that may decrease oxidative damage to various tissues.[2] The effects of green tea on exercise-induced oxidative stress was investigated in young male Sprague Dawley rats. For 6.5 weeks, the rats received one of the following treatments: (a) water only, (b) water plus aerobic exercise, (c) green tea only and (d) green tea plus aerobic exercise. Prior to sacrifice, rats that received aerobic exercise ran at an average speed of 25m/min for 30 minutes on a treadmill. There was no change in malonaldehyde equivalents, a biomarker of oxidative stress, in the kidney tissue of rats fed tea. Rats in the water consumption group had a 290% increase of malonaldehyde equivalents. The transport of EGCG to the kidneys for excretion enhanced antioxidant protection, thereby preventing kidney tissue from exercise-induced oxidative stress.[12]

Short-term intervention studies on human subjects with elevated blood pressure have indicated ingestion of either black or green tea had no effect on urinary F_2- isoprostanes, an indicator of lipid peroxidation.[13] Thus, it appears the antioxidant properties of tea may be beneficial in preventing oxidative damage, which may serve as an ergogenic aid. Though positive results were noted in animals, future research needs to focus on tea consumption as an ergogenic aid in athletes.

IV. Tea and Blood Pressure

It has been suggested that oxidative stress is involved in the development of raised blood pressure.[14] Previous research by Uchida et al.[15] had demonstrated that persimmon tannin was effective in maintaining blood pressure and prolonging life expectancy of stroke-prone spontaneously hypertensive rats. Since green tea contains abundant amounts of EGCG, and to determine whether EGCG would be involved in hypertensive lesions via lipid peroxidation, stroke-prone spontaneously hypertensive rats were treated with four variations of fluids, namely distilled water, 0.1% EGCG, 0.5% EGCG and 0.5% persimmon tannin. Results revealed long-term administration of either 0.5% EGCG or 0.5% persimmon tannin had no effect on blood pressure. EGCG-like persimmon tannin prevented lipid peroxidation, inhibiting the occurrence of stroke, and prolonged the life expectancy of the rats.[16]

Similar observations were noted in a 4-week randomized crossover study that evaluated the effect of black tea consumption on blood lipids, bowel habits and blood pressure in humans. There were no significant differences in diastolic and systolic blood pressures between tea consumers (6 mugs or 48 fluid ounces of tea/day) and nonconsumers.[17]

Yokozawa et al.[18] found oral administration of 2 and 4 mg of green tea tannin resulted in lowered diastolic and systolic blood pressure in rats. Compared with the control group, rats on 2 mg of tea showed a 7% reduction in blood pressure. A significant decrease in mean blood pressure was noted in rats fed 4 mg of green tea (163.0 to 141.9 mm Hg). The diastolic (144.0 to 122.2 mm Hg) and systolic (144.0 to 122.2 mm Hg) blood pressure was also significantly lowered in rats given 4 mg of green tea. Increased urinary excretion of sodium and prostaglandin E_2, inhibitor of sodium reabsorption were also observed, with higher excretions in rats fed 4 mg green tea tannin.

Because γ-aminobutyric acid (GABA) is considered as a neurotransmitter inhibitor, the role of GABA in regulating blood pressure was investigated. For 4 weeks, 11-month-old Dahl S rats received water, ordinary tea or GABA-rich tea. Although blood pressure was not different among the three groups at the start of the research, after 4 weeks blood pressure was significantly lower in the GABA group than in the other two groups. Even though blood pressure was lower in the tea group than in the water group; this was not found to be statistically significant. The investigators concluded that GABA may have vasodilation abilities and as such GABA-rich tea might be effective in preventing essential hypertension in humans.[19]

Unlike coffee, tea consumption in humans has been positively associated with healthy behaviors such as decreased intake of meats, increased physical activity and higher consumption of water.[20] Since there is evidence that tea is a relaxing beverage, inducing vasodilation, association between tea and positive mood states were assessed in 49 volunteers (males n = 18; females n = 31) in two working groups (teachers and nurses). For 8 weeks, participants kept daily records of fluids consumed and ratings of positive moods. There was no association between tea consumption and mood. Interestingly, more positive moods were associated with alcohol consumption than any other beverage. When gender differences were assessed, women who had enjoyed greater social aspects of their work drank more tea on days they felt relaxed. On the other hand, men with little to no social support drank more tea on days they were relaxed. [20]

In Norway, mortality from cardiovascular disease was found to be higher in persons drinking no tea or less than one cup of tea/day. Irrespective of gender, an inverse relationship existed between tea consumption and systolic blood pressure.[21] Because of the differences in polyphenols between green and black tea, their effects on blood pressure examined with 20 normotensive men showed a significant increase in systolic and diastolic blood pressure after 30 minutes of consumption of black tea. However, there were no significant changes in diastolic or systolic blood pressure 60 minutes after consumption of green or black tea. In addition, heart rate remained unaffected at 30 and 60 minutes after drinking green or black tea. Similarly, in the second part of their research, the authors found 1 week consumption of either type of tea had no significant effect on ambulatory blood pressure in subjects with mild and high systolic blood pressure.[22] Anomalous blood pressure as a result of postprandial hypotension has been observed with aging. Short-term consumption of five cups of tea/day in 63 hypertensive adults demonstrated that tea did not lead to elevations in fasting systolic blood pressure and may be used as an alternative to drug treatment.[23]

V. Tea and Cardiovascular Disease

Studies from the UK and USA have demonstrated that tea consumption had no effect on ischemic heart disease.[24,25] In spite of controversies over the effects of tea consumption on cardiovascular disease, Geleijnse et al.[26] in their study population of >6000 men and women showed an inverse association between tea intake and severity of atherosclerosis. Interestingly, women had a greater protective effect than men. Research in Norway also found mean serum cholesterol decreased with increasing tea consumption. Differences in serum cholesterol ranged from 5.8 mg/dl–9.3 mg/dl between drinkers of less than one cup and those of five or more cups/day.[21] Similarly, Sesso et al.[27] found that subjects who drank one or more cups (> 250 ml) of black tea per day had only a 50% greater risk of a heart attack than those who refrained from tea consumption. Although the full potential of tea as an antioxidant has been attributed to tea without milk,[28] research has demonstrated that the presence of milk failed to interfere with gastrointestinal absorption or excretion of catechins.[29,30]

Since hypercholesterolemia is a contributing factor to heart disease, several researchers investigated tea consumption on serum cholesterol levels in humans. Unlike animal studies (e.g., rats and hamsters) where green tea, black tea and tea polyphenols had a positive effect on serum cholesterol and lipid levels,[31-34] research on humans failed to repeatedly exhibit a serum-cholesterol-lowering effect from either green or black tea consumption.[17,35-37] Studies in the late '80s demonstrated that green tea catechins increased plasma levels of HDL cholesterol and reduced concentrations of LDL cholesterol possibly through reverse cholesterol transport, reduced intestinal absorption or increased bile acid excretion.[38]

Even though inconsistencies regarding the protective effect of tea against cardiovascular disease exist in the literature,[30,35,39] there are a few proposed mechanisms for its effectiveness. It appears that tea polyphenols have the ability to inhibit the oxidation of LDL[9] and to decrease ADP-induced platelet aggregation.[40] Although, to date, no research has looked at the effectiveness of tea in preventing heart disease in athletes, the ability to inhibit LDL oxidation and decrease platelet aggregation may encourage physically active individuals to speculate the use of tea as an alternative to other ergogenic aids. This coupled with healthy eating may not only enhance athletic performance but may also alleviate the onset of other diseases.

VI. Tea and Cancer

Epidemiological studies on the effects of tea consumption on human cancer have been inconclusive.[5,41-45] In certain regions of the world (e.g., China, Italy, Japan, Turkey and Sweden) a protective effect or lower incidence on various types of cancers (e.g., gastric, oral, pharyngeal, gastric, breast, digestive) has been associated with tea consumption. The cancer-preventive effect of green tea is through the major polyphenolic component EGCG. According to Mukthar et al.,[46] there is 200 mg of EGCG in one cup of brewed green tea (240 mL). In females, daily consumption of greater than 10 cups (2 L) of green tea was associated with decreased risk of breast cancer.[47]

In most circumstances, preventive effects were from studies of Asians, who drink predominantly green tea. On the contrary, many studies, especially in Europe, have been unable to support tea in cancer prevention.[48] A possible explanation to this could be that, generally, Europeans drink black tea, which is lower in catechin content than green tea. Unlike black tea, the cancer-preventive activity in green tea appears to be not only catechins but also theaflavins. Other reasons for the discrepancy are that there have been very few heavy tea drinkers in the Western study population, and the differences in dietary practices between Asian and Western countries may have contributed to variances in the results.

Animal studies on rats, mice and hamsters have demonstrated the effectiveness of tea in preventing various types of cancers (e.g., skin, esophagus, small intestine, colon, bladder, stomach, mammary glands, lung, liver, pancreas and prostate). [5,41,42,49,50] Most experimental studies on animals were conducted with green tea and very few studies have assessed the effectiveness of black tea.[46] In mice, administration of tea during any stage of cancer development was effective in inhibiting tumor activity. Similarly, other researchers have shown tannins to reduce hepatomas in mice.[51] In addition to oral feeding, topical application of tea polyphenols have been effective in inhibiting tumor initiation and promotion in rats.[52] In most circumstances, the major polyphenol that served as a potent inhibitor of cancer was EGCG. However, in a majority of animal studies, it is important to note that high amount of tannins (EGCG and catechins) that would be inappropriate for daily human consumption were utilized.

Although the exact mechanism as to how tea inhibits the occurrence of cancer has not been studied in depth, there have been many proposed models. Besides the frequently cited antioxidant activities, polyphenols, catechins and theaflavins in tea have been shown to inhibit cell transformation, cell growth and tumor-promotion enzymes in animals and cell cultures.[41,42,53,54] Most human studies on the effectiveness of tea in inhibiting the development of cancer have been inconclusive. However, there is supporting evidence from animal research that shows a strong association between tea and the immune system. Several researchers have demonstrated tea consumption to enhance immunity, thereby decreasing the occurrence of cancer. The type of tea (black versus green), method of preparation (with or without milk) and the amount consumed have all affected the effectiveness of tea, which may possibly increase the use of tea as an ergogenic aid.

VII. Conclusion

Although many beneficial health effects of tea have been discussed, inconsistencies exist in the literature. There is a strong possibility that the antioxidant properties of tannins may enable them to exert an ergogenic effect, thereby preventing the onset of vascular diseases and cancer.

Based on the literature, discrepancies in the effectiveness of tannins in preventing and treating various disease states is due to the high doses administered in animal studies that are impractical and unrealistic for daily human consumption.[49] In humans, since there is limited information on the bioavailability of tea polyphenols, future research needs to focus on the absorption and metabolism of black and green tea polyphenols. Additionally, long-term clinical and intervention trials should be conducted in different population groups (e.g., athletes and nonathletes) of varying ages, and using larger samples.

References

1. Bravo, L. Polyphenols: Chemistry, dietary sources, metabolism and nutritional significance. *Nutr. Rev.* 56, 317–333, 1998.
2. Riedl, K.M., Carando, S., Alessio, H.M., McCarthy, M. and Hagerman, A.E., Antioxidant Activity of Tannins and Tannin-Protein Complexes: Assessment *in Vitro* and *in Vivo*. American Chemical Society 88–200, 2002, chap 14.
3. Chung, K.-T., Tannins and human health: A review. *Crit. Rev. Food Sci. Nutr.* 14, 325–335, 1998.
4. Sakanaka, S., Kim, M, Taniguchi, M. and Yamamoto, T., Prevention of dental caries by consumption of green tea. *Agric. Biol. Chem.* 53, 2307-2311 1989.
5. Katiyar, SK., Mukthar, H. Tea in chemoprevention of cancer: Epidemiologic and experimental studies, *Int. J. Oncol.* 8, 221–38, 1996.
6. Lin, J.K., Liang, Y.C. and Lin-Shiau, S.Y., Survey of catechins, gallic acid and methylxanthines in green, oolong, pu-erh and black teas. *J. Agric. Food Chem,* 46, 3635–3642, 1998.
7. Balentine, D.A., Wiseman, S.A. and Bouwens, L.C.M., The chemistry of tea flavonoids. *Crit. Rev. Food Sci. Nutr.* 37, 693–704, 1997.
8. Harbowy, M.E., Balentine, D.A., Tea chemistry. *Crit. Rev. Plant Sci.*16, 415–480, 1997.
9. Wiseman, S. A., Balentine, D.A., Kuo, M.C. and Schantz,S.P., Antioxidants in tea. *Crit. Rev. Food Sci. Nutr.* 37, 705–718, 1997.
10. Zhu, Q.Y., Huang, Y., Tsang, D., Chen, Z.Y., Regeneration of α-tocopherol in human low-density lipoprotein by green tea catechin. *J. Agric. Food Chem.* 47, 2020–2025, 1999.
11. Packer, L., Vitamin E and the antioxidant network: Protection of human low density lipoprotein from oxidation, in Ohigashi, H., *Food Factors for Cancer Prevention.* Tokyo: Springer-Verlag, 452–459, 1997.
12. Alessio, H., Hagerman, A.E., Romanello, M, Carando, S., Threlkeld, S., Rogers, J., Dimitrova, Y., Muhammed, S. and Wiley, R.L., Consumption of green tea protects rats from exercise-induced oxidative stress in kidney and liver. *Nutr. Res.*, 22, 1177–1188, 2002.
13. Hodgson, J.M., Croft, K.D., Mori, A., Burke, V., Beilin, L.J. and Puddey, I.B., Regular ingestion of tea does not inhibit *in vivo* lipid peroxidation in humans *J. Nutr.* 132, 55–58, 2002.
14. Romero-Alvira, D. and Roche, A., High blood pressure, oxygen radicals and antioxidants: Etiological relationships. *Med. Hypoth.* 46, 414–420, 1996.
15. Uchida, S., Ohta, H., Niwa, M., Mori, A., Nonaka, G., Nishioka, I. and Ozaki, M., Prolongation of life span of stroke-prone spontaneously hypertensive rats (SHRSP) ingesting persimmon tannin. Chemical and Pharmaceutical Bulletin, 38, 1049–1050, 1990.
16. Uchida, S., Ozaki, M., Akashi, T., Yamashita, K., Niwa, M. and Taniyama, K., Effects of (-)-epigallocatechin-3-0-gallate (green tea tannin) on the life span of stroke-prone spontaneously hypertensive rats. *Clin. Exp. Pharmacol. Physiol.,* Suppl. 1, S302–S303, 1995.
17. Bingham, S.A., Vorster, H., Jerling, C., Magee, E., Mulligan, A., Runswick, S.A. and Cummings, J.H., Effect of black tea drinking on blood lipids, blood pressure and aspects of bowel habit, *Brit. J. Nutr.* 78, 41-55, 1997.

18. Yokozawa, T., Oura, H., Sakanaka, S., Ishigaki, S. and Mujo. K. Depressor effect of tannin in green tea on rats with renal hypertension. *Biosci. Biotech. Biochem.* 58, 5, 855–858, 1994.
19. Yasuhiko, A., Umemura, S., Sugimoto, K-I, Hirawa, N., Kato, Y., Yokoyama,N. and Yokoyama, T., Effect of green tea rich in ϒ-aminobutyric acid on blood pressure of dahl salt-sensitive rats. *Am. J. Hyperten*s 8, 74–79, 1995.
20. Schwarz, B., Bischoff, H-P and Kunze, M., Coffee, tea and lifestyle. *Prev. Med.* 23, 377–384, 1994.
21. Stensvold, I., Tverdal, A., Solvoll, K. and Foss, O.P., Tea consumption relationship to cholesterol, blood pressure and coronary and total mortality. *Prev. Med.* 21, 546–553, 1992.
22. Hodgson, J.M., Puddey, I.B., Burke, V., Beilin, L.J. and Jordan, N., Effects on blood pressure on drinking green and black tea. *J. Hyperten.* 17, 457–463, 1999.
23. Rakic, V., Effect of coffee and tea drinking on postprandial hypotension in older men and women. *Clin. Exp. Pharmacol.Physiol.*, 23, 559–563, 1996.
24. Klatsky, A.L., Friedman, G.D. and Armstrong, M.A., Coffee use prior to myocardial infarction restudied: heavier intake may increase the risk. *Am. J. Epidemiol.* 132, 479–88, 1990.
25. Woodward, M. and Tunstall-Pedoe, H., Coffee and tea consumption in the Scottish Heart Health Study follow up: Conflicting relations with coronary risk factors, coronary disease and all cause mortality. *J. Epidemiol. Comm. Health* 53, 481–7, 1999.
26. Geleijnse, J.M., Launer, L.J., Hofman, A., Pols, H.A. and Witterman, J.C. Tea flavonoids may protect against atherosclerosis: The Rotterdam Study. *Arch. Intern. Med.* 159, 2170–2174, 1999.
27. Sesso, H.D., Gaziano, J.M., Buring, J.E., Hennekens, C.H., Coffee and tea intake and the risk of myocardial infraction. *Am. J. Epidemiol.* 149, 162–167, 1999.
28. Hertog MGL, Antioxidant flavonols and ischemic heart disease in a Welsh population of men: The Caerphilly Study. *Am. J. Clin. Nutr.* 65,1489–94,1997.
29. van het Hof, K.H., Kivits, G.A.A. and Weststrate, J.A., Bioavailability of catechins from tea: The effect of milk. *Eur. J. Clin. Nutr.* 52, 356–9, 1998.
30. van het Hof, K.H., Wiseman, K.H., Yang, C.S. and Tijbung, L.B.M., Plasma and lipoprotein levels of tea catechins following repeated tea consumption. *Proc. Soc. Exp. Bio.* 220, 203–209, 1999.
31. Vinson, J.A. and Dabbagh, Y.A., Effect of green and black tea supplementation on lipids, lipid oxidation and fibrinogen in the hamster: mechanisms for the epidemiological benefits of tea drinking. *FEBS Lett.* 433, 44–46, 1998.
32. Yang, T.T. and Koo, M.W., Hypocholesterolemic effects of Chinese tea. *Pharmacol. Res.* 35, 505–515. 1997.
33. Matsumoto, N., Okushio, K. and Hara, Y. Effect of black tea polyphenols on plasma lipidsin cholesterol-fed rats. *J. Nutr, Sci. Vitaminol.* 44, 337-342. 1998.
34. Chan, P.T., Fong, W.P., Cheung, Y.L., Huang, Y., Ho, W.K. and Chen, Z.Y., Jasmine green tea epicatechins are hypolipidemic in hamsters (*Mescoricetus auratus*) fed a high fat diet. *J. Nutr.* 129, 1094–1101. 1999.
35. Tijburg, L.B.M., Mattern, T., Folts, J.D., Weisgerber, U.M. and Katan, M.B., Tea flavonoids and cardiovascular diseases: a review. *Crit. Rev. Food Sci. Nutr.* 37, 771–785, 1997.
36. Princen, H.M., van Duyvenvoorde, W., Buyterhek, R., Blonk, C., Tijburg, L.B., Languis, J.A., Meinders, A.E. and Pijl, H., No effect of consumption of green and black tea on plasma lipid and antioxidant levels and on LDL oxidation in smokers. *Arterioscler. Thromb. Vasc.* Biol. 18, 833–841, 1998.

37. Van het Hot, K.H., De Boer, H.S.M., Wiseman, S.A., Weststrate, J.A. and Tijburg, L.B.M. Consumption of green tea or black tea does not increase the resistance of LDL oxidation in humans. *Am. J. Clin. Nutr.* 66, 1125–1132, 1997.
38. Muramatsu, K., Fukuyu, M., Hara, Y., Effect of green tea catechins on plasma cholesterol level in cholesterol-fed rats. *J. Nutr. Sci. Vitaminol.* 32, 613–22, 1986.
39. Freese, R., Basu, S., Hietanen, E., Nair, J., Nakachi, K., Bartsch, H. and Mutanen, M., Green tea extract decreases plasma malondialdehyde concentration but does not affect other indicators of oxidative stress, nitric oxide production, or hemostatic factors during a high linoleic acid diet in healthy females. *Eur. J. Nutr.*38, 149–157, 1999.
40. Tijburg, L.B.M., Wiseman, S.A., Meijer, G.W. and Westsrate, J.A., Effects of green tea, black tea and dietary lipophilic antioxidants on LDL oxidizability and atherosclerosis in hypercholesterolemic rabbits. *Atherosclerosis* 135, 37–47, 1997.
41. Yang, C.S. and Wang, Z-Y., Tea and cancer: A review. *J. Natl. Cancer Inst.* 58, 1038–1049, 1993.
42. Yang C.S., Chung, J.Y., Yang, G-Y.,Tea and tea polypnenols in cancer prevention *J. Nutr.* 130, 472S–478S, 2000.
43. Blot, W.J., McLaughlin, J.K. and Chow, W-H., Cancer rates among drinkers of black tea. *Crit. Rev, Food Sci. Nutr.* 37, 739–760. 1997.
44. Kohlmeier, L., Weterings, K.G.C., Steck, S. and Kok, F.J., Tea and cancer prevention: An evaluation of the epidemiology literature. *Nutr. Cancer* 27, 1–13. 1997.
45. Buschman, J.L., Green tea and cancer in humans: A review of the literature. *Nutr. Cancer* 31, 151–159, 1998.
46. Mukthar, H. and Ahmad, N., Tea polyphenols: Prevention of cancer and optimizing health. *Am. J.Clin. Nutr.* 71 (suppl), 1698S–1702S, 2000.
47. Nakachi, K., Suemasu, K., Suga, K., Takeo, T., Imai, K. and Higashi, Y., Influence of drinking green tea on breast cancer malignancy among Japanese patients. *Jpn. J. Cancer Res.* 89, 254–261, 1998.
48. Goldbohm, R.A., Hertog, M.G.L.,Brants, H.A.M., van Poppel, G. and van den Brandt, P.A., Consumption of black tea and cancer risk: A prospective cohort study. *J. Natl. Cancer Inst.* 88, 93–100, 1996.
49. Nepka, C., Asprondini, E. and Kouretas, D., Tannins, xenobiotic metabolism and cancer chemoprevention in experiemental animals. *Eur. J. Drug. Metab. Pharmacol.* 24, 183–189, 1999.
50. Chung, F.L., Wang, M., Rivenson, A., Iatrpoulos, M.J., Reinhardt, J.C., Pittman, B., Ho, C.T. and Amin, S.G., Inhibition of lung carcinogenesis by black tea in Fischer rats treated with a tobacco-specific carcinogen: Caffeine as an important constituent. *Cancer Res.,* 58, 4096–4101, 1998.
51. Nishida, H., Omori, M., Fukutomi, Y., Inhibitory effects of EGCG on spontaneous hepatoma in $C_3H/HeNCrj$ mice and human hepatoma derived PLC/PRF/5 cells. *Jpn. J. Cancer Res.*, 85, 221–225, 1994.
52. Gensler, H.L., Timmerman, B.N., Valcic, S., Wachter, G.A., Dorr, R., Dvorkova, K. and Alberts, D.S., Prevention of photocarcinogenesis by topical administration of pure epigallocatechin gallate isolated from green tea. *Nutr. Cancer* 26, 325–335, 1996.

53. Chen, G., Perchellet, E.M., Gao, X.M., Ability of M-chloroperoxybenzoic acid to induce the ornithine decarboxylase marker of skin tumor promotion and inhibition of this response by gallotannins, oligomeric proanthocyanidins and their monomeric units in mouse epidermis *in vivo*. *Anticancer Res.*, 15, 1183–1189, 1995.
54. Yang C.S. and Chung, J.Y., Growth inhibition of human cancer cell lines by tea polyphenols. *Curr Pract. Med.* 2, 163–166, 1999.
55. Chung, K-T., Wei, C-I., Johnson, M.G., Are tannins a double-edged sword in biology and health? *Trends Food Sci.Technol.*9, 168–175, 1998.
56. Kinlen, L.L., Willows, A.N., Goldblatt, P. and Yudkin, J., Tea consumption and cancer. *Brit. J. Cancer* 58, 397–401, 1988.
57. Ramanathan, R., Tab, C.H. and Das, N.P., Tannic acid promotes benzo(a)pyrene-induced mouse skin carcinogenesis at low concentrations. *Med. Sci. Res.* 20, 711–712, 1992.

Part V

Evaluation of Effectiveness

25

Systematic and Critical Evaluation of Benefits and Possible Risks of Nutritional Ergogenic Aids

Amy A. Turpin, Shawn M. Talbott, Jeff Feliciano and Luke R. Bucci

CONTENTS

I. Introduction 469
II. Long-Term Endurance Exercise Performance 470
III. Muscle Mass and Strength 472
IV. Body Fat Loss 474
V. Sports Skills and Exercise-Associated Health 474
VI. Current Product Trends 476
VII. Possible Risks and Safety Issues of Ergogenic Aids 477
VIII. Summary 481
Acknowledgments 482
References 482

I. Introduction

This chapter will summarize what works and what does not work for the four major areas of ergogenicity affected by nutritional ergogenic aids (sports nutrition and dietary supplement products). This chapter will briefly consider what is acknowledged to be efficacious or not by scientific researchers, based mostly on recent reviews for: (1) long-term endurance exercise, (2) muscle mass and strength (anaerobic exercise), (3) body fat loss and (4) sports skills and exercise-associated health (mental performance, muscle soreness, immune function, musculoskeletal health, sleep or jet lag).

Tables for efficacy ratings have been divided into three categories dependent on the degree of efficacy: (1) Class A for consensus of efficacy (sometimes conditional), (2) Class B for equivocal or limited supportive data and (3) Class C for consensus on lack of efficacy. Nutritional ergogenic aids are listed in alphabetical order within each class. Ratings are derived primarily from a consensus of recent reviews. Also, the authors considered the weight of evidence based on experience in this field and individual articles.

Supporting evidence (as a list of references) will be presented only for Class A nutritional ergogenic aids. Class A nutrients have either reached a consensus of efficacy from most reviewers or a majority of studies have reproducibly shown ergogenicity. Nutrients in Class B usually have some support for ergogenic actions in well-controlled human clinical trials. However, there are also trials that did not find ergogenic actions, meaning their support is equivocal and, at this point, is not agreed upon by reviewers. It is possible that, with further research, some nutrients may have reproducible and consistent ergogenic effects under certain conditions. Some compounds in this category have an insufficient (less than three) number of reports showing ergogenic effects. Class C nutrients are not considered ergogenic, nor is their status likely to change under any condition. A majority or consensus of research has not found improvements in performance from these ingredients as single agents. Ergogenic aids untested in human clinical trials are not listed.

Discussion of guidelines for use and mechanisms of action entail more time and detail than this chapter permits, and can be found in other chapters of this book, as well as other reviews that are referenced in each table and the following text.

As with any rating system, the conclusions are only as valid as the input. Important research issues such as dose-response, statistical power (subject number), applicability of measurements, study design details, statistical analysis, investigator bias and reproducibility of results affect an overall determination of efficacy. Usually, these issues are suboptimal, leading to Type II errors (not finding a significant difference) and thus, any sweeping conclusion is probably pessimistic in nature. However, the available data, stressing controlled human clinical trials and reviews, have been expressed in these tables.

II. Long-Term Endurance Exercise Performance

The nutritional ergogenic aids useful for improving long-term endurance exercise performance are listed in Table 25.1. Sports drinks, carbohydrates and caffeine are the most efficacious ergogenic aids for endurance performance. Antioxidants, mostly as tocopherols and combinations containing tocopherols, have conditional efficacy for maintaining endurance performance at high altitudes or hypoxic conditions.[1-17] Caffeine dosed at 3–6 mg/kg 1 hour before exercise improved endurance performance, while also staying below the IOC cutoff for doping with caffeine.[1-3,16,18-46] Higher

TABLE 25.1
What Works for Long-Term Endurance Exercise Performance

Class A: Majority of studies support maintenance of or improvements in physical performance:

1. Antioxidants for high altitudes (tocopherols, specific combinations) — 400 mg tocopherol
2. Caffeine — 3–6 mg/kg 1 hour before exercise
3. Carbohydrates (pre-, during, post-exercise feedings) — 300 g low-fiber, low-fat, complex carbohydrate meal 3–4 hours before exercise; see sports drinks for during exercise; 50 g high-glycemic index carbohydrates within one hour after exercise and every 2 hours (500–700 g carbohydrates in 24 hours)
4. L-Carnitine and salts (only for cardiopulmonary disease patients) — 2–3 g/day
5. Coenzyme Q_{10} (only for cardiovascular disease patients) — 100–300 mg/day
6. Ephedrine/caffeine combinations (efficacy evident, but other issues confound use)
7. Iron (if anemia or low ferritin stores present) — 18–100 mg/day
8. Sports drinks (water, electrolytes, carbohydrates) — 200–400 ml before exercise, 100–150 ml every 15 minutes

Class B: Evidence for effects are equivocal, conditional or too few for adequate conclusions:

1. Antioxidants (N-acetyl-L-cysteine, ascorbate, beta carotene, carotenoids, coenzyme Q_{10}, glutathione, lipoic acid, selenium, soy isoflavones, tocopherols and combinations)
2. Aspartate salts
3. Branched chain amino acids (BCAAs)
4. L-Carnitine and salts
5. Coenzyme Q_{10}
6. Glycerol
7. Herbals (singly or in blends — *Cordyceps sinensis, Eleutherococcus senticosus, Panax ginseng, Rhodiola* spp.)
8. beta-Hydroxy-beta-methylbutyrate (HMB)
9. Hyperhydration
10. Medium chain triglycerides (MCTs)
11. Minerals (iron, magnesium, zinc)
12. Nicotinamide Adenine Dinucleotide (NADH)
13. Omega-3 fatty acids
14. Phosphates and other buffers (citrate)

Class C: Majority of evidence does not support improvements in physical performance

1. Bee Pollen
2. Calcium
3. Choline, Lecithin, Phosphatidylcholine
4. Creatine
5. Dimethylglycine (DMG) and pangamic acid
6. Fat loading
7. L-Glutamine
8. Multiple vitamin/minerals (200% daily value amounts or less)
9. Niacin
10. Octacosanol
11. Ribose
12. Royal jelly
13. Selenium
14. Single amino acids
15. Spirulina (blue-green algae)
16. Succinate salts
17. Vitamin B combinations
18. Wheat germ oil

doses may be needed for caffeine-habituated persons, but diuretic concerns have been lessened.[23] Carbohydrates before, during and after exercise remain the most-studied nutritional ergogenic aid for endurance performance.[1-3,16-22,32,38,46,48-75] Carnitine salts[1,3,76-109] and coenzyme Q_{10}[1,3,109-121] improve exercise performance in persons with cardiovascular diseases, a point largely ignored by most reviewers. Since people with cardiovascular conditions are often prescribed exercise as a treatment, it is important to know what may help these persons comply with exercise and possibly derive more therapeutic benefits from it. Ephedrine and caffeine combinations are not included in this book because of limited safety data in exercising persons and banned status by most sports-governing agencies. However, ephedrine and caffeine combinations enhance exercise performance over caffeine or ephedrine alone.[3,18-20,29,33,122-131] Iron repletion to persons with iron-deficiency anemia or low ferritin stores can restore exercise performance to normal levels.[1-3,40,69,132-148] Sports drinks simultaneously administer water, electrolytes and carbohydrates before and during exercise and have become a mainstay of nutritional ergogenic aids for endurance performance.[1-3,16,19,20,31,35,49,58,68,73,149-175] Under conditions of high rates of body water loss, sports drinks can avert serious or life-threatening consequences.

III. Muscle Mass and Strength

The nutritional ergogenic aids that have been shown to provide beneficial effects for increasing muscle mass and strength (anaerobic exercise performance) are shown in Table 25.2. Bicarbonate is mostly used to enhance short-term intense anaerobic exercise performance.[1-3,17-19,22,25,33,35,46-48,73,176-182] Calories (usually carbohydrate and protein combinations with other ingredients sometimes added in the form of powders or drinks) enhanced body and muscle mass gain in resistance training programs.[3,19,25,35,183-190] Creatine can increase body weight, lean mass, power, torque, strength and repetitive short-term anaerobic exercise performance.[3,18,19,21,22,25,33,35,39,40,46-48,73,128,176,178,189,191-237] Beta-hydroxy-beta-methylbutyrate (HMB), a leucine metabolite, has exhibited consistent effects on strength and anaerobic exercise performance.[3,19,33,35,40,69,73,128,176,186,189,200,225,238-241] Protein (and amino acid mixtures resembling protein) from foods and supplements maintained nitrogen balance and muscle protein synthesis during intense training periods better than control groups.[1-3,17-19,21,22,33,48,73,176,184,185,187-189,200,242-267] Replenishment of deficient status for magnesium and zinc has been patented and marketed as ZMA, restoring serum testosterone levels decreased by overtraining.[40,268] Deficient status of magnesium and zinc is not uncommon among exercising individuals.[1-3,40,138,139,142,144,268-281]

TABLE 25.2
What Works for Strength and Muscle Mass

Class A: Majority of studies support maintenance of or improvements in muscle mass, strength, protein synthesis or anaerobic exercise performance

1. Bicarbonate — 300-500 mg/kg body weight/day
2. Calories (usually Carbohydrate/protein combinations, sometimes with other ingredients) — dose varies with desired result
3. Creatine — loading dose 20 g/day for 5 days; maintenance dose 5 g/day
4. HMB — 1.5 to 3 g/day
5. Protein and mixtures of amino acids — 1.6 to 1.7 g/kg body weight/day
6. Replenishing deficiencies of essential minerals (magnesium, zinc) — 450 mg magnesium and 30 mg zinc/day

Class B: Evidence for effects are equivocal, or too few for adequate conclusions

1. Antioxidants (N-acetyl-L-cysteine, ascorbate, carotenoids, coenzyme Q_{10}, glutathione, lipoic acid, soy isoflavones, selenium, tocopherols and combinations)
2. Caffeine
3. L-Carnitine
4. Conjugated linoleic acid (CLA)
5. L-Glutamine and L-Ornithine-alpha-ketoglutarate (OKG)
6. Herbals (singly or in blends — e.g., *Eleutherococcus, Panax ginseng, Rhodiola*)
7. Omega-3 fatty acids
8. Vitamin C

Class C: Majority of evidence does not support improvements in muscle mass or strength, or unstudied in human outcome trials

1. Amino acid somatotropin secretagogues (arginine, arginine aspartate, arginine pyroglutamate, combinations, glutamine, glycine, 5-hydroxytryptophan, ornithine, tryptophan, tyrosine)
2. Antiestrogenic herbal compounds (chrysin, diindolylmethane, indole-3-carbinol)
3. Boron
4. Calcium
5. Carnosine
6. Chromium
7. Desiccated beef liver
8. Glandulars (adrenal, liver, orchic, pituitary)
9. Herbals (*Coleus*, combinations, *Commiphora, Gymnema, Lagerstroemia, Tribulus*)
10. Inosine and other nucleotides (RNA, Brewer's yeast)
11. MCTs
12. Multiple vitamin/minerals (200% RDA amounts or less)
13. Myostatin inhibitors (*Cystoseira canariensis* algal extracts)
14. Nitric oxide inducers (arginine, arginine alpha-ketoglutarate, various herbals)
15. Octopamine (*Citrus aurantium* extracts, synthetic)
16. Phosphatidylserine
17. Prohormones (DHEA, androstenedione, norandrostenediones, norandrostenediols, others)
18. Ribose
19. Single antioxidants (selenium, vitamin C, vitamin E, herbal phenolics)
20. Sterols (beta-sitosterol, gamma oryzanol, Sarsaparilla, *Smilax* spp.)
21. Taurine
22. Vanadium
23. Vitamin B combinations

IV. Body Fat Loss

Ergogenic Aids that promote body fat loss are listed in Table 25.3. Inadequate calcium intake appeared to blunt weight loss from caloric restriction, but calcium supplementation restored weight loss.[282-288] Caloric restriction is by far the most common, viable and successful means of reducing body weight.[289-301] Often, meal-replacement drinks, powdered drink mixes, bars or other drinks are used to replace foodstuffs or meals during periods of caloric restriction. This practice covers a wide range of food and dietary supplement products, which can be considered ergogenic acids when used to reduce body weight to enhance exercise performance or appearance. Conjugated linoleic acid (CLA) appeared to improve body composition (maintain lean muscle mass) during periods of caloric restriction, more than reducing body fat.[302-306] Ephedrine and caffeine combinations have consistently exhibited increased loss of weight and body fat during periods of caloric restriction in overweight subjects.[3,18,19,33,58,128,307-320] It is less clear whether such combinations reduce body fat in persons who are not overweight. 7-keto DHEA has exhibited increased loss of body fat with retention of muscle mass during periods of caloric restriction, without the ability to form other steroid hormones.[321-327]

V. Sports Skills and Exercise-Associated Health

There are many ways to enhance performance other than improving physiological capabilities. One factor not considered in this review is mood. Although there are substances that are known to improve mood, application toward exercising individuals is sorely lacking and thus, presentation of this concept here would be premature. Nevertheless, the reader is reminded that use of some nutrients to improve mood may also improve compliance with exercise or provide a placebo effect that remains unstudied but potentially ergogenic.

Another potential ergogenic aid is improved repair of musculoskeletal injuries, a topic covered in another book.[328] Another chapter in this book describes glucosamine and chondroitin sulfates, which in theory may allow thousands of persons suffering from arthritis to start, resume or increase exercise training.

Neuromuscular control and reaction times are presumed to be important for certain skill sports and there is some evidence that certain nutrients can enhance these capabilities in healthy individuals and sportspersons (Table 25.4). B Vitamins,[1-3,329-334] *Panax ginseng* extracts[1-3,335,336] and octacosanol[1-3,19,176,337,338] have reproducibly shown improvements in neuromotor skills, primarily reaction times to auditory and visual stimuli. Another interesting development has been use of melatonin for ameliorating jet lag,

TABLE 25.3
Body Fat Loss or Body Composition (Leanness)

Class A: Majority of studies support improvements in body fat loss or leanness

1. Calcium — 1,000 mg/day
2. Caloric restriction — amount varies depending upon desired effect
3. Conjugated linoleic acid (CLA) — 3.4 to 6 g/day
4. Ephedrine/Caffeine combinations (efficacy evident, but other issues confound use)
5. 7-ketoDHEA — 150 mg/day

Class B: Evidence for effects are equivocal, or too few for adequate conclusions

1. Caffeine (herbal [green tea, Guarana, Kola Nut, Yerba Mate] or synthetic sources)
2. Chitosan (only certain sources)
3. Chromium
4. Ephedrine (*Ephedra sinensis* extracts)
5. Forskolin (*Coleus forskholii* extracts)
6. Green tea polyphenols plus caffeine
7. Guggul lipids (*Commiphora mukul* extracts)
8. Herbals (singly or in blends — *Cordyceps sinensis, Eleutherococcus senticosus, Panax ginseng, Rhodiola* spp.)
9. HMB
10. Hydroxycitric acid (HCA or Malabar tamarind rind extract)
11. Octopamine (Norsynephrine, *Citrus aurantium* extracts)
12. Phaseolamin® (Amylase inhibitor)
13. DL-Phenylalanine
14. Phosphate Salts
15. Pyruvate & dihydroxyacetone (DHA)
16. Para-Synephrine (*Citrus aurantium* extracts)
17. Yohimbine (Yohimbe (*Pausinystalia yohimbe*) bark extracts)

Class C: Majority of evidence does not support improvements in body fat loss or leanness

1. L-Carnitine
2. Chitosan
3. Creatine
4. DHEA
5. Herbal combinations (non-*Ephedra*)
6. 5-Hydroxytryptophan (5HTP)
7. MCTs
8. Multiple vitamin/minerals (200% RDA amounts or less)
9. Single antioxidants (selenium, vitamin C, vitamin E, herbal phenolics)
10. Vitamin B complexes

a problem for international competitors. Enough research has been conducted with melatonin to find reproducible and meaningful amelioration of jet lag, speeding acclimatization, in a Cochrane Review meta-analysis.[339] Relatively few studies on prevention of delaying onset of muscle soreness with nutrients have been published, but chronic high doses of vitamin C appear to have significant benefits.[340–348]

Immune function is perhaps the most visible nonperformance means to enhance or maintain performance. However, even though several review

TABLE 25.4

Sports Skills and Exercise-Associated Health

Effect Improved	Nutrient	Effective Doses	References
Delayed onset muscle soreness	Vitamin C	400–3000 mg/d for at least 14 d	1–3, 340–348
Immune function — prevention of upper respiratory tract illness in distance runners	Vitamin C	>1000 mg/d	349–351
Jet lag	Melatonin	0.5–5 mg before local bedtime	339
Neuromuscular — fine motor control, mood, reaction time	B Vitamin complex	>500% Daily value	1–3, 329–334
Neuromuscular — reaction time	Panax ginseng extracts	400 mg 4% ginsenosides	1–3, 335, 336
Neuromuscular — reaction time	Octacosanol	1–10 mg/d	1–3, 337, 338

books, many review chapters and an entire journal (*Exercise Immunology Reviews*) are devoted to addressing this topic, there are very few studies testing whether specific nutrients reduce incidence of upper respiratory tract illness and infections, especially in long-distance runners. Prevention of lost training time or missed events (rather than other measures of immune function) is important for serious athletes as well as recreational persons. Vitamin C stands as the nutrient with the most data on preventing infectious events in exercising individuals.[349–351]

VI. Current Product Trends

Significant changes in the array of dietary supplements available to consumers since DSHEA was passed show several trends. First, since 2000, prohormones and ephedrine-containing products have greatly decreased in number and visibility. Most large companies do not sell either anymore. On the other hand, a few very small companies are selling newly seen prohormones in a continuous procession. Combination products, usually containing caffeine and synephrine alkaloids (from *Citrus aurantium*) and a host of other nutrients, have largely replaced ephedrine-containing products for the category of energy and weight management.

Endurance products continue to be dominated by mass market CHO/electrolyte drinks and high-carbohydrate energy bars, but an increasing number of herbal combinations have appeared, usually with *Rhodiola*, *Cordyceps* or other adaptogenic herbs.

The strength and muscle mass market segment continues to be dominated by protein powders, meal replacements and creatine. However, the level of sophistication of each has increased. For example, numerous patents have illustrated the addition of other nutrients with creatine to improve results. Another trend is the proliferation of complicated herbal/vitamin/metabolite blends for regulating insulin as an anabolic vector. Protein powders increasingly have additional nutrients added and advances in protein food techology have made high-dose (40–50 gram) ready-to-drink protein products practical. High-protein foodstuffs, such as high-protein pasta, have not been commercially successful. Timing of ingestion is receiving increased attention. Other trends are nutrient combinations to block myostatin receptors and combinations to regulate other growth factors involved in muscle growth. Nitric oxide biochemistry has become a fashionable trend, and numerous nutrient combinations exist with the intent of improved circulation (pump) for muscles. Manipulating growth hormone with amino acids is passé, except in Internet spam advertisements.

The weight management segment has seen the rapid demise of ephedrine-containing products, in spite of favorable scientific studies. A wide variety of herbal and nutrient combinations have rushed to fill the void, most without the level of substantiation enjoyed by ephedrine and caffeine combinations. A rapidly growing trend is the proliferation of low-carbohydrate, high-protein foodstuffs, especially bars. New technologies have made high-protein bars that remain moist for their shelf life, making these products acceptable to a wider range of consumers. Entire programs for weight management, including diet plans, exercise plans and combinations of several dietary supplements, seem to be the trend in this category, rather than touting a single ingredient for weight loss.

Although the dietary supplement industry in general has many nutrients with potential for immune system health and antiinflammatory actions, few have been applied to sports or athletes. One thing is for certain — any new concept with any possibility of efficacy will have a nutritional approach and dietary supplements to match.

VII. Possible Risks and Safety Issues of Ergogenic Aids

Safety is always of paramount importance, but there will never be enough data on any food, drug or supplement to conclusively prove safety or lack of safety. The real question is: are products being sold for the sake of profit over legitimate concerns for consumer safety? Examples of such practices were gamma hydroxybutyrate (GHB) and its related compounds (gamma butyrolactone and butanediol) and tiratricol (a thyroid hormone). Both were sold as dietary supplements by a small number of companies to the weight-loss and weightlifter market niches, in spite of well-known published adverse effects, and use as prescription drugs prior to 1994 (which DSHEA specifically forbade). Both types of products were shown to cause or to be associated with expected adverse effects, and use spread to nonathletic groups. The FDA and Department of Justice quickly took action and removed these products from the marketplace. Judicial agencies investigated and brought criminal charges against manufacturers or distributors of these products. In other words, the current system did not prevent unscrupulous persons from introducing prohibited substances to the marketplace, but it did react and remove these products.

Except for a few controversial dietary supplements that are not considered in detail in this publication (ephedrine-containing herbal extracts and pro-hormones), sports nutrition products and dietary supplements have a very low potential for causing harm to consumers (see Table 25.5).[352-354] Overdoses are possible with anything, as exemplified by a handful of water-induced deaths during long-term endurance events.[355] Fortunately, very few ingredients used in nutritional ergogenic aids are potentially lethal in overdose amounts. Caffeine perhaps has the greatest risk of lethal overdosing and there are some concerns about ingestion by pregnant women, but in general, widespread use of caffeine and coffee has conditioned chronic users from caffeine abuse.[43]

Creatine safety has been a popular subject in the media, with rampant misinformation. Concerns about creatine and dehydration, electrolyte imbalance, excess injuries, muscle cramps, renal damage, liver toxicity, muscle damage, compartment syndrome, glucose metabolism, formation of nitroso compounds, cancer, contaminants, down-regulation of endogenous synthesis and unforeseen adverse effects have been based on animal studies (at very high doses), supposition, anecdotal reports or hearsay.[191, 194, 356-370] For example, creatine was implied as causal (without any evidence) in the unfortunate deaths of wrestlers making weight by a USA Today news report on December 18, 1997, when, in fact, there was no involvement of creatine, as found by an expert panel convened by the FDA. The weight of evidence from retrospective and carefully controlled, prospective human studies heavily favors a lack of adverse effects from creatine (5 g/d) in any age group, including children and for long time periods.[19,189,211,230,233,234,356,358,371-426] Long-term safety has been questioned, but long-term use in athletes for almost 2 years[215] and in children with gyrate atrophy for 5–10 years[386,416,421,420] have not shown any adverse effects. At this point, suggesting that creatine supplementation at normal usage levels causes adverse side effects is anachronistic, unsupported and counter to the weight of scientific evidence. Normal usage levels are 5 grams four times daily for 5–7 days, followed by up to 5 grams daily indefinitely.

Since most nutritional ergogenic aids are foods, are derived from foods or are already present in the human body, potential for serious adverse effects are relatively low. The low incidence of reported adverse effects from non-prohibited substances and the very large number of doses used argue strongly that nutritional ergogenic aids are, in general, safe. In the case of sports drinks, it can be argued that it is unsafe not to use them.

TABLE 25.5

Potential Risks Associated with Nutritional Ergogenic Aids

Substance	Dose Ranges	Safety Issues
Androstenedione	300 mg/d for 4 wks or more	Males: • HDL cholesterol reduction • Estrogen increase (temporary) Females: • Unknown
Ascorbate (Vitamin C)	LOAEL: unknown	Acute: (>2000 mg) • Osmotic gastrointestinal effects Chronic: (>2000 mg/d) • Rebound scurvy, oxalate kidney stones, prooxidant effects, excess iron uptake, vitamin B12 destruction, negative copper balance all unsupported and/or refuted
Beta Carotene	LOAEL: unknown	• Carotenemia (orange skin) — harmless • Conflicting data on increased incidence of lung cancer from high dose (>100,000 IU/d) only in smokers drinking alcohol — not seen when other antioxidants co-administered
Calcium	LOAEL: 2500 mg	• Gastrointestinal upset, constipation • Hypercalciuric males: kidney stone risk increased • Possible suppression of copper, iron, manganese, zinc uptake (dependent on many conditions)
Caffeine	LOAEL: varies with individual	• LD_{50} 150-200 mg/kg • Anxiety, tremors, nausea, vomiting • Overdose can be lethal
L-Carnitine salts	LOAEL: unknown	• Osmotic diarrhea in massive doses (over 6 grams at once) possible
Cholecalciferol (Vitamin D_3)	LOAEL: 2000 IU (children) >5000 IU (adults)	• Hypercalcemia • Idiopathic hypercalcemia, tuberculosis, sarcoidosis at increased risk for adverse effects • Most adverse effects from high-dose, parenteral administration (not applicable to dietary supplements)
Chondroitin sulfates	LOAEL: unknown	• Occasional gastrointestinal upset
Chromium (Cr^{3+})	LOAEL: unknown	• None documented
Coenzyme Q_{10}	LOAEL: unknown	• None documented
Creatine	LOAEL: unknown	• Osmotic diarrhea at doses of 5–10 grams • Gastrointestinal upset and bloating with doses over 5 grams • Possible compartment syndrome • Muscle cramps, dehydration, renal damage, liver toxicity, carcinogenicity and downregulation of endogenous synthesis have all been shown to be of little or no concern

(continued)

TABLE 25.5 (CONTINUED)
Potential Risks Associated with Nutritional Ergogenic Aids

Substance	Dose Ranges	Safety Issues
DHEA	LOAEL: unknown	• Mild androgenization (skin oiliness, facial hair) in minority of female users at doses >100 mg/d • Transient, short-term increases in estrogen/testosterone in doses >100 mg/d (long-term effects unknown)
Ephedrine		• Anxiety, nervousness • Heart rate increased 10–20% • Blood pressure increased 10% • Overdose can be lethal
Ephedrine + Caffeine		• Anxiety, nervousness • Heart rate increased 10–20% • Blood pressure increased 10% • Overdose can be lethal
Glucosamine salts	LOAEL: unknown	• Occasional gastrointestinal upset • Effects on blood sugar and insulin resistance are not attainable after oral administration, even in doses over 5 grams/d • No adverse effects on blood lipid levels
HMB	LOAEL: unknown	• No adverse effects reported for subchronic use (< 8 wks) • Possible GI upset with high doses (>6 g)
Iron	LOAEL: 100 mg	Acute: (>900 mg in children) • Overdose can be lethal Chronic: • Hereditary hemochromatosis • Liver disease (cirrhosis) • Heart disease link not supported • Gastrointestinal upset
Magnesium	LOAEL: unknown	• gastrointestinal upset, diarrhea, laxative (>15 g)
Melatonin	LOAEL: unknown	• Possible interactions with warfarin and other oral anticoagulants • Possible interactions in epilepsy — consult physician
Niacin	LOAEL: 1000 mg	• Flushing (redness, tingling of face, neck) • Gastrointestinal upset, diarrhea, vomiting • Liver enzyme elevation (time-release products)
Niacinamide	LOAEL: unknown	• None documented
Phosphorus (usually as phosphate)	LOAEL: >2500 mg	• Gastrointestinal upset, osmotic diarrhea with very large doses (over 4–5 grams at once)
Phylloquinone (Vitamin K)	LOAEL: unknown	• Only concern is in persons taking coumarin anticoagulants — consult physician

(continued)

TABLE 25.5 (CONTINUED)

Potential Risks Associated with Nutritional Ergogenic Aids

Substance	Dose Ranges	Safety Issues
Pyridoxine (Vitamin B6)	LOAEL: 500 mg	• Sensory neuropathy (numbness, tingling)
Pyruvate salts	LOAEL: unknown	• Gastrointestinal upset at very large doses (over 5 grams at once), osmotic diarrhea possible at higher doses • Possible breakdown or polymerization into parapyruvate and other ketones with unknown safety profiles
Retinol (Vitamin A)	LOAEL: 21600 IU	• Liver damage • Possible teratogenicity • Overdoses can be lethal if continued
Selenium	LOAEL: 910 mcg	• Adverse effects on nails, hair, skin
Tocopherols (Vitamin E)	LOAEL: unknown	• Extensive human study data do not support adverse effects on coagulation at doses <2400 IU/d
Zinc	LOAEL: 60 mg	Acute: • Nausea, vomiting, diarrhea
	>150 mg daily	Chronic: • Suppression of copper status • Hypochromic, microcytic anemia • HDL cholesterol decreased • Lymphocyte stimulation decreased • Cu/Zn SOD decreased

Note: Any substance can induce osmotic diarrhea if taken in excess, usually in doses of 10 grams or more at one time.

LOAEL = Lowest observed adverse effect level

Source: Safety data taken from refs. 352–354

VIII. Summary

Exercising individuals universally want to perform better and are willing to try almost anything that offers a promise of better performance. A barometer of practical efficacy can be found in ingredients that stay in the marketplace. Product sales are the ultimate arbiter of efficacy and ergogenicity. Segmentation of ergogenic aids into classes of efficacy remains the responsibility of the authors and we relied on peer-reviewed review articles or book chapters from the same authors that wrote reviews. A thorough listing of ingredients that are actually in current and past products was presented so that the reader can form opinions about products that consumers already have access to. With time and further research, Class B ergogenic aids will end up in either Class A or C, depending on their merit. Class B nutrients represent a larger proportion of newer products, since it is faster to sell products than conduct human clinical trials.

In terms of efficacy, it can be seen that for endurance exercise, sports drinks and carbohydrates are effective, safe and, indeed, are commonly used. The majority of sales of ergogenic aids constitute these two product categories (see Chapter 1). For strength and muscle mass, eating more calories, including protein, is commonly practiced and effective. Creatine and HMB can augment anaerobic muscular performance and have been shown to be tolerable and safe with normal usage. Caloric restriction is still the foremost way to lose body fat and body weight. Only a few ingredients have been shown to produce results beyond caloric restriction in spite of many products and ingredients being sold for this purpose. Since many persons exercise to improve their appearance rather than compete in sports, body-fat loss has become the second most important commercial niche of ergogenic aids. Much research remains to be done in this arena and many people anxiously await news of efficacy or lack thereof for components. Conditions that would improve ergogenicity by maintaining training and preventing downtime from illness and injuries are less well studied than actual performance enhancers. Nevertheless, a few nutrients have accumulated enough data to suggest they may aid sports performance in nonphysical ways.

Excluding ephedrine and caffeine combinations and prohormones, there are few, if any, nutritional ergogenic aids that represent a significant hazard. Overall, the more food-like products (sports drinks, meal replacements, protein powders) can be considered as safe as foods, and most other ingredients are already present in the human body. Questions to explore further include use by children and adolescents, monitoring systems to observe long-term effects, possible effects from combinations not seen from individual nutrients and drug testing of products to prevent unintentional positive tests.

In conclusion, there is a consensus among reviewers of ergogenic aids that some nutrients can successfully improve exercise performance. Many nutrients remain insufficiently or improperly studied to reach firm conclusions.

Acknowledgments

The authors wish to thank Brooke Bouwhuis, Camilla Kragius, Charlotte Oler, Christina Beer, Janet Sorensen, Stephanie Blum and Trudy Day for their excellent technical assistance.

References

1. Bucci, L., *Nutrients as Ergogenic Aids for Sports and Exercise*, CRC Press, Boca Raton, 1993.
2. Bucci, L.R., Nutritional ergogenic aids, in *Nutrition in Exercise and Sport*, 2nd ed., Wolinsky, I. and Hickson, J.F., Eds., CRC Press, Boca Raton, FL, 1994, 295.

3. Bucci, L.R., Dietary supplements as ergogenic aids, in *Nutrition in Exercise and Sport*, 3rd ed., Wolinsky, I., Ed., CRC Press, Boca Raton, FL, 1998, 315–360.
4. Clarkson, P.M. and Thompson, H.S., Antioxidants: What role do they play in physical activity and health? *Am. J. Clin. Nutr.*, 72, 637S, 2000.
5. Gerster, H., Function of vitamin E in physical exercise: A review, *Z. Ernahrungswiss*, 30, 89, 1991.
6. Goldfarb, A.H., Antioxidants: Role of supplementation to prevent exercise-induced oxidative stress, *Med. Sci. Sports Exerc.*, 25, 232, 1993.
7. Kagan, V.E., Spirichev, V.B., Serbinova, E.A., Witt, E., Erin, A.M. and Packer, L., The significance of vitamin E and free radicals in physical exercise, in *Nutrition in Exercise and Sport*, 2nd ed., Wolinsky, I. and Hickson, J.F., Eds., CRC Press, Boca Raton, FL, 1994, 185.
8. Kanter, M., Nutritional antioxidants and physical activity, in *Nutrition in Exercise and Sport*, 3rd ed., Wolinsky, I., Ed., CRC Press, Boca Raton, FL, 1998, 245–255.
9. Meydani, M., Fielding, R.A. and Fotouhi, N., Vitamin E, in *Sports Nutrition. Vitamins and Trace Elements*, Wolinsky, I. and Driskell, J.A., Eds., CRC Press, Boca Raton, FL, 1997, 119.
10. Packer, L. and Singh, V.N., Nutrition and exercise introduction and overview, *J. Nutr.*, 122, 758, 1992.
11. Packer, L., Oxidants, antioxidant nutrients and the athlete, *J. Sports Sci.*, 15, 353, 1997.
12. Packer, L., Reznick, A.Z., Simon-Schnass, I. and Landvik, S.V., Significance of Vitamin E for the athlete, in *Vitamin E in Health and Disease*, Packer, L. and Fuchs, J., Eds., Marcel Dekker, New York, 1993, 465.
13. Simon-Schnass, I., Vitamin requirements for increased physical activity: Vitamin E, *World Rev. Nutr. Diet.*, 71,144, 1993.
14. Simon-Schnass, I., Vitamin E and high-altitude exercise, in *Vitamin E in Health and Disease*, Packer, L. and Fuchs, J., Eds., Marcel Dekker, New York, 1993, 455.
15. Singh, V.N., A current perspective on nutrition and exercise, *J. Nutr.*, 122, 760, 1992.
16. Williams, M.H., Ergogenic and ergolytic substances, *Med. Sci. Sports Exer.*, 24, S344, 1992.
17. Williams, M.H., Nutritional ergogenics: help or hype? Some "performance enhancers" may leave marathoners running on empty, *J. Am. Diet. Assoc.*, 92, 1213, 1992.
18. Ahrendt, D.M., Ergogenic aids: Counseling the athlete, *Am. Fam. Physician*, 63, 913, 2001.
19. Antonio, J. and Stout, J.R., *Sports Supplements*, Lippincott Williams & Wilkins, Philadelphia, 2001.
20. Antonio, J. and Stout, J.R., *Supplements for Endurance Athletes*, Human Kinetics, Champaign, IL, 2002.
21. Applegate, E.A. and Grivetti, L.E., Search for the competitive edge: A history of dietary fads and supplements, *J. Nutr.*, 127, 869S, 1997.
22. Applegate, E., Effective nutritional ergogenic aids, *Int. J. Sport Nutr.*, 9, 229, 1999.
23. Armstrong, L.E., Caffeine, body fluid-electrolyte balance and exercise performance, *Int. J. Sport Nutr. Exerc. Metab.*, 12, 189, 2002.
24. Clarkson, P.M., Nutritional ergogenic aids: Caffeine, *Int. J. Sport Nutr.*, 3, 103, 1993.

25. Clarkson, P.M., Nutrition for improved sports performance. Current issues on ergogenic aids, *Sports Med.*, 21, 292, 1996.
26. Conlee, R.K., Amphetamine, caffeine and cocaine, in *Perspectives in Exercise Science and Sports Medicine Volume 4: Ergogenics — Enhancement of Performance in Exercise and Sport*, Lamb, D.R. and Williams, M.H., Eds., Brown & Benchmark, Ann Arbor, MI, 1991, 285.
27. Dodd, S.L., Herb, R.A. and Powers, S.K., Caffeine and exercise performance. An update, *Sports Med.*, 15, 14, 1993.
28. Graham, T.E., Caffeine, *Can. J. Appl. Physiol.*, 23, 323, 1998.
29. Graham, T.E., Caffeine, coffee and ephedrine: Impact on exercise performance and metabolism, *Can. J. Appl. Physiol.*, 26, S103, 2001.
30. Graham, T.E., Caffeine and exercise: Metabolism, endurance and performance, *Sports Med.*, 31, 785, 2001.
31. Jeukendrup, A.E. and Martin, J., Improving cycling performance: How should we spend our time and money? *Sports Med.*, 31, 559, 2001.
32. Juhn, M.S., Ergogenic aids in aerobic activity, *Curr. Sports Med. Rep.*, 1, 233, 2002.
33. Krcik, J.A., Performance-enhancing substances: What athletes are using, *Cleve. Clin. J. Med.*, 68, 283, 2001.
34. Lamarine, R.J., Caffeine as an ergogenic aid, in *Caffeine*, Spiller, G.A., Ed., CRC Press, Boca Raton, FL, 1998, 233.
35. Maughan, R., The athlete's diet: Nutritional goals and dietary strategies, *Proc. Nutr. Soc.*, 61, 87, 2002.
36. Nehlig, A. and Debry, G., Caffeine and sports activity: A review, *Int. J. Sports Med.*, 15, 215, 1994.
37. Paluska, S.A., Caffeine and exercise, *Curr. Sports Med. Rep.*, 2, 213, 2003.
38. Peters, E.M., Nutritional aspects in ultra-endurance exercise, *Curr. Opin. Clin. Nutr. Metab. Care*, 6, 427, 2003.
39. Rigassio Radler, D., Nutritional supplements, ergogenic aids and herbals, *Dent. Clin. North Am.*, 47, 245, 2003.
40. Schwenk, T.L. and Costley, C.D., When food becomes a drug: Nonanabolic nutritional supplement use in athletes, *Am. J. Sports Med.*, 30, 907, 2002.
41. Sinclair, C.J. and Geiger, J.D., Caffeine use in sports. A pharmacological review, *J. Sports Med. Phys. Fitness*, 40, 71, 2000.
42. Smith, D.A. and Perry, P.J., The efficacy of ergogenic agents in athletic competition. Part II: other performance-enhancing agents, *Ann. Pharmacother.*, 26, 653, 1992.
43. Spiller, G.A., *Caffeine*, CRC Press, Boca Raton, 1998.
44. Spriet, L.L., Caffeine and performance, *Int. J. Sport Nutr.*, 5, S84, 1995.
45. Tarnopolsky, M.A., Caffeine and endurance performance, *Sports Med.*, 18, 109, 1994.
46. Williams, M.H., *Nutrition for Fitness and Sport*, 4th ed., Brown & Benchmark, 1995.
47. Williams, M.H., Nutritional ergogenics in athletics, *J. Sports Sci.*, 13, S63, 1995.
48. Williams, M.H., *The Ergogenics Edge: Pushing the Limits of Sports Performance*, Human Kinetics Publishers, Champaign, IL, 1998.
49. Burke, L.M. and Read, R.S., Dietary supplements in sports, *Sports Med.*, 15, 43, 1993.
50. Burke, L.M., Cox, G.R., Culmmings, N.K. and Desbrow, B., Guidelines for daily carbohydrate intake: Do athletes achieve them? *Sports Med.*, 31, 267, 2001.

51. Clark, N., *Nancy Clark's Sports Nutrition Guidebook: Eating to Fuel Your Active Lifestyle*, 2nd ed., Human Kinetics Publishers, Champaign, IL, 1997.
52. Coleman, E., *Eating for Endurance*, 3rd ed., Bull Publishing Company, Palo Alto, CA, 1997.
53. Coleman, E. and Steen, S.N., *The Ultimate Sports Nutrition Handbook*, Bull Publishing Company, Palo Alto, CA, 2000.
54. Costill, D.L. and Hargreaves, M., Carbohydrate nutrition and fatigue, *Sports Med.*, 13, 86, 1992.
55. Coyle, E.F., Carbohydrate supplementation during exercise, *J. Nutr.*, 122, 788, 1992.
56. Coyle, E.F. and Coyle, E., Carbohydrates that speed recovery from training, *Phys. Sportsmed.*, 21, 111, 1993.
57. Doyle, J.A. and Papadopoulos, C., Simple and complex carbohydrates in exercise and sport, in *Energy-Yielding Macronutrients and Energy Metabolism in Sports Nutrition*, Driskell, J.A. and Wolinsky, I., Eds., CRC Press, Boca Raton, FL, 2000, 57.
58. Fillmore, C.M., Bartoli, L., Bach, R., Park, Y., Nutrition and dietary supplements, *Phys. Med. Rehabil. Clin. N. Am.*, 10, 673, 1999.
59. Friedman, J.E., Neufer, P.D. and Dohm, G.L., Regulation of glycogen resynthesis following exercise. Dietary considerations, *Sports Med.*, 11, 232, 1991.
60. Guezennec, C.-Y., Oxidation rates, complex carbohydrates and exercise. Practical recommendations, *Sports Med.*, 19, 365, 1995.
61. Hargreaves, M., Pre-exercise nutritional strategies: Effects on metabolism and performance, *Can. J. Appl. Physiol.*, 26 Suppl, S64, 2001.
62. Hawley, J.A., Dennis, S.C. and Noakes, T.D., Oxidation of carbohydrate ingested during prolonged endurance exercise, *Sports Med.*, 14, 27, 1992.
63. Ivy, J.L. Dietary strategies to promote glycogen synthesis after exercise, *Can. J. Appl. Physiol.*, 26 Suppl, S236, 2001.
64. Jacobs, K.A. and Sherman, W.M., The efficacy of carbohydrate supplementation and chronic high-carbohydrate diets for improving endurance performance, *Int. J. Sport Nutr.*, 9, 92, 1999.
65. Jonnalagadda, A.A., Carbohydrate supplements in exercise and sport, in *Energy-Yielding Macronutrients and Energy Metabolism in Sports Nutrition*, Driskell, J.A. and Wolinsky, I., Eds., CRC Press, Boca Raton, FL, 2001, 163.
66. Kiens, B., Diet and training in the week before competition, *Can. J. Appl. Physiol.*, 26, S56, 2001.
67. Liebman, M. and Wilkinson, J.G., Carbohydrate metabolism and exercise, in *Nutrition in Exercise and Sport*, 2nd ed., Wolinsky, I. and Hickson, J.F., Eds., CRC Press, Boca Raton, FL, 1994, 15.
68. Manore, M.M., Barr, S.I. and Butterfield, G.E., Joint position statement: nutrition and athletic performance. American College of Sports Medicine, American Dietetic Association and Dietitians of Canada, *Med. Sci. Sports Exerc.*, 32, 3120, 2000.
69. Maughan, R., Sports nutrition: An overview, *Hosp. Med.*, 63, 136, 2002.
70. Miller, G.D., Carbohydrate in ultra-endurance exercise and athletic performance, in *Nutrition in Exercise and Sport*, 2nd ed., Wolinsky, I. and Hickson, J.F., Eds., CRC Press, Boca Raton, FL, 1994, 50.
71. Petibois, C., Cazorla, G., Poortmans, J.R. and Deleris, G. Biochemical aspects of overtraining in endurance sports: A review, *Sports Med.*, 32, 867, 2002.

72. Sherman, W.M., Carbohydrate feedings before and after exercise, in *Perspectives in Exercise Science and Sports Medicine Volume 4: Ergogenics — Enhancement of Performance in Exercise and Sport*, Lamb, D.R. and Williams, M.H., Eds., Brown & Benchmark, Ann Arbor, MI, 1991, 1.
73. Talbott, S.M., Sports supplements and ergogenic aids, *A Guide to Understanding Dietary Supplements*, Haworth Press, New York, 2003, 101.
74. Walberg-Rankin, J., Dietary carbohydrate as an ergogenic aid for prolonged and brief competitions in sport, *Int. J. Sport Nutr.*, 5, S13, 1995.
75. Wilkinson, J.G. and Liebman, M., Carbohydrate metabolism in sport and exercise, in *Nutrition in Exercise and Sport*, 3rd ed., Wolinsky, I., Ed., CRC Press, Boca Raton, FL, 1998, 63.
76. Ahmad, S. Robertson, H.T., Golper, T.A., Wolfson, M., Kurtin, P., Katz, L.A., Hirschberg, R., Nicora, R., Ashbrook, D.W. and Kopple, J.D., Multicenter trial of L-carnitine in maintenance hemodialysis patients. II. Clinical and biochemical effects, *Kidney Int.*, 38, 912, 1990.
77. Anand, I., Chandrashekhan, Y., De Giuli, F., Pasini, E., Mazzoletti, A., Confortini, R. and Ferrari, R., Acute and chronic effects of propionyl-L- carnitine on the hemodynamics, exercise capacity and hormones in patients with congestive heart failure, *Cardiovasc. Drugs Ther.*, 12, 291, 1998.
78. Anonymous, Study on propionyl-L-carnitine in chronic heart failure, *Eur. Heart J.*, 20, 70, 1999.
79. Barker, G.A., Green, S., Askew, C.D., Green, A.A. and Walker, P.J., Effect of propionyl-L-carnitine on exercise performance in peripheral arterial disease, *Med. Sci. Sports Exerc.*, 33, 1415, 2001.
80. Bartels, G.L., Remme, W.J., Holwerda, K.J. and Kruijssen, D.A., Anti-ischemic effect of L-propionylcarnitine — a promising novel metabolic approach to ischaemia? *Eur. Heart J.*, 17, 414, 1996.
81. Bartels, G.L., Remme, W.J., den Hartog, F.R., Wielenga, R.P. and Kruijssen, D.A., Additional antiischemic effects of long-term L-propionylcarnitine in anginal patients treated with conventional antianginal therapy, *Cardiovasc. Drugs Ther.*, 9, 746, 1995.
82. Brass, E.P., Supplemental carnitine and exercise, *Am. J. Clin. Nutr.*, 72, 618S, 2000.
83. Brevetti, G., Perna, S., Sabba, C., Rossini, A., Scotto di Uccio, V., Berardi, E. and Godi, L., Superiority of L-propionylcarnitine vs L-carnitine in improving walking capacity in patients with peripheral vascular disease: an acute, intravenous, double-blind, cross-over study, *Eur. Heart J.*, 13, 251, 1992.
84. Brevetti, G., Perna, S., Sabba, C., Martone, V.D. and Condorelli, M., Propionyl-L-carnitine in intermittent claudication: double-blind, placebo-controlled, dose titration, multicenter study, *J. Am. Coll. Cardiol.*, 25, 1411, 1995.
85. Brevetti, G., Perna, S., Sabba, C., Martone, V.D., Di Iorio, A. and Barletta, G., Effect of propionyl-L-carnitine on quality of life in intermittent claudication, *Am. J. Cardiol.*, 15, 777, 1997.
86. Brevetti, G., Chiariello, M., Ferulano, G., Policicchio, A., Nevola, E., Rossini, A., Attisano, T., Ambrosio, G., Siliprandi, N. and Angelini, C., Increases in walking distance in patients with peripheral vascular disease treated with L-carnitine: A double-blind, cross-over study, *Circulation*, 77, 767, 1988.

87. Cacciatorre, L., Cerio, R., Ciarimboli, M., Cocozza, M., Coto, V., D'Alessandro, A., D'Alessandro, L., Grattarola, G., Imparato, L. and Lingetti, M., The therapeutic effect of L-carnitine in patients with exercise-induced stable angina: A controlled study, *Drugs Exp. Clin. Res.*, 17, 225, 1991. 1991.
88. Canale, C., Terrachini, V., Biagini, A., Vallebons, A., Masperone, M.A., Valice, S. and Castellano, A., Bicycle ergometer and echocardiographic study in healthy subjects and patients with angina pectoris after administration of L-carnitine: Semiautomatic computerized analysis of M-mode tracing, *Int. J. Clin. Pharmacol. Ther. Toxicol.*, 26, 221, 1988.
89. Cherchi, A., Lai, C., Angelino, F., Trucco, G., Caponnetto, S., Meretto, P.E., Rosolen, G., Manzoli, U., Schiavoni, A. and Reale, A., Effects of L-carnitine on exercise tolerance in chronic stable angina: A multicenter, double-blind, randomized, placebo controlled crossover study, *Int. J. Clin. Pharmacol. Ther. Toxicol.*, 23, 569, 1985.
90. Cherchi, A., Lai, C., Onnis, E., Orani, E., Pirisi, R., Pisano, M.R., Soro, A. and Corsi, M., Propionyl carnitine in stable effort angina, *Cardiovasc. Drugs Ther.*, 4, 481, 1990.
91. Dal Negro, R., Pomari, G., Zoccatelli, O and Turco, P., L-Carnitine and rehabilitative physiokinesitherapy: Metabolic and ventilatory response in chronic respiratory insufficiency, *Int. J. Clin. Pharmacol. Ther. Toxicol.*, 24, 452, 1986.
92. Dal Negro, R., Turco, P., Pomari, G. and De Conti, F., Effects of L-carnitine on physical performance in chronic respiratory insufficiency, *Int. J. Clin. Pharmacol. Ther. Toxicol.*, 26, 269, 1988.
93. Dean, S.M., Pharmacologic treatment for intermittent claudication, *Vasc. Med.*, 7, 301, 2002.
94. Ferrari, R. and De Giuli, F., The propionyl-L-carnitine hypothesis: An alternative approach to treating heart failure, *J. Card. Fail.*, 3, 217, 1997.
95. Gleim, G.G. and Glace, B., Carnitine as an ergogenic aid in health and disease, *J. Am. Coll. Nutr.*, 17, 203, 1998.
96. Goa, K.L. and Brogden, R.N., L-Carnitine. A preliminary review of its pharmacokinetics and its therapeutic use in ischaemic heart disease and primary and secondary carnitine deficiencies in relationship to its role in fatty acid metabolism, *Drugs*, 34, 1, 1987.
97. Hiatt, W.R., Regensteiner, J.G., Creager, M.A., Hirsch, A.T., Cooke, J.P., Olin, J.W., Gorbunov, G.N., Isner, J., Lukjanov, Y.V., Tsitsiashvili, M.S., Zabelskaya, T.F. and Amato, A., Propionyl-L-carnitine improves exercise performance and functional status in patients with claudication, *Am. J. Med.*, 110, 616, 2001.
98. Iyer, R.N., Khan, A.A., Gupta, A., Vajifdar, B.U. and Lokhandwala, Y.Y., L-Carnitine moderately improves the exercise tolerance in chronic stable angina, *J. Assoc. Physicians India*, 48, 1050, 2000.
99. Jackson, G., Combination therapy in angina: A review of combined haemodynamic treatment and the role for combined haemodynamic and cardiac metabolic agents, *Int. J. Clin. Pract.*, 55, 256, 2001.
100. Kamikawa, T., Suzuki, Y., Kobayashi, A., Hayashi, H., Masumura, Y., Nishihara, K., Abe, M. and Yamazaki, N., Effects of L-carnitine on exercise tolerance in patients with stable angina pectoris, *Jpn. Heart J.*, 25, 587, 1984.
101. Kobayashi, A., Masumura, Y. and Yamazaki, N., L-Carnitine treatment for congestive heart failure — experimental and clinical study, *Jpn. Circ. J.*, 56, 86, 1992.

102. Kosolcharoen, P., Nappi, J., Peducci, P., Shug, A., Patel, A., Filipek, T. and Thomsen, J.H., Improved exercise tolerance after administration of L-carnitine, *Curr. Ther. Res.*, 30, 753, 1981.
103. Lagioia, R., Scrutinio, D., Mangini, S.G., Ricci, A., Mastropasqua, F., Valentini, G., Ramunni, G., Totaro Fila, G. and Rizzon, P., Propionyl-L-carnitine: A new compound in the metabolic approach to the treatment of effort angina, *Int. J. Cardiol.*, 34, 167, 1992.
104. Mancini, M., Rengo, F., Lingeti, M., Sorrentino, G.P. and Nolfe G., Controlled study on the therapeutic efficacy of propionyl-L-carnitine in patients with congestive heart failure, *Arzneimittelforschung*, 42, 1101, 1992.
105. Neumann, G., Effect of L-carnitine on athletic performance, in *Carnitine — Pathobiochemical Basics and Clinical Applications*, Seim, H. and Loster, H., Eds., Ponte Press Verlags-GmbH, Bochum, Germany, 1996, 61–72.
106. Pauly, D.F. and Pepine, C.J., The role of carnitine in myocardial dysfunction, *Am. J. Kidney Dis.*, 41, S35, 2003.
107. Pucciarelli, G., Mastursi, M., Latte, S., Sacra, C., Setaro, A., Lizzado, A. and Nolfe, G., Effetti clinici ed emodinamici della propionil-L-carnitina nel trattamento dello scompenso cardiaco congestizio, *Clin. Ter.*, 141, 379, 1992.
108. Terranova, R. and Luca, S., Trattamento con L-propionilcarnitina dell'arteriopatia obliterante degli arti inferiori in pazienti anziani, *Minerva Med.*, 92, 61, 2001.
109. Witte, K.K., Clark, A.L. and Cleland, J.G., Chronic heart failure and micronutrients, *J. Am. Coll. Cardiol.*, 37, 765, 2001.
110. Greenberg, S.M. and Frishman, W.H., Coenzyme Q_{10}: A new drug for cardiovascular disease, *J. Clin. Pharmacol.*, 30, 596, 1990.
111. Hofman-Bang, C., Rehnqvist, N., Swedberg, K., Wiklund, I. and Astrom, H., Coenzyme Q_{10} as an adjunctive in the treatment of chronic congestive heart failure. The Q_{10} Study Group, *J. Card. Fail.* 1, 101, 1995.
112. Kamikawa, T., Kobayashi, A., Yamashita, T., Hayashi, H. and Yamazaki, N., Effects of coenzyme Q_{10} on exercise tolerance in chronic stable angina pectoris, *Am. J. Cardiol.*, 56, 247, 1985.
113. Khatta, M., Alexander, B.S., Krichten, C.M., Fisher, M.L., Freudenberger, R., Robinson, S.W. and Gottlied, S.S., The effect of coenzyme Q_{10} in patients with congestive heart failure, *Ann. Intern. Med.*, 132, 636, 2000.
114. Langsjoen, P.H. and Langsjoen, A.M., Overview of the use of CoQ_{10} in cardiovascular disease, *Biofactors*, 9, 273, 1999.
115. Langsjoen, P.H., Vadhanavikit, S. and Folkers, K., Effective treatment with coenzyme Q_{10} of patients with chronic myocardial disease, *Drugs Exp. Clin. Res.*, 11, 577, 1985.
116. Morisco, C., Nappi, A., Argenziano, L., Sarno, D., Fonatana, D., Imbriaco, M., Nicolai, E., Romano, M., Rosiello, G. and Cuocolo, A., Noninvasive evaluation of cardiac hemodynamics during exercise in patients with chronic heart failure: Effects of short-term coenzyme Q_{10} treatment, *Mol. Aspects Med.*, 15, S155, 1994.
117. Munkholm, H., Hansen, H.H. and Rasmussen, K., Coenzyme Q_{10} treatment in serious heart failure, *Biofactors*, 9, 285, 1999.
118. Permanetter, B., Rossy, W., Klein, G., Weingartner, F., Seidl, K.F. and Blomer, H., Ubiquinone (coenzyme Q_{10}) in the long-term treatment of idiopathic dilated cardiomyopathy, *Eur. Heart J.*, 13, 1528, 1992.

119. Sacher, H.L., Sacher, M.L., Landau, S.W., Kersten, R., Dooley, F., Sacher, A., Sacher, M., Dietarick, K. and Ichkhan, K., The clinical and hemodynamic effects of coenzyme Q_{10} in congestive heart failure, *Am. J. Ther.*, 4, 66, 1997.
120. Satta, A., Grandi, M., Landoni, C.V., Migliori, G.B., Spanavello, A., Vocaturo, G. and Neri, M., Effects of ubidecarenone in an exercise training program for patients with chronic obstructive pulmonary diseases, *Clin. Ther.*, 13, 754, 1991.
121. Tran, M.T., Mitchell, T.M., Kennedy, D.T.,a nd Giles, J.T., Role of coenzyme Q_{10} in chronic heart failure, angina and hypertension, *Pharmacotherapy*, 21, 797, 2001.
122. Bell, D.G., Jacobs, I. and Zamecnik, J., Effects of caffeine, ephedrine and their combination on time to exhaustion during high-intensity exercise, *Eur. J. Appl. Physiol.*, 77, 427, 1998.
123. Bell, D.G. and Jacobs, I., Combined caffeine and ephedrine ingestion improves run times of Canadian Forces Warrior Test, *Aviat. Space Environ. Med.*, 70, 325, 1999.
124. Bell, D.G., Jacobs, I., McLellan, T.M., Miyazaki, M. and Sabiston, C.M., Thermal regulation in the heat during exercise after caffeine and ephedrine ingestion, *Aviat. Space Environ. Med.*, 70, 583, 1999.
125. Bell, D.G., Jacobs, I., McLellan, T.M. and Zamecnik, J., Reducing the dose of combined caffeine and ephedrine preserves the ergogenic effect, *Aviat. Space Environ. Med.*, 71, 415, 2000.
126. Bell, D.G., McLellan, T.M. and Sabiston, C.M., Effect of caffeine and ephedrine ingestion on 10 km run, *Med. Sci. Sports Exerc.*, 32, S117, 2000. [abstract #463].
127. Bohn, A.M., Khodaee, M. and Schwenk, T.L., Ephedrine and other stimulants as ergogenic aids, *Curr. Sports Med. Rep.*, 2, 220, 2003.
128. Fomous, C.M., Costello, R.B. and Coates, P.M., Symposium: Conference on the science and policy of performance-enhancing products, *Med. Sci. Sports Exerc.*, 34, 1685, 2002.
129. Nevola, V.R., Weller, A.S. and Harrison, M.H., Cardiovascular effects of a combined dose of caffeine and ephedrine in man, *Med. Sci. Sports Exerc.*, 31, S117, 1999. [abstract #448].
130. Shekelle, P.G., Hardy, M.L., Morton, S.C., Maglione, M., Mojica, W.A., Suttorp, M.J., Rhodes, S.L., Jungvig, L. and Gagne, J., Efficacy and safety of ephedra and ephedrine for weight loss and athletic performance: a meta-analysis, *J.A.M.A.*, 289, 1537, 2003.
131. Talbott, S.M., Supplements for boosting energy levels, in *A Guide to Understanding Dietary Supplements*, Haworth Press, New York, NY, 2003, 181–216.
132. Bartsch, P., Mairbaurl, H. and Friedmann, B., Pseudoanamie durch Sport, [Pseudo-anemia caused by sports], *Ther. Umsch.*, 55, 251, 1998.
133. Beard, J.L. and Tobin, B.W., Iron deficiency affects exercise and exercise affects iron metabolism: Is this a chicken and egg argument? in *Sports Nutrition. Minerals and Electrolytes*, Kies, C.V. and Driskell, JA., Eds., CRC Press, Boca Raton, FL, 1995, 33.
134. Beard, J and Tobin, B., Iron status and exercise, *Am. J. Clin. Nutr.*, 72, 594S, 2000.
135. Beard, J.L., Iron biology in immune function, muscle metabolism and neuronal functioning, *J. Nutr.*, 131, 568S, 2001.
136. Beltz, S.D. and Doering, P.L., Efficacy of nutritional supplements used by athletes, *Clin. Pharm.*, 12, 900, 1993.
137. Chatard, J.C., Mujika, I., Guy, C., Lacour, J.R., Anaemia and iron deficiency in athletes. Practical recommendations for treatment, *Sports Med.*, 27, 229, 1999.

138. Clarkson, P.M., Vitamins and trace minerals, in *Perspectives in Exercise Science and Sports Medicine Volume 4: Ergogenics — Enhancement of Performance in Exercise and Sport*, Lamb, D.R. and Williams, M.H., Eds., Brown & Benchmark, Ann Arbor, MI, 1991, 123.
139. Clarkson, P.M. and Haymes, E.M., Exercise and mineral status of athletes: Calcium, magnesium, phosphorus and iron, *Med. Sci. Sports Exerc.*, 27, 831, 1995.
140. Eichner, E.R., Sports anemia, iron supplements and blood doping, *Med. Sci. Sports Exerc.*, 24, S315, 1992.
141. Ekblom, B., Micronutrients: effects of variation in [Hb] and iron deficiency on physical performance, *World Rev. Nutr. Diet.*, 82,122, 1997.
142. Gleeson, M., Minerals and exercise immunology, in *Nutrition and Exercise Immunology*, Nieman, D.C. and Pedersen, B.K., Eds., CRC Press, Boca Raton, FL, 2000, 137.
143. Haas, J.D. and Brownlie, T., Iron deficiency and reduced work capacity: A critical review of the research to determine a causal relationship, *J. Nutr.*, 131; 676S, 2001.
144. Haymes, E.M., Trace minerals and exercise, in *Nutrition in Exercise and Sport*, 2nd ed., Wolinsky, I. and Hickson, J.F., Eds., CRC Press, Boca Raton, FL, 1994, 223.
145. Manore, M.M., Dietary recommendations and athletic menstrual dysfunction, *Sports Med.*, 32, 887, 2002.
146. Newhouse, I.J and Clement, D.B., The efficacy of iron supplementation in iron depleted women, in *Sports Nutrition. Minerals and Electrolytes*, Kies, C.V. and Driskell, JA., Eds., CRC Press, Boca Raton, FL, 1995, 47.
147. Nielsen, P. and Nachtigall, D., Iron supplementation in athletes. Current recommendations, *Sports Med.*, 26, 207, 1998.
148. Weaver, C.M. and Rajaram, S., Exercise and iron status, *J. Nutr.*, 122, 782, 1992.
149. American College of Sports Medicine, Position stand on exercise and fluid replacement, *Med. Sci. Sports Exer.*, 28, i, 1996.
150. Brouns, F., Saris, W and Schneider, H., Rationale for upper limits of electrolyte replacement during exercise, *Int. J. Sport Nutr.*, 2, 229, 1992.
151. Burke, L.M., Nutritional needs for exercise in the heat, *Comp. Biochem. Physiol Part A*, 128, 735, 2001.
152. Buskirk, E.R. and Puhl, S.M., *Body Fluid Balance: Exercise and Sport*, CRC Press, Boca Raton, FL, 1996.
153. Galloway, S.D., Dehydration, rehydration and exercise in the heat: Rehydration strategies for athletic competition, *Can. J. Appl. Physiol.*, 24, 188, 1999.
154. Gisolfi, C.V. and Duchman, S.M., Guidelines for optimal replacement beverages for different athletic events, *Med. Sci. Sports Exerc.*, 24, 679, 1992.
155. Grandjean, A.C. and Ruud, J.S., Nutrition for cyclists, *Clin. Sports Med.*, 13, 235, 1994.
156. Hawley, J.A., Dennis, S.C. and Noakes, T.D., Carbohydrate, fluid and electrolyte requirements during prolonged exercise, in *Sports Nutrition. Minerals and Electrolytes*, Kies, C.V. and Driskell, J.A., Eds., CRC Press, Boca Raton, FL, 1995, 235.
157. Hickey, M.S. and Israel, R.G., Fluid replacement and exercise-thermal stress: an update, *AMMA Quarterly*, Winter, 10, 1997.
158. Holzheimer, L.A., Sports nd energy drinks, *Diabetes Self Manag.*, 20, 96, 2003.
159. Horswill, C.A., Effective fluid replacement, *Int. J. Sport Nutr.*, 8, 175, 1998.

160. Johnson, H.L., The requirements for fluid replacement during heavy sweating and the benefits of carbohydrate and minerals, in *Sports Nutrition. Minerals and Electrolytes*, Kies, C.V. and Driskell, J.A., Eds., CRC Press, Boca Raton, FL, 1995, 215.
161. Kay, D. and Marino, F.E., Fluid ingestion and exercise hyperthermia: Implications for performance, thermoregulation, metabolism and the development of fatigue, *J. Sports Sci.*, 18, 71, 2000.
162. Latzka, W.A. and Montain, S.J., Water and electrolyte requirements for exercise, *Clin. Sports Med.*, 18, 513, 1999.
163. Maughan, R.J. and Naokes, T.D., Fluid replacement and exercise stress. A brief review of studies on fluid replacement and some guidelines for the athlete, *Sports Med.*, 12, 16, 1991.
164. Maughan, R.J., Fluid and electrolyte loss and replacement in exercise, *J. Sports Sci.*, 9, 117, 1991.
165. Maughan, R., Carbohydrate-electrolyte solutions during prolonged exercise, in *Perspectives in Exercise Science and Sports Medicine Volume 4: Ergogenics — Enhancement of Performance in Exercise and Sport*, Lamb, D.R. and Williams, M.H., Eds., Brown & Benchmark, Ann Arbor, MI, 1991, 35.
166. Maughan, R.J., Food and fluid intake during exercise, *Can. J. Appl. Physiol.*, 26 Suppl, S71, 2001.
167. Maughan, R.J., *Sports Drinks: Basic Science and Practical Aspects*, CRC Press, Boca Raton, 2001.
168. Millard-Stafford, M., Fluid replacement during exercise in the heat. Review and recommendations, *Sports Med.*, 13, 223, 1992.
169. Puhl, S.M. and Buskirk, E.M., Nutrient beverages for physical performance, in *Nutrition in Exercise and Sport*, 3rd ed., Wolinsky, I., Ed., CRC Press, Boca Raton, FL, 1998, 277.
170. Rehrer, N.J., Fluid and electrolyte balance in ultra-endurance sport, *Sports Med.*, 31, 701, 2001.
171. Sawka, M.N. and Coyle, E.F., Influence of body water and blood volume on thermoregulation and exercise performance in the heat, *Exerc. Sport Sci. Rev.*, 27, 167, 1999.
172. Sawka, M.N. and Montain, S.J., Fluid and electrolyte supplementation for exercise heat stress, *Am. J. Clin. Nutr.*, 72, 564S, 2000.
173. Senay, L.C., Water and electrolytes during physical activity, in *Nutrition in Exercise and Sport*, 3rd ed., Wolinsky, I., Ed., CRC Press, Boca Raton, 1998, 257-276.
174. Shirreffs, S.M., Effects of ingestion of carbohydrate-electrolyte solutions on exercise performance, *Int. J. Sports Med.*, 19, S117, 1998.
175. Shirreffs, S.M., Restoration of fluid and electrolyte balance after exercise, *Can. J. Appl. Physiol.*, 26, S228, 2001.
176. Antonio, J. and Stout, J.R., *Supplements for Strength-Power Athletes*, Human Kinetics, Champaign, IL, 2002.
177. Gledhill, N., Bicarbonate ingestion and anaerobic performance, *Sports Med.*, 1, 177, 1984.
178. Gomes, M.R. and Tirapegui, J., [Relation of some nutritional supplements and physical performance], *Arch. Latinoam. Nutr.*, 50, 317, 2000.

179. Heigenhauser, G.J.F. and Jones, N.L., Bicarbonate loading, in *Perspectives in Exercise Science and Sports Medicine Volume 4: Ergogenics — Enhancement of Performance in Exercise and Sport*, Lamb, D.R. and Williams, M.H., Eds., Brown & Benchmark, Ann Arbor, MI, 1991, 183.
180. Horswill, C.A., Effects of bicarbonate, citrate and phosphate loading on performance, *Int. J. Sport Nutr.*, 5, S111, 1995.
181. Linderman, J. and Fahey, T.D., Sodium bicarbonate ingestion and exercise performance. An update, *Sports Med.*, 11, 71, 1991.
182. Matson, L.G. and Tran Z.V., Effects of sodium bicarbonate ingestion on anaerobic performance: A meta-analytic review, *Int. J. Sport Nutr.*, 3, 2, 1993.
183. Anding, J., Body weight regulation and energy needs: weight loss, in *Energy-Yielding Macronutrients and Energy Metabolism in Sports Nutrition*, Driskell, J.A. and Wolinsky, I., Eds., CRC Press, Boca Raton, 2001, 309.
184. Bazzarre, T.L., Nutrition and strength, in *Nutrition in Exercise and Sport*, 2nd ed., Wolinsky I. and Hickson, J.F., Eds., CRC Press, Boca Raton, FL, 1994, 417.
185. Bazzarre, T.L., Nutrition and strength, in *Nutrition in Exercise and Sport*, 3rd ed., Wolinsky I., Ed., CRC Press, Boca Raton, FL, 1998, 369.
186. Grandjean, A., Nutritional requirements to increase lean mass, *Clin. Sports Med.*, 18, 623, 1999.
187. Grunewald, K.K. and Bailey, R.S., Commercially marketed supplements for bodybuilding athletes, *Sports Med.*, 15, 90, 1993.
188. Kleiner, S.M, Performance-enhancing aids in sport: Health consequences and nutritional alternatives, *J. Am. Coll. Nutr.*, 10, 115, 1991.
189. Kreider, R.B., Dietary supplements and the promotion of muscle growth with resistance exercise, *Sports Med.*, 27, 97, 1999.
190. Volek, J.S., Houseknecht, K. and Kraemer, W.J., Nutritional strategies to enhance performance of high-intensity exercise, *J. Strength Cond. Res.*, 11, 11, 1997.
191. Andres, L.P.A., Sachek, J. and Tapia, S., A review of creatine supplementation: Side effects and improvements in athletic performance, *Nutr. Clin. Care*, 2, 73, 1999.
192. Antonio, J., Street, C., Kalman, D. and Colker, C., Dietary supplements used by athletes, *AMAA Quarterly*, 13, 6, 1999.
193. Balsom, P.D., Soderlund, K. and Ekblom, B., Creatine in humans with special reference to creatine supplementation, *Sports Med.*, 18, 268, 1994.
194. Benzi, G. Is there a rationale for the use of creatine either as nutritional supplementation or drug administration in humans participating in a sport? *Pharmacol. Res.*, 41, 255, 2000.
195. Benzi, G. and Ceci, A., Creatine as nutritional supplementation and medicinal product, *J. Sports Med. Phys. Fitness*, 41, 1, 2001.
196. Branch, J.D., Effect of creatine supplementation on body composition and performance: a meta-analysis, *Int. J. Sport Nutr. Exerc. Metab.*, 13, 198, 2003.
197. Casey, A and Greenhaff, P.L., Does dietary creatine supplementation play a role in skeletal muscle metabolism and performance? *Am. J. Clin. Nutr.*, 72, 607S, 2000.
198. Clark, J.F., Creatine and phosphocreatine: A review of their use in exercise and sport, *J. Athlet. Train.*, 32, 45, 1997.
199. Clark, J.F., Creatine: A review of its nutritional applications in sport, *Nutrition*, 14, 322, 1998.
200. Clarkson, P.M. and Rawson, E.S., Nutritional supplements to increase muscle mass, *Crit. Rev. Food Sci. Nutr.* 39, 317, 1999.

201. Conway, M.A. and Clark, J.F., Eds., *Creatine and Creatine Phosphate: Scientific and Clinical Perspectives*, Academic Press, San Diego, CA, 1996.
202. Demant, T.W. and Rhodes, E.C., Effects of creatine supplementation on exercise performance, *Sports Med.*, 28, 49, 1999.
203. Derave, W., Eijinde, B.O. and Hespel, P., Creatine supplementation in health and disease: what is the evidence for long-term efficacy? *Mol. Cell. Biochem.*, 244, 49, 2003.
204. Di Pasquale, M., Essential amino acids, in *Amino Acids and Proteins for the Athlete*, CRC Press, Boca Raton, FL, 1997, 121.
205. Feldman, E.B. Creatine: A dietary supplement and ergogenic acid, *Nutr. Rev.*, 57, 45, 1999.
206. Francaux, M. and Poortmans, J.R., Effects of training and creatine supplement on muscle strength and body mass, *Eur. J. Appl. Occup. Physiol.*, 80, 165, 1999.
207. Graham, A.S. and Hatton, R.C., Creatine: A review of efficacy and safety, *J. Am. Pharm. Assoc. (Wash.)*, 39, 803, 1999.
208. Greenhaff, P.L., Creatine and its application as an ergogenic aid, *Int. J. Sport Nutr.*, 5, S100, 1995.
209. Hespel, P., Eijinde, B.O., Derave, W. and Richter, E.A., Creatine supplementation: exploring the role of the creatine kinase/phosphocreatine system in human muscle, *Can. J. Appl. Physiol.*, 26, S79, 2001.
210. Jacobs, I., Dietary creatine monohydrate supplementation, *Can. J. Appl. Physiol.*, 24, 503, 1999.
211. Juhn, M.S. and Tarnopolsky, M., Oral creatine supplementation and athletic performance: A critical review, *Clin. J. Sport Med.*, 8, 286, 1998.
212. Kraemer, W.J. and Volek, J.S., Creatine supplementation. Its role in human performance, *Clin. Sports Med.*, 18, 651, 1999.
213. Krämer, K., Weiss, M. and Liesen, H., Creatine: physiology and exercise performance, in *Nutraceuticals in Health and Disease Prevention*, Krämer, K., Hoppe, P.-P. and Packer, L., Eds., Marcel Dekker, New York, 2001, 165.
214. Kreider, R.B., Creatine supplementation in exercise and sport, in *Energy-Yielding Macronutrients and Energy Metabolism in Sports Nutrition*, Driskell, J.A. and Wolinksy, I., Eds., CRC Press, Boca Raton, 2000, 213-242.
215. Kreider, R.B., Effects of creatine supplementation on performance and training adaptations, *Mol. Cell. Biochem.*, 244, 89, 2003.
216. Lawrence, M.E. and Kirby, D.F., Nutrition and sports supplements: Fact or fiction, *J. Clin. Gastroenterol.*, 35, 299, 2002.
217. Lemon, P.W., Dietary creatine supplementation and exercise performance: Why inconsistent results? *Can. J. Appl. Physiol.*, 27, 663, 2002.
218. Maughan, R.J., Creatine supplementation and exercise performance, *Int. J. Sport Nutr.*, 5, 94, 1995.
219. McChesney, P., Smith, M. and Emmerton, L., Creatine. The sports supplement of the decade, *N.Z. Pharmacy*, March, 31, 2000.
220. Mendes, R.R. and Tirapegui, J., Considerações sobre exercício físico, creatina e nutrição, *Br. J. Pharm. Sci.*, 35, 195, 1999.
221. Mendes, R.R. and Tirapegui, J., Creatina: O suplemento nutricional para a atividade física — conceitos atuais, *Arch. Latinoam. Nutr.*, 52, 117, 2002.
222. Mesa, J.L., Ruiz, J.R., Gonzalez-Gross, M.M., Gutierrez, Sainz, A. and Castillo Garzon, M.J., Oral creatine supplementation and skeletal muscle metabolism in physical exercise, *Sports Med.*, 32, 903, 2002.

223. Metzl, J.D., Strength training and nutritional supplement use in adolescents, *Curr. Opin. Pediatr.*, 11, 292, 1999.
224. Mujika, I. and Padilla, S., Creatine supplementation as an ergogenic aid for sports performance in highly trained athletes: a critical review, *Int. J. Sports Med.*, 18, 491, 1997.
225. Nissen, S.L. and Sharp, R.L., Effect of dietary supplements on lean mass and strength gains with resistance exercise: A meta-analysis, *J. Appl. Physiol.*, 94, 651, 2003.
226. Paoletti, R., Poli, A. and Jackson, A.S., *Creatine. From basic science to clinical application*, Kluwer Academic Publishers, Dordrecht, 2000.
227. Persky, A.M. and Brazeau, G.A., Clinical pharmacology of the dietary supplement creatine monohydrate, *Pharmacol. Rev.*, 53, 161, 2001.
228. Rubinstein, M.L. and Federman, D.G., Sports supplements. Can dietary additives boost athletic performance and potential? *Postgrad. Med.*, 108, 130, 2000.
229. Saint-Pierre, M.-A., Poortmans, J. and Léger, L., Supplémentation en créatine – État de la question, *Science Sports*, 17, 55, 2002.
230. Terjung, R.L., Clarkson, P., Eichner, E.R., Greenhaff, P.L., Hespel, P.J., Israel, R.G., Kraemer, W.J., Meyer, R.A., Spriet, L.L., Tarnopolsky, M.A., Wagenmakers, A.J.M., Williams, M.H., American College of Sports Medicine roundtable. The physiological and health effects of oral creatine supplementation, *Med. Sci. Sports Exerc.*, 32, 706, 2000.
231. Toler, S.M., Creatine is an ergogen for anaerobic exercise, *Nutr. Rev.*, 55, 21, 1997.
232. Volek, J.S. and Kraemer, W.J., Creatine supplementation: Its effect on human muscular performance and body composition, *J. Strength Cond. Res.*, 10, 200, 1996.
233. Williams, M.H. and Branch, J.D., Creatine supplementation and exercise performance: an update, *J. Am. Coll. Nutr.*, 17, 216, 1998.
234. Williams, M.H., Kreider, R. and Branch, J. *Creatine: The Power Supplement*, Human Kinetics, Champaign, 1999.
235. Wyss, M. and Kaddurah-Daouk, R., Creatine and creatinine metabolism, *Physiol. Rev.*, 80, 1107, 2000.
236. Yang, Z.Y., Creatine supplementation and exercise performance: An overview, *J. Sport Sci.*, 20, 76, 2000.
237. Dempsey, R.L., Mazzone, M.F. and Meurer, L.N., Does oral creatine supplementation improve strength? A meta-analysis, *J. Fam. Pract.*, 51, 945, 2002.
238. Di Pasquale, M., Essential amino acids, in *Amino Acids and Proteins for the Athlete*, CRC Press, Boca Raton, FL, 1997, 112.
239. Gallagher, P.M., Carrithers, J.A., Godard, M.P., Schulze, K.E. and Trappe, S.W., Beta-hydroxy-beta-methylbutyrate ingestion, Part I: Effects on strength and fat free mass, *Med. Sci. Sports Exerc.*, 32, 2109, 2000.
240. Nissen, S.L. and Abumrad, N.N., Nutritional role of the leucine metabolite beta-hydroxy beta-methylbutyrate (HMB), *Nutr. Biochem.*, 8, 300, 1997.
241. Slater, G.J. and Jenkins, D., Beta-hydroxy-beta-methylbutyrate (HMB) supplementation and the promotion of muscle growth and strength, *Sports Med.*, 30, 105, 2000.
242. Bucci, L.R. and Unlu, L.M., Proteins and amino acids supplements in exercise and sport, in *Energy-Yielding Macronutrients and Energy Metabolism in Sports Nutrition*, Driskell, J.A. and Wolinksy, I., Eds., CRC Press, Boca Raton, 2000, 191.

243. Butterfield, G., Amino acid and high protein diets, in *Perspectives in Exercise Science and Sports Medicine Volume 4: Ergogenics — Enhancement of Performance in Exercise and Sport*, Lamb, D.R. and Williams, M.H., Eds., Brown & Benchmark, Ann Arbor, MI, 1991, 87.
244. Di Pasquale, M., Dietary protein and amino acids, in *Amino Acids and Proteins for the Athlete. The Anabolic Edge*, CRC Press: Boca Raton, FL, 1997, 63.
245. Di Pasquale, M., Protein foods vs. protein and amino acid supplements, in *Amino Acids and Proteins for the Athlete. The Anabolic Edge*, CRC Press: Boca Raton, FL, 1997, 89.
246. Di Pasquale, M.G., Proteins and amino acids in exercise and sport, in *Energy-Yielding Macronutrients and Energy Metabolism in Sports Nutrition*, Driskell, J.A. and Wolinsky, I., Eds., CRC Press, Boca Raton, 2000, 119.
247. Evans, W.J., Protein nutrition and resistance exercise, *Can. J. Appl. Physiol.*, 26, S141, 2001.
248. Fielding, R.A., Parkington, J., What are the dietary protein requirements of physically active individuals? New evidence on the effects of exercise on protein utilization during post-exercise recovery, *Nutr. Clin. Care*, 5, 191, 2002.
249. Gibala, M.J., Nutritional supplementation and resistance exercise: What is the evidence for enhanced skeletal muscle hypertrophy? *Can. J. Appl. Physiol.*, 25, 524, 2000.
250. Houston, M.E. Gaining weight: The scientific basis of increasing skeletal muscle mass, *Can. J Appl. Physiol.*, 24, 305, 1999.
251. Kreider, R.B., Miriel, V. and Bertun, E., Amino acid supplementation and exercise performance. Analysis of the proposed ergogenic value, *Sports Med.*, 16, 190, 1993.
252. Lambert, C.P. and Flynn, M.G., Fatigue during high-intensity intermittent exercise: application to bodybuilding, *Sports Med.*, 32, 511, 2002.
253. Lemon, P.W., Protein and amino acid needs of the strength athlete, *Int. J. Sport Nutr.*, 1, 127, 1991.
254. Lemon, P.W. and Proctor, D.N., Protein intake and athletic performance, *Sports Med.*, 12, 313, 1991.
255. Lemon, P.W., Do athletes need more dietary protein and amino acids? *Int. J Sport Nutr.*, 5, S39, 1995.
256. Lemon, P.W., Is increased dietary protein necessary or beneficial for individuals with a physically active lifestyle? *Nutr. Rev.*, 54, S169, 1996.
257. Lemon, P.W.R., Dietary protein requirements in athletes, *J. Nutr. Biochem.*, 8, 52, 1997.
258. Lemon, P.W., Effects of exercise on dietary protein requirements, *Int. J. Sport Nutr.*, 8, 426, 1998.
259. Lemon, P.W., Beyond the zone: Protein needs of active individuals, *J. Am. Coll. Nutr.*, 19, 513S, 2000.
260. Millward, D.J., Protein and amino acid requirements of adults: Current controversies, *Can. J. Appl. Physiol.*, 26, S130, 2001.
261. Paul, G.L., Gautsch, T.A. and Layman, D.K., Amino acid and protein metabolism during exercise and recovery, in *Nutrition in Exercise and Sport*, 3rd ed., Wolinsky, I., Ed., CRC Press, Boca Raton, FL, 1998, 125.
262. Rankin, J.W., Role of protein in exercise, *Clin. Sports Med.*, 18, 499, 1999.
263. Rasmussen, B.B., Tipton, K.D., Miller, S.L., Wolf, S.E. and Wolfe, R.R., An oral essential amino acid-carbohydrate supplement enhances muscle protein anabolism after resistance exercise, *J. Appl. Physiol.*, 88, 386, 2000.

264. Ratamess, N.A., Kraemer, W.J., Volek, J.S., Rubin, M.R., Gomez, A.L., French, D.N., Sharman, M.J., McGuigan, M.M., Scheett, T., Hakkinen, K., Newton, R.U. and Diogaurdi, F., The effects of amino acid supplementation on muscular performance during resistance training overreaching, *J. Strength Cond. Res.*, 17, 250, 2003.
265. Tipton, K.D. and Wolfe, R.R., Exercise, protein metabolism and muscle growth, *Int. J. Sport. Nutr. Exerc. Metab.*, 11, 109, 2001.
266. Wolfe, R.R., Protein supplements and exercise, *Am. J. Clin. Nutr.*, 72, 551S, 2000.
267. Wolfe, R.R., Regulation of muscle protein by amino acids, *J. Nutr.*, 132, 3219S, 2002.
268. Brilla, L.R. and Conte, V., Effects of a novel zinc-magnesium formulation on hormones and strength, *J. Exerc. Physiol. Online*, 3, 26, 2000.
269. Brilla, L.R. and Lombardi, V.P., Magnesium in sports physiology and performance, in *Sports Nutrition. Minerals and Electrolytes*, Kies, C.V. and Driskell, JA., Eds., CRC Press, Boca Raton, FL, 1995, 139.
270. Campbell, J.D., Lifestyle, minerals and health, *Med. Hypotheses*, 57, 521, 2001.
271. Clarkson, P.M., Minerals: Exercise performance and supplementation in athletes, *J. Sports Sci.*, 9, 91, 1991.
272. Cordova, A. and Alvarez-Mon, M., Behavior of zinc in physical exercise: A special reference to immunity and fatigue, *Neurosci. Biobehav. Rev.*, 19, 439, 1995.
273. Konig, D., Weinstock, C., Keul, J., Northoff, H. and Berg, A., Zinc, iron and magnesium status in athletes — Influence on the regulation of exercise-induced stress and immune function, *Exerc. Immunol. Rev.*, 4, 2, 1998.
274. Lukaski, H.C., Magnesium, zinc and chromium nutriture and physical activity, *Am. J. Clin. Nutr.*, 72, 585S, 2000.
275. Lukaski, H.C., Magnesium, zinc and chromium nutrition and athletic performance, *Can. J. Appl. Physiol.*, 26, S13, 2001.
276. Micheletti, A., Rossi, R. and Rufini, S., Zinc status in athletes: Relation to diet and execise, *Sports Med.*, 31, 577, 2001.
277. Newhouse, I.J., Finstad, E.W., The effects of magnesium supplementation on exercise performance, *Clin. J. Sport. Med.*, 10, 195, 2000.
278. Ohno, H., Sato, Y., Kizaki, T., Yamashita, H., Ookawara T. and Ohira, Y., Physical exercise and zinc metabolism, in *Sports Nutrition. Minerals and Electrolytes*, Kies, C.V. and Driskell, JA., Eds., CRC Press, Boca Raton, FL, 1995, 129.
279. Rayssiguier, Y., Geuzennec, C.Y. and Durlach, J., New experimental and clinical data on the relationship between magnesium and sport, *Magnes. Res.*, 3, 93, 1990.
280. Shephard, R.J. and Shek, P.N., Immunological hazards from nutritional imbalance in athletes, *Exerc. Immunol. Rev.*, 4, 22, 1998.
281. Speich, M., Pineau, A. and Ballereau, F., Mineral, trace elements and related biological variables in athletes and during physical activity, *Clin. Chim. Acta*, 312, 1, 2001.
282. Barr, S.I., Increased dairy product or calcium intake: Is body weight or composition affected in humans?, *J. Nutr.*, 133, 245S, 2003.
283. Heaney, R.P., Davies, K.M., Barger-Lux, M.J., Calcium and weight: Clinical studies, *J. Am. Coll. Nutr.*, 21, 152S, 2002.
284. Parikh, S.J., Yahovski, J.A., Calcium intake and adiposity, *Am. J. Clin. Nutr.*, 77, 281, 2003.
285. Teegarden, D., Calcium intake and reduction in weight or fat mass, *J. Nutr.*, 133, 249S, 2003.

286. Zemel, M.B., Calcium modulation of hypertension and obesity: Mechanisms and implications, *J. Am. Coll. Nutr.*, 20, 428S, 2001.
287. Zemel, M.B., Regulation of adiposity and obesity risk by dietary calcium: Mechanisms and implications, *J. Am. Coll. Nutr.*, 21, 146S, 2002.
288. Zemel, M.B., Mechanisms of dairy modulation of adiposity, *J. Nutr.*, 133, 252S, 2003.
289. Anderson, J.W., Konz, E., Frederich, R.C. and Wood, C.L., Long-term weight maintenance: A meta-analysis of US studies, *Am. J. Clin. Nutr.*, 74, 579, 2001.
290. Astrup, A., Dietary approaches to reducing body weight, *Bailleres Best Pract. Res. Clin. Endocrinol. Metab.*, 13, 109, 1999.
291. Grace, C.M., Dietary management of obesity, in *The Management of Obesity and Related Disorders*, Kopelman, P.G., Ed., Martin Dunitz, London, 2001, 129.
292. Hyman, F.N., Sempos, E., Saltsman, J., Glinsmann, W.H., Evidence for success of caloric restriction in weight loss and control. Summary of data from industry, *Ann. Intern. Med.*, 119, 681, 1993.
293. Jakicic, J.M., Clark, K., Coleman, E., Donnelly, J.E., Foreyt, J., Melanson, E., Volek, J., Volpe, S.L., American College of Sports Medicine, American College of Sports Medicine position stand. Appropriate intervention strategies for weight loss and prevention of weight regain for adults, *Med. Sci. Sports Exerc.*, 33, 2145, 2001.
294. Miller, W.C., How effective are traditional dietary and exercise interventions for weight loss? *Med. Sci. Sports Exerc.*, 31, 1129, 1999.
295. Miller, W.C., Effective diet and exercise treatments for overweight and recommendations for intervention, *Sports Med.*, 31, 717, 2001.
296. Moloney, M., Dietary treatments of obesity, *Proc. Nutr. Soc.*, 59, 601, 2000.
297. Nawaz, H. and Katz, D.L., American College of Preventive Medicine practice policy statement. Weight management counseling of overweight adults, *Am. J. Prev. Med.*, 21, 73, 2001.
298. Rolls, B.J. and Bell, E.A., Dietary approaches to the treatment of obesity, *Med. Clin. N. Am.*, 84, 401, 2000.
299. Wadden, T.A., Treatment of obesity by moderate and severe caloric restriction. Results of clinical research trials, *Ann. Intern. Med.*, 119, 688, 1993.
300. Wadden, T.A. and Stunkard, A.J., Eds., *Handbook of Obesity Treatment*, Guilford Press, New York, 2003.
301. Wing, R.R. and Hill, J.O., Successful weight maintenance, *Annu. Rev. Nutr.*, 21, 323, 2001.
302. Belury, M.A., Dietary conjugated linoleic acid in health: Physiological effects and mechanisms of action, *Annu. Rev. Nutr.*, 22, 505, 2002.
303. DeLany, J.P., West, D.B., Changes in body composition with conjugated linoleic acid, *J. Am. Coll. Nutr.* 19, 487S, 2000.
304. Evans, M.E., Brown, J.M., McIntosh, M.K., Isomer-specific effects of conjugated linoleic acid (CLA) on adiposity and lipid metabolism, *J. Nutr. Biochem.*, 13, 508, 2002.
305. Ntambi, J.M., Choi, Y., Park, Y., Peters, J.M., Pariza, M.W., Effects of conjugated linoleic acid (CLA) on immune responses, body composition and stearoyl-CoA desaturase, *Can. J. Appl. Physiol.*, 27, 617, 2002.
306. Pariza, M.W., Park, Y., Cook, M.E., Mechanisms of action of conjugated linoleic acid: Evidence and speculation, *Proc. Soc. Exp. Biol. Med.*, 223, 8, 2000.
307. Astrup, A., Toubro, S., Christensen, N.J., Quaade, F., Pharmacology of thermogenic drugs, *Am. J. Clin. Nutr.*, 55, 246S, 1992.

308. Astrup, A., Lundsgaard, C., What do pharmacological approaches to obesity management offer? Linking pharmacological mechanisms of obesity management agents to clinical practice, *Exp. Clin. Endocrinol Diabetes*, 106, 29, 1998.
309. Astrup, A., Thermogenic drugs as a strategy for treatment of obesity, *Endocrine*, 13, 207, 2000.
310. Atkinson, R.L., The herbal ephedra and caffeine debate continues, *Int. J. Obes.*, 26, 589, 2002.
311. Bray, G.A., Drug treatment of obesity, *Baillieres Best Pract. Res. Clin. Endocrinol. Metab.*, 13, 131, 1999.
312. Bray, G.A., Reciprocal relation of food intake and sympathetic activity: Experimental observations and clinical implications, *Int. J. Obes. Relat. Metab. Disord.*, 24, S8, 2000.
313. Bray, G.A., A concise review on the therapeutics of obesity, *Nutrition*, 16, 953, 2000.
314. Boozer, C.N., Daly, P.A., Homel, P., Solomon, J.L., Blanchard, D., Nasser, J.A., Strauss, R. and Meredith, T., Herbal ephedra/caffeine for weight loss: a 6-month randomized safety and efficacy trial, 26, 593, 2002.
315. Clarkson, P.M. and Thompson, H.S., Drugs and sport. Research findings and limitations, *Sports Med.*, 24, 366, 1997.
316. Dulloo, A.G. and Miller, D.S., Ephedrine, caffeine and aspirin: "Over-the-Counter" drugs that interact to stimulate thermogenesis in the obese, *Nutr.*, 5, 7, 1989.
317. Dulloo, A.G., Ephedrine, xanthines and prostaglandin-inhibitors: Actions and interactions in the stimulation of thermogenesis, *Int. J. Obes.*, 17, S35, 1993.
318. Dulloo, A.G., Herbal stimulation of ephedrine and caffeine in treatment of obesity, *Int. J. Obes.*, 26, 590, 2002.
319. Pasquali, R., Casimirri, F., Clinical aspects of ephedrine in the treatment of obesity, *Int. J. Obes.*, 17, S65, 1993.
320. Talbott, S.M., Supplements for weight loss, in *A Guide to Understanding Dietary Supplements*, Haworth Press, New York, NY, 2003, 57.
321. Kalman, D.S., Colker, C.M., Swain, M.A., Torina, G.C., Shi, Q., A randomized, double-blind, placebo-controlled study of 3-acetyl-7-oxo-dehydroepiandrosterone in healthy overweight adults, *Curr. Ther. Res.*, 61, 435, 2000.
322. Lardy, H., Stratman, F., Eds., *Hormones, Thermogenesis and Obesity*, Elsevier, New York, 1989.
323. Lardy, H., Kneer, N., Bellei, M. and Bobyleva, V., Induction of thermogenic enzymes by DHEA and its metabolites, *Annal. N.Y. Acad. Sci.*, 774, 171, 1995.
324. Lardy, H.A., Reich, I.L. and Wei, Y., Delta 5-androstenes useful for promoting weight maintenance or weight loss and treatment process, U.S. Patent 5424463, 1995.
325. Lardy, H.A., Reich, I.L. and Wei, Y., Delta 5-androstenes useful for promoting weight maintenance or weight loss and treatment process, U.S. Patent 5506223, 1996.
326. Partridge, B.E., Lardy, H.A., Treatment process for promoting weight loss employing a substituted delta-5-androstene, U.S. Patent 5296481, 1994.
327. Zenk, J.L., Helmer, T.R., Kassen, L.J. and Kuskowski, M.A., The effect of 7-Keto Naturalean™ on weight loss: A randomized, double-blind, placebo-controlled trial, *Curr. Ther. Res.*, 63, 263, 2002.
328. Bucci, L.R., *Nutrition Applied to Injury Rehabilitation and Sports Medicine*, CRC Press, Boca Raton, FL, 1994.

329. Benton, D., Haller, J. and Fordy, J., Vitamin supplementation for 1 year improves moods, *Neuropsychobiology*, 32, 98, 1995.
330. Benton, D., Fordy, H. and Haller, J., The impact of long-term vitamin supplementation on cognitive functioning, *Psychopharmacology*, 117, 298, 1995.
331. Benton, D., Griffiths, R. and Haller, J., Thiamine supplementation, moor and cognitive functioning, *Psychopharmacology* 129, 66, 1997.
332. Bonke, D. and Nickel, B., Improvement of fine motorific movement control by elevated dosages of vitamin B_1, B_6 and B_{12} in target shooting, *Int. J. Vitam. Nutr. Res.*, Suppl. 30, 198, 1989.
333. Carroll, D., Ring, C., Suter M. and Willemsen, G., The effects of an oral multivitamin combination with calcium, magnesium and zinc on psychological wellbeing in healthy young male volunteers: a double-blind placebo-controlled trial, *Psychopharmacology*, 150, 220, 2000.
334. Heseker, H., Kubler, W., Pudel, V. and Westenhofer, J., Interaction of vitamins with mental performance, *Bibl. Nutr. Dieta*, 43, 1995.
335. Bucci, L.R., Selected herbals and exercise performance, *Am. J. Clin. Nutr.*, 72, 624S, 2000.
336. Bucci, L.R., Turpin, A.A., Beer, C., Feliciano, J., Ginseng, *Nutritional Ergogenic Aids*, Driskell, J.A. and Wolinsky, I., Eds., CRC Press, Boca Raton, FL, in press.
337. Brozek, B., Soviet studies on nutrition and higher nervous activity, *Ann. N.Y. Acad. Sci.*, 93, 665, 1963.
338. Saint-John, M. and McNaughton, L., Octacosanol ingestion and its effects on metabolic responses to submaximal cycle ergometry, reaction time and chest and grip strength, *Int. Clin. Nutr. Rev.*, 6, 81, 1986.
339. Herxheimer, A. and Petrie, K.J., Melatonin for the prevention and treatment of jet lag, *Cochrane Database Syst. Rev.*, (2), CD001520, 2002.
340. Staton, W.M., The influence of ascorbic acid in minimizing post-exercise muscle soreness in young men, *Res. Quart.*, 23, 356, 1952.
341. Syed, I.H., Muscle stiffness and vitamin C, *Br. Med. J.*, 2, 304, 1966.
342. Kaminski, M. and Boal, R., An effect of ascorbic acid on delayed-onset muscle soreness, *Pain*, 50, 317, 1992.
343. Jakeman, P. and Maxwell, S., Effect of antioxidant vitamin supplementation on muscle function after eccentric exercise, *Eur. J. Appl. Physiol.*, 67, 426, 1993.
344. Thompson, D., Nicholas, C.W., McGregor, S.J., McArdle, F., Jackson, M.J. and Williams, C., Muscle soreness and damage following two weeks vitamin C supplementation, *Med. Sci. Sports Exerc.*, 32, S171, 2000.
345. Thompson, D., Williams, C., Kingsley, M., Nicholas, C.W., Lakomy, H.K., McArdle, F. and Jackson, M.J., Muscle soreness and damage parameters after prolonged intermittent shuttle-running following acute vitamin C ingestion, *Int. J. Sports Med.*, 22, 68, 2001.
346. Thompson, D., Williams, C., McGregor, S.J., Nicholas, C.W., McArdle, F., Jackson, M.J. and Powell, J.R., Prolonged vitamin C supplementation and recovery from demanding exercise, *Int. J. Sport Nutr. Exerc. Metab.*, 11, 466, 2001.
347. Bryer, S.C. and Goldfarb, A.H., The effect of vitamin C supplementation on blood glutathione status, DOMS & creatine kinase, *Med. Sci. Sports Exerc.*, 33, S122, 2001.
348. Bailey, D.M., Williams, C., Hurst, T. and Powell, J., Recovery from downhill running following ascorbic acid supplementation, *Med. Sci. Sports Exerc.*, 33, S122, 2001.

349. Peters, E.M., Vitamin C, neutrophil function and upper respiratory tract infection risk in distance runners: the missing link, *Exerc. Immunol. Rev.*, 3, 32, 1996.
350. Nieman, D.C. and Pedersen, B.K., Exercise and immune function. Recent developments, *Sports Med.*, 27, 73, 1999.
351. Pedersen, B.K., Bruunsgaard, H., Jensen, M., Krzywkowski, K. and Ostrowski, K., Exercise and immune function: Effect of ageing and nutrition, *Proc. Nutr. Soc.*, 58, 733, 1999.
352. Hathcock, J.N., Safety limits for nutrients, *J. Nutr.*, 126, 2386S, 1996.
353. Hathcock, J.N., Vitamin and mineral safety, Council for Responsible Nutrition, Washington, DC, 1997.
354. Subcommittee on the Tenth Edition of the RDAs, Food and Nutrition Board, Commission on Life Sciences, National Research Council, *Recommended Dietary Allowances*, 10th ed., National Academy Press, Washington, DC, 1989.
355. Gardner JW. Death by water intoxication. *Mil. Med.* 167(5), 432, 2002.
356. Archer, M.C., Use of oral creatine to enhance athletic performance and its potential side effects, *Clin. J. Sport Med.*, 9, 119, 1999.
357. Culpepper, R.M., Creatine supplementation: Safe as steak? *South. Med. J.*, 91, 890, 1998.
358. Guerrero-Ontiveros, M.L. and Wallimann, T., Creatine supplementation in health and disease. Effects of chronic creatine ingestion *in vivo*: Down-regulation of the expression of creatine transporter isoforms in skeletal muscle, *Mol. Cell. Biochem.*, 184, 427, 1998.
359. Juhn, M.S. and Tarnopolsky, M., Potential side effects of oral creatine supplementation: A critical review, *Clin. J. Sport Med.*, 8, 298, 1998.
360. Juhn, M.S., O'Kane, J.W. and Vinci, D.M., Oral creatine supplementation in male collegiate athletes: A survey of dosing habits and side effects, *J. Am. Diet. Assoc.*, 99, 593, 1999.
361. Juhn, M.S. Does creatine supplementation increase the risk of rhabdomyolysis? *J. Am. Board Fam. Pract.*, 13, 150, 2000.
362. Koshy, K.M., Griswold, E. and Schneeberger, E.E., Interstitial nephritis in a patient taking creatine, *N. Engl. J. Med.*, 340, 814, 1999.
363. Loud, K.J., Rozycki, A.A. and Chobanian, M.C., Creatine nephropathy — lacrosse, *Med. Sci. Sports Exerc.*, 33, S10, 2001.
364. Pritchard, N.R. and Kalra, P.A., Renal dysfunction accompanying oral creatine supplements, *Lancet*, 351, 1252, 1998.
365. Robinson, S.J., Acute quadriceps compartment syndrome and rhabdomyolysis in a weight lifter using high-dose creatine supplementation, *J. Am. Board Fam. Pract.*, 13, 134, 2000.
366. Rossi, R., Gambelunghe, C., Lepri, E., Micheletti, A., Sommavilla, M., Parisse, I. and Rufini, S., Integrazione orale di creatine e sport valutazione critica dei rischi e benefici, *Med. Sport*, 51, 349, 1998.
367. Schroeder, C., Potteiger, J., Randall, J., Jacobsen, D., Magee, L., Benedict, S. and Hulver, M., The effects of creatine dietary supplementation on anterior compartment pressure in lower leg during rest and following exercise, *Clin. J. Sport Med.*, 11, 87, 2001.
368. Silber, M.L., Scientific facts behind creatine monohydrate as sport nutrition supplement, *J. Sports Med. Phys. Fitness*, 39, 179, 1999.
369. Sullivan, J.C., McGuine, T.M. and Bernhardt, D.A., The benefits and risks of creatine supplementation as perceived by high school athletes, *Med. Sci. Sports Exerc.*, 33, S203, 2001.

370. Yu, P.H. and Deng, Y., Potential cytotoxic effect of chronic administration of creatine, a nutrition supplement to augment athletic performance, *Med. Hypotheses*, 54, 726, 2000.
371. Almada, A.L., Kreider, R., Melton, C., Rasmussen, C., Lundberg, J., Ransom, J., Greenwood, M., Stroud, T., Cantler, E., Milnor, P. and Fox, J., Long-term creatine supplementation does not affect markers of renal function, *J. Strength Cond. Res.*, 14, 259, 2000.
372. Bemben, M.G., Bemben, D.A., Loftiss, D.D., Knehans, A.W., Creatine supplementation during resistance training in college football athletes, *Med. Sci. Sports Exerc.*, 33, 1667, 2001.
373. Bermon, S., Venembre, P., Sachet, C., Valour, S., Dolisi, C., Effects of creatine monohydrate ingestion in sedentary and weight-trained older adults, *Acta Physiol. Scand.*, 164, 147, 1998.
374. Earnest, C.P., Almada, A.L. and Mitchell, T.L., Influence of chronic creatine supplementation on hepatorenal function, *FASEB J.*, 10, A790, 1996.
375. Eijinde, B.O. and Hespel, P., Short-term creatine supplementation does not alter the hormonal responses to resistance training, *Med. Sci. Sports Exerc.*, 33, 449, 2001.
376. Graham, R.E., Bell, R.W., Lambert, B. and Bailey, M., Effects of brief creatine monohydrate loading on responses to prolonged exercise in the heat, *Med. Sci. Sports Exerc.*, 34, S145, 2002.
377. Greenhaff, P., Renal dysfunction accompanying oral creatine supplements, *Lancet*, 352, 233, 1998.
378. Greenwood, M., Kreider, R., Rasmussen, C., Ransom, J., Melton, C., Stroud, T., Cantler, E. and Milnor, P., Creatine supplementation does not increase incidence of cramping or injury during college football training II, *J. Strength Cond. Res.*, 13, 425, 1999.
379. Greenwood, M., Farris, J., Kreider, R., Greenwood, L. and Byars, A., Creatine supplementation patterns and perceived effects in select division I collegiate athletes, *Clin. J. Sport Med.*, 10, 191, 2000.
380. Greenwood, M., Kreider, R., Melton, C., Rasmussen, C., Lundberg, J., Stroud, T., Cantler, E., Milnor, P. and Almada, A.L., Short- and long-term creatine supplementation does not affect hematological markers of health, *J. Strength Cond. Res.*, 14, 362, 2000.
381. Greenwood, M., Greenwood, L.D., Kreider, R. and Byars, A., Effects of creatine supplementation on the incidence of cramping/injury during college football three a day training, *Med. Sci. Sports Exerc.*, 32, S136, 2000.
382. Greenwood, M., Greenwood, L.D., Kreider, R. and Byars, A., Effects of creatine supplementation on the incidence of cramping/injury during college fall baseball, *Med. Sci. Sports Exerc.*, 32, S136, 2000.
383. Greenwood, L., Greenwood, M., Kreider, R., Earnest, C.P., Brown, L.E., Farris, J. and Byars, A., Effects of creatine supplementation on the incidence of cramping/injury during eighteen weeks of division I football training/competition, *Med. Sci. Sports Exerc.*, 34, S146, 2002.
384. Greenwood, M., Kreider, R., Greenwood, L., Earnest, C.P., Farris, J., Brown, L.E., Comeau, M. and Byars, A., Effects of creatine supplementation on the incidence of cramping/injury during eighteen weeks of collegiate baseball training/competition, *Med. Sci. Sports Exerc.*, 34, S146, 2002.

385. Greenwood, M., Kreider, R.B., Melton, C., Rasmussen, C., Lancaster, S., Cantler, E., Milnor, P. and Almada, A., Creatine supplementation during college football training does not increase the incidence of cramping or injury, *Mol. Cell. Biochem.*, 244, 83, 2003.
386. Heinanen, K., Nanto-Salonen, K., Komu, M., Erkintalo, M., Alanen, A., Heinonen, O.J., Pulkki, K., Nikoskelainen, E., Sipila, I. and Simell, O. Creatine corrects muscle 31P spectrum in gyrate atrophy with hyperornithinaemia, *Eur. J. Clin. Invest.*, 29, 1060, 1999.
387. Hulver, M.W., Campbell, A., Haff, G., Schroeder, C., Comeau, M. and Potteiger, J.A., The effects of creatine supplementation on total body fluids, performance and muscle cramping during exercise, *Med. Sci. Sports Exerc.*, 32, S133, 2000.
388. Hunt, J., Kreider, R., Melton, C., Rasmussen, C., Stroud, T., Cantler, E. and Milnor, P., Creatine does not increase incidence of cramping or injury during pre-season college football training II, *Med. Sci. Sports Exerc.*, 31, S355, 1999.
389. Incledon, T. and Kreider, R.B., Creatine alpha-ketoglutarate is experimentally unproven, *J. Sports Med. Phys. Fitness*, 40, 373, 2000.
390. Kreider, R., Rasmussen, C., Ransom, J. and Almada, A.L., Effects of creatine supplementation during training on the incidence of muscle cramping, injuries and GI distress, *J. Strength Cond. Res.*, 12, 275, 1998.
391. Kreider, R., Ransom, J., Rasmussen, C., Hunt, J., Melton C., Stroud, T., Cantler, E. and Milnor, P., Creatine supplementation during pre-season football training does not affect markers of renal function, *FASEB J.*, 13, A543, 1999.
392. Kreider, R., Melton C., Ransom, J., Rasmussen, C., Stroud, T., Cantler, E., Greenwood, M. and Milnor, P., Creatine supplementation does not increase incidence of cramping or injury during college football training, *J. Strength Cond. Res.*, 13, 428, 1999.
393. Kreider, R., Melton C., Hunt, J., Rasmussen, C., Ransom, J., Stroud, T., Cantler, E. and Milnor, P., Creatine supplementation does not increase incidence of cramping or injury during college football training II, *Med. Sci. Sports Exerc.*, 31, S355, 1999.
394. Kreider, R., Melton, C., Rasmussen, C., Greenwood, E., Cantler, E., Milnor, P. and Almada, A., Effects of long-term creatine supplementation on renal function and muscle & liver enzyme efflux, *Med. Sci. Sports Exerc.*, 33, S207, 2001.
395. Kreider, R., Greenwood, M., Melton, C., Rasmussen, C., Cantler, E., Milner, P. and Almada, A., Creatine supplementation during training/competition does not increase perceptions of fatigue or adversely affect health status, *Med. Sci. Sports Exerc.*, 34, S146, 2002.
396. Kreider, R.B., Melton, C., Rasmussen, C.J., Greenwood, M., Lancaster, S., Cantler, E.C., Milnor, P., Almada, A.L., Long-term creatine supplementation does not significantly affect clinical markers of health in athletes, *Mol. Cell. Biochem.*, 244, 95, 2003.
397. Kuehl, K., Soehler, S. Kulacki, K., Goldberg, L., Elliot, D., Bennett, W. and Haddock, B., Effects of oral creatine monohydrate supplementation on renal function in adults, *Med. Sci,. Sports Exerc.*, 32, S168, 2000.
398. Maher, M.A., Selby, A.K., Bouffleur, K.A., Porcari, J.P. and Wilson, T., Creatine supplementation does not increase exercise-induced hyperthermia in rats, *FASEB J.*, 14, A320, 2000.
399. Mayhew, D.L., Mayhew, J.L. and Ware, J.S., Effects of long-term creatine supplementation on liver and kidney functions in American college football players, *Int. J. Sport Nutr. Exerc. Metab.*, 12, 453, 2002.

400. McArthur, P.D., Webster, M.J., Body, J.C., May, R.A., Eschbach, L.D., Eimer, A.J., Angelosoulos, T.J., Zoeller, R.F. and Krebs, G.V., Creatine supplementation and acute dehydration, *Med. Sci. Sports Exerc.*, 31, S263, 1999.
401. McDonough, P., Biggerstaff, K.D., Bearden, S.E., Bergen, S.N., Cheuvront, S.N., Moffatt, R.J. and Haymes, E.M., Creatine monohydrate and cycle ergometry: Effects on hydration status, *Med. Sci. Sports Exerc.*, 31, S122, 1999.
402. Mendel, R.W., Cheatham, C.C. and Sinning, W.E., Creatine supplementation effects on cardiovascular and thermal responses during exercise in the heat, *Med. Sci. Sports Exerc.*, 32, S195, 2000.
403. Mertschenk, B., Gloxhuber, C. and Wallimann, T., Gesundheitliche Bewertung von Kreatin als Nahrungsergänzungsmittel, *Deutsche Lebensmittel-Rundschau*, 97, 250, 2001.
404. Papadopoulos, C., Imamura, R. and Bandon, L.J., The effect of creatine supplementation on repeated bouts of high-intensity exercise in the heat, *Med. Sci. Sports Exerc.*, 33, S203, 2001.
405. Poortmans, J.R., Auquier, H., Renaut, V., Durussel, A. Saugy, M. and Brisson, G.R., Effect of short-term creatine supplementation on renal responses in men, *Eur. J. Appl. Physiol. Occup. Physiol.*, 76, 566, 1997.
406. Poortmans, J.R. and Francaux, M., Renal dysfunction accompanying oral creatine supplements, *Lancet*, 351, 1252, 1998.
407. Poortmans, J.R. and Francaux, M., Long-term creatine supplementation does not impair renal function in healthy athletes, *Med. Sci. Sports Exerc.*, 31, 1108, 1999.
408. Poortmans, J.R. and Francaux, M., Adverse effects of creatine supplementation. Fact or fiction? *Sports Med.*, 30, 155, 2000.
409. Powers, M.E., The effects of creatine supplementation on intracellular and extracellular water content, Ph.D. thesis, University of Virginia, 2000.
410. Rasmussen, C., Kreider, R., Ransom, J., Hunt, J., Melton, C., Stroud, T., Cantler, E. and Milnor, P., Creatine supplementation during pre-season football training does not affect fluid or electrolyte balance, *Med. Sci. Sports Exerc.*, 31, S299, 1999.
411. Rizzo, C.S., Glover, T.A. anderson, M.J., Edwards, J.E., Ingersoll, C.D. and Merrick, M.A., Creatine monohydrate has no effect on the incidence of skeletal muscle cramping induced through electrically assisted maximal voluntary contractions, *J. Athletic Training*, 34, S33, 1999.
412. Robinson, T.M., Sewell, D.A., Casey, A., Steenge, G. and Greenhaff, P.L., Dietary creatine supplementation does not affect some haematological indices, or indices of muscle damage and hepatic and renal function, *Br. J. Sports Med.*, 34, 284, 2000.
413. Rosene, J.M. and Whitman, S.A., Effects of short-term, high-dosage creatine supplementation on thermoregulation in females in a thermoneutral environment, *Med. Sci. Sports Exerc.*, 34, S146, 2002.
414. Schilling, B.K., Stone, M.H., Uter, A., Kearnery, J.T., Johnson, M., Coglianese, R., Smith, L., O'Bryant, H.S., Fry, A.C., Starks, M., Keith, R. and Stone, M.E., Creatine supplementation and health variables: A retrospective study, *Med. Sci. Sports Exerc.*, 33, 183, 2001.
415. Sheppard, H.L., Raichada S.M., Kouri, K.M., Stenson-Bar-Maor, L. and Branch, J.D., Use of creatine and other supplements by members of civilian and military health clubs: A cross-sectional survey, *Int. J. Sport Nutr. Exerc. Metab.*, 10, 245, 2000.

416. Sipila, I, Rapola, J., Simell, O. and Vannas, A., Supplementary creatine as a treatment for gyrate atrophy of the choroid and the retina, *N. Engl. J. Med.* 304, 867, 1981.
417. Smith, J. and Dahm, D.L., Creatine use among a select population of high school athletes, *Mayo Clin. Proc.*, 75, 1257, 2000.
418. Stephens, M.B. and Olsen, C., Ergogenic supplements and health risk behaviors, *J. Fam. Pract.*, 50, 696, 2001.
419. Tarnopolsky, M., Parise, G., Fu, M.H., Brose, A., Parshad, A., Speer, O. and Wallimann, T., Acute and moderate-term creatine monohydrate supplementation does not affect creatine transporter mRNA or protein content in either young or elderly humans, *Mol. Cell. Biochem.*, 244, 159, 2003.
420. Valtonen, M., Nanto-Salonen, K., Jaaskelainen, S., Heinanen, K., Alanen, A., Heinonen, O.J., Lundbom, N., Erkintalo, M. and Simell, O., Central nervous system involvement in gyrate atrophy of the choroid and retina with hyperornithinaemia, *I. Inherit. Metab. Dis.*, 22, 855, 1999.
421. Vannas-Sulonen, K., Sipila, I., Vannas, A., Simell, O. and Rapola, J., Gyrate atrophy of he choroid and retina. A five year follow-up of creatine supplementation. *Ophthalmology*, 92, 1719, 1985.
423. Volek, J.S., Duncan, N.D., Mazzetti, S.A., Putukian, M., Gomez, A.L. and Kraemer, W.J., No effect of heavy resistance training and creatine supplementation on blood lipids, *Int. J. Sport Nutr. Exerc. Metab.*, 10, 144, 2000.
424. Volek, J.S., Mazzetti, S.A., Farquhar, W.B., Barnes, B.R., Gomez, A.L. and Kraemer, W.L., Physiological responses to short-term exercise in the heat after creatine loading, *Med. Sci. Sport Exerc.*, 33, 1101, 2001.
425. Webster, M.J., Vogel, R.A., Erdmann, L.D. and Clark, R.D., Creatine supplementation: effect on exercise performance at two levels of acute dehydration, *Med. Sci. Sports Exerc.*, 31, S263, 1999.
426. Young, J.C., Young, R.E. and Young, C.J., The effect of creatine supplementation on glucose transport in rat skeletal muscle, *FASEB J.*, 14, A91, 2000.

Part VI

Summary

26

Summary and Implications — Nutritional Ergogenic Aids

Ira Wolinsky and Judy A. Driskell

CONTENTS

I. Introduction 507
II. Amino Acid Derivatives 508
III. Lipid Derivatives 512
IV. Other Substances in Foods not Classified as Essential ...513
V. Evaluation of Effectiveness 515
VI. Summary and Implications 516
References 517

I. Introduction

Many commercially available nutritional ergogenic aids are currently marketed and are being used to improve or supposedly improve athletic performance and help build muscle. Ergogenic aids have been utilized for centuries to enhance athletic performance, although research supporting their use for this purpose frequently has been lacking. Many of these ergogenic aids are nutritional in nature. Some of these nutritional ergogenic aids are beneficial, while others are not — and some may even be harmful. Athletes are continually looking for an "edge" so they can win competitions. Recreational athletes also frequently take various nutritional ergogenic aids.

Nutritional ergogenic aids are dietary supplements and, as such, are regulated by the U.S. Food and Drug Administration under the Dietary Supplement Health and Education Act.[1] Around $18 billion was spent in 2002 on dietary supplements in the United States, and much of this was for nutritional ergogenic aids.[2] Most of these supplements are purchased in familiar retail outlets.

This volume provides detailed information on the efficacy of nutritional ergogenic aids that are used extensively by athletes, professional or recreational, to enhance their physical performance or build muscle. The use of vitamins and minerals as ergogenic aids has been reviewed recently.[3-4] Several books summarize the importance of all nutrients in sport and exercise.[5-8]

Nutritional ergogenic aids are defined as being dietary manipulations. These dietary manipulations are alterations of food choices and the addition of macronutrients or micronutrients for specified uses in sports and exercise. This includes adding nutrients to drinking water to make a sports beverage. Most nutritional ergogenic aids are used for endurance exercise or for building muscles and losing body fat.

Nutritional ergogenic aids have an advantage over other food or dietary supplements in being able to make factual claims without making disease or drug claims, in that physical performance is not considered a disease or abnormal condition. The ability to craft claims from available substantiations remains the major means of communicating potential product benefits. Continued research on product quality, doping, safety, certification, industry credibility and efficacy of the nutritional ergogenic aids themselves is needed.

II. Amino Acid Derivatives

Arginine is a nonessential amino acid in healthy humans. It appears to be important in ammonia detoxification, hormone secretion, cardiovascular function and immune modulation. It has been suggested to be an immunonutrient important in physiological stress, in wellness as a dietary supplement and as an ergogenic aid in enhancing athletic performance. Research on arginine and its purported benefits expanded with recognition of its role as precursor for nitric acid. Arginine does appear to be beneficial in specific clinical conditions, but the use of arginine supplements as ergogenic aids is unsupported. Arginine appears to be safe for use in humans in doses up to 30 grams daily.

Aspartate is a nonessential amino acid formed by the transamination of oxaloacetate from the Krebs cycle. Supplemental aspartate has been theorized to have an anti-fatigue effect on skeletal muscle. The evidence for this is equivocal in both human and animal studies. Evidence exists that aspartate offers no benefit to the athlete performing sub-maximal exercise, nor does it exhibit any body composition altering properties. Large doses of aspartate have been shown to be detrimental to laboratory rats; however, no evidence of toxicity exists in humans. The most common dose of aspartate (magnesium/potassium salt) is 10 grams daily. Not enough evidence currently exists that aspartate has ergogenic effects.

About two decades ago, the **branched-chain amino acids** (BCAA) emerged as a potential ergogenic aid and currently enjoy much popularity.

BCAA are a combination of the essential amino acids leucine, isoleucine and valine. BCAA have been touted to improve physical performance by reducing fatigue that is associated with sustained moderately high-intensity activity. Also, BCAA supplementation may have a beneficial influence on immune status following strenuous exercise, thus possibly reducing the incidence of sickness in athletes. BCAA supplementation may influence plasma glutamine levels, thus affecting the immune response. The primary site of BCAA oxidation seems to be the skeletal muscle, where these amino acids may be involved in protein synthesis at the translational level. BCAA may influence the transport of tryptophan and thus have an effect on serotonin synthesis. Thus, BCAA may function as an ergogenic aid by inhibiting fatigue, enhancing protein synthesis and favorably influencing the immune system following strenuous exercise; however, the data are not yet considered conclusive.

Carnitine, a short-chain carboxylic acid containing nitrogen, is a cofactor for carnitine translocase and acylcarnitine transferases I and II, which function in the transfer of activated long-chain fatty acids from the cytoplasm to the mitochrondria. This translocation is critical for the synthesis of ATP via β-oxidation of fatty acids in skeletal muscle. Thus, carnitine potentially influences endurance performance. L-carnitine is synthesized from lysine and methionine in the human body. Dietary sources of carnitine include red meats and dairy products. Most of the research indicates that carnitine supplements are effective, chiefly in a primary or secondary carnitine deficiency. There seems to be little to no effect of carnitine supplementation on total muscle carnitine levels during acute exercise. The vast majority of published research indicates that taking supplemental carnitine does not enhance physical performance. There appears to be no toxicity and few side effects when carnitine is taken in doses as high as 15 grams daily.

Creatine is a very popular nutritional supplement among athletes. The body can synthesize creatine from arginine, glycine and methionine and appears to produce about 1 gram daily. Reportedly, creatine supplementation increases strength, enhances work performed during repetitive sets of muscle contraction, improves repetitive sprint performance and increases body mass/fat-free mass. Increasing dietary creatine boosts both total creatine and phosphocreatine concentrations in muscle. Anaerobic or intense exercise performance improves when muscle creatine levels are increased. The proposed mechanism for improvements in anaerobic and aerobic capacity resulting from creatine supplementation is the transferring of high-energy phosphate in the mitochondria and cytosol of cells. Less supportive evidence exists that creatine supplementation enhances exercise performance during moderate- to high-intensity prolonged exercise. Most of the published studies relate to short-term (less than 2 months) rather than long-term creatine supplementation. However, a number of studies have reported that long-term creatine supplementation may promote even greater gains in strength, sprint performance and fat-free mass. Creatine supplements may also be beneficial to patients with a variety of neuromuscular conditions. Twenty grams daily for 5 days is the usual creatine dosage, followed by ingesting 3–5 grams daily thereafter. Weight gain is a common side effect of creatine supplementation. Creatine supplementation seems to be a safe and effective nutritional strategy to enhance physical performance.

Gelatin is a protein that is found in many foods, including protein bars. It is derived from collagen. Gelatin is a common food additive. The pharmaceutical industry uses gelatin in making capsules, tablets and for microencapsulation. Gelatin is composed of about 33% glycine, 22% proline/hydroxyproline and 45% other amino acids. Some evidence indicates that hydrolyzed gelatin and hydrolyzed collagen may be beneficial in relieving symptoms of osteoarthritis, and stimulating change and enhancement of cartilage tissue. Preliminary findings indicate that certain types of gelatin may be beneficial to athletes with osteoarthritis. Gelatin is an incomplete protein; therefore, gelatin or collagen supplementation as a protein source is not recommended.

Glucosamine and **chondroitin sulfate** are publicized in the popular media as being capable of protecting joints and tendons and positively influencing osteoarthritis. Glucosamine is an amino monosaccharide found in cartilage, tendons and ligaments. Glucosamine is synthesized in the body from glucose and glutamic acid. It is available in supplemental form as glucosamine hydrochloride, glucosamine sulfate and N-acetyl-glucosamine. Chrondroitin sulfate, a glycoaminoglysan, found in articular cartilage, is available in supplemental form usually as mixtures of chrondroitin sulfate A and C. Both glucosamine and chondroitin sulfate have been reported to stimulate chondrocyte growth in animals, though only equivocal reports exist for chondroprotection in humans. Evidence exists that combined preparations of glucosamine and chondroitin sulfate may be synergistic. Both supplements can be safely consumed, with some individuals experiencing mild digestive problems; the usual dosage is 1500 mg glucosamine and 1200 mg chrondroitin sulfate. Preliminary evidence exists that glucosamine and chrondroitin sulfate may improve symptoms of osteoarthritis at least as effectively as authorized nonsteriodal anti-inflammatory drugs. Thus, it is too early to recommend use of glucosamine and chondroitin sulfate to athletes as nutritional ergogenic aids.

Glutamine, a popular supplement and nonessential amino acid, plays a major role in maintenance of a healthy immune system and body energy levels. Glutamine can become a conditionally essential amino acid under certain physiological conditions such as infection, stress and trauma. Then, exogenous sources, dietary or supplements, appear to be necessary to meet body needs. Glutamine supplements are available as L-glutamine, either as an individual amino acid or a protein supplement. Glutamine may be beneficial to endurance athletes because of its influence on the synthesis of glycogen in muscle, and it may decrease the incidence of infection secondary to overtraining and improve immune response; however, these findings are not considered conclusive. Glutamine supplements may prevent post-exercise fluctuations in plasma glutamine concentrations. Intakes up to 40 grams daily appear to have no adverse effects other than mild digestive discomfort in some individuals. Well controlled long- and short-term studies on the efficacy of glutamine supplementation are needed.

β-hydroxy-β-methylbutyrate (HMB), a leucine metabolite, has been shown to have a positive effect on muscle protein. HMB is synthesized in the body from the essential amino acid leucine. The benefits claimed for HMB include magnification of the strength and fat-free mass gains associated with resistance training, reduction of muscle damage occurring during intense exercise and enhancement of endurance performance indicators. HMB theoretically functions in improving cell membrane integrity by supplying substrate for cholesterol synthesis, which is needed by cells for synthesis of new cell membranes as well as the repair of damaged membranes in maintaining proper cell function and growth. Research findings indicate that HMB supplementation decreases the muscle damage that occurs from repeated bouts of intense resistance exercise, as well as improving endurance. HMB supplementation appears to be safe, particularly at the recommended adult dose of 3 grams daily. Rather strong evidence exists that HMB supplementation is beneficial to athletes.

Lysine, an essential amino acid, functions in general protein synthesis, especially bone collagen formation, and in carnitine synthesis. Most athletes likely have adequate dietary intakes of lysine. Lysine deficiency could impair the body's adaptive responses to both strength and endurance training and could also compromise immune functioning. The thinking behind lysine supplementation in sports and fitness training is to stimulate growth hormone release and to favorably alter anabolic actions that govern skeletal muscle hypertrophy and regulation of energy substrate utilization. However, it appears that supplementation with lysine or lysine/arginine mixtures has no significant effect on training outcomes. Even if such supplementation did raise growth hormone levels beyond the level attributable to exercise, there is no apparent added value of elevated growth hormone levels. Adverse effects of L-lysine supplementation have not been reported other than possibly increasing the toxicities of the drugs streptomycin and gentamycin. Lysine supplementation does not seem to have ergogenic benefits.

Ornithine, ornithine α-ketoglutarate (OKG), and taurine have been the focus of some research relative to their potential anabolic and anticatabolic properties. Research in trauma patients indicates that OKG has some anabolic and anticatabolic properties; however, similar results were not observed in athletes. Taurine may come from dietary sources or it can be synthesized by the body from methionine and cysteine. Improvement has been reported in congestive heart patients given taurine supplements. Endurance-trained athletes supplemented with taurine and a caffeine-containing drink reportedly had improvements in their maximal performances and those supplemented with taurine and glucurovolactone, improved their endurances; however, these results may not be due to taurine alone. Toxicity data are not available on ornithine, OKG, or taurine.

III. Lipid Derivatives

Conjugated linoleic acid (CLA) is a collective term for a series of linoleic isomers. CLA are found in meat and milk products from ruminant animals; commercially prepared dietary supplements are available. Beneficial effects of CLA on adiposity and lean mass have been reported in animals and humans. Hence, CLA have been of special interest to resistance-trained athletes and body builders. Human data are limited at the present time and additional research is warranted. In both animals and humans there have been no reports of adverse health incidences from CLA ingestion.

Medium-chain triglycerides (MCT), composed of glycerol and fatty acids with chain lengths of 6–12 carbons, have attracted attention as a nutritional ergogenic aid. Glycerol, a triglyceride component, has also been examined as a nutritional ergogenic aid. MCT are quickly absorbed and metabolized by the liver, therefore providing a more immediate energy source than long-chain triglycerides that are commonly found in foods. MCT ingestion before and during exercise seems to have no effect on muscle glycogen levels and little effect on the utilization of carbohydrates for energy formation, but positive effects have only been reported when MCT are consumed along with an adequate quantity of carbohydrates. Glycerol supplementation increases the body's fluid retention. Benefits in physical performance were reported in five studies when individuals were "hyperhydrated" with glycerol and water, but no effects were observed in five other studies. So, it is not clear whether MCT alone or glycerol alone improves physical performance. MCT ingestion up to 1 g/kg body weight appears to be safe; however, diabetics should avoid using this supplement because of its ketogenic effects. Glycerol is well tolerated at 1 g/kg body weight if ample water is also ingested.

Octacosanol, a long-carbon-chain alcohol, is the primary active component of wheat germ oil. Small quantities of octacosanol are available from dietary plant waxes, leaves, skins and seeds. Wheat germ oil, rather than octacosanol, is the form generally used as a dietary supplement. Research studies have indicated that octocosanol or wheat germ oil supplementation seems to be a promising lipid-lowering agent, especially in lowering cholesterol and platelet antiaggregation. Wheat germ oil ingestion is claimed to enhance physical performance and delay fatigue. Unfortunately, well controlled research studies on wheat germ oil as an ergogenic aid are not available. Few side effects have been reported from ingestion of wheat germ oil or octacosanol supplements.

IV. Other Substances in Foods not Classified as Essential

Buffers, whether sodium bicarbonate, citrate, or phosphate, elicit a physiological response by increasing the buffering capacity of blood and muscle, thus maintaining a contraction-conductive myocellular environment. This type of environment enables muscle fibers to contract at higher forces or for longer periods of time, allowing an athlete to enhance physical performance. For both sodium bicarbonate and sodium citrate ingestion, physical performance outcomes are mixed. Phosphate loading effects on exercise performance may occur due to increased capacity to store inorganic phosphate, permitting a greater quantity of creatine phosphate to be stored for use during short-duration exercise bursts, but more explication is needed to understand its effects. With current information at hand, it is clear that the effects of ingestion of sodium bicarbonate, citrate and phosphate buffers on performance are inconsistent and, if administered or dosed inappropriately, may actually hinder performance.

Caffeine, the most widely consumed psychoactive drug worldwide, is used by both professional and recreational athletes in an attempt to enhance exercise performance and delay fatigue. This substance may improve alertness, concentration, reaction time and energy levels. The primary actions of caffeine are to stimulate the central nervous system and to amplify lipolysis. *In vivo*, caffeine has been observed to increase circulating levels of catacholamines, especially adrenalin, both at rest and during exercise; it also increases plasma free fatty acid concentrations. Strong evidence exists that caffeine improves aerobic endurance. Ergogenic effects of caffeine are less frequently observed during short-term than long-term intensive exercise. Caffeine is widely used, not only for its perceived ergogenic efficacy but also because it is inexpensive, can be consumed in generally legal amounts, has few acute adverse health effects and is a socially acceptable practice. Its use with ephedrine, by all means, should be avoided.

Carotenoids are long-chain hydrocarbons, all terpenes, originating in plants that are yellow, orange or red in color. Some carotenoids can serve as precursors to vitamin A. Supplements, synthetic or natural, are available which contain β-carotene, lycopene, zeaxanthin, lutein, β-cryptoxanthin and α-carotene, individually or in various combinations. *In vivo*, carotenoids can act as antioxidants and, in high dietary amounts, they may assist the body in reducing damage resulting from high levels of oxidative stress due to increases in oxygen consumption during intense physical activity. No clear evidence exists of a beneficial role for antioxidants in exercise-induced lipid peroxidation, nor in their ability to improve physical performance. Although dietary recommendations are not available for the individual carotenoids, the Institute of Medicine[9] firmly supported the five or more servings of fruits and vegetables daily recommendation and indicates that the general population should not take β-carotene. Specific daily intakes for athletes have not been proposed. Several studies indicate that physically active people appear to consume more fruits and vegetables than individuals in the normal population.

Coenzyme Q_{10} (CoQ), a fat-soluble compound found in all cells, is also known as ubiquinone or ubidecarenone. CoQ_{10} is synthesized by the body. The crucial roles of CoQ_{10} in the body are as an antioxidant and as a redox electron carrier in the mitochondria that is coupled to energy transfer. CoQ_{10} has been used effectively as a complementary therapeutic agent in certain health disorders, including some cancers, cardiovascular disease and neurodegenerative diseases. There is some evidence of deficiency during intensive physical activity. Equivocal evidence has been reported about the ergogenic effect of CoQ_{10} supplements in athletic populations with regard to enhancement of physical performance or reducing exercise-induced oxidation. Usual daily dosages are 30–150 mg. Few adverse effects of CoQ_{10} supplementation have been reported.

Ginseng is a popular herbal remedy traditionally used by the Chinese. Asian (*Panax*) ginseng usage appears to positively influence exercise and mental performance measurements, with the evidence being stronger for its influence on the latter, which, in turn, affects exercise performance. To date, American ginseng usage has not been reported to influence exercise parameters. Cogent arguments for efficacy or lack of efficacy can be made depending on which studies or outcomes are emphasized. Overall, review of the literature on ergogenic effects of ginseng has yielded equivocal conclusions. There may be some key experimental factors that explain discrepancies in results on ginseng. No adverse side effects resulting from the usage of commercially prepared ginseng products have been reported, and one study showed that the use of an American ginseng product did not cause positive doping results. It may be that some, but not all, ginseng products are capable of favorably influencing physical performance in humans; strong evidence exists that Asian ginseng does favorably influence physical performance via its effect on mental performance. Asian ginseng products are currently slowly being replaced in the marketplace by ginseng-like herbs, and even less research is available on their efficacies.

α-lipoic acid, also known as thioctic acid, and its reduced form, dihydrolipoic acid, are unique and vital antioxidants. Their functions include free radical quenching, metal chelation, interactions with other antioxidants and modulation of gene expression. Experimental and clinical trials indicate potential usefulness of α-lipoic acid as a promoter of glucose utilization. Evidence in rodent models strongly suggests a beneficial effect of α-lipoic acid supplementation on protection against oxidative damage, but there is a lack of data examining similar hypothesis in humans. α-lipoic acid supplements appear to be well tolerated by humans with few reported side effects. Until convincing data are available, it is only speculative that administration of α-lipoic or dihydrolipoic acids enhances physical performance in humans.

Myo-inositol, an isomer of glucose, is sometimes incorrectly referred to as vitamin B_8. Pangamic acid, inappropriately called vitamin B_{15}, is composed of dimethylglycine and sorbitol. There is growing literature that may support clinical and therapeutic uses of myo-insitol or pangamic acid, but no evidence yet exists to support any role(s) as an ergogen.

Pyruvate and **dihydroxyacetone**, both 3-carbon carbohydrate metabolites and trioses, when used in combination in animals and humans have been reported to provide beneficial effects including reduction of body weight and fat, lowering of blood lipids and improved endurance during aerobic exercise. These studies often employed very high dosages of pyruvate. Side effects of pyruvate and dihydroxyacetone ingestion have been reported. Studies using modest or moderate dosages of pyruvate alone are needed. Data emanating from a very small number of laboratories on potential ergogenic properties of pyruvate/hydroxyacetone combinations are highly equivocal.

Tannins are intermediate- to high-molecular-weight polyphenols. In recent years, tea polyphenols have gained much recognition in the nutrition and health areas. There is evidence indicating that tea consumption can mitigate oxidative stress and protect against a number of diseases, including high blood pressure and cardiovascular disease. Its ability to relax individuals is well known. These observations suggest a role for tea and its high tannin content in improvement of physical performance. To date, however, this remains a *terra incognita* of nutrition science.

V. Evaluation of Effectiveness

Which nutritional ergogenic aids are efficacious? The chapter written by the well established sports nutrition group Turpin, Talbott, Feliciano and Bucci rated the efficacies of these nutritional ergogenic aids based primarily on a consensus of recent reviews, but also the weight of the available evidence. Of the nutritional ergogenic aids discussed in the current volume, these authors indicated that for caffeine the majority of studies supported improvement of long-term endurance-exercise performance, with carnitine and coenzyme Q_{10} showing improvement in this regard in individuals with cardiovascular diseases. The nutritional ergogenic aids covered in this volume in which the majority of studies support muscle mass and strength according to these authors were bicarbonates (as in buffers), creatine and HMB. The nutritional ergogenic aids covered in this volume in which the majority of studies support improvements in body fat loss according to these authors were CLA and caffeine/ephedrine combinations; other issues with ephedrine confound the use of these combinations. Neuromuscular control and reaction time seem to be positively influenced by Asian (*Panax*) ginseng extracts and octacosanol, according to these authors.

VI. Summary and Implications

Athletes of all types want to enhance their physical performance and frequently are willing to try almost anything that offers a promise of better performance. Some of the ergogenic aids are nutritional in nature. Professional, collegiate and recreational athletes, as well as some of the general public, are spending billions of dollars annually on nutritional ergogenic aids. In that most of the nutritional ergogenic aids are foods, food derivatives or body metabolites, the potential for serious side effects for most of these substances is relatively low, especially in healthy individuals.

Generally, the evaluations of Turpin, Talbott, Feliciano and Bucci (hereafter referred to as Turpin and coauthors), contributors of the chapter on evaluation of benefits and possible risks of nutritional ergogenic aids (Chapter 25), were similar to those of the contributor(s) of the chapter on a specific ergogenic aid. In agreement with Turpin and coauthors, the chapter contributors for creatine (Kreider, Leutholtz and Greenwood) and HMB (Nissen) indicated that the usage of each of these two nutritional ergogenic aids was beneficial with regard to muscle mass and strength. Nissen and Sharp,[10] in a meta-analysis utilizing peer-reviewed studies published on dietary supplements that met certain quality control criteria, also concluded that the use of creatine as well as HMB supplements enhanced one's muscle mass and strength. Magkos and Kavouras, the contributors of the chapter on caffeine, agreed with Turpin and coauthors that there was strong evidence that caffeine positively influenced aerobic endurance in humans. The use of Asian ginseng extracts affects mental performance, which, in turn, influences physical performance according to the contributors of the chapter on ginseng, as well as the contributors of the evaluation chapter. The conclusions of some of the chapter contributors regarding the efficacy of some of the nutritional ergogenic aids were occasionally different from that of Turpin and coauthors; however, many times, experts in a specific area deem evidence to be conclusive while others do not. Occasionally, different experts may have different interpretations of available data. Sometimes too much credence is given to poorly designed studies, and even good studies may use insensitive techniques that have limited ability to find statistical significance. Some reviewers look for consensus —or near consensus — before they "stick their necks out" and say that something has an effect.

More research, particularly long-term well-controlled research, is needed on the effects of nutritional erogenic aids on physical performance. There is insufficient well-conducted research on most of the nutritional ergogenic aids with regard to their efficacy and safety.

Individual authors or groups of authors, primarily experts in their areas, wrote chapters in this volume on the popular nutritional ergogenic aids. Those that were deemed by these experts to positively influence physical performance and to be safely used were creatine and HMB. Equivocal evidence seems to exist for several of these nutritional ergogenic aids with regard to their having a beneficial effect on physical performance. Insufficient evidence is available with regard to some of these nutritional ergogenic aids as to their having beneficial effects on physical performance in humans. Many of these nutritional ergogenic aids function as antioxidants, favorably influence immune functioning or have a beneficial effect on mood, and thus may be beneficial with regard to physical performance.

References

1. Dietary Supplement Health and Education Act of 1994, Public Law 103-417, *Federal Register*, 59, 4325, 1994.
2. National Nutritional Foods Association, *Facts and Stats*, http://www.nnfa.org/facts, accessed February 20, 2004.
3. Wolinsky, I. and Driskell, J.A. (Eds.), *Sports Nutrition: Vitamins and Trace Elements*, CRC Press, Boca Raton, FL, 1997, 235 p.
4. Driskell, J.A. and Wolinsky, I. (Eds.), *Macroelements, Water and Electrolytes in Sports Nutrition*, CRC Press, Boca Raton, FL, 1999, 256 p.
5. Driskell, J.A., *Sports Nutrition*, CRC Press, Boca Raton, FL, 2000, 280 p.
6. Wolinsky, I. (Ed.), *Nutrition in Exercise and Sport*, CRC Press, Boca Raton, FL, 1998, 685 p.
7. Driskell, J.A. and Wolinsky, I. (Eds.), *Nutritional Assessment of Athletes*, CRC Press, Boca Raton FL, 2002, 410 p.
8. Wildman, R.E.C. (Ed.), *Handbook of Nutraceuticals and Functional Foods*, CRC Press, Boca Raton, FL, 2001, 568 p.
9. Institute of Medicine, *Dietary Reference Intakes for Vitamin C, Vitamin E, Selenium and Carotenoids*, National Academy Press, Washington, DC, 2000, pp. 325–382.
10. Nissen, S.L. and Sharp, R.L., Effect of dietary supplementation on lean mass and strength gains with resistance exercise: A meta-analysis. *J. Appl. Physiol.*, 94, 651, 2003.

Index

A

AA, *see* Amino acid
Acetic acid, diluted, 226
Acid–base balance
    sodium bicarbonate loading on, 261
    sodium citrate loading on, 265
    stress on, 257
Acidosis, force-depleting effects of, 260
Actin cytoskeletal functioning, 433
Active energy releasing factor, octacosanol as, 249
Active lifestyle, HMB supplementation and, 164
Acute exercise, carnitine levels and, 68
Acylcarnitines, 66, 68, 71
Adenine nucleotide degradation, 87
Adenosine A_3 receptors, 284
Adenosine diphosphate (ADP), 81, 357
Adenosine receptors
    antagonism of, 282
    blockade of, 285, 311–312
Adenosine triphosphate (ATP), 42, 62, 81
    as body's energy broker, 228
    concentrations, pre-exercise, 83
    generation of, 62
    replenishment of, 85
    resynthesis of, 38
    synthesis, role of CoQ_{10} in mitochondrial, 357
ADP, *see* Adenosine diphosphate
Adriamycin, CoQ_{10} and, 366
Adverse-event questionnaires, 161
Advertising claims, FTC regulation of, 12
Aerobic capacity, increased, 151
Aerobic exercise, 133
Aerobic performance, improved, 311
AHPA, *see* American Herbal Products Association
AIDS
    DMG and, 438, 439
    MCT use and, 224
    patients, muscle-wasted, 159
    treatment of with CoQ_{10}, 356

Air-braked ergometer, 267
Alanine, 106
Alanine aminotransferase (ALT), 43
Alcohol consumption, 365
AllSport®, 6
ALT, *see* Alanine aminotransferase
Alzheimer's disease, 432
American College of Rheumatology Subcommittee on Osteoarthritis, 124
American College of Sports Medicine, 13, 342
American Dietetic Association, 342
American Herbal Products Association (AHPA), 13
Amino acid (AA), 173
    branched-chain, *see* Branched-chain amino acids
    classification of, 22
    combinations, marketing claims, 181
    concentrations in muscle, 175
    consumption of among athletes, 21
    cyclic, 105–106
    destruction of, 142
    dicarboxylic, 37
    dispensable, 139
    essential, 64, 75, 173, 510
    excitatory, 44
    HMB and, 159
    imbalance, 178
    increased release of, 51
    metabolism, changes in, 198
    neutral polar, 130
    nonessential, 38, 508
    oral ingestion of, 179
    preferential uptake of, 50
    somatotrophic, 42
    supplementation, overview of in sports and fitness, 172
Ammonia
    detoxification, arginine and, 508
    production, aspartate supplementation and, 41
Amyotrophic lateral sclerosis, 92
Anabolic androgen steroids, CoQ_{10} and, 366, 367

Anabolism, ingredients applied to, 15
Anaerobic exercise, 133, 469
Anaerobic performance, improved, 311
Androstenedione, 11
Anorexia, 26
Anterior pituitary (AP), 187
Antigelatin antibodies, 111
Antioxidant(s)
    carotenoids as, 341
    defense systems, 412
    dihydrolipoic acid and, 415
    effects of on gene expression, 416
    lipid-soluble, 359
    lung function and, 344
    nonspecific, 358
    novel, 369
    nutrients, 359
    water-soluble, 415
AP, *see* Anterior pituitary
Arginine, 21–35, 159, 508
    chemical structure and classification, 22
        classification, 22
        structure, 22
    clinical applications, 27–31
        hormone secretion, 27–28
        immune functions, 28–29
        physical activity, 29–31
    GH release and, 179
    metabolic functions, 22–24
        arginine, 22–23
        nitric oxide, 23–24
    nutrient assessment, 24–27
        body reserves, 24–25
        dietary supplementation, 25–27
        requirements, 25
        toxicity, 27
    oral ingestion of, 182
    synthesis of, 198
Arthritis, 92
Asian ginseng, 384, 389
Aspartate, 37–45, 508
    aminotransferase (AST), 43
        activity, neural tissue, 44
        production of, 43
    athletic uses, 43
    delivery agent, 43
    ergogen in animals, 38–40
    ergogen in humans, 41–42
    -glutamate transporter, 42
    link to liver function, 43
    malate-aspartate shuttle, 42–43
    role as nonessential amino acid, 38
Aspartic acid, 37, 38
AST, *see* Aspartate aminotransferase
Astaxanthin, industrial synthesis of, 337

Asthenia, 395–399
Atherosclerosis, 117
Athlete(s)
    causal, HMB supplementation and, 164
    consumption of amino acids among, 21
    CoQ_{10} supplementation of, 360
    high-performance, glutamine
        concentrations among, 137–138
    incidence of sickness in, 509
    mental performance in, 399
    Olympic, lasting fatigue of, 134
    popular nutritional supplement
        marketed to, 82
    potential uses of gelatin for, 111
    power-trained, 52
    resistance-trained, 88, 210
    risk of infection of, 133
    weight-consciousness of, 340
    well-trained, HMB and, 163
Athletic performance, lysine
    supplementation to enhance, 188
ATP, *see* Adenosine triphosphate
Attention deficit disorder, 432
Autism, DMG and, 441

B

Bacterial culture media, use of gelatin in, 108
Basal metabolic rate (BMR), 173
BCAA, *see* Branched-chain amino acids
BCKADH, *see* Branch Chain α-ketoacid
    Dehydrogenase
Bench press, 88
BIA, *see* Bioelectrical impedance analysis
Bicarbonate buffers, *see* Buffers
Biocell™ group, 110
Biodex Multi-Joint System B2000, 109
Bioelectrical impedance analysis (BIA), 212
Black tea, 460
Blood clotting factors, 24
Blood pressure, tea and, 458
Blood urea nitrogen, 163, 212–213
B-lymphocytes, 29
BMR, *see* Basal metabolic rate
Body composition, determination of, 449
Body fat
    foremost way to lose, 482
    loss, 469, 474
Body reserves
    arginine, 24, 25
    carotenoids, 337
    CoQ_{10}, 364
    ornithine, 198
    taurine, 202

Index

Body weight, reducing effect of CLA on, 214
Bone(s)
 chipping of, 116
 pathologies, human, 179
Branched-chain amino acids (BCAA), 47–59, 508
 absorption of, 50
 exercise-induced changes in immune status, 55–56
 exercise-induced fatigue and, 52–54
 food and supplementation of, 49–50
 muscle fiber hypertrophy and pathology in response to exercise, 56–57
 supplementation, 49, 172, 509
 tissue distribution and metabolism of, 50–52
Branch Chain α-ketoacid Dehydrogenase (BCKADH), 52
Breast cancer patients, arginine administered to, 26
Breast implants, capsular contracture in, 117
Brussels marathon, glutamine supplementation and, 140
Buffers, 257–273
 areas of future research, 268–269
 chronic buffer supplementation, 268
 ethical issue, 269
 hydrogens and muscle fatigue, 258–259
 lactate production and hydrogen accumulation, 259–261
 phosphate loading on physiology and performance, 267–268
 physiological response elicited by, 512
 recommendations for coaches and athletes, 269
 sodium bicarbonate loading on acid–base balance and performance, 261–265
  administration and dose, 261–264
  performance effects, 264–265
  potential mechanism, 261
 sodium citrate loading on acid–base balance and performance, 265–267
  administration and dose, 265–266
  combining bicarbonate and citrate, 267
  performance effects, 266–267
  potential mechanism, 265
Butyric acid, 227

C

Caffeine, 275–323, 513
 abstinence, 287
 abuse, 277
 caffeine and exercise performance, 296–311
  caffeine and ephedrine, 309–311
  incremental exercise, 309
  long-term submaximal exercise, 296–308
  short-term intense exercise, 308
 caffeine use in sports, 276–277
 cellular and molecular mechanisms of caffeine action *in vitro*, 280–285
  antagonism of adenosine receptors, 282–284
  calcium release, 281
  inhibition of glycogen phosphorylase, 282
  inhibition of phosphodiesterase, 281
  inhibition of phosphoinositide kinases, 285
  stimulation of Na^+/K^+-ATPase, 284
 -containing drink, 203
 –ephedrine treatments, 310
 -naive individuals, 286
 pharmacokinetics of, 277–280
  absorption, 277
  distribution, 278
  excretion, 279–280
  metabolism, 278
 physiological and metabolic effects of caffeine *in vivo*, 285–295
  catecholamine release, 286–287
  glycogen breakdown, 292–294
  lipolysis and fat oxidation, 287–292
  neuromuscular function, 294–295
 tolerance, 311
 withdrawal, 302
Calcium
 absorption, lysine and, 179
 phosphate ingestion, 268
 release, caffeine and, 281
Caloric restriction, 482
Camellia Sinensis, 456
Cancer
 DMG and, 438, 441
 herbal pharmaceutical under investigation for treatment of, 390–391
 inhibitor of, 461
 lycopene and lower risk of, 333
 myo-inositol and, 437
 patients, muscle-wasted, 159, 160
 tea and, 460
 therapy, glutamine supplements and, 142
 treatment of with CoQ_{10}, 356
Canthaxanthin, 337, 344

Carbohydrate(s)
  endurance performance and, 472
  indigestible, 225–226
  ingestion of during endurance activity, 52–53
  oxidation, 231, 232
  tannins and, 455
Carbomyl phosphate synthetase I, 22
Cardiomyopathy, treatment of with CoQ_{10}, 356
Cardiovascular disease
  herbal pharmaceutical under investigation for treatment of, 390–391
  tea and, 459
  treatment of with CoQ_{10}, 356
Cardiovascular function, arginine and, 508
Carnitine, 61–79, 509
  biosynthesis of, 65
  discovery of, 62
  history of, 62
  interactions during exercise, 65–66
    acylcarnitines, 66
    pyruvate dehydrogenase complex, 65–66
  metabolic actions of, 62–64
  metabolism, 64–65
    distribution and excretion of carnitine, 65
    synthesis of carnitine, 64
  rationale for carnitine supplementation, 66–67
    effectiveness of supplementation, 67
    reason for using carnitine, 66–67
  recommendations, 74–75
  supplementation and performance, 67–74
    body carnitine levels with supplementation, 67–68
    carnitine levels and acute exercise, 68–70
    effect of chronic training on carnitine status, 70–72
    summary of carnitine and training effects, 72
    supplementation and performance, 72–74
  synthesis of, 175
D-Carnitine, 63
L-Carnitine, 63, 74
β-Carotene, industrial synthesis of, 337
Carotenoids, 325–353, 513
  animal absorption of from food supply, 325
  apolar, 336
  body reserves, intake and athletic performance, 337–342
    epidemiological studies, 339–340
    human intervention studies, 340–342
    measurement of carotenoid levels in tissues and plasma, 337–339
  content of foods, 328–329
  -deficient diet, 345
  dietary sources of carotenoids, 334–337
    amounts in various foodstuffs, 334–335
    carotenoid supplements, 337
    cooking techniques and bioavailability, 335–337
  functions of carotenoids in human body, 327–334
    β-carotene and other provitamin A carotenoids, 327–331
    lutein and zeaxanthin, 331–332
    lycopene, 332–334
  recommended intake, 342–344
Cartilage
  amino monosaccharide found in, 510
  poor-quality, 116
  shark, 117
Catabolism
  exercise-induced, 182
  false signal of, 186–187
Cataracts, 332, 422
  gastrointestinal absorption of, 459
  LDL oxidation and, 457
Catecholamine(s)
  release, 286, 303
  role of, 305
CDC, *see* Centers for Disease Control and Prevention
Cell damage
  exercise-induced, 371
  minimized, 152
Cell death
  inhibition of, 355
  signaling, 433
Cellular signaling, function of CoQ_{10} in, 371
Center for Food Safety and Nutrition (CFSAN), 7
Centers for Disease Control and Prevention (CDC), 111
Central nervous system (CNS), 433, 435
Certification programs, 14
CFSAN, *see* Center for Food Safety and Nutrition
Chemotherapy, 26, 369
Cholesterol
  HDL, 26, 162, 251, 337
  LDL, 367, 368, 457, 460

Index
523

lowering of with octacosanol, 250, 252
synthesis pathway, 250
Chondroitin sulfate, 116, 117, 510, *see also* Glucosamine and chondroitin sulfate
Chondroprotection, 120
Chromoplasts, carotenes present as, 336
Chronic training, effect of on carnitine status, 70
Cigarette smoke, myo-inositol and, 437
Circumplex test of emotion, 162
Citrate buffers, *see* Buffers
Citrulline, synthesis of, 198
Citrus aurantium, 476
CK, *see* Creatine kinase
CLA, *see* Conjugated linoleic acid
Clarifying agent, use of gelatin as, 108
Clif Bar®, 6
^{13}C NMR spectroscopy, 293
CNS, *see* Central nervous system
Coenzyme Q_{10} (CoQ_{10}), 355–378, 513
  body reserves, 364–365
  chemical structure, 357
  deficiency, 360
  dietary and supplemental sources, 365–366
  functions, 357–364
    antioxidant, 358–360
    effects of supplementation on exercise performance, 360–364
    energy coupling in ATP synthesis, 357–358
    other functions, 360
  future research directions, 370–371
  interactions with other nutrients and drugs, 366–368
  recommendations for dietary supplementation, 368–369
Cola beverages, 276
Cold-FX®, 384
Collagen, 105
  pharmaceutical-grade, 109
  production, lysine and, 179
  recovery of gelatin from, 106
  synthesis, lysine and, 175
Colorectal adenoma, 343
Colorful vegetables, human preference for, 326–327
Competence factor, 131
Confectionary products, gelatin used in, 107
Congenital defect, 331
Congestive heart failure
  taurine supplementation and, 203, 204
  treatment of with CoQ_{10}, 356
Conjugated linoleic acid (CLA), 209–219, 512

body composition and, 474
content of in selected foods, 211
effects of on body composition, 213–215
effects of on energy intake and expenditure, 215–216
effects of on muscle mass and strength, 210–213
health properties of, 210
isomeric content of, 212
ConsumerLabs.com's Athletic Banned Substances Screening Program, 14
Controlled Substances Act, 11
Cooking
  techniques, carotenoid loss occurring in, 335
  use of healthy amount of fat in, 343
CoQ_{10}, *see* Coenzyme Q_{10}
Cordyceps sinensis, 380, 390
Coronary heart disease
  CoQ_{10} concentration and, 365
  DMG and, 441
  platelet aggregation and, 250
Council for Responsible Nutrition (CRN), 12
CPK, *see* Creatine phosphokinase
Cramping, 92, 233, 261
C-reactive protein, 331
Creatine, 81–104, 509
  effects of short and long-term creatine supplementation, 85–90
  HMB and, 157
  kinase (CK), 84, 363
  muscle creatine content and phosphocreatine resynthesis, 82–85
  phosphokinase (CPK), 154, 158, 165
  potential medical uses of creatine, 92
  research, criticism of, 85
  safety, 91–92, 478
  synthesis
    biochemical pathway for, 83
    deficiencies, 92
CRN, *see* Council for Responsible Nutrition
Crohn's disease, 117, 224
Cross-country skiing time, effects of caffeine on, 305
β-Cryptoxanthin, main dietary sources of, 335
Cycling, supramaximal, 310
Cysteine availability, 415
Cystic fibrosis, 331

D

DAG, *see* Diacylglycerol

DEA, see Drug Enforcement Agency
Dehydration, 91, 233–236, 238
Dehydroepiandrosterone (DHEA), 157
Delivery agent, aspartate as, 43
Depression, 432, 441
DEXA, see Dual energy x-ray absorptiometry
Dexamethasone, 437
DHEA, see Dehydroepiandrosterone
Diabetes
    cataract formation in, 422
    complications, prevention of, 422
    herbal pharmaceutical under investigation for treatment of, 390–391
    MCT supplementation and, 238
    treatment of with CoQ_{10}, 356
    Type II, 156
Diabetic neuropathy, 419
Diacylglycerol (DAG), 434
Diastolic dysfunction, treatment of with CoQ_{10}, 356
Dietary manipulations, major efforts, 4
Dietary supplement(s)
    definition of, 8
    industry
        economic competition, 4
        self-regulation, 12
    mislabeling, 9
Dietary Supplement Education Alliance (DSEA), 13, 15
Dietary Supplement Health and Education Act (DSHEA), 5, 117, 182, 476, 507
    FDA enforcement powers limited by, 9
    major intent of, 10
    pertinent issues with ergogenic aids, 8
    provision of, 9
    regulations, breach of, 9
Dieting, extreme, 340
Dietitians of Canada, 342
Dihydrolipoic acid
    complexation of metals by, 414
    regenerated oxidized antioxidants and, 415
Dihydroxyacetone, see Pyruvate and dihydroxyacetone
N,N-Dimethylglycine (DMG), 438, 441
Diphtheria-tetanus-acellular pertussis (DTP), 111
Disease
    prevention, 11
    -treatment claims, 11
DMG, see N,N-Dimethylglycine
DNA, oxidative damage to, 333, 345, 359, 365
Drug Enforcement Agency (DEA), 11

DSEA, see Dietary Supplement Education Alliance
DSHEA, see Dietary Supplement Health and Education Act
DTP, see Diphtheria-tetanus-acellular pertussis
Dual energy x-ray absorptiometry (DEXA), 213, 449
Duchannes Muscular Dystrophy, 92

E

EDRF, see Endothelium-derived relaxing factor
EFA, see Essential fatty acids
Elderly population, common cause of injury in, 156
Electrolytes, 472
Eleutherococcus senticosus, 380, 390, 391
Emotion, Circumplex test, 162
Endothelium-derived relaxing factor (EDRF), 23
Endurance
    activity, ingestion of carbohydrate during, 52–53
    caffeine-induced increase in, 304
    -cycling performance, 266
    exercise
        long-term, 469, 470, 471
        prolonging of, 72
    trial completion time, BCAA supplementation and, 54
Endurox®, 391
Energy
    bars, 5, 231
    broker, ATP as body's, 228
    coupling, ATP synthesis, 357
    drink, 276
    expenditure, role of CLA in, 216
    intake, effects of CLA on, 215
    source
        glycerol supplementation as, 222
        protein stores used as, 24
    supply, glutamine and, 134
    utilization of fat for, 391
Enforcement Policy Statement on Food Advertising, 12
Ephedrine, 276
    caffeine and, 309
    effect of on muscle endurance, 310
Epicatechins, 456
Epilepsy medications, anticonvulsant effects of, 142
Erogen, aspartate as, 38, 41

Ergogenic aids, hidden agendas surrounding, 3–4
Erythropoietic protoporphyria, 331
Essential fatty acids (EFA), 238
Excitatory neurotransmitters, synthesis of, 130
Exercise(s)
  acute, carnitine levels and, 68
  aerobic, 133
  amino acid release and, 51
  anaerobic, 133, 469
  -associated health, 474, 476
  bench-press, 311
  chondroitin sulfate and, 123
  duration, 72
  effects of on immune system, 133
  effects of wheat germ oil on humans during, 247
  endurance
    anaerobic metabolism testing, 44
    long-term, 469, 470, 471
    prolonging of, 72
  -to-exhaustion, animal studies, 39
  FFA concentrations during, 288
  fluid-electrolyte imbalances during, 296
  handgrip, 360
  incremental, caffeine and, 309
  -induced fatigue, 39, 52
  intensity, HMB supplementation and, 158
  metabolism, lipoic acid and, 421
  muscle glycogen catabolism during, 312
  NSAIDS vs., 124
  performance
    caffeine and, 296
    CoQ_{10} supplementation and, 360
    hyperhydration and, 222
    octacosanol and, 248
  prolonged, 133, 233–236
  resistance-type, 28
  short-term intense, caffeine and, 308
  sprint anaerobic, 340
  strenuous, free radicals and, 359
  sub-maximal, 508
  treadmill, lipoic acid and, 421
  type of, 302
  working skeletal muscle during sustained, 53
Exhaustion
  cycling to, 310
  muscle glycogen concentrations at, 450
  run time to, 395
  time to, caffeine and, 308
  voluntary, 264
Extracellular buffers, 260
Extreme dieting, 340

F

FAD, *see* Flavin adenine dinucleotide
Fair trade practices, 12
Fat-free mass (FFM), 88, 93, 148, 173, 213
Fatigue
  central, reduction of, 47
  exercise-induced, 39, 52
  lasting, 134
  muscle, hydrogens and, 258
  peripheral, 47, 52
  protein catabolism and, 132
  resistance, 89
Fat oxidation, lipolysis and, 287
Fat-soluble vitamins, 326
Fatty acid(s)
  absorption of, 227
  commonly found in nature, 224
  metabolism, abnormal, 64
  partitioning of, 225
FDA, *see* Food and Drug Administration
Federal Trade Commission (FTC), 11, 12
FFA, *see* Free fatty acid
FFM, *see* Fat-free mass
Fitness enthusiast, HMB supplementation and, 164
Flavin adenine dinucleotide (FAD), 358
Flicker-photometry, 339
Fluid-electrolyte imbalances, 296
Fluvastatin, 251
Food
  choices, alteration of, 4, 5
  industry, applications of gelatin in, 105, 108
  -intake interviews, 327
Food and Drug Administration (FDA), 7
  authority, 8
  enforcement powers, 9
  final ruling on structure/function claims, 10
  good manufacturing processes, 9
  Health Claims approved by, 7
Food, Drug and Cosmetic Act, 7
Fortifying health booster, claim of, 433, 437
Free fatty acid (FFA), 39, 287
  concentrations
    during exercise, 288, 289
    effects of caffeine on, 289–291
  uptake, muscle, 292
  use of carnitine to transport, 41
Free radical(s), 358–359
  damage, transition mineral elements and, 413
  production, 341
  quenching, 413, 415–416

signaling processes and, 359
strenuous exercise and production of, 359
FTC, *see* Federal Trade Commission

G

GABA, *see* Gamma-aminobutyric acid
GAGs, *see* Glycosaminoglycans
Gamma-aminobutyric acid (GABA), 179, 434, 458
Gamma hydroxybutyrate (GHB), 477
Gas chromatography (GC), 416–417
Gastric irritation, 250
Gastrointestinal cramping, 233
Gastrointestinal distress, triose dosages and, 451
Gatorade®, 6
GC, *see* Gas chromatography
GC mass spectroscopy (GC-MS), 416–417
GC-MS, *see* GC mass spectroscopy
Gelatin, 105–113, 510
    clinical trials in bone and joint disease, 109–110
    composition of, 106
    delivery systems, 112
    financials of worldwide use, 108
    gelatin production, 106–107
    history of human consumption, 107–108
    kosher, 106
    manufacture of, 105
    potential anaphylactic reaction to gelatin, 111
    potential uses for athlete, 111
    uses in food and pharmaceutical industry, 108–109
GelFilm™, 108
Gene expression
    CoQ$_{10}$ and, 371
    effects of antioxidants on, 416
    modulation of, 413
Generally Recognized as Safe (GRAS), 231
Geriatric Pharmaton®, 384
Gericomplex®, 384
Get-up-and-go performance time, 160
Get-up-and-go (GUG) test, 156
GH, *see* Growth hormone
GHB, *see* Gamma hydroxybutyrate
GHRH, *see* Growth hormone releasing hormone
Gincosan®, 384
Ginkgo biloba, 384
Ginkoba®, 384
Ginsana G115®, 399
Ginseng, 379–410, 514
    alternatives, 379, 390
    Chinese literature on medical uses of, 380
    human studies on mental performance related to, 385–388
    mental effects, 384
    other adaptogenic herbs, 390–399
        cordyceps, 390–391
        eleutherococcus, 391–395
        Rhodiola species, 395–399
    physical performance, 380–384
    unresolved issues and guide for future research, 399–400
    variables delineating conditions of efficacy, 384–390
Gluconeogenesis, precursor for, 138
Glucosamine, 116, 510
    production of, 118
    side effects, 122
Glucosamine and chondroitin sulfate, 115–127
    description of products, 117–118
        chondroitin sulfate, 117–118
        glucosamine, 117
    mechanisms, 118–119
        chondroitin sulfate, 118–119
        glucosamine, 118
    recommendations, 124
    review of research studies and clinical trials, 119–122
        chondroitin sulfate, 120–121
        combined glucosamine and chondroitin sulfate, 121–122
        glucosamine, 119–120
    side effects, 122–123
        chondroitin sulfate, 123
        glucosamine, 122
    use in sport and exercise, 123
Glucose
    ingestion, HMB absorption and, 150
    intolerance, 92
Glutamine, 129–145, 159, 510
    absorption of, 137
    acute ingestion of, 141
    body stores and regulation, 135–136
        liver and kidney, 135–136
        skeletal muscle, 135
    chemical structure, 130
    dietary intake, 136–137
        digestion and absorption, 137
        food sources, 136
        supplemental sources, 136–137
    drug–nutrient interactions, 141–142
    ergogenic effects, 139–141
        endurance activities, 139–140
        nonendurance activities, 140–141

Index

first isolation of, 129
food sources of, 136
immune system-enhancing effects of, 142
long-term ingestion of, 141
metabolic functions, 130–134
    energy supply, 134
    glucose regulation, 132
    immune regulation, 132–134
    protein synthesis and degradation, 131–132
nutritional status assessment, 137–139
physiological functions of, 131
reduction, postexercise, 55
safety and toxicity, 142
supplements, 136, 137
Glycerol, *see also* Medium-chain triglycerides and glycerol
    ingestion, 236, 238
    supplementation, metabolic effects of, 237
Glycogen
    breakdown, 292
    metabolism, effects of caffeine on, 293
    phosphorylase, inhibition of, 282
    sparing, 72
    synthesis, stimulation of, 135
    utilization, rate of, 450
Glycosaminoglycans (GAGs), 118, 119
GMPs, *see* Good manufacturing processes
GNC health food store, 181
Golden Age Games, 340
Gold's Gym, 182
Good Clinical Practices, 400
Good manufacturing processes (GMPs), 9
G protein-coupled receptors, 282
GRAS, *see* Generally Recognized as Safe
Great Earth health food store, 181
Green bone, 106
Green tea, 457, 460
Growth hormone (GH), 172
    concentration, increase in, 183
    -deficient subjects, 187
    inhibitor, 181
    release
        enhancement of, 179
        GABA influence on, 180
    releasing hormone (GHRH), 180
    secretion
        diagnostic test for, 178
        high protein intakes and, 187
Growth-retarded children, 200
GUG test, *see* Get-up-and-go test

H

Handgrip exercise, CoQ_{10} and, 360
HDL, *see* High-density lipoprotein
Health claims, 11
Heart rate recovery post-exercise, 395
Hemoglobin, 260
Hepatocytes, synthesis of arginine via, 23
Herbal extracts, ephedrine-containing, 478
Herpes simplex virus, treatment of, 172, 188
High-density lipoprotein (HDL), 26, 162, 251, 337
High-performance athletes, glutamine concentrations among, 137–138
High-performance liquid chromatography (HPLC), 416–417
HIV, MCT use and, 224
HMB, *see* β-Hydroxy-b-methylbutyrate
Hormone secretion, 27, 508
HPLC, *see* High-performance liquid chromatography
Human growth hormone
    release, enhancement of, 27
    secretion, increased, 28
Huntington's disease, 92, 368
Hydrogen accumulation, lactate production and, 259
Hydroxyl radical scavenging assay, 413–414
β-Hydroxy-b-methylbutyrate (HMB), 48, 89, 147–169, 511
    absorption, 150
    applications, 152–160
        endurance, 157–158
        muscle damage, 158–159
        resistance training, 152–157
        reversing unwanted muscle loss, 159–160
        wound healing, 160
    butyric acid and, 227
    dietary and supplemental sources, 150–151
    effect of on emotion, 162
    effect of on muscle strength, 472
    endogenous production, 148–149
    fate, 149–150
    future research, 164–165
    mechanism of action, 151–152
    recommendations, 163–164
        active lifestyle, 164
        casual athlete and fitness enthusiast, 164
        well trained athletes, 163
    safety, 161–163
        higher dosages, 162–163
        recommended dosages, 161–162
Hypertension, treatment of with CoQ_{10}, 356

I

IGF-I, *see* Insulin-like growth factor-I
Illness, downtime from, 482
Immune deficiencies, herbal pharmaceutical under investigation for treatment of, 390–391
Immune function
    performance and, 475
    T-cell-mediated, 29
Immune modulation, arginine and, 508
Immune responses, effect of arginine on, 28
Immune suppression, chemotherapy-induced, 26
Immune system
    effects of exercise on, 133
    glutamine and, 142
    lysine deficiency and, 189
Immuno-nutrient, 133
Increase in NO theory, 180
Inflammatory disease, herbal pharmaceutical under investigation for treatment of, 390–391
Ingestible supplement, least expensive, 264
Inositol retailers, 433
Institute of Medicine, Food and Nutrition Board, 175
Insulin
    -like growth factor-I (IGF-1), 182, 200
    receptors, 203
    resistance, 156, 214
International Olympic Committee (IOC), 13, 280, 384
Intraocular dehydration, 238
IOC, *see* International Olympic Committee
Ischemia heart disease
    patients, reduced blood viscosity in, 367
    treatment of with CoQ_{10}, 356
Isoleucine, physical performance improved by, 47

J

Jet lag, 474–475
Joint
    injuries, noncontact, 92
    pain, chronic, 115
    space narrowing, 120

K

Kashruth religious law, gelatin and, 106
α-Ketoisocaproate (KIC), 148, 149
Ketone bodies, 227, 228, 238
KIC, *see* α-Ketoisocaproate
Kidney disease, glutamine supplements and, 142
Killer-cell activity, lymphokine-activated, 133
Knee extension strength, 85
Knox NutraJoint™, 109
Krebs cycle, 73, 228, 413
    acetyl-CoA delivered to, 66
    generation of ATP via, 62
    intermediate, 41, 445

L

Labile protein stores, 25
Lactate
    dehydrogenase (LDH), 158, 363
    disposal of, 260
    production, hydrogen accumulation and, 259
    threshold (LT), 39, 69, 86, 360
Lactic acidosis, congenital, 439
LAK cells, *see* Lymphokine-activated killer cells
Lauric acid, 222
Laws, food regulation, 7
LCFA, *see* Long-chain fatty acids
LCT, *see* Long-chain triglycerides
LDH, *see* Lactate dehydrogenase
LDL, *see* Low-density lipoprotein
Lean mass
    effects of CLA on, 210
    gain, 155
Lean-tissue wasting, 177
Leg noradrenaline spillover, 287
Legumes, 188
Leptin infusion, 214
Leucine
    physical performance improved by, 47
    turnover studies, 149
Leukotriene, synthesis of, 132
Linoleic acid, *see* Conjugated linoleic acid
Lipid
    derivatives, 512
    digestion, 223, 225
    peroxidation, 368
        copper-induced, 414
        exercise-induced, 364
        prevention of, 458
    -soluble antioxidant, 359
Lipoic acid, 411–429
    dietary sources, 416–419
        absorption and bioavailability, 417–419

Index

analytical methods, 416–417
animal and plant sources, 417
α-lipoic acid, 412–416, 514
   antioxidant functions, 413–416
   chemical structure and biochemical functions, 412–413
   recommendations for future research, 423–424
   tolerance and safety, 423
   uses of supplemental α-lipoic acid, 419–423
      exercise metabolism, 421–422
      glucose disposal, 419–421
      prevention of diabetes complications, 422–423
α-Lipoic Acid Diabetic Neuropathy, 423
Lipoyllysine, 417, 418
Liver
   damage, alcohol-induced, 343
   disease, glutamine supplements and, 142
   inhibition of glycogen phosphorylase enzymes in, 311
   oxidation of MCFA in, 221–222
   radioactivity in, 251, 252
   reversal of electron transport in, 449
LMH, *see* L-Lysine monohydrochloride salt
Long-chain fatty acids (LCFA), 221
   delivery of, 225
   enzymes responsible for transport of, 75
   transport of, 66
   triglycerides composed of, 223
Long-chain triglycerides (LCT), 221
Lou Gehrig's Disease, 92
Lovastatin, 367
Low-density lipoprotein (LDL), 162, 251, 331
   CoQ_{10} and, 367
   oxidation
      catechins and, 457
      inhibition of, 460
   protection of from lipid peroxidation, 368
LT, *see* Lactate threshold
Lung
   function, antioxidants and, 342
   tumor formation, myo-inositol and, 437, 438
Lutein, 331
   cataract development and, 332
   documented action of, 332
   foods rich in, 335
Lycopene, 326, 332
   food sources of, 334
   industrial synthesis of, 337
   lower risk of cancer and, 333
   plants rich in, 327

Lymphocyte proliferation, nitrogen-stimulated, 134
Lymphokine-activated killer (LAK) cells, 55
Lysine, 64, 75, 171–196, 511
   amino acid supplementation in sports and fitness, 172
   chemical structure and metabolism, 173–178
      clinical applications, 176–177
      dietary and supplemental sources, 175–176
      interactions with other nutrients and drugs, 178
      nutrient assessment, 177–178
      toxicity and precautions, 178
   deficiency, 177, 189
   oral ingestion of, 182
   proposed sports and fitness applications, 172–173
   RDA values for, 188
   use for sports and fitness, 178–188
      current arginine and lysine use in sport and fitness, 182
      future research, 188
      lysine sports- and fitness-related research, 182–188
      marketing claims, 181–182
      rationale for including lysine in amino acid mixtures to raise growth hormone, 179–181
      sports origin, 178–179
L-Lysine monohydrochloride salt (LMH), 176

M

Macronutrient manipulations, primary benefits of, 5
Macular pigment, 332
Magnetic resonance spectroscopy (MRS), 87
Malate-aspartate shuttle, 39, 40, 42
MAOD test, *see* Maximal oxygen deficit test
Marathon training program, sedentary individuals completing, 71
Marketing claims, amino acid combinations, 181
Maximal oxygen deficit (MAOD) test, 310
Maximal voluntary contractions (MVC), 85
MCA, *see* Methylcrotenoic acid
MCFA, *see* Medium-chain fatty acids
MCT, *see* Medium-chain triglycerides
Measles, mumps, rubella (MMR), 111
Medium-chain fatty acids (MCFA), 221
   proportion of in milk fat, 229
   rapid oxidation of in liver, 221–222

triglycerides composed of, 223
Medium-chain triglycerides (MCT), 221, 231, 512
   FDA classification of, 231
   first development of, 229
   hydrolysis, 225
   ingestion, 232
   oil, 229
   water-solubility of, 224
Medium-chain triglycerides and glycerol, 221–246
   dietary and supplemental sources, 229–231
   ergogenic benefits, 231–236
      glycerol, 233–236
      medium-chain triglycerides, 231–233
   metabolism, 222–229
      chemical and physical properties, 222–223
      fatty acid and glycerol utilization, 227–229
      lipid digestion and absorption, 223–227
   recommendations, 238–239
   toxicity, 236–238
Mental performance, 399, 469
Metabolic alkalosis, 258
Metabolife®, 6
Metanephrine, urinary excretion of, 286
Methionine, 64, 75, 201
Methylcrotenoic acid (MCA), 149
Methylxanthines, 284, 448
MHC, see Myosin heavy chain
Migraines, 117
Milk fat, proportion of MCFA in, 229
MMR, see Measles, mumps, rubella
Momordica cochinchinensis, 337
Monarthritis, 112
Mood, improved, 163
MRS, see Magnetic resonance spectroscopy
Muscle
   active, glycerol fluxes across, 292
   amino acid concentration in, 175
   atrophy, rates of, 90
   cardiac, 75
   carnitine, 68, 69
   catabolism, 200
   cells, CPK leakage from, 165
   compound action potentials, 295
   creatine content, 82
   damage, 158, 159
      antioxidant effects and, 364
      minimized, 152, 163
   endurance, effects of caffeine and ephedrine on, 310
   energy metabolism, improved, 370
   extracts, discovery of carnitine in, 62
   fatigue, 64, 258
   fatty acids taken up by, 227
   FFA uptake, 292
   fiber(s)
      diameter, increased, 90
      hypertrophy, 56
      slow-twitch, 365
   glucose uptake, altered, 450
   glycogen, 66
      breakdown, 293
      catabolism, during exercise, 312
      sparing of, 231
   loss, reversing unwanted, 159
   magazines, 15
   mass, 472
      CLA effects on, 210
      effects of HMB on, 154
      mechanism of action of HMB on, 151
      role of CLA to alter, 212
   protein
      breakdown, decreasing, 152
      HMB and, 148
      turnover, favorable changes in, 56
   proteolysis, 151
   resting, 293
   skeletal, glutamine and, 135
   soreness, 475
   strength, 472
   tightness, 91
   wasting, 90
   weight, loss of, 200
Muscular dystrophy, treatment of with CoQ_{10}, 356
MVC, see Maximal voluntary contractions
Myocardial infarction risk, carotenoid levels and, 333
Myo-inositol and pangamic acid, 431–444
   myo-inositol, 432–438
      chemical structure and purification or synthesis, 433
      current research on myo-inositol, 435–437
      dietary sources, 435
      fortifying health booster, 437–438
      metabolism, 433–435
   pangamic acid, 438–441
      chemical structure and purification or synthesis, 439–440
      dosage, 440
      metabolic enhancer, 440–441
      sources, 440
Myosin heavy chain (MHC), 90
Myristic acid, 222, 229

Index

N

NAD, see Nicotinamide adenine dinucleotide
Na^+/K^+-ATPase, stimulation of, 284
National Center for Complementary and Alternative Medicine (NCCAM), 124
National Institute of Arthritis and Musculoskeletal and Skin Diseases (NIAMS), 124
National Institutes of Health (NIH), 7
  Office of Alternative Medicine, 7
  Office of Dietary Supplements, 7
National Nutritional Foods Association (NNFA), 12
National Sanitation Foundation (NSF)
Natural killer cells, 29
NCCAM, see National Center for Complementary and Alternative Medicine
NDI submissions, see New dietary ingredient submissions
Nerve damage, 432
Neurodegenerative diseases, treatment of with CoQ_{10}, 356
Neuropathy impairment score, 423
Neurotransmitter(s)
  inhibitor, 458
  proteins and, 24
  synthesis of excitatory, 130
Neutrophils, 29, 134
New dietary ingredient (NDI) submissions, 9
NIAMS, see National Institute of Arthritis and Musculoskeletal and Skin Diseases
Nicotinamide adenine dinucleotide (NAD), 358
NIH, see National Institutes of Health
Nitric oxide synthase (NOS), 24, 180
Nitrogen transport, inter-organ, 130
NLEA, see Nutrition Labeling Education Act of 1990
NNFA, see National Nutritional Foods Association
Nonessential amino acid, role of aspartate as, 38
Nonsteroidal anti-inflammatory drugs (NSAIDS), 110, 116, 120
  alternatives to, 121
  exercise vs., 124
Norandrostenediones, 11
Normetanephrine urinary excretion of, 286
NOS, see Nitric oxide synthase
NSAIDS, see Nonsteroidal anti-inflammatory drugs
NSF, see National Sanitation Foundation
Nuclear protein biosynthesis, arginine and, 22
Nursing women, chondroitin sulfate and, 123
Nutrient(s)
  analysis databases, 202
  classes of, 470
  deficiency diseases, 10
Nutritional ergogenic aids, introduction, definitions and regulatory issues, 1–17
  regulatory aspects, 6–14
    certification programs, 14
    dietary supplement industry self-regulation, 12–13
    laws regulating foods and dietary supplements, 7–12
    sports governing agencies, 13–14
  sports nutrition marketplace update, 5–6
  working definitions, 4–5
Nutritional ergogenic aids, summary and implications, 507–517
  amino acid derivatives, 508–511
  evaluation of effectiveness, 515
  implications, 515–516
  lipid derivatives, 512
  other substances in foods not classified as essential, 512–515
Nutritional ergogenic aids, systematic and critical evaluation of benefits and possible risks of, 469–504
  body fat loss, 474
  current product trends, 476–477
  long-term endurance exercise performance, 470–472
  muscle mass and strength, 472–473
  possible risks and safety issues, 477–481
  sports skills and exercise-associated health, 474–476
Nutritional supplements, comparison of HMB with, 157
Nutrition Labeling Education Act of 1990 (NLEA), 7

O

OA, see Osteoarthritis
OAA, see Oxaloacetate
OAM, see Office of Alternative Medicine
Obese population, changes in body fat in, 449
Obesity, animal models of, 210
Obsessive-compulsive disorder (OCD), 432, 436
OCD, see Obsessive-compulsive disorder

Octacosanol, *see* Wheat germ oil and octacosanol
ODS, *see* Office of Dietary Supplements
Office of Alternative Medicine (OAM), 7
Office of Dietary Supplements (ODS), 7
OKG, *see* Ornithine alpha-ketoglutarate
Older adults, HMB supplementation of, 155
Oleoresin-containing carotenoids, 337
Olympic athletes, lasting fatigue of, 134
Olympic Triathlon, 55, 56
Oolong tea, manufacturing of, 457
Ornithine alpha-ketoglutarate (OKG), 197, 199, 511
Ornithine, ornithine alpha-ketoglutarate and taurine, 197–206
    metabolic function, 198–200, 201–202
        body reserves, 198–199, 202
        interaction with other nutrients or drugs, 200, 202
        nutrient assessment, 199–200, 202
        toxicity, 200, 202
    ornithine and ornithine alpha-ketoglutarate, 197–198
    recommendations, 201, 203
    summary of research on athletes and active individuals, 200–201, 203
    summary of research on general population, 200, 202–203
    taurine, 201
Osteoarthritis (OA), 109
    diseased joints of, 124
    knee-based, 109
    prevalence of, 115
    symptoms, 118
    treatment of, 110, 112
Oxaloacetate (OAA), 41, 42
Oxidative damage, CoQ supplementation and, 356
Oxidative stress
    biothiols and, 412
    conditions associated with increased, 331
    CoQ_{10} biosynthesis and, 363
    definition of, 359
    diabetes complications and, 422
    exercise-induced, 457
    tea and, 457
Oxygen
    consumption, submaximal, 441
    extraction, 268
    quencher, 333

P

Palm oil, food cooked with, 337

Panax
    ginseng, 379, 380, 474
    japonicus, 401
    notoginseng, 401
    quinquefolius L., 380
Pancreatic insufficiency, MCT use and, 224
Pangamic acid, *see* Myo-inositol and pangamic acid
Panic disorder, 432, 436
Paraxanthine, 278, 284
PCr, *see* Phosphocreatine
PDC, *see* Pyruvate dehydrogenase complex
PDH, *see* Pyruvate dehydrogenase
Peer behavior, caffeine consumption and, 277
Performance
    effects of carnitine supplementation on, 75
    mental, 399, 400, 469
    phosphate loading on, 267
    time, get-up-and-go, 160
    treadmill running, 268
Peripheral fatigue, 52
Peripheral neuropathy, myo-inositol and, 437
Permeability transition pore (PTP), 360
PFK, *see* Phosphofructokinase
Pharmaceutical industry, uses of gelatin in, 108
Phenylpropanolamine, 309
Phosphate
    buffers, *see* Buffers
    loading, 267, 268
Phosphatidylinositol (PI), 434
Phosphatidylinositol 4-phosphate (PIP), 434
Phosphocreatine (PCr), 81, 82
Phosphodiesterase
    activity, methylxanthine-induced suppression of, 281
    inhibitor, 282
Phosphofructokinase (PFK), 259
Phosphoinositide kinases, inhibition of, 285
Photosynthetic plants, role of carotenoids in, 326
PI, *see* Phosphatidylinositol
D-Pinitol, 82
PIP, *see* Phosphatidylinositol 4-phosphate
Placebo effect, buffer research and, 265
Plant foods, sources of glutamine in, 136
Plasma proteins, 260
Platelet aggregation, 250
Policosanol, 251
Pork skin, gelatin and, 107
PowerBar®, 6
Power lifters, 88
Pregnancy
    FDA ruling on claims about, 10

Index

glucosamine consumption during, 122
use of CoQ$_{10}$ during, 369
Premenstrual syndrome, 10
Product quality assurance, 15
Prohormones, regulatory status of, 15
Prostate-specific antigen (PSA), 333
Proteases, genes expressing, 165
Protective colloid, use of gelatin as, 108
Protein(s)
   animal, 188
   building block of body, 173
   catabolism, fatigue and, 132
   -containing foods, BCAA found in, 49
   C-reactive, 331
   degradation, decreased, 152
   dietary, 186
   GLUT-4, 420, 450
   high-protein foodstuffs, 477
   hypercatabolism, 200
   incomplete, 111
   kinase B phosphorylation-activation, 285
   kinase C, activation of, 434
   lens, 332
   lysine content of, 176
   muscle, HMB and, 148
   oxidative damage to, 365
   plasma, 260
   powders, 5
   sources, 177
   stores
      labile, 25
      use of as energy source, 24
   synthesis
      amino acids available for, 50
      glutamine, 131
   tannins and, 455
   turnover, 24–25, 56, 57
   whey, 49, 89
PSA, *see* Prostate-specific antigen
Pseudoephedrine, 309
Psychoactive drug, world's most widely consumed, 276
PTP, *see* Permeability transition pore
Pyruvate, 514
Pyruvate dehydrogenase (PDH), 66
   complex (PDC), 65
   subunits of, 412
Pyruvate and dihydroxyacetone, 445–454
   athletic performance, 449–452
      human studies, 450–452
      rationale and possible mechanisms, 449–450
   description and reported effects, 445–446
   metabolism, 446–447
   recommendations, 452

   weight loss effects, 447–449
      human studies, 448–449
      initial animal studies, 447–448

R

RAE, *see* Retinol activity equivalent
Raman spectroscopy, 339
RBC, *see* Red blood cells
RDA, *see* Recommended dietary allowance
RE, *see* Retinol equivalent
Reactive oxygen species (ROS), 413
Recommended dietary allowance (RDA), 49, 177, 330
Red blood cells (RBC), 161
Red Bull®, 6, 203
Renal function, assessment of, 163
Renal glutaminase activity, 136
Repetitive cycling sprint performance, 87
Reproductive disorders, treatment of with CoQ$_{10}$, 356
Research Triangle Institute Center for Economics Research, 5
Reserpine drug, hypoactivity effects of, 438
Resistance training
   fat-free mass gains associated with, 148
   HMB and, 152, 157
   muscle mass gains from, 154
   sessions, enhanced recovery following, 140
Respiratory exchange ratio, 42
Respiratory quotient (RQ), 72, 73
Retinal degeneration, treatment of with CoQ$_{10}$, 356
Retinol activity equivalent (RAE), 329
Retinol equivalent (RE), 330
Rhodiola species, 380, 390, 395, 396–398
ROS, *see* Reactive oxygen species
RQ, *see* Respiratory quotient
Rumenic acid, 209

S

Saccharum officinarum L., 247
Saguenay-Lac-Saint-Jean cytochrome-c oxidase deficiency (SLSJ-COX), 439
Salt, aspartate delivered as, 41
SAMe, *see* Sulfur-adenosylmenthionine
Sarcoplasmic reticulum (SR), 281, 311
SCFA, *see* Short-chain fatty acids
Selective serotonin reuptake inhibitors (SSRI), 435, 436
Serum
   amino acid concentration in, 175

growth hormone, integrated
concentrations of, 186
Shark cartilage, 117
Short bowel syndrome, MCT use and, 224
Short-chain fatty acids (SCFA), 225, 226
Siberian ginseng, 391, 392–394
Skeletal muscle
glutamine and, 135
hypertrophy, enhanced, 172, 173
Skin photoprotector, 344
SlimFast®, 6
Slow-twitch (ST) muscle fibers, 365
SLSJ-COX, see Saguenay-Lac-Saint-Jean cytochrome-c oxidase deficiency
Smoking, β-carotene and, 343
Smooth muscle relaxation, 24
Sodium
bicarbonate, citrate and, 267
myo-inositol transporter, 435
Sports
drinks, 472
governing agencies, 13
supplement market, U.S., 49
Sprint
performance, 82, 87
test, 86
SR, see Sarcoplasmic reticulum
SSRI, see Selective serotonin reuptake inhibitors
Statins, CoQ_{10} and, 366
Steroids
anabolic androgen, 366, 367
animal production of, 326
ST muscle fibers, see Slow-twitch muscle fibers
Strength
gain, 156
HMB and, 152
Stress, chronic physiological, 25
Sulfur-adenosylmenthionine (SAMe), 439
Supramaximal cycling, 310
Sympathomimetic drug, 309
Symptomatic slow-acting drugs in osteoarthritis (SYSADOAs), 121
SYSADOAs, see Symptomatic slow-acting drugs in osteoarthritis

T

Tanning pill, 344
Tannins, 455–465, 515
chemistry and constituents of tea, 456–457
condensed, 455
hydrolyzable, 455
overview of plant tannins, 455–456

tea and blood pressure, 458–459
tea and cancer, 460–461
tea and cardiovascular disease, 459–460
tea and oxidative stress, 457
Taurine, see also Ornithine, ornithine alpha-ketoglutarate and taurine
occurrence of, 202
oral supplementation of, 202
production of, 201
supplementation, congestive heart failure and, 203, 204
TCA cycle pathway, see Tricarboxylic acid cycle pathway
T-cell-mediated immune function, 29
T-cell mitogens, lymphocyte proliferative response to, 139
TCr, see Total creatine
Tea, see also Tannins
black, 460
blood pressure and, 458
cancer and, 460
cardiovascular disease and, 459
constituents of, 456
green, 457, 460
oolong, 457
oxidative stress and, 457
TEE, see Total energy expenditure
Tenebrio molitor, 62
Tennis players, sprint performance in, 87
Terpenes, 326
Test-to-test reliability, 87
Theobromine, 278
Theophylline, 278, 281, 284, 288
Thromboxane A_2 synthesis, 250
Time to exhaustion, caffeine and, 308
Time trials, 258
Tissue proteins, incorporation of arginine in, 25
TNF-α, see Tumor-necrosis factor
Total creatine (TCr), 81, 82
Total energy expenditure (TEE), 173
Training
adaptations, creatine supplementation and, 89
bouts, poor recovery from, 177
chronic, effect of on carnitine status, 70
status, HMB and, 154
Transition mineral elements, 413
Treadmill, 341
exercise, lipoic acid and, 421
motorized Quinton, 342
running performance, 268
Tricarboxylic acid (TCA) cycle pathway, 358
Triglycerides, see also Medium-chain triglycerides and glycerol

Index

hydrolysis of, 227
lipolysis, 337
medium-chain, 231
mobilization, 228
synthesis of, 228
Triose supplementation, 446, 450
Tryptophan
    transport, 57
    uptake, reduced, 54
Tumor
    -necrosis factor (TNF-α), 416
    protein synthesis rate, 26
Type II diabetes, 156

U

Ubidecarenone, see Coenzyme Q_{10}
Ubiquinone, see Coenzyme Q_{10}
Ulcerative colitis, 117
United Kingdom Medicines Act, 248
United States Anti-Doping Agency, 13–14, 264
United States Department of Agriculture (USDA), 334
    database, carotenoid content calculated using, 334
    National Nutrient Database for Standard Reference, 199
United States Olympic Committee (USOC), 13
United States Pharmacopeia (USP), 14
Urea
    -cycle disorders, 199
    synthesis, glutamine and, 132
USDA, see United States Department of Agriculture
U.S. Marines, cold-weather field training of, 342
USOC, see United States Olympic Committee
U.S. Olympic Cycling Team (1984), caffeine suppositories used by, 277
USP, see United States Pharmacopeia

V

Vaccines, gelatin portion of, 111
Valine, physical performance improved by, 47
VANT, see Ventilatory anaerobic threshold
Vascular arterial disease, platelet aggregation and, 250
Vegetarian diets, 63, 64
Ventilatory anaerobic threshold (VANT), 86

Ventricular arrhythmia, treatment of with CoQ_{10}, 356
Very-low-density lipoprotein cholesterol (VLDL), 162
Vision protection, 332
Vitamin A
    carotenoids and, 344
    formation of, 327
    toxicity of, 330
Vitamin C
    deficiency, 414
    regeneration of, 359
Vitamin E
    CoQ_{10} and, 366
    deficiency, 414
Vitamin Q, 356
Vitamins, fat-soluble, 326
VLDL, see Very-low-density lipoprotein cholesterol
Volitional exhaustion, 264

W

Warfarin, 366, 367
WBC, see White blood cells
Weightlifting
    muscle endurance during, 310
    performance, glutamine supplementation and, 141
Weight loss
    carnitine and, 74
    NADH and, 447
Weight training regimen, HMB-supplemented subjects in, 159
Well-being, ginseng and, 384
Wheat germ oil and octacosanol, 247–254
    biodistribution, 251–252
    mechanism of action, 248–251
        antiaggregatory properties, 249–250
        cholesterol-lowering effects, 250–251
        exercise performance, 248–249
Whey protein, 49, 89
White blood cells (WBC), 161
Wingate test, 308, 340
World Anti-Doping Agency, 13, 280
Wound healing, 134, 160

X

Xanthine derivatives, 280
Xanthophylls, 327, 336–337
Xenobiotics, detoxification of, 415

Y

Yale-Brown Obsessive Compulsive Scale
(YBOCS), 436
YBOCS, *see* Yale-Brown Obsessive
Compulsive Scale
Yeast infections, 226

Z

Zeaxanthin, 331
  cataract development and, 332
  documented action of, 332
  foods rich in, 335
  industrial synthesis of, 337
ZMA, 472

American River College Library
4700 College Oak Drive
Sacramento, CA 95841